Nanostructured Materials in Electrochemistry

Edited by
Ali Eftekhari

Further Reading

Zehetbauer, M. J., Zhu, Y. T. (Eds.)

Bulk Nanostructured Materials

2008
ISBN: 978-3-527-31524-6

Bard, A. J., Stratmann, M. (Eds.)

Encyclopedia of Electrochemistry

11 Volume Set
2007
ISBN: 978-3-527-30250-5

Hamann, C. H., Hamnett, A., Vielstich, W.

Electrochemistry

2007
ISBN: 978-3-527-31069-2

Wang, J.

Analytical Electrochemistry

2006
ISBN: 978-0-471-67879-3

Samori, P. (Ed.)

Scanning Probe Microscopies Beyond Imaging

Manipulation of Molecules and Nanostructures
2006
ISBN: 978-3-527-31269-6

Vollath, D.

Nanomaterials

An Introduction to Synthesis, Charaterization and Processing
2008
ISBN: 978-3-527-31531-4

Rao, C. N. R., Müller, A., Cheetham, A. K. (Eds.)

Nanomaterials Chemistry

Recent Developments and New Directions
2007
ISBN: 978-3-527-31664-9

Nanostructured Materials in Electrochemistry

Edited by
Ali Eftekhari

WILEY-VCH Verlag GmbH & Co. KGaA

The Editor

Dr. Ali Eftekhari
Avicenna Institute of Technology
Cleveland, Ohio
USA

Cover Description
Center: In solid-state electrochemistry, the shape of a cyclic voltammogram depends on numerous complicated processes occurring simultaneously, and the cyclovoltammogram is its statistical representative. Bottom right: Fullerenes have also garnered interest for electrochemical nanotechnology as cages in which ions can be trapped. Top right: The classic scheme of the tunneling effect as the basis of scanning tunneling microscopy (STM) is shown, which is capable of visualizing the occurrence of various electrochemical processes at the electrode surface and has also inspired scanning electrochemical microscopy (SECM).
Original artwork by Ali Eftekhari, adapted by Bernd Adam.

■ All books published by Wiley-VCH are carefully produced. Nevertheless, authors, editors, and publisher do not warrant the information contained in these books, including this book, to be free of errors. Readers are advised to keep in mind that statements, data, illustrations, procedural details or other items may inadvertently be inaccurate.

Library of Congress Card No.: applied for

British Library Cataloguing-in-Publication Data
A catalogue record for this book is available from the British Library.

Bibliographic information published by the Deutsche Nationalbibliothek
Die Deutsche Nationalbibliothek lists this publication in the Deutsche Nationalbibliografie; detailed bibliographic data are available in the Internet at <http://dnb.d-nb.de>.

© 2008 WILEY-VCH Verlag GmbH & Co. KGaA, Weinheim

All rights reserved (including those of translation into other languages). No part of this book may be reproduced in any form – by photoprinting, microfilm, or any other means – nor transmitted or translated into a machine language without written permission from the publishers. Registered names, trademarks, etc. used in this book, even when not specifically marked as such, are not to be considered unprotected by law.

Typesetting Thomson Digital, Noida, India
Printing Strauss GmbH, Mörlenbach
Binding Litges & Dopf GmbH, Heppenheim
Cover Design Adam-Design, Weinheim

Printed in the Federal Republic of Germany
Printed on acid-free paper

ISBN: 978-3-527-31876-6

Foreword by R. Alkire

Electrochemical phenomena control the existence and movement of charged species in the bulk phases as well as across interfaces between ionic, electronic, semiconductor, photonic, and dielectric materials. During the past several decades, the study of electrochemical phenomena has advanced rapidly owing, in large part, to the invention of a suite of new scientific tools. Electrochemical processes have thus provided the ability to create precisely characterized systems for fundamental study. Usage includes the monitoring of behavior at unprecedented levels of sensitivity, atomic resolution, and chemical specificity and the prediction of behavior using new theories and improved computational abilities. These capabilities have revolutionized fundamental understanding, as well as contributed to the rapid pace of discovery of novel material structures, devices, and systems.

This volume, "*Nanostructured Materials in Electrochemistry*," focuses on the importance of electrochemistry in the fabrication as well as the functional capabilities of a great many nanostructured materials, processes and devices. It provides an authoritative overview of a dozen key topics contributed by leading experts from academic, industrial, federal and private research institutions located around the world. The viewpoints of the authors arise from a variety of disciplines that span science, mathematics, and engineering. Of particular note are the references to the recent literature, more than 2100, which has grown exponentially since the mid-1990's.

Interwoven throughout the chapters of this work are several overarching themes that, taken together, provide a strategic framework for closing the gap between nanoscience and nanotechnology. These include:

Open-ended Discovery and Targeted Design

Open-ended curiosity-driven research and discovery at the nanoscale has established a spectacular record of success, based in large part to the availability of a multiplicity of experimental methods and data assimilation/visualization tools that provide broad access and the development of informed intuition. Targeted design builds on the foundation of curiosity-driven discoveries, but involves working backward from the desired function or product to perfect the underlying material and the process conditions by which it can be fabricated. This volume expands the common ground between these approaches, both of which are essential. A particularly valuable contribution of this volume is the identification of numerous "model" systems which

---

Nanostructured Materials in Electrochemistry. Edited by Ali Eftekhari
Copyright © 2008 WILEY-VCH Verlag GmbH & Co. KGaA, Weinheim
ISBN: 978-3-527-31876-6

have been found to provide consistent results suitable for developing refined scientific experiments, as well as for establishing robust well-engineered systems that work.

The Flow of Information between Individuals both Within and Amongst Disciplines

The ease with which new results and insights are used by specialists working on other aspects of a related problem is extremely important for the integration of shared purposes across disciplines. Clearly evident in this volume are many examples of how critical knowledge is shared amongst specialists working together to transfer innovative ideas and insights into new products and processes. These examples serve to emphasize the importance of reporting results in ways that others can not only use them, but can also modify them for other purposes.

Multi-scale Phenomena

New applications of novel materials and devices are being discovered where the critical functional events depend upon the control of structure at the nanoscale, while the product fabrication is controlled by macroscopic variables. The examples in this volume can serve to inspire the creation of new process engineering methods to ensure product quality in complex, multi-scale, multi-phenomena systems.

Collaborative Environment

The scientific discoveries described in this volume are leading inexorably to new technological advances, the manufacturability of which requires precise quantitative understanding at a magnitude, sophistication, and completeness that is extraordinarily difficult to assemble today. Methods for collaborative discovery and problem-solving across disciplines today are in their infancy. The pioneering efforts reported in this volume provide the breeding ground for the particular applications at hand.

The editor has assembled a group of chapters that provide excellent coverage of the literature and tutorial background in addition to details direct from the authors' own experience. The chapters include numerous examples of methods and merits of various electrochemical and other procedures for the formation of nanostructured materials which have a wide range of forms and combinations of properties; the effect of electrochemical processing conditions on morphology, structure, reactivity and properties with numerous discussions of mechanistic aspects and the novel devices which result. These include devices based on ultra-small electrodes and sensors; nanocomposites and alloys for energy storage; photoelectrochemically active nanoparticles for batteries and solar cells; nanostructured interfaces for biosensors; and noble metal nanoparticles for electroanalytical applications.

The volume represents a benchmark for the current state-of-the-art, and provides numerous paths by which nanoscale science and technology is moving from an art form into the science and technology of well-engineered devices and products. The contents describe generic approaches that have the potential to contribute well beyond the specific systems used here.

December 2007

Richard Alkire
Charles and Dorothy Prizer Professor
Department of Chemical and Biomolecular Engineering
University of Illinois, Urbana, USA

Foreword by Y. Gogotsi and P. Simon

Nanostructured materials or nanomaterials – which are materials with structural units on a nanometer scale in at least one direction – have received much attention worldwide over the past decade. Indeed, for just one single class of nanostructured materials – the carbon nanotubes – the number of published reports has increased from less than 500 per year in 2000 to almost 3000 in 2007 (ISI Web of Science). Since material properties become different on the nanoscale, much effort is currently being dedicated to the synthesis, structure control and property improvement of nanomaterials. For example, the deformation mechanisms of nanocrystalline metals are different from those of microcrystalline metals. One-dimensional nanomaterials such as nanotubes and nanowires possess many attractive properties which can be fine-tuned by controlling their diameter. Today, the industrial applications of nanomaterials continue to grow in number, with hundreds of products now available worldwide. Nanostructured materials are widely used in many applications where people do not expect to see nanotechnology. An average reader of this book will probably know that the huge storage capacity of a computer hard drive is achieved thanks to nanosized magnetic particles, and that a diamond-like carbon coatings of a few nanometers thickness protects the surface of the magnetic head which is reading the hard drive. However, few people realize that when they make an omelet on an aluminum frying pan they benefit from an anodic alumina coating of less than 100 nm thickness with well-aligned cylindrical pores of less than 20 nm diameter, whether they use a non-stick Teflon-coated frying pan or professional-grade cookware with no Teflon coating. These coatings are produced by an electrochemical process – the anodization of alumina – and are widely used to provide protection and/or give specific colors to aluminum surfaces. On the other hand, anodic alumina films (membranes) are widely used as templates for producing metal nanowires, carbon nanotubes and other elongated nanostructures. These examples demonstrate the clear synergy that exists between nanomaterials and electrochemistry.

Nanostructured Materials in Electrochemistry. Edited by Ali Eftekhari
Copyright © 2008 WILEY-VCH Verlag GmbH & Co. KGaA, Weinheim
ISBN: 978-3-527-31876-6

Nanostructured materials have also led to major advances in electrochemical energy storage areas and, more specifically, to lithium-ion batteries. The recent discovery of the universal "conversion reaction" mechanism, which involves the formation of nanosized metal particles, is an excellent example of a breakthrough that has been achieved due to nanotechnology. Moreover, today the nanosize dimension makes it possible to use materials that once were considered useless for battery applications. The best illustration of this is carbon-coated nano $LiFePO_4$, which, today, is one of the most widely studied materials for positive electrodes of lithium-ion batteries.

Nanostructured materials also promise to revolutionize the field of supercapacitors, thus opening the doors to many applications, such as Hybrid Electric Vehicles (HEV) and portable electronic devices. Unfortunately, however, the fields of materials and electrochemistry are not necessarily well connected – many materials scientists do not receive any formal training in electrochemistry during their undergraduate (and even graduate) education, while most chemists have only a limited knowledge of the structure–property relationships of materials. Therefore, a book addressing both communities – and written by research scientists with backgrounds in both chemistry and materials – should prove to be very useful for those who produce nanomaterials using electrochemistry methods, who study the electrochemical behavior of nanomaterials, or develop materials for electrochemical applications.

The chapters of this book describe the preparation of anodic alumina membranes, the preparation of nanopatterned electrodes, the use of porous alumina and polycarbonate templates for synthesis of nanowires, and the electrochemical deposition of nanostructured oxide and metal coatings with different morphologies, as well as the use of nanoparticles and nanomaterials in lithium-ion batteries, hydrogen storage, solar cells, biosensors, and electroanalysis. Whilst the area addressed by the book is very broad, it is hardly possible to cover all nanomaterials for all electrochemical applications in a single volume. Hence, metals have received more attention than carbon nanomaterials; while materials for batteries are described in two chapters materials for supercapacitors and fuel cells did not receive equal attention. However, overall, a wide range of topics has been addressed and the book content certainly corresponds to its title.

This book has been written for today's scientists, graduate students and engineering professionals in order to provide an overview of nanostructured materials and electrochemical techniques and applications. The book consists of 12 chapters written by researchers representing a wide geographic spectrum. It provides coverage of the latest developments in the United States, Western Europe, and Japan, as well as of investigations conducted in Brazil and Eastern European countries, which received less attention in previous volumes on nanomaterials, such as the *Nanomaterials Handbook*, edited by Y. Gogotsi (CRC Press, 2006). We are confident that many readers will find interesting reviews covering a broad range of subjects in this interdisciplinary volume.

November 2007

Yury Gogotsi[1]
Professor of Materials Science and Engineering
Drexel University
and
Director, A.J. Drexel Nanotechnology Institute,
Philadelphia, PA, USA

Patrice Simon[2]
Professor of Materials Science
Paul Sabatier University
Toulouse, France

---

[1] Y.G. has a MS degree in metallurgy, PhD in physical chemistry and DSc in materials engineering, but considers himself a material scientist.

[2] P.S. has a BS and MS in Chemistry and Physics, and a PhD in materials science, but considers himself an electrochemist.

They work together to solve problems at the interface between materials science and electrochemistry.

Contents

Foreword by R. Alkire *V*
Foreword by Y. Gogotsi and P. Simon *VII*
Preface *XIX*
List of Contributors *XXIII*

1 **Highly Ordered Anodic Porous Alumina Formation by Self-Organized Anodizing** *1*
  Grzegorz D. Sulka
1.1 Introduction *1*
1.2 Anodizing of Aluminum and Anodic Porous Alumina Structure *6*
1.2.1 Types of Anodic Oxide Film *7*
1.2.2 General Structure of Anodic Porous Alumina *8*
1.2.2.1 Pore Diameter *9*
1.2.2.2 Interpore Distance *12*
1.2.2.3 Wall Thickness *13*
1.2.2.4 Barrier Layer Thickness *14*
1.2.2.5 Porosity *17*
1.2.2.6 Pore Density *19*
1.2.3 Incorporation of Anions *20*
1.2.4 Cell-Wall Structure *23*
1.2.5 Crystal Structure of Oxide *26*
1.2.6 Density and Charge of Oxide Film *26*
1.2.7 Miscellaneous Properties of Anodic Porous Alumina *27*
1.3 Kinetics of Self-Organized Anodic Porous Alumina Formation *28*
1.3.1 Anodizing Regimes and Current/Potential-Time Transient *28*
1.3.2 Pores Initiation and Porous Alumina Growth *32*
1.3.2.1 Historical Theories *32*
1.3.2.2 Field-Assisted Mechanism of Porous Film Growth *34*
1.3.2.3 Steady-State Growth of Porous Alumina *36*
1.3.2.4 Growth Models Proposed by Patermarakis and Colleagues *39*

1.3.2.5	Other Phenomenological Models of Porous Alumina Growth *41*	
1.3.2.6	Other Theoretical Models of Porous Alumina Growth *44*	
1.3.3	Volume Expansion: The Pilling–Bedworth Ratio (PBR) *45*	
1.3.4	Rates of Oxide Formation and Oxide Dissolution *46*	
1.4	Self-Organized and Prepatterned-Guided Growth of Highly Ordered Porous Alumina *50*	
1.4.1	Aluminum Pre-Treatment *53*	
1.4.2	Self-Organized Anodizing of Aluminum *58*	
1.4.2.1	Structural Features of Self-Organized AAO *60*	
1.4.2.2	Order Degree and Defects in Nanopore Arrangement *74*	
1.4.3	Post-Treatment of Anodic Porous Alumina *81*	
1.4.3.1	Removal of the Aluminum Base *81*	
1.4.3.2	Removal of the Barrier Layer *82*	
1.4.3.3	Structure and Thinning of the Barrier Layer *85*	
1.4.3.4	Re-Anodization of Anodic Porous Alumina *87*	
1.5	AAO Template-Assisted Fabrication of Nanostructures *88*	
1.5.1	Metal Nanodots, Nanowires, Nanorods, and Nanotubes *89*	
1.5.2	Metal Oxide Nanodots, Nanowires, and Nanotubes *91*	
1.5.3	Semiconductor Nanodots, Nanowires, Nanopillars, and Nanopore Arrays *91*	
1.5.4	Polymer, Organic and Inorganic Nanowires and Nanotubes *93*	
1.5.5	Carbon Nanotubes *94*	
1.5.6	Photonic Crystals *95*	
1.5.7	Other Nanomaterials (Metallic and Diamond Membranes, Biomaterials) *95*	
	References *97*	
2	**Nanostructured Materials Synthesized Using Electrochemical Techniques** *117*	
	Cristiane P. Oliveira, Renato G. Freitas, Luiz H.C. Mattoso, and Ernesto C. Pereira	
2.1	Introduction *117*	
2.2	Anodic Synthesis *119*	
2.2.1	Electropolishing and Anodization *119*	
2.2.2	Porous Anodic Alumina *128*	
2.2.2.1	Porous Anodic Alumina as Template *135*	
2.2.2.2	Porous Anodic Alumina to Create Nanodevices *137*	
2.3	Cathodic Synthesis *144*	
2.3.1	Nanowires *144*	
2.3.1.1	Template Procedures to Prepare Nanowires *145*	
2.3.1.2	Magnetic Nanowires *147*	
2.3.1.3	Nanotubes *152*	
2.3.2	Multilayers *158*	

2.3.3	Other Materials *162*	
2.3.3.1	Semiconductors *165*	
2.3.3.2	Oxides *168*	
2.3.3.3	Metals *170*	
2.4	Final Remarks *173*	
	References *174*	

3 Top-Down Approaches to the Fabrication of Nanopatterned Electrodes *187*
Yvonne H. Lanyon and Damien W.M. Arrigan

3.1	Introduction *187*
3.2	Considerations for Choosing a Nanoelectrode Fabrication Strategy *189*
3.3	Nanoelectrode Fabrication Using Top-Down Approaches *190*
3.3.1	E-Beam Lithography *191*
3.3.2	Focused Ion Beam Lithography *196*
3.3.3	Nano-Imprint Lithography *199*
3.3.4	Nanogap Electrodes *203*
3.3.5	Non-High-Resolution Techniques *205*
3.4	Applications *206*
3.5	Conclusions *207*
	References *209*

4 Template Synthesis of Magnetic Nanowire Arrays *211*
Sima Valizadeh, Mattias Strömberg, and Maria Strømme

4.1	Introduction *211*
4.2	Electrochemical Synthesis of Nanowires *213*
4.2.1	Fabrication of Nanoelectrodes *213*
4.2.2	Reactions, Diffusion, and Nucleation in the Electrochemical Deposition of Co Nanowires *214*
4.2.2.1	Theoretical Considerations of Spherical Diffusion at a Nanode Array *214*
4.2.3	Electrodeposition of Magnetic Multilayered Nanowire Arrays *222*
4.2.3.1	Electrodeposition of 8 nm Ag/15 nm Co Multilayered Nanowire Arrays (Wire Diameter 120 nm) *224*
4.2.3.2	Template Synthesis of 2 nm Au/4 nm Co Multilayered Nanowire Arrays (Wire Diameter 110 nm) *225*
4.3	Physical Properties of Electrodeposited Nanowires *231*
4.3.1	Magnetic Properties of Nanowire Arrays *231*
4.3.2	Electrical Transport Measurements on Single Nanowires Using Focused Ion Beam Deposition *234*
4.4	Summary *238*
	References *238*

5 Electrochemical Sensors Based on Unidimensional Nanostructures 243
Arnaldo C. Pereira, Alexandre Kisner, Nelson Durán, and Lauro T. Kubota

5.1 Introduction 243
5.2 Preparation of Nanowires and Nanotubes by Template-Based Synthesis 243
5.2.1 Template-Based Mesoporous Materials 244
5.2.1.1 The Memorable Marks of Electrochemical Nanowires 247
5.2.2 Nanowires as Nanoelectrodes 247
5.2.2.1 Electrochemical Aspects of Nanoelectrodes 248
5.2.2.2 Nanoelectrodes Based on Chemically Modified Surface 249
5.3 An Electrochemical Step Edge Approach 251
5.3.1 The Predeterminant Mechanism 251
5.3.2 Nanowire-Based Gas Sensors 253
5.4 Atomic Metal Wires from Electrochemical Etching/Deposition 255
5.4.1 Sensing Molecular Adsorption with Quantized Nanojunction 257
5.5 Future Prospects and Promising Technologies 259
5.6 Concluding Remarks 261
References 262

6 Self-Organized Formation of Layered Nanostructures by Oscillatory Electrodeposition 267
Shuji Nakanishi

6.1 Introduction 267
6.1.1 Self-Organized Formation of Ordered Nanostructures 267
6.1.2 Dynamic Self-Organization in Electrochemical Reactions 268
6.1.3 The Important Role of Negative Differential Resistance (NDR) in Electrochemical Oscillations 271
6.1.4 Outline of the Present Chapter 272
6.2 Current Oscillation Observed in H_2O_2 Reduction on a Pt Electrode 273
6.3 Nanoperiod Cu–Sn Alloy Multilayers 275
6.4 Nano-Scale Layered Structures of Iron-Group Alloys 279
6.5 Other Systems 283
6.5.1 Nano-Multilayers of Cu/Cu_2O 283
6.5.2 Ag–Sb Alloy with Periodical Modulation of the Elemental Ratio 285
6.6 Summary 286
References 286

7 Electrochemical Corrosion Behaviour of Nanocrystalline Materials 291
Omar Elkedim

7.1 Introduction 291
7.2 Electrochemical Corrosion Behavior of Nanocrystalline Materials 292
7.3 Conclusions 315
References 315

8	**Nanoscale Engineering for the Mechanical Integrity of Li-Ion Electrode Materials** *319*
	Katerina E. Aifantis and Stephen A. Hackney
8.1	Introduction *319*
8.2	Electrochemical Cycling and Damage of Electrodes *320*
8.2.1	Fracture Process of Planar Electrodes *320*
8.2.2	Electrochemical Cycling of Particulate Electrodes *323*
8.3	Electrochemical Properties for Nanostructured Anodes *330*
8.3.1	Nanostructured Metal Anodes *331*
8.3.1.1	Sn and Sn-Sb Anodes at the Nanoscale *331*
8.3.1.2	Si Anodes at the Nanoscale *331*
8.3.1.3	Bi Anodes at the Nanoscale *333*
8.3.2	Embedding/Encapsulating Active Materials in Less-Active Materials *334*
8.3.2.1	Sn-Based Anodes *335*
8.3.2.2	Si-Based Anodes *337*
8.4	Modeling Internal Stresses and Fracture of Li-anodes *339*
8.4.1	Stresses Inside the Matrix *339*
8.4.2	Stable Crack Growth *341*
8.4.3	Griffith's Criterion *342*
8.4.4	No Cracking *344*
8.5	Conclusions and Future Outlook *345*
	References *345*
9	**Nanostructured Hydrogen Storage Materials Synthesized by Mechanical Alloying** *349*
	Mieczyslaw Jurczyk and Marek Nowak
9.1	Introduction *349*
9.1.1	The Aim of the Research *349*
9.1.2	Types of Hydride *352*
9.1.3	The Absorption–Desorption Process *353*
9.1.4	Hydrides Based on Intermetallic Compounds of Transition Metals *354*
9.1.5	Prospects for Nanostructured Metal Hydrides *355*
9.2	The Fundamental Concept of the Hydride Electrode and the Ni-MH Battery *357*
9.2.1	The Hydride Electrode *357*
9.2.2	The Ni-MH Battery *357*
9.2.2.1	Normal Charge–Discharge Reactions *357*
9.2.2.2	Overcharge Reactions *357*
9.2.2.3	Over-Discharge Reaction *358*
9.3	An Overview of Hydrogen Storage Systems *358*
9.3.1	The TiFe-Type System *359*
9.3.2	The ZrV_2-Type System *364*
9.3.3	The $LaNi_5$-Type System *366*

9.3.4	The Mg_2Ni-Type System 369
9.3.5	Nanocomposites 371
9.4	Electronic Properties 376
9.5	Sealed Ni-MH Batteries 381
9.6	Conclusions 382
	References 383

10 Nanosized Titanium Oxides for Energy Storage and Conversion 387
Aurelien Du Pasquier

10.1	Introduction 387
10.2	Preparation of Nanosized Titanium Oxide Powders 387
10.2.1	Wet Chemistry Routes 387
10.2.2	Chemical Vapor Deposition 389
10.2.3	Vapor-Phase Hydrolysis 389
10.2.4	Physical Vapor Deposition 390
10.3	Other TiO_2 Nanostructures 390
10.4	Preparation of Nano-$Li_4Ti_5O_{12}$ 390
10.5	Nano-$Li_4Ti_5O_{12}$ Spinel Applications in Energy Storage Devices 393
10.5.1	Asymmetric Hybrid Supercapacitors 394
10.5.2	High-Power Li-Ion Batteries 396
10.6	Nano-TiO_2 Anatase for Solar Energy Conversion 398
10.6.1	TiO_2 Role in Dye-Sensitized Solar Cells 398
10.6.2	Trap-Limited Electron Transport in Nanosized TiO_2 399
10.6.3	Electron Recombination in Dye-Sensitized Solar Cells 400
10.6.4	Preparation of Flexible TiO_2 Photoanodes 401
10.6.4.1	Sol–Gel Additives 402
10.6.4.2	Mechanical Compression 403
10.6.4.3	Metallic Foils 403
10.7	Conclusions 404
	References 405

11 DNA Biosensors Based on Nanostructured Materials 409
Adriana Ferancová and Ján Labuda

11.1	Introduction 409
11.2	Nanomaterials in DNA Biosensors 410
11.2.1	Carbon Nanotubes 410
11.2.1.1	Electronic Properties and Reactivity of CNTs 411
11.2.1.2	CNT–DNA Interaction 412
11.2.1.3	CNTs in DNA Biosensors 413
11.2.2	Fullerenes 422
11.2.3	Diamond and Carbon Nanofibers 423
11.2.3.1	Diamond 423
11.2.3.2	Carbon Nanofibers 424
11.2.4	Clays 424

11.2.5	Metal Nanoparticles	*425*
11.3	Conclusions	*428*
	References	*430*

12 Metal Nanoparticles: Applications in Electroanalysis *435*
Nathan S. Lawrence and Han-Pu Liang

12.1	Introduction	*435*
12.2	Electroanalytical Applications	*439*
12.2.1	Gold Nanoparticles	*439*
12.2.2	Platinum Nanoparticles	*441*
12.2.3	Silver Nanoparticles	*442*
12.2.4	Palladium Nanoparticles	*443*
12.2.5	Copper Nanoparticles	*448*
12.2.6	Nickel Nanoparticles	*449*
12.2.7	Iron Nanoparticles	*449*
12.2.8	Nanoparticles of Other Metallic Species	*450*
12.3	Future Prospectives	*451*
	References	*451*

Index *459*

Preface

The interaction of electrochemistry and nanotechnology has two sides, namely the applications of nanotechnology in electrochemistry, and vice versa. Although, as inferred by the title, this book deals with the former subject, the basic concept behind it was to unite the two sides of this newly born field, which we can then refer to as Electrochemical Nanotechnology. Due to vast range of topics in this field, there was a clear obligation to focus on only a part of the field, and hence this book is not considered to be an exhaustive resource on the subject, but rather to provide some important information on a variety of topics that will attract the attention of readers to current issues in the subject. In my opinion, such a united volume is indeed capable of providing a comprehensive perspective of the whole field. After undergoing rapid and growing specialization during the past few decades, now is the time for interdisciplinary studies and collaboration between the various fields. Today, the successful research groups are those which conduct studies that are significant and important not only for the people working within the field, but also for those working in other areas. Here, nanotechnology represents a vivid example, as the extreme success of this newborn field is due as much to the generality of its findings as to the interest of the research teams working in its various areas.

The reason why such emphasis is placed on Electrochemical Nanotechnology is due not only to the existence of so many interesting topics within the category, but also to its important concepts. Today, many research groups working in nanotechnology also have wider interests in electrochemistry, as electrochemical methods are typically low-cost and also highly effective for the preparation of nanostructures. This newfound attention is due largely to the methodology employed, which may also be used for fundamental studies. In fact, rather than electrochemistry being considered as a branch of chemistry, its footprints can be seen in a variety of fields for both methodological applications and fundamental studies. For example, when studying chaotic dynamics in chemical systems, electrochemical oscillators provide the best means of proposing general models, as both controllable parameters and system response form part of the electrochemical set-up. Indeed, this is also the case for nanotechnology.

The reason why I first came to the field of nanotechnology stemmed in fact from my studies in electrochemistry – it was not the "fame" of nanotechnology, because in

Nanostructured Materials in Electrochemistry. Edited by Ali Eftekhari
Copyright © 2008 WILEY-VCH Verlag GmbH & Co. KGaA, Weinheim
ISBN: 978-3-527-31876-6

those days the subject was not famous! My first encounter with the subject occurred while studying electrochemical oscillations, when I noticed a classic theory that the distribution of potential is inhomogeneous across the electrode surface. So, I thought that it might be very interesting to identify a way in which the local currents on an electrode surface could be inspected. Subsequently, the invention of the scanning electrochemical microscope (SECM) paved the path to this goal. My second encounter occurred when I tried to use carbon nanotubes as the anode material of a lithium battery, and I had considered preparing separate sheets of graphene (not rolled as nanotubes), as solid-state diffusion within graphite interlayers occurred so slowly. Although neither of these topics has yet been fully addressed, these early calls for nanotechnology within the realms of electrochemistry were due to the essential role of nanoscale in electrochemical systems.

The SECM is commonly considered as a form of scanning probe microscopy (SPM), and is of major interest to electrochemists. In fact, opinion suggests SECM is an advanced form of SPM, as it provides the great opportunity to control not only (electro)chemical processes but also the common applications of SPM (here, I am not talking about the features of currently available commercial microscopes, but rather the concepts involved). Unfortunately, non-electrochemists are often afraid to use the SECM due to the existence of strange electrochemical processes that may affect their results. There is, therefore, a clear need for scientific collaboration, rather than simply ignoring these great opportunities. In the second case, as well as using electrolysis to prepare graphene sheets by simply cutting a graphite electrode layer-by-layer, the opportunity exists to examine these nanomaterials by using electrochemical methods, rather than by their applications. In this respect, recent advances in methods such as fast voltammetry have provided new opportunities in surface electrochemistry, mainly in the identification of nanostructures.

Richard Alkire has well described the journey of electrochemistry towards nanotechnology and summarized the contents of this book upon this connection. The present book deals with the area of Electrochemical Nanotechnology where nanomaterials are applied in electrochemical systems. Yury Gogotsi and Patrice Simon expressed the rapidly growing applications of nanotechnology in our everyday life as electrochemistry is an important part of such industries, and also the server need of nanotechnology in modern electrochemistry (e.g., electrochemical power sources). However, such mutual involvements are not vivid to both parties. In fact, the book's contents describe the importance of nanostructured materials in electrochemical systems, and the value of electrochemical methods in the preparation of nanostructures.

At this point the reader may wonder why I so frequently place emphasis on Electrochemical Nanotechnology, when in fact the book does not comprehensively cover all aspects of the subject. The main mission of books such as this is to review certain "hot" topics within specified areas of research – something that review articles in scholarly journals often cannot do because they are published in a too-general or too-specialized medium. In this regard, electrochemical materials science is of particular interest due to a very broad readership since, within the electrochemical literature, most studies are associated with materials science, and numerous electrochemical studies are also reported in the materials science literature. Yet, according to the similarities of the electrochemical processes (both in applications and synthesis), it is very useful for the different research groups to know similar systems. Consequently, in order to address so many different aspects of the subject under consideration, a variety of current topics that should be of interest to all readers are discussed.

Today, perhaps the main emphasis in the rapidly growing field of Electrochemical Nanotechnology is to identify a new way of thinking. However, whilst all fields of science have their own "jargon", it is clearly more important to devise a consistent method of thinking rather than a unique terminology. Moreover, such concerted effort should lead to a united scientific community, which is essential for the advancement of any field of investigation. Within the realms of Electrochemical Nanotechnology, researchers of different training and thought methods are becoming increasingly involved, and this can surely only prove to be advantageous for the subject in the long term.

It is hoped that, although similar volumes have been produced in the past, this book will attract the attention of many research groups, who hopefully will unite in their studies of the general features of this new area. Undoubtedly, such a situation will not only result in a more comprehensive realization of the subject, but also lead to improved problem-solving capabilities in the field of Electrochemical Nanotechnology.

The realm of Electrochemical Nanotechnology in fact consists of a broad range of topics, hence leading researchers from various areas of study were involved in this book project. They address the most fascinating current issues and challenges that have presented themselves at the interface of electrochemistry and nanotechnology. Though coming from various different backgrounds in electrochemistry or materials science, the authors share a joint belief that the essential link between electrochemistry and nanotechnology has previously been missing, and must now be tackled.

It is my great pleasure and good fortune to have two invaluable forewords by three highly esteemed scientists. As a leading electrochemist, the fame of Richard Alkire is due to his considerable contributions to the fundamentals of electrochemistry, and in this capacity he has also contributed brilliantly to the fundamental aspects of Electrochemical Nanotechnology, in particular electrodeposition.

Yury Gogotsi is one of the leading scientists in nanomaterials, and has carried out groundbreaking work on numerous types of nanomaterials, especially carboneous ones. His collaboration with Patrice Simon is an example of the need for the combination of electrochemistry with nanotechnology, which cannot be emphasized often enough.

Last but not least, I would like to note my appreciation of the Wiley-VCH editors' foresight in picking out this particular topic and their kind efforts which made the publication of this book possible. I wish to thank them for their essential roles.

I sincerely hope that the readers find the contents of this work useful for their scientific research.

January 2008   *Ali Eftekhari*

List of Contributors

Katerina E. Aifantis
Aristotle University of Thessaloniki
Lab of Mechanics and Materials
Box 468
54124 Thessaloniki
Greece

Damien W. M. Arrigan
University College
Tyndall National Institute
Lee Maltings
Cork
Ireland

Aurelien Du Pasquier
Rutgers, The State University of
New Jersey
Department of Materials Science and
Engineering
Energy Storage Research Group
671, Highway 1
North Brunswick NJ 08902
USA

Nelson Durán
Universidade Estadual de Campinas
Instituto de Química
P.O. Box 6154
13083-970, Campinas, SP
Brazil

Omar Elkedim
Université de Technologie de
Belfort-Montbéliard
Génie Mécanique et Conception
Rue du Château
90010 Belfort cedex
France

Adriana Ferancová
Slovak University of Technology
Faculty of Chemical and Food
Technology
Institute of Analytical Chemistry
Radlinského 9
81237 Bratislava
Slovakia

Renato G. Freitas
Universidade Federal de São Carlos
Departamento de Química
CX.P.: 676
13560-970. São Carlos, SP
Brazil

Stephen A. Hackney
Michigan Technological University
Department of Materials Science and
Engineering
1400 Townsend Drive
Houghton, MI 49931
USA

Nanostructured Materials in Electrochemistry. Edited by Ali Eftekhari
Copyright © 2008 WILEY-VCH Verlag GmbH & Co. KGaA, Weinheim
ISBN: 978-3-527-31876-6

Mieczyslaw Jurczyk
Poznan University of Technology
Institute of Materials Science and Engineering
Sklodowska-Curie 5 Sq.
60-965 Poznan
Poland

Ján Labuda
Slovak University of Technology
Faculty of Chemical and Food Technology
Institute of Analytical Chemistry
Radlinského 9
81237 Bratislava
Slovakia

Yvonne H. Lanyon
University College
Tyndall National Institute
Lee Maltings
Cork
Ireland

Current Address

Stirling Medical Innovations Ltd.
FK9 4NF
United Kingdom

Nathan S. Lawrence
Schlumberger Cambridge Research
High Cross
Madingley Road
Cambridge CB3 0EL
United Kingdom

Han-Pu Liang
Schlumberger Cambridge Research
High Cross
Madingley Road
Cambridge CB3 0EL
United Kingdom

Luiz H.C. Mattoso
EMBRAPA-CNPDIA
Rua XV de Novembro, 1452
São Carlos
13560-970 São Carlos, SP
Brazil

Shuji Nakanishi
Osaka University
Graduate School of Engineering Science
Division of Chemistry
1-3 Machikaneyama, Toyonaka
Osaka 560-8631
Japan

Marek Nowak
Poznan University of Technology
Institute of Materials Science and Engineering
Sklodowska-Curie 5 Sq.
60-965 Poznan
Poland

Cristiane P. Oliveira
Universidade Federal de São Carlos
Departamento de Química
CX.P.: 676
13560-970 São Carlos, SP
Brazil

Arnaldo C. Pereira
Universidade Estadual de Campinas
Instituto de Química
P.O. Box 6154
13083-970, Campinas, SP
Brazil

Ernesto C. Pereira
Universidade Federal de São Carlos
Departamento de Química
CX.P.: 676
13560-970 São Carlos, SP
Brazil

Mattias Strömberg
Uppsala University
Department of Engineering Sciences
Division of Solid State Physics
The Ångström Laboratory
Box 534
Lägerhyddsvägen 1
751 21 Uppsala
Sweden

Maria Strømme
Uppsala University
Department of Engineering Sciences
Division of Solid State Physics
The Ångström Laboratory
Box 534
Lägerhyddsvägen 1
751 21 Uppsala
Sweden

Grzegorz D. Sulka
Jagiellonian University
Department of Physical Chemistry and Electrochemistry
Ingardena 3
30060 Krakow
Poland

Sima Valizadeh
Uppsala University
Department of Engineering Sciences
Division of Solid State Physics
The Ångström Laboratory
Box 534
Lägerhyddsvägen 1
751 21 Uppsala
Sweden

1
Highly Ordered Anodic Porous Alumina Formation by Self-Organized Anodizing
Grzegorz D. Sulka

1.1
Introduction

Nanotechnology, in combination with surface engineering focused on a fabrication of various nanostructures and new materials, has recently attracted a vast amount of research attention, and has become a subject of intense scientific interest. Particularly, the inexpensive formation of periodically ordered structures (e.g., nanopore, nanotubes and nanowire arrays) with a periodicity lower than 100 nm, has triggered extensive activities in research. The present, huge progress in nanotechnology is a direct result of the modern trend towards the miniaturization of devices and the development of specific instrumentation that could visualize the nanoworld and allow surfaces to be studied at nanoscale resolution. Among various technologies that allow the visualization and characterization of nanomaterials and nanosystems, scanning probe microscopy (SPM) techniques and especially, scanning tunneling microscopy (STM) or atomic force microscopy (AFM), must definitely be quoted [1,2]. Recently, the basis of STM/AFM instrumentation, including a near-field imaging process and the use of piezoelectric actuators, has been successfully adapted to several techniques [3]. Consequently, scanning near-field optical microscopy (SNOM), photon scanning tunneling microscopy (PSTM), magnetic force microscopy (MFM) or scanning thermal profiling (STP) have begun to be used widely, not only for the surface imaging and characterization of materials in nanometer scale but also for providing additional information on surface properties [3].

The strong reduction of the dimensions and precise control of the surface geometry of nanostructured materials has resulted in the occurrence of novel and unique catalytic, electronic, magnetic, optoelectronic and mechanical properties. The unique properties of the nanostructure, or even of an integral functional unit consisting of multiple nanostructures, are the result of the collective behavior and interaction of a group of nano-elements acting together and producing responses of the system as a whole [4]. The potential of highly ordered nanomaterials for future technological applications lies mainly in the field of various nanophotonic,

Nanostructured Materials in Electrochemistry. Edited by Ali Eftekhari
Copyright © 2008 WILEY-VCH Verlag GmbH & Co. KGaA, Weinheim
ISBN: 978-3-527-31876-6

photocatalytic, microfluidic and sensing devices, as well as functional electrodes and magnetic recording media.

A huge variety of nanodevices based on nanostructured materials has been reported recently in the literature. For instance, two-dimensional (2D) photonic crystals are seriously taken into account as very useful nanostructures used for the construction of various important functional devices [5]. Photonic crystals are periodic dielectric structures having a band gap that prohibits the propagation of a certain frequency range of electromagnetic waves. The integration of photonic crystals made from patterned semiconductor nanostructures with active optoelectronic devices has been studied [6,7]. Photonic crystals have been proposed as mirrors in a single-mode semiconductor laser with small cavity length, and also as a tunable laser with a tuning range of over 30 nm [7]. Recently, three-dimensional (3D) photonic crystals have been prepared by the electrochemical etching of silicon and subsequent pore-widening treatment [8]. The array of nanochannels with a diameter of 30 nm fabricated on silicon can be used for separation, cell encapsulation, and drug release [9]. Silicon nanowires as highly sensitive biosensor devices, allowing the electrical detection of selective adsorption biomolecules, such as specific proteins related to certain types of cancer, has also been reported [10]. Nanoparticles, nanowires and nanotubes have been extensively studied by Vaseashta *et al.* [11] in order to determine their biocompatability and their further possible use in the detection of molecular binding for molecules such as DNA, RNA, proteins, cells, and small molecules. Semiconductor quantum dots (QDs), in conjugation with biomolecules, have been used to produce a new class of fluorescent probes and QD–antibody complexes [12]. These nanoscale semiconductor QDs, which have affinities for binding with selected biological structures, have been used to study the dynamics of various biological processes, including neuronal processes [12]. Self-assembly QDs have also been used for manufacturing a quantum-dot field-effect and quantum-dot memory devices [13]. Electrochemical biosensors with a unique electrocatalytic properties used for electro-analytical purposes have been fabricated on a basis of carbon nanotubes [14–16] and other nanoporous materials [17,18]. Nanoporous, well-ordered materials have been used for gas moisture measurement and the successful fabrication of humidity sensors [19,20]. Currently, there is a great demand in the microelectronics industry for the ideal magnetic medium consisting of a 2D array of ordered islands with nanometer dimensions. Patterned magnetic media fabricated directly by various lithography techniques and regular arrays of magnetic dots and wires obtained by a template synthesis approach have been one of the more widely discussed possibilities for useful devices to extend the density of recording and information storage [21–26]. Magnetic structures have been also used for the fabrication of a nanoscale single domain magnetoresistive bridge sensor, and for a new ultra-high-resolution tip of the magnetic force microscope [27]. A single-electron memory device, in which one bit of information can be stored by one electron [28], an organic photovoltaic device based on polymer nanowires [29], or a fast-response hydrogen sensor based on arrays of palladium wires [30,31] have also been presented. Only limited applications of nanostructured materials for the manufacture of nanodevices have been presented in this chapter. However, major reviews of various devices based on nanowires, nanotubes and nanoporous materials have been reported in the literature [2,32–36].

Lithography patterning techniques can be used directly to create various nanoparticles, nanowires and nanotubes arranged with highly ordered arrays [37]. By using lithography techniques, all of the processing steps to accomplish the pattern transfer from a mask to a resist and then to devices, can be successfully performed with ultra-high precision, and even nanosize resolution. Lithography techniques can be also employed for the preparation of nanoporous membranes and various templates used for the subsequent deposition of metals [38]. The fabrication of high-ordered nanostructures with the periodicity less than 50 nm is beyond what a conventional optical lithography could afford [39,40]. Advanced non-optic lithographic techniques, such as electron-beam [41–45], ion-beam [46,47], X-ray [48], interference or holographic lithographies [49–51], can replicate patterns with a sufficient resolution of few nanometers, but required sophisticated facilities. Moreover, the high cost of lithographic equipment makes these techniques unavailable to many researchers. Notwithstanding the undeniable advantages of these techniques, certain major drawbacks exist. For example, a low aspect ratio (the ratio of length to diameter) of the formed nanostructure and the high cost of its preparation limit the applications of lithography techniques to the laboratory scale.

Therefore, in order to overcome such drawbacks of conventional lithographic methods, nanoimprint lithography (NIL) as a high-throughput and low-cost method has been developed for fabricating nanometer-scale patterns [52,53]. In nanoimprint lithography, a nanostructured mold is pressed into a resist film existing on a substrate, and in this way a thickness contrast pattern in the resin is created. The duplicated nanostructure in the resist film is then transferred to the substrate by reacting ion etching. Today, the use of this method has rapidly spread quickly and it is now used for the preparation of arrays of various materials [27,54,55].

Among several nanolithographic techniques, scanning probe lithography employing STM and AFM has been considered one of the best tools for atomic level manipulation and forming nanostructures [56–59]. The surface nanostructuring within this method is based on the chemical vapor deposition (CVD) process, local electrodeposition or voltage pulses.

Nanosphere lithography (NSL) employs a hexagonal, close-packed monolayer or bilayer of spheres formed on a supporting substrate (e.g., Au, Si, glass). Spheres on the substrate are arranged by a self-organized process upon solvent evaporation or drying [37,60–62]. Close-packed monolayers of submicron-diameter monodisperse polymer spheres are then formed from a chemical solution spread over the substrate surface. NSL is a simple and inexpensive process which can generate structures even on a curved surface with a high throughput. Self-assembled, hexagonally arranged nano-sized latex, polystyrene or silica spheres have been used as a lithography shadow mask to generate a variety of 2D metal nanoparticle arrays. Au [18], Ni [63], Fe_2O_3 and In_2O_3 [64] hexagonal, close-packed nanoarrays have been obtained by a subsequent deposition or etching through the nanosphere mask. Well-ordered arrays of nanostructures of complex oxides including $BaTiO_3$ and $SrBi_2Ta_2O_9$ has been obtained by NSL in a combination with pulsed laser deposition [65].

Surfaces with periodic structure can be produced by a self-organized process occurring during the ion irradiation of surface, optical or electron-beam lithography, molecular beam epitaxy, arc discharge or evaporation, and laser-focused atomic deposition [22,23,37]. Facets and steps appear spontaneously, or strain relief can create a self-organized nanopattern on the metal or semiconductor surface [22,66–68]. The local periodicity of nanostructure and limited surface area of ordered domains are insuperable disadvantages of the method.

Enormous progress has been made in the preparation of various highly ordered and uniform nanostructures through wet chemical processes employing microemulsion systems, reverse micelles and self-assembled close-packed monolayers or bilayers [23,69,70]. The self-assembled growth of nanoparticles with highly monodisperse diameters or nanowires occurs by the decomposition or reduction of precursors in a solution [71–76], solid matrix [77], and reverse micelles [78]. For example, Co monodisperse nanoparticles with a diameter of 6 nm and self-ordered into spherical superstructures, as well as a hexagonal close-packed self-organized array of $Fe_{50}Pt_{50}$, have been prepared by reduction in solution [71]. The nanoscale patterning and fabrication of nanostructured materials do not necessarily require sophisticated methods and equipment, and remarkable examples of nanoparticles and nanowires synthesis have been presented. Extraordinary hopper- and flower-shaped silver oxide micro- and nanoparticles have been prepared electrochemically by the anodization of silver wires [79]. Ag nanowires have been successfully synthesized from a solution containing surfactant by a simple ultraviolet (UV) photochemical reduction method [80]. Lamellar mono- and bilayers have been used as microreaction matrices to form nanoparticles and nanowires. Selective copper and cobalt deposition on e-beam patterned thiol self-assembled monolayers on a solid support has also been studied [81]. The smallest structures produced in this way were 30 nm-wide Co or Cu lines and 30 nm dots. The ability of surfactants and lipids to form self-assembled monolayers or bilayers on a solid support has been used for the nanostructuring of materials [82,83]. The Langmuir–Blodgett film of surfactant-stabilized Si and SiO_2 nanowires, parallel aligned with a controlled pitch, has been used as a template for the transfer of its arrangement into a chromium substrate over areas up to 20 cm^2 [84]. Soft chemical routes based on sol–gel techniques and wet impregnation methods, often with modern modifications, have been employed successfully to synthesize a large number of inorganic nanotubes, nanorods such as: CeO_2, ZrO_2, HfO_2, TiO_2, Al_2O_3 [85–89] and hierarchically ordered mesoporous SiO_2, Ni_2O_5 [90] or nanocomposite materials [91].

Based on the electrochemical techniques used for nanostructure manufacture, a great number of sophisticated nanomaterials have been prepared on various substrates. Highly oriented pyrolytic graphite (HOPG) with a low-energy electrode surface has been used extensively for nanoparticle and nanowire electrodeposition. Monodispersed in size Ni [92,93], Cd, Cu, Au, Pt and MoO_2 nanoparticles [93,94] have been deposited on the HOPG surface. The electrochemical step edge decoration method has been applied for the selective deposition of metal at step edges existing at HOPG. A great variety of nanomaterials, including Mo [95,96], Cu, Ni, Au, Ag, Pd [97–102], MoO_2 [95,96], MnO_2 [103], MoS_2 [104], Bi_2Te_3 [105] nanowires,

3D Co nanocrystal array [106], Ag nanoflakes and nanorods [107], have been prepared using this method. The synthesis of nanostructures through electrochemical means has also been studied on different substrates. Copper nanowires with a diameter of between 10 and 20 nm have been prepared on a glass substrate covered with a thin layer of CuI under a direct current electric field [108]. Spontaneous displacement reactions have been successfully employed for electroless Ag deposition into patterned Si [109], Pt, Pd, Cu and Ag on a HOPG surface with crystals of a ferrocene derivative [110], and Cu on AlN substrate with a patterned copper seed layer [111].

A template synthesis has recently proved to be an elegant, inexpensive and technologically simple approach for the fabrication of various nanoscale sophisticated materials. These alternative methods overcome many of the drawbacks of lithographic techniques, and exploit templates which differ in material, pattern, range of order and periodicity, feature size and overall size of templates. Templating methods employing a variety of porous membranes and films are commonly used for the synthesis of high-density, ordered arrays of nanodots, nanotubes and nanowires. Recently, scientific attention has been focused on two types of template: track-etched polymeric and porous alumina membranes [36]. Although the origin of the template, used for nanofabrication, may also be lithographic, the high cost of its preparation drastically limits the applications. Although the aspect ratio achievable is not high, lithographically prepared templates have the undeniable advantage of exploiting a much wider spectrum of substrate choice than do porous membranes. The most significant drawback of membrane-templating methods is a poor long-range order in the formed nanostructure, even though a short-range order is maintained satisfactorily. Nickel nanowire arrays have been prepared by electrodeposition into the porous single-crystal mica template with a uniform size of pores and orientation [112,113]. Mesoporous thin films (MTFs) as templates have been used for the successful synthesis of ordered arrays of Ge [114], Co and Fe_3O_4 [115] nanowires. Cobalt dots [116,117], monodisperse Au microwire arrays [118] and Sn nanotubes [119] have also been deposited on the porous silicon substrate. During the past decade, an enormous variety of metallic nanowires has been synthesized electrochemically within the pores of the nuclear track-etched polycarbonate membrane. Since it has been discovered that the deposition of materials in porous membranes proceeds preferentially on the pore walls [120,121], polycarbonate membranes are widely used for the fabrication of Co [122–132], Ni [122,133,134], Cu [124,134,135], Pb [136] nanowires, ferromagnetic NiFe [133,137] or CoNiFe [137] nanowire arrays and Co/Cu [124,138–142], NiFe/Cu [138,140–143], Ag/Co [144] multilayered nanowires. An electrochemical deposition method using the polycarbonate membrane, has been also employed for a successful preparation of Co, Fe, Ni [145], Bi [146] rods, conducting polypyrrole nanotubes and polypyrrole–copper nanocomposites [146]. The physical process of solvent volatilization from the polycarbonate membrane filled with an appropriate solution has been used for the synthesis of manganite oxide-based ($La_{0.325}Pr_{0.300}Ca_{0.375}MnO_3$) nanowires and nanotubes [147]. Cobalt nanoparticles with a very fine diameter of 3.5 nm have been synthesized in the perfluorinated sulfo-cation exchange polymeric membrane by using the ion-exchange method [148]. Recently, ordered thin films of diblock copolymer have been used to fabricate periodic arrays of holes with nanometer dimensions [46,149–151]. The template strategy employing a thin film of the self-assembly diblock copolymer in combination with electrodeposition, has been used to prepare Co nanowires with a diameter of 12 or 24 nm [152] and highly ordered nanoelectrode arrays [153]. The concept of using anodic porous alumina membranes as templates for fabrication-ordered nanowire and nanotube arrays seems to be the most widespread among research groups, and will be discussed in Section 1.5.

A considerable effort has been also placed on developing alternative nanopatterning methods. For example, a variety of methods have been developed for the fabrication of ordered structures on a nanometer length scale, including physical vapor deposition (PVD) [154,155], catalyzed vapor-phase transport processes [156,157], CVD based on the vapor–liquid–solid (VLS) growth mechanism [4,36,158], a direct self-assembly route that exploits a mask with protruding patterns [159], laser-assisted direct imprinting [160], self-patterning methods based on the instability of ultrathin films during a high-temperature treatment [161], and template synthesis in conjunction with the pressure injection filling technique [36]. An extraordinarily novel method based on the unidirectional solidification of monotectic alloys such as Al–In and NiAl–Re and subsequent electrochemical etching has been developed for the preparation of Al and NiAl nanoporous arrays [162–164].

1.2
Anodizing of Aluminum and Anodic Porous Alumina Structure

Currently, much effort has been undertaken to develop an effective and technologically simple method used for the synthesis of nanostructures over a macroscopic surface area. Today, the research spotlight is especially focused on self-organized nanostructured materials with a periodic arrangement of nanopores due to the high expectations regarding their applications. A highly desired densely packed hexagonal array nanoporous structure can be obtained by anodization, which is a relatively easy process for nanostructured material fabrication. The electrochemical formation of self-organized nanoporous structures produced by the anodic oxidation of semiconductors or metals, except aluminum, has been reported for only a few materials such as Si [165–172], InP [173,174], Ti [175–183], Zr [181,184–186], Nb [187–190], Hf [191] and Sn [192]. During recent years, the anodization of aluminum, due to its great commercial significance, represents one of the most important and widespread method used for the synthesis of ordered nanostructures consisting of close-packed cells in a hexagonal arrangement with nanopores at their centers.

Over the past decades, the anodizing of aluminum has raised substantial scientific and technological interest due to its diverse applications which include dielectric film production for use in electrolytic capacitors, increasing the oxidation resistance of materials, decorative layers by incorporation of organic or metallic pigments during the sealing of anodized materials, and increasing abrasion wear resistance

[193–196]. Recently, a self-organized process that occurs during the anodization of aluminum in acidic electrolytes has became one of the most frequently employed method for the synthesis of highly ordered nanostructures.

1.2.1
Types of Anodic Oxide Film

In general, the anodizing of aluminum can result in two different types of oxide film: a barrier-type anodic film, and a porous oxide film. It was generally accepted that the nature of an electrolyte used for anodizing aluminum is a key factor which determines the type of oxide grown on the surface [197–199]. Recent progress in the anodizing of aluminum in various media highlights the depth of the problem of electrolyte influence on porous oxide layer formation. The capability of porous anodic oxide development from the initially formed barrier-type film was reported elsewhere [200–204]. Nevertheless, it has long been believed that a strongly adherent, non-porous and non-conducting barrier-type of anodic film on aluminum can be formed by anodizing in neutral solutions (pH = 5–7) in which the anodic oxide layer is not chemically affected and stays practically insoluble. These films are extremely thin and dielectrically compact. The group of electrolytes used for this barrier-type film formation includes boric acid, ammonium borate, ammonium tartrate and aqueous phosphate solutions, as well as tetraborate in ethylene glycol, perchloric acid with ethanol and some organic electrolytes such as citric, malic, succinic, and glycolic acids [200,202,205,206]. Recently, a compact, and dense anodic oxide layer has been obtained by the anodization of aluminum in 5-sulfosalicylic acid [207]. In contrast, porous oxide films were reported mainly for the anodizing of aluminum in strongly acidic electrolytes, such as sulfuric, oxalic, phosphoric and chromic acid solutions, where the resulting oxide film can be only sparingly soluble [197,208]. For anodic porous alumina, the film growth is associated with localized dissolution of the oxide, as a result of which pores are formed in the oxide film. However, the porous anodic layer formation has been recently reported for various acidic electrolytes such as malonic [209–211], tartaric [211–213], citric [214–217], malic [213], glycolic [213] and even chromic acid [218]. Anodic porous oxide films have been also obtained in unpopular electrolytes, including a mixed solution of phosphoric and organic acids with cerium salt [219], or in a mixture of oxalic acid, sodium tungstate, phosphoric and hypophosphorous acids [220]. The porous oxide film was also observed for anodization carried out in a fluoride-containing oxalic acid electrolyte [221]. The formation of nanopores by self-organized anodization has also been studied in a mixture of sulfuric and oxalic acids [222]. In fact, these results show that is no distinct difference in the selection of an electrolyte used for the formation of barrier or porous films during the anodization of aluminum. Moreover, the transition from a barrier-type film to porous oxide occurs easily. The anodizing time is a key factor responsible for the development of porous oxide structure on previously formed barrier-type film [202,222]. Some excellent reviews on barrier-type oxide films have been produced [200,202,203,205], whilst details of the porous oxide layers are reported elsewhere [195,198,200,202,203,205,208,224]. The porous oxide film formation resulting in a self-organized, highly ordered nanopore array is one subject of this chapter.

The anodic behavior of aluminum in electrolytes used for anodizing is not limited to the formation of a barrier-type film or a porous oxide layer. A fiber-like porous oxide morphology has been reported for the anodization of aluminum in chromic [225] and sulfuric acid [213,226]. The completely different morphology of the oxide layer formed by anodizing in oxalic or sulfuric acid has been observed by Palibroda et al. [227]. A cross-sectional view of an anodized layer exhibits hemispherical and spherical patterns that appear during anodization as a result of simultaneous gas evolution and the formation of a gel-like, nascent oxide. In addition, other anodic behaviors such as localized pitting corrosion in monobasic carboxylic acids (e.g., formic, acetic) [200] and in the presence of halide ions [200,202,228,229], burning [211,216,230] or crystallographic dissolution [231] have also been reported.

1.2.2
General Structure of Anodic Porous Alumina

The specific nature of a porous oxide layer on aluminum has attracted scientific attention over several decades, and determined fully the present applications of anodized aluminum in nanotechnology. Self-organized anodic porous alumina grown by the anodization of aluminum can be represented schematically as a closed-packed array of hexagonally arranged cells containing pores in each cell-center (Figure 1.1). High-ordered nanostructures are often characterized by given parameters such as a pore diameter, wall thickness, barrier layer thickness and interpore distance (cell diameter). The uniform pore diameter, which is easily controllably by altering the anodizing conditions, can range from a few nanometers to hundreds of nanometers. The depth of fine parallel channels can even exceed 100 μm, a characteristic which makes anodic porous alumina one of the most desired nanostructures with a high aspect ratio and high pore density. A parallel growth of controlled dimensional pores can proceed throughout the complete anodized material.

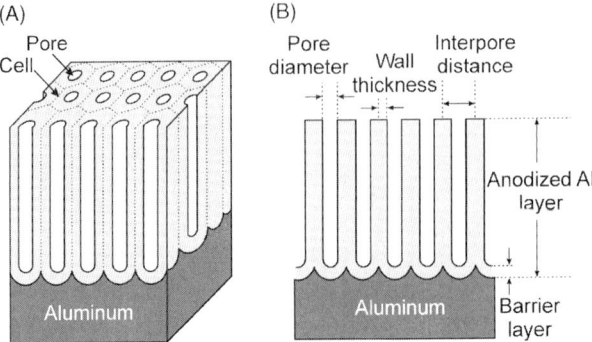

Figure 1.1 Idealized structure of anodic porous alumina (A) and a cross-sectional view of the anodized layer (B).

Growth of the oxide layer takes place at the metal/oxide interface at the pore bottoms, and involves the conversion of a preexisting, naturally occurring film on the surface into the barrier-type film and further into a porous oxide layer. During the porous oxide growth, a thin and compact barrier layer at the pore bottom/electrolyte interface is continuously dissolved by locally increased field, and a new barrier layer at the metal/oxide interface is rebuilt. For steady-state film growth, there is a dynamic equilibrium between the rate of film growth and its field-assisted dissolution. All of the major anodic film parameters are directly dependent on forming a voltage attendant, steady-state growth of the porous oxide layer. During the steady state of oxide growth, current density under potentiostatic anodizing or anodizing potential under galvanostatic anodizing remains almost unchanged. The cylindrical, in-section pores appear as a result of such steady-state growth.

1.2.2.1 Pore Diameter

Generally, for the anodic porous alumina structure, the pore diameter is linearly proportional to the anodizing potential with a proportionality constant λ_p of approximately 1.29 nm V^{-1} [208]:

$$D_p = \lambda_p \cdot U \tag{1}$$

where D_p is a pore diameter (nm) and U denotes an anodizing potential (V). The dependence of the diameter on the voltage is not sensitive to the electrolyte. In describing the anodic porous alumina structure, researchers usually list the outer layer of oxide close to the surface, and the inner layer close to the pore bottoms. The diameter of pores in the inner oxide layer does not change significantly with anodizing time [197,203]. A higher diameter of pores observed in a region close to the film surface is the result of an irregular initial growth of the pores during the very early stages of pore development, and their further reorganization in an hexagonal arrangement. It should be noted that an enhanced chemical dissolution of oxide, resulting in the development of widened pores, may occur also during anodizing at a sufficiently high temperature, or in strong acidic solutions [200,203]. The solvent chemical action along the cell walls, and especially in the outer oxide layer, increases the diameter of pores measured at the surface of the anodic film [232].

During the early stage of collecting research data concerning the structure of anodic porous alumina, it was believed that pore diameter was independent of the forming potential [197]. However, later studies have reported the pore diameter as a parameter which depends upon the anodizing potential or current density [233]. According to O'Sullivan and Wood [208], the pore diameter for anodizations conducted at a constant anodizing potential can be calculated as follows:

$$D_p = D_c - 2 \cdot W = D_c - 1.42 \cdot B = D_c - 2 \cdot W_U \cdot U \tag{2}$$

where D_c is the cell diameter, interpore distance (nm), W is the wall thickness (nm), B is the barrier layer thickness (nm), and W_U is the wall thickness per volt (nm/V).

For potentiostatic conditions of anodization, wall thickness divided by the anodizing voltage gives the thickness of oxide-wall per applied volt (W_U). The empirical dependence of pore diameter on the anodizing potential or the ratio of anodizing potential to a critical value of the potential (U_{max}) has been reported by Palibroda [234]:

$$D_p = 4.986 + 0.709 \cdot U = 3.64 + 18.89 \cdot \frac{U}{U_{max}} \qquad (3)$$

The critical potential is an experimentally determined maximum anodizing potential which can be applied during anodization, without triggering intense gas evolution on the aluminum. In describing a 10% porosity rule, Nielsch et al. [235] suggested that a diameter of pores formed by the anodizing aluminum under optimum self-ordering conditions, leading to a quasiperfect hexagonal arrangement of pores, can be calculated from the following equation:

$$D_p = \sqrt{\frac{2\sqrt{3} \cdot P}{\pi}} k \cdot U \qquad (4)$$

where P is porosity ($P = 0.1 = 10\%$) and k is the proportionality constant ($k \cong 2.5$).

Obviously, the temperature of the electrolyte and the hydrodynamics conditions in the electrolytic cell affects the pore diameter. At higher temperatures of anodizing (e.g., near room temperature) a significant acceleration of chemical dissolution of the outer oxide layer, especially in a strong acidic solution is expected. On the other hand, stirring of the electrolyte during anodizing under the constant anodizing potential causes a significant increase in the local temperature at the inner oxide layer, and the recorded current density increases [236]. As a result of the increasing local temperature, a chemical dissolution of oxide in the inner layer, as well as the electrochemical formation of anodic oxide layer, are accelerated. The effect of electrolyte stirring on a pore diameter is shown schematically in Figure 1.2.

These experimental data show that the effect of electrolyte temperature and electrolyte concentration on pore diameter can be totally different for potentiostatic and galvanostatic conditions of anodizing [208,237,238]. For instance, the pore

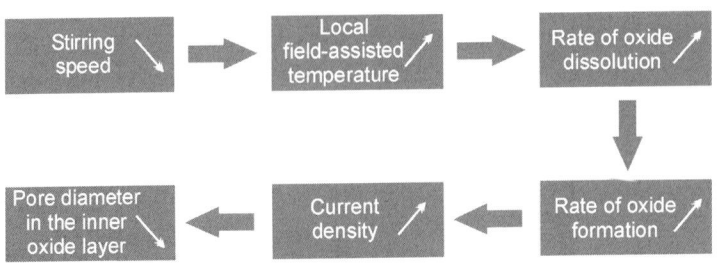

Figure 1.2 Effect of stirring on the pore diameter of nanostructures obtained by anodization of aluminum.

Table 1.1 Relationship between pore diameter and potential for anodization conducted in 2.4 M H_2SO_4 at various anodizing temperatures [240].

Potential range (V)	Temperature (°C)	$D_c = f(U)$ (nm)
15–25	−8	$1.06 + 0.80 \cdot U$
	1	$12.35 + 0.53 \cdot U$
	10	$9.34 + 0.72 \cdot U$

diameter has been found to depend on temperature for the constant potential anodizing in a phosphoric acid solution [208]. With increasing temperature of the electrolyte, an increase in pore diameter in the outer oxide layer has been noted. At the same time, a decrease in pore diameter in the inner oxide layer has been observed. It was found that the concentration of electrolyte does not significantly influence the pore diameter [208], although more recently it has been suggested that the pore diameter decreases with decreasing pH of the solution [239] and temperature [240]. A decrease in pore diameter with increasing concentration of the acidic electrolyte or decreasing pH can be attributed to the decreasing threshold potential for a field-assisted dissolution of oxide at pore bottoms, resulting in an enhanced rate of anodic oxide formation [235]. The temperature influence on the pore diameter has been reported recently in detail for sulfuric acid [240]. The linear relationships have been established for a variety of anodizing temperatures (Table 1.1).

A summary of the influence of anodizing parameters on the pore diameter of a nanostructure formed under potentiostatic conditions is presented schematically in Figure 1.3.

When anodizing under a galvanostatic regime, the pore diameter is affected by current density, electrolyte temperature, and acid concentration. As might be

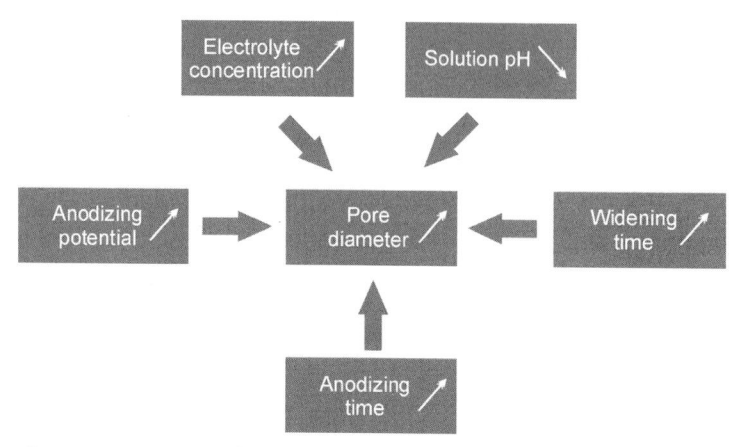

Figure 1.3 Parameters influence on the pore diameter of nanostructure formed by anodization of aluminum at potentiostatic regime.

expected, with increasing current density the pore diameter increases [233]. An increase in the temperature of phosphoric acid, when used as an electrolyte, results in a decrease in the pore diameter measured in the inner oxide layer, although only slight changes in pore diameter are observed in the outer oxide layer [208]. Additionally, the pore diameter decreases with increasing concentration of phosphoric acid. On the other hand, there was no observable evidence of the influence of anodizing temperature on pore diameter for anodization conducted in oxalic and sulfuric acids [237,238]. The final pore diameter of the nanostructure formed by the anodization of aluminum is a superposition of those parameters influences. Therefore, a significant variation in pore diameter determined for similar anodizing conditions and electrolytes is reported in the literature.

1.2.2.2 Interpore Distance

It is generally accepted that the interpore distance of anodic porous alumina is linearly proportional to the forming potential of the steady-state growth of anodic porous alumina with a proportionality constant λ_c of approximately 2.5 nm V^{-1} [235]:

$$D_c = \lambda_c \cdot U \tag{5}$$

According to Keller *et al.* [197], the cell diameter can be calculated precisely from the following equation:

$$D_c = 2 \cdot W + D_p = 2 \cdot W_U \cdot U + D_p \tag{6}$$

The linear dependence of interpore distance on anodizing potential is assumed on the hypothesis that pore diameter is independent of anodizing voltage. O'Sullivan and Wood [208] have found that the wall thickness is about 71% of the barrier layer thickness. Taking into account this fact, the following expression can be proposed:

$$D_c = 1.42 \cdot B + D_p \tag{7}$$

An in-depth study of aluminum anodization in sulfuric and oxalic acid has been conducted by Ebihara *et al.* [237,238]. Experimental equations describing the relationship between the interpore distance and anodizing potentials have been established as follows:

$$H_2SO_4 \text{ [238]}: \quad D_c = 12.1 + 1.99 \cdot U \quad (U = 3-18 \text{ V}) \tag{8}$$

$$C_2H_2O_4 \text{ [237]}: \quad D_c = 14.5 + 2.00 \cdot U \quad (U \leq 20 \text{ V}) \tag{9}$$

$$D_c = -1.70 + 2.81 \cdot U \quad (U \geq 20 \text{ V}) \tag{10}$$

The dependence of anodizing potential on interpore distance has been observed by Hwang *et al.* [241]. For anodization in oxalic acid conducted under anodizing

Table 1.2 Relationship between interpore distance and potential for anodization conducted in 2.4 M H_2SO_4 at various anodizing temperatures [240].

Potential range (V)	Temperature (°C)	$D_c = f(U)$ (nm)
15–25	−8	$12.23 + 1.84 \cdot U$
	1	$12.20 + 2.10 \cdot U$
	10	$12.72 + 1.87 \cdot U$

potential ranges between 20 and 60 V, the linear equation can be expressed as follows:

$$D_c = -5.2 + 2.75 \cdot U \tag{11}$$

The interpore distance has been found to be slightly dependent or independent on the temperature of anodization at the constant anodizing potential regime [208]. For anodizing carried out in oxalic acid, the interpore diameter has been found to be independent of the electrolyte temperature [241]. Contrary to this, for self-organized anodizing in sulfuric acid, an influence of temperature on interpore distance has been observed [240]. For the highest studied temperature of anodization, 10 °C, the calculated interpore distance increased by about 8–10% in comparison to that at a temperature of 1 or −8 °C. Equations describing the interpore distance as a function of anodizing potential are collected in Table 1.2 for a variety of temperatures.

The study performed by O'Sullivan and Wood [208] showed that increasing the concentration of electrolyte decreases the interpore distance. In the case when anodizing is carried out in phosphoric acid at constant current density, increasing the temperature of anodizing as well as increasing the electrolyte concentration causes a decrease in the interpore distance in nanostructures [208]. An unchanged cell size was observed when anodizing was conducted in phosphoric and oxalic acids [237,238].

The interpore distance can be also calculated from 2D Fourier transforms of anodic alumina triangular lattice. The 2D Fourier transform provides unique information about the structure periodicity in the inverse space. According to Marchal and Demé [242], the interpore distance for the triangular lattice with the hexagonal arrangement of nanopores can be calculated from the following expression:

$$D_c = \frac{4\pi}{\sqrt{3} \cdot Q_{10}} \tag{12}$$

where Q_{10} is the position of the first Bragg reflection of the hexagonal arrangement of nanopores.

1.2.2.3 Wall Thickness

Among other morphological features of anodic porous oxide described by Keller et al. [197], a wall thickness per volt for some acidic electrolytes frequently used for anodization of aluminum (Table 1.3) appears to be of importance.

Table 1.3 Wall thickness per volt for various anodizing electrolytes [197].

Electrolyte	0.42 M H_3PO_4	0.22 M $H_2C_2O_4$	1.7 M H_2SO_4	0.35 M H_2CrO_4
Anodizing temperature (°C)	24	24	10	38
Wall thickness per volt (nm V^{-1})	1.00	0.97	0.80	1.09

The transformation of Eq. (1) gives the following form for the wall thickness calculation:

$$W = \frac{D_c - D_p}{2} \tag{13}$$

According to O'Sullivan and Wood [208], the wall thickness built during anodizing in phosphoric acid is related to the barrier layer thickness as follows:

$$W = 0.71 \cdot B \tag{14}$$

For anodization conducted in the oxalic acid solution, Ebihara et al. [237] found that the proportionality between wall thickness and barrier layer thickness varies slightly with anodizing potential in the range of 5 to 40 V. For anodizing potentials between 5 and 20 V, a relationship between W and B with a proportionality constant of about 0.66 has been observed, while for higher anodizing potential a gradual increase in the proportionality constant to the final value of about 0.89 has been reported.

1.2.2.4 Barrier Layer Thickness

During the anodization of aluminum, a very thin, dense and compact dielectric layer is formed at pore bases. The barrier layer has the same nature as an oxide film formed naturally in the atmosphere, and allows the passage of current only due to existing faults in its structure. The existing compact barrier layer at the pore bottoms makes the electrochemical deposition of metals into pores almost impossible. On account of this limit, the thickness of the barrier layer is extremely important and can determine any further applications of nanostructures formed by the anodization of aluminum. The thickness of the barrier layer depends directly on the anodizing potential. The dependence is about 1.3–1.4 nm V^{-1} for barrier-type coatings, and 1.15 nm V^{-1} for porous structures [195].

Some variations have been reported in the barrier layer thickness with anodizing potential or the concentration of electrolyte. Evidence of experimental values of B_U, known as an anodizing ratio and defined as a ratio between the thickness of the barrier layer and anodizing potential, are listed in Table 1.4 for a variety of electrolytes.

An inspection of the data in Table 1.4 shows that variation of the barrier layer thickness per volt depends on whether oxide films are formed at a constant potential or at constant current density regimes [208]. The increasing temperature

Table 1.4 Anodizing ratio (B_U) for various anodizing electrolytes.

Electrolyte	Current density[a] (mA cm^{-2}) or anodizing potential[b] (V)	Concentration (M)	Temperature (°C)	B_U (nm V^{-1})	Reference(s)
H_3PO_4	100[a]; 80[b]	0.4 (3.8%)	20	0.89[a]; 1.14[b]	[208]
			25	0.90[a]; 1.09[b]	
			30	1.05[a]; 1.04[b]	
		1.5 (13%)	25	1.10[a]; 1.04[b]	
		2.5 (21%)		1.17[a]; 0.82[b]	
	(20–60)[b]	0.42 (4%)	24	1.19[b]	[197,243]
	87[b]		25	0.99[b]	[232]
	103[b]			0.96[b]	
	117.5[b]			1.08[b]	
	87[b]	1.70 (15%)		0.99[b]	
	87[b]	3.10 (25%)		0.97[b]	
$H_2C_2O_4$	(20–60)[b]	0.22 (2%)	24	1.18[b]	[197,243]
	3[b]	0.45 (4%)	30	1.66[b]	[237]
	10[b]			1.40[b]	
	20[b]			1.19[b]	
	30[b]			1.10[b]	
	40[b]			1.06[b]	
H_2SO_4	15[b]	1.70 (15%)	10	1.00	[243]
		1.10 (10%)	21	1.00[b]	[244]
		1.70 (15%)		0.95[b]	
		5.6–9.4 (40–60%)		0.80[b]	
		12.8 (75%)		0.95[b]	
		16.5 (90%)		0.10[b]	
	3[b]	2.0 (17%)	20	1.45[b]	[238]
	10[b]			1.23[b]	
	15[b]			1.05[b]	
	18[b]			0.92[b]	
H_2CrO_4	(20–60)[b]	0.26 (3%)	38	1.25[b]	[244]
$Na_2B_4O_7$	60[b]	0.25 (pH = 9.2)	60	1.3[b]	[201]
$(NH_4)_2C_4H_4O_6$	(25–100)[b]	0.17 (3%, pH = 7.0)	–	1.26[b]	[223]
Citric acid	(260–450)[b]	0.125	21	1.1[b]	[217]

[a]Constant current density anodizing.
[b]Constant potential anodizing.

of anodization decreases the thickness of the barrier layer for anodizing at the constant potential. An opposite relationship is observed, however, for the constant current density anodization. The increasing concentration of phosphoric acid at the constant temperature either decreases or increases the barrier layer thickness for the potentiostatic or galvanostatic anodizing regime, respectively. This can be attributed to the fact that, at constant current density, a constant electric field is maintained across the given barrier layer. The increasing thickness of the barrier layer with increasing electrolyte concentration suggests that ionic conduction becomes easier under the set current density, and that most of the ionic current passes through microcrystallites in the barrier layer. On the other hand, the observed decrease in barrier layer thickness with increasing electrolyte concentration and anodizing temperature is a direct result of an enhanced field-assisted dissolution of oxide at the oxide/electrolyte interface.

Recently, the barrier layer thickness has been determined for other less-popular anodizing electrolytes such as glycolic, tartaric, malic and citric acids [213]. The effect of anodizing potential on the thickness of the barrier layer for the anodic porous alumina formed in various electrolytes is shown in Figure 1.4.

In general, the anodizing ratio (B_U) determined for various anodizing electrolytes is very close to $1\,\text{nm}\,\text{V}^{-1}$ (the diagonal, dotted line in Figure 1.4) over the whole range of anodizing potential. These results suggest a general constant relationship between the anodizing ratio and anodizing potential.

Nielsch et al. [235] suggested that for optimum self-ordering conditions of anodizing, leading to a 10% porosity of the nanostructure and perfect hexagonal arrangement of nanopores, the barrier layer thickness is proportional to the interpore distance as follows:

$$B \cong \frac{D_c}{2} \tag{15}$$

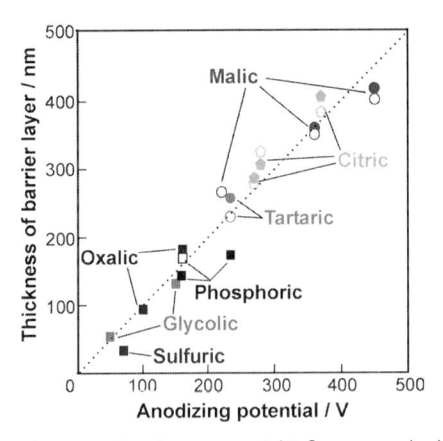

Figure 1.4 Anodizing potential influence on the barrier layer thickness for anodic porous alumina formed in sulfuric, oxalic, glycolic, phosphoric, tartaric, malic, and citric acid solutions. (Solid symbols: measured values; open symbols: calculated values from the half-thickness of the pore walls). (Reproduced with permission from Ref. [213], © 2006, The Electrochemical Society.)

Table 1.5 Effect of anodizing potential on the barrier layer thickness formed at 20 °C.

Electrolyte	Anodizing potential (V)	$B = f(U)$ (nm)	Reference
Potentiostatic anodising			
0.42 M H_3PO_4	$U < 38$	$0.28 + 1.10 \cdot U$	[245]
	$U > 38$	$1.58 + 1.03 \cdot U$	
0.45 M $H_2C_2O_4$	$U < 57$	$5.90 + 0.90 \cdot U$	[246]
	$U > 57$	$0.96 + 0.97 \cdot U$	
Galvanostatic anodising			
0.45 M $H_2C_2O_4$	$U < 55$	$0.92 + 1.09 \cdot U$	[247]
	$U \geq 55$	$5.02 + 0.92 \cdot U$	

Recently, Vrublevsky et al. [245–247] reported empirical equations used for calculating the barrier layer thickness in anodic alumina formed by anodizing in 0.42 M H_3PO_4 and 0.45 M $C_2H_2O_4$ solutions. By using a re-anodizing technique and assuming that the anodizing ratio equals 1.14 nm V^{-1}, the thickness of the barrier layer has been evaluated for anodic film formation at 20 °C. The equations expressing the thickness of the barrier layer as a function of anodizing potential, $B = f(U)$, are presented in Table 1.5.

The wall thickness, as well as the barrier layer thickness, in nanostructures formed by anodization can be easily altered by post-treatment procedures involving chemical etching; this is known as a process of isotropic pore widening.

1.2.2.5 Porosity

The porosity of nanostructures formed by aluminum anodizing depends heavily on the rate of oxide growth, the rate of chemical dissolution of oxide in acidic electrolyte, and anodizing conditions such as: the type of electrolyte, the concentration of electrolyte, time of anodization, anodizing potential, and temperature. The most important factor governing the porosity of the structure is the anodizing potential and pH of the solution. There is a great inconsistency among experimental data on the porosity of nanostructures, with the estimated porosity of anodic porous alumina varying from about 8% to 30%, and even more. An exponential decrease in porosity with increasing anodization potential has been reported for anodizing in sulfuric acid [238,248] and oxalic acid [237]. A decrease in the porosity of nanostructures with increasing anodizing potential has been observed for constant potential anodizations conducted in sulfuric, oxalic, phosphoric, and chromic acids [216,249,250]. On the other hand, a slight increase in porosity is observed with increasing anodizing potential for anodization carried out in sulfuric acid [251]. As might be expected, the porosity of nanostructures may also be affected by the anodizing time, an extension of which usually results in increasing porosity of the nanostructure formed in tetraborate [201] and phosphoric acid [252] solutions. Increasing the anodizing temperature decreases the porosity of the nanostructure formed in oxalic acid [253]; the opposite effect has been observed in sulfuric acid [240].

The porosity is defined as a ratio of a surface area occupied by pores to the whole surface area. For a single regular hexagon with one pore inside, the porosity formulation can be written as follows:

$$\alpha = \frac{S_{pores}}{S} = \frac{S_p}{S_h} \quad (16)$$

Assuming that each single pore is a perfect circle, the following equations for S_p and S_h can be further evolved:

$$S_p = \pi \cdot \left(\frac{D_p}{2}\right)^2 \quad (17)$$

$$S_h = \frac{\sqrt{3} \cdot D_c^2}{2} \quad (18)$$

Substitution of Eqs. (17) and (18) into Eq. (16) leads to the following expression for the porosity of a nanostructure with hexagonally arranged cells:

$$\alpha = \frac{\pi}{2\sqrt{3}} \cdot \left(\frac{D_p}{D_c}\right)^2 = 0.907 \cdot \left(\frac{D_p}{D_c}\right)^2 \quad (19)$$

The porosity of the hexagonally arranged nanostructure can be also calculated from the expression given by Ebihara et al. [237,238]:

$$\alpha = 10^{-14} n \cdot \pi \cdot \left(\frac{D_p}{2}\right)^2 \quad (20)$$

where n is a pore density (1 cm^{-2}) being an overall number of pores per cm^2 and D_p is a pore diameter (in nm).

Nielsch et al. [235] reported that, for a perfect hexagonal arrangement of nanopores formed by self-organized anodization under optimum anodizing conditions, the ratio between pore diameter and interpore distance is almost constant and is equal to ∼0.33–0.34. Consequently, the optimum porosity for the best chosen anodizing conditions should be 10%. The optimal anodizing conditions depend mainly on the applied anodizing potential; for anodizing conducted in sulfuric, oxalic and phosphoric acids, the anodizing potentials that guarantee the perfect hexagonal order in a nanostructure are limited to values of 25, 40, and 195 V, respectively. The applied anodizing potential (which is different from the optimal value for a certain electrolyte) results in a significantly larger or smaller porosity of the nanostructure. The porosity rule has been derived only for self-ordering of alumina at the optimum anodizing conditions. In general, the self-organized anodic porous alumina requires a porosity of 10% independently of the anodizing potential, type of electrolyte and anodizing conditions.

Taking into account the ratio of pore diameter to cell diameter as a constant value for the self-ordering anodizing, Ono et al. [216,249,250] have suggested the following expression for porosity:

$$\alpha = \left(\frac{D_p}{D_c}\right)^2 \tag{21}$$

Most importantly, it should be noted that the relationship in Eq. (21) is only a rough approximation of Eq. (19) with a conceptual error of about 10%.

A different equation for porosity has been proposed by Bocchetta et al. [253]:

$$\alpha = \frac{S_{pores}}{S} = \frac{S - S_{ox}}{S} = 1 - \frac{m_p}{\rho h S} \tag{22}$$

where S_{ox} is the oxide surface, and m_p, h and ρ are the mass, thickness, and density of the porous layer, respectively. The density of anodic porous alumina formed by anodizing at 70 V in 0.15 M oxalic acid has been estimated at 3.25 g cm^{-3}. A wide variety of experimental values of porosity result from the different anodizing conditions used, in addition to the problem of completely different equations being used for the calculation.

It was mentioned previously that pore diameter can be altered by the widening process, which is based on the chemical etching of porous alumina walls in acidic solutions. A very interesting formula has been proposed for the porosity of nanostructure formed by anodizing in 0.42 M phosphoric acid at 140 V and 20 °C [252]. After anodization, the pores were widened in a 2.4 M H_2SO_4 solution at 20 °C, with the widening time being varied from 60 to 300 min:

$$\alpha = 0.196 + 6.54 \cdot 10^{-4} t_a + 2.62 \cdot 10^{-4} t_w + 7.27 \cdot 10^{-7} t_a^2 \\ + 4.36 \cdot 10^{-7} t_a t_w + 8.73 \cdot 10^{-8} t_w^2 \tag{23}$$

where t_a (min) and t_w (min) denote the anodization time and widening time, respectively.

1.2.2.6 Pore Density

The highly ordered nanomaterial with a close-packed arrangement of nanopores or nanotubes is seen as an "object of desire" for the microelectronics industry. Due to the hexagonal symmetry of the cells, anodic porous alumina is a nanostructure with the highest packing density, and consequently the number of pores created during anodization represents one of the most important features of porous alumina.

For the hexagonal distribution of cells in the nanostructure, the density of pores defined as a total number of pores occupying the surface area of 1 cm^2 is expressed by

$$n = \frac{10^{14}}{P_h} = \frac{2 \cdot 10^{14}}{\sqrt{3} \cdot D_c^2} \tag{24}$$

where P_h is a surface area of a single hexagonal cell (in nm^2) and D_c is given in nm. The substitution of D_c by Eq. (5) leads to the expression as follows:

$$n = \frac{2 \cdot 10^{14}}{\sqrt{3} \cdot \lambda_c^2 \cdot U^2} \cong \frac{18.475 \cdot 10^{12}}{U^2} \qquad (25)$$

A different approach to the pore density calculation has been proposed by Palibroda [234,254]:

$$n = 1.6 \cdot 10^{12} \exp\left(-\frac{4.764 \cdot U}{U_{max}}\right) \qquad (26)$$

where U_{max} has the same meaning as in Eq. (3). For certain anodizing conditions, this critical value of potential (U_{max}) can easily be estimated from Eq. (3), when the pore diameter and anodizing potential are known.

As might be expected from Eqs. (24) and (25), an increasing anodizing potential or interpore distance leads to a decrease in the number of pores formed within the structure [237,238,255]. For anodizing in oxalic acid, increasing the temperature of anodizing decreases the pore density [253]. Pakes et al. [201] studied a variation of pore density with anodizing time in disodium tetraborate solution (pH 9.2) at 60 V and 60 °C, and found pore density to decrease slightly with anodizing time at the initial stages of anodizing. This behavior was attributed to the rearrangement of pores, this being a consequence of the transformation of incipient pores into the true pores.

1.2.3
Incorporation of Anions

The incorporation of anions into the structure of anodic oxide layer depends heavily on the film type of the formed oxide. The formation of porous alumina during anodizing leads to a higher anion content in the structure than for barrier-type coatings. It is generally accepted that incorporated electrolyte species are present in the oxide films in a form of acid anion derived from the electrolyte used for anodizing [200]. Published data show an agreement with respect to the content of incorporated acid anions in the bulk of the oxide layer. The typical content of incorporated anions observed for some popular electrolytes are presented in Table 1.6.

Over the past few decades, much research effort has been focused on the determination of profiles of incorporated anions along the width of the oxide layer thickness. A wide variety of techniques has been employed for a depth profiling analysis of the barrier-type oxide layer, including Auger electron spectroscopy (AES) [256], impedance measurements [206], Rutherford backscattering spectroscopy (RBS) [257–259], secondary ion mass spectrometry (SIMS) [260], electron probe microanalysis (EPMA) [261], glow discharge optical emission spectroscopy (GDOES) [201,261–267], and X-ray photoelectron spectroscopy (XPS) [268]. An

Table 1.6 Percentage of incorporated anions in the porous oxide layer [195,200].

Electrolyte	H_2CrO_4	H_3PO_4	$H_2C_2O_4$	H_2SO_4
Anion content (%)	0.1–0.3	6–8	2–3	10–13

excellent review on methods used to study the distribution of incorporated anions in anodic oxide layers was prepared by Despić and Parkhutik [202], in which the kinetics of anion incorporation into the growing alumina films was also discussed.

Schematic profiles of anion concentration in the barrier-type oxide layer are presented in Figure 1.5 for electrolytes frequently used in the formation of highly ordered nanostructures. Sulfate and chromate profiles are drawn on basis of data obtained by the GDOES method [261,262,265,266], whilst the oxalate profile is derived from the AES method [256]. The profile of phosphate anions concentration is a result of analysis of profiles obtained by both methods [256,262,265].

For the steady-state growth of anodic barrier-type film, oxide formation occurs simultaneously at the electrolyte/oxide and oxide/metal interfaces, and is associated with the opposite-direction-migration of Al^{3+} and O^{2-}/OH^- ions. However, a portion of the migrating Al^{3+} ions are ejected directly into solution and do not take a part in formation of the solid oxide film [203,256]. Barrier-type anodic films on aluminum are amorphous, and the transport numbers of Al^{3+} and O^{2-} are 0.44 and 0.56, respectively [256,259,266,267]. Therefore, for film growth at high efficiency, about 40% of the film material is formed at the film surface, and the remainder is formed at the oxide/metal interface. The electrolyte anions are adsorbed at the electrolyte/oxide interface at the pore bases. For steady-state porous film growth, the incorporation of anions into the oxide layer occurs at pores bases as a direct result of the migration of electrolyte species. Electrolyte species can be negatively, positively or not charged, and consequently can either be immobilized on the oxide surface or migrate inwards and outwards at a constant rate, which differs for the various electrolytes. For example, phosphate, sulfate and oxalate anions migrate inwards within the oxide film under the electric field, while chromate anions are characterized by an outward migration [257]. The migration rate of phosphate, oxalate and sulfate anions related to the migration rate of O^{2-}/OH^- ions have been determined

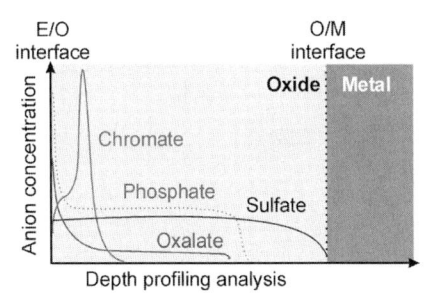

Figure 1.5 A typical depth profile of various anions concentration in the barrier-type oxide film formed by anodization of aluminum.

Table 1.7 Oxide composition and movement of some electrolyte species in the barrier-type amorphous alumina films.

Electrolyte	pH	Anion	Direction of migration	Relative rate of migration	$\dfrac{N_X}{N_{Al}} \cdot 10^{-2}$		Methods of determination	Reference(s)
0.1 M Na$_2$CrO$_4$	10.0	CrO$_4^{2-}$	Out	0.74	0.50 ± 0.01		RBS	[257,269]
0.1 M Na$_2$HPO$_4$	9.4	HPO$_4^{2-}$	In	0.50 ± 0.05	3.6 ± 0.3		RBS	[257]
0.1 M (COONH$_4$)$_2$	6.4	COO$^-$	In	0.67	–		AES	[256]
0.1 M Na$_2$SO$_4$	5.8	SO$_4^{2-}$	In	0.32 ± 0.07	7.4 ± 0.6		RBS	[257]
1M Na$_2$SO$_4$	–			0.62	–		GDOES	[266]

$\dfrac{N_X}{N_{Al}}$ is a ratio of the total number of X and Al atoms in the film.

AES: Auger electron spectroscopy; GDOES: glow discharge optical emission spectroscopy; RBS: Rutherford backscattering spectroscopy.

[195,200]. The migration of chromate anions is outward similarly to Al^{3+} ions, and therefore its rate of migration was compared to Al^{3+} ions. The migration rate and composition of film for some anodizing electrolytes used for anodic porous alumina formation are listed in Table 1.7.

The incorporation of boron species was also reported for anodizing conducted under a constant anodizing potential of 60 V in 0.25 M Na$_2$B$_4$O$_7$ solution (pH 9.2) at 60 °C [201]. About 40% of the total anodic oxide layer thickness, when monitored from the top of the film, was found to be associated with incorporated electrolyte species.

There is a significant difference in the porous alumina film growth in comparison to the barrier-type coating formed on aluminum. For porous alumina film growth, film formation occurs only at the oxide/metal interface, and anions migrate into the barrier layer according to the electric field. The electric field at the barrier layer is not uniform due to the semi-spherical shape of the pore base, and is much higher near the pore base close to the electrolyte/oxide interface than at the cell base close to the oxide/metal interface [270]. For this reason, the incorporation of electrolyte anions proceeds more easily. The higher content of incorporated anions in porous alumina layers is also a direct consequence of a long-term exposure of oxide walls for an acid active penetration. The concentration of incorporated SO$_4^{2-}$ increases with increasing current density and temperature [271]. Moreover, the incorporation of SO$_4^{2-}$ anions along pores, as well as across the barrier layer and pore walls near the pore bases in anodic porous alumina formed in sulfuric acid, was described in detail [272,273]. The parabolic distributions of the incorporated species determined is shown schematically in Figure 1.6.

A significant amount of incorporated anions can be found on the pore bases, such that across the barrier layer a local maximum is reached and a gradual decrease is then observed. The analysis of incorporated anions in the cell walls showed that negligible amounts are observed at cell-wall boundaries, but that the concentration of SO$_4^{2-}$ increases steadily, reaches a maximum, and then decreases slightly in a region close to the cell-wall/electrolyte interface. The distribution of anions along the cell wall (along the oxide thickness) is similar to that observed in the barrier layer.

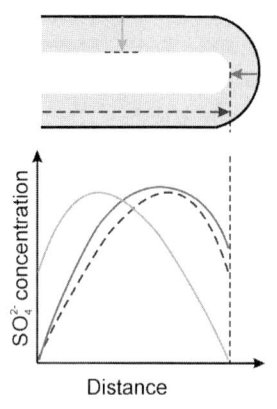

Figure 1.6 Schematic distribution diagram of SO_4^{2-} concentration in the anodic porous alumina formed in sulfuric acid. (After Ref. [272].)

1.2.4
Cell-Wall Structure

The properties of porous alumina films formed by anodizing are related to the electrolyte species incorporated into the oxide walls. For instance, the incorporation of anions modifies space charge accumulation in the porous and barrier-type alumina films [202]. Moreover, the mechanical properties of anodic alumina films, including flexibility, hardness and abrasion resistance, are greatly influenced by the incorporation of anions [200]. The content of incorporated species, and their distribution, depend on the anodizing conditions such as anodizing potential/current density and temperature. Consequently, different wall structures can be expected at different anodizing conditions. The duplex structure of the cell walls (Figure 1.7A) was proposed by Thompson et al. [200,274], whereby two different regions – the inner layer containing relatively pure alumina and outer layer with incorporated electrolyte anions – were distinguished. It was reported that the thickness of the inner layer increases in the order:

$$H_2SO_4 < C_2H_2O_4 < H_3PO_4 < H_2CrO_4 \tag{27}$$

According to Thompson [274], there is a transition from solid to gel-like material in moving across the cell walls towards the pore interior. It was also found that the ratio of the inner to outer layer thickness depends on the electrolyte, and equals 0.05, 0.1 and 0.5 for sulfuric, oxalic and phosphoric acids, respectively [274]. Recently, the triplex structure of the cell walls (Figure 1.7B) was reported for anodized alumina formed by anodization conducted in phosphoric acid [275]. The outer and intermediate layers are contaminated by electrolyte species, mainly anions and protons. The outer layer is rich in anions and protons, whereas the intermediate layer contains mainly anions. The inner layer consists of pure alumina.

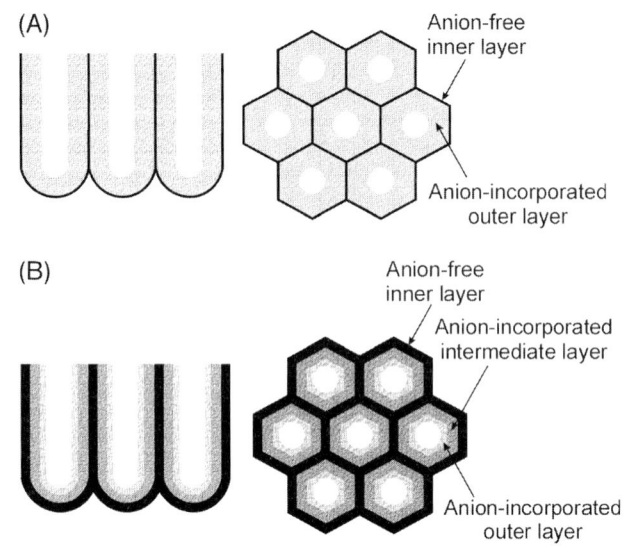

Figure 1.7 Schematic representations of the sectional and plan views of the duplex (A) and triplex (B) structures of porous alumina cell-walls formed in sulfuric and phosphoric acid, respectively. (After Refs. [274,275].)

Although the water content in the porous alumina film can vary between 1 and 15% [200,205], it is generally accepted that the amount of water in porous alumina depends on the anodizing conditions, sample handling, and the measuring technique. The porous oxide films are essentially dry when H_2SO_4 electrolyte is used for their preparation [276]. The porous oxide grown in acidic electrolytes does not contain any bonded water in the film bulk [202], but chemisorption of OH groups and water molecules can occur on the porous oxide layer. The adsorption of water onto the anodic porous alumina formed by anodizing in sulfuric acid was studied by Palibroda and Marginean [277], who found that the amount of adsorbed water on the porous oxide layer was equal to about 100 OH groups per nm^2 and was constant, independent of the sulfuric acid concentration used for anodizing, the temperature of anodizing, and the current density.

The formation of voids in the anodic alumina layer has been reported elsewhere [278–282]. Ono *et al.* [279,280] reported voids on the apexes of aluminum protrusions at the oxide/metal interface in the inner layer of cell walls (see Figure 1.8). These authors suggested that this occurred due to oxygen evolution, to existing tensile stress in the film, or to electrostriction pressure. The size of voids formed in the oxide layer were also found to increase with increasing anodizing potential [279,280]. Moreover, the formed voids could enlarge and merge under electron beam irradiation [281].

A clear and exhaustive explanation of the process of breakdown of anodic passive film and formation of voids was provided by Macdonald [283–285]. The proposed vacancy condensation mechanism of void formation involves a localized

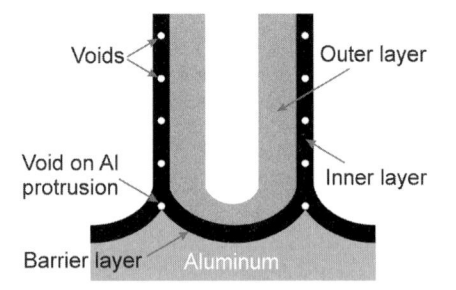

Figure 1.8 Voids in anodic porous alumina film. (After Ref. [285].)

condensation of cation and/or metal vacancies at the oxide/metal interface, and a subsequent detachment of the formed void. A schematic representation of voids formation in the anodic porous alumina is shown in Figure 1.9. When the anodizing process begins, vacancies are produced at the oxide/metal interface as a result of enhanced field-assisted ejection of Al^{3+} directly into the electrolyte. The condensation of vacancies begins at the defected area at the intersection of metal grain boundaries, and a void is formed. The growing void is detached from the apex of

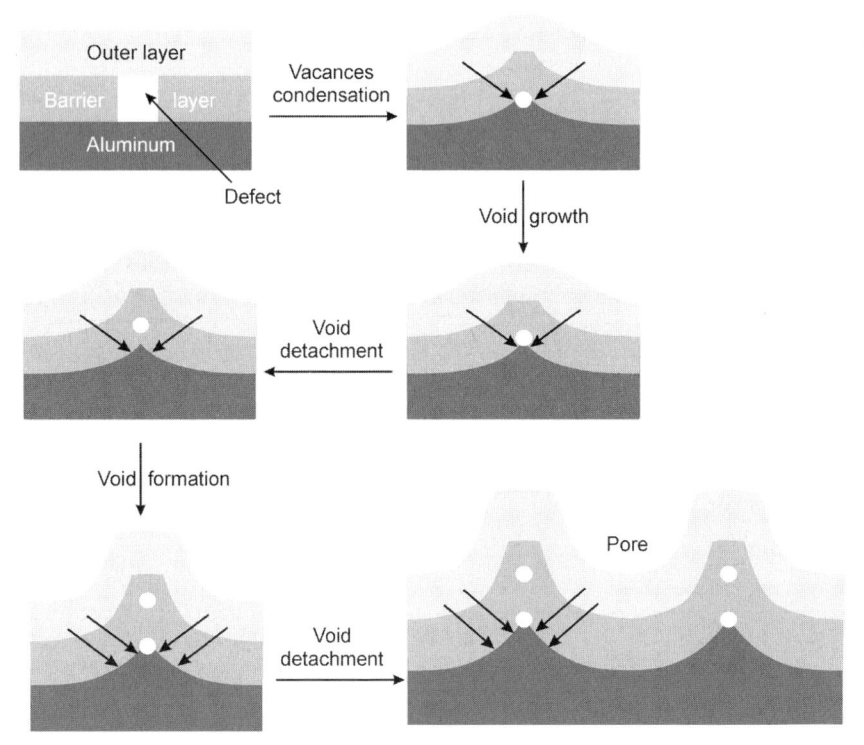

Figure 1.9 Schematic diagrams of the vacancy condensation mechanism for the formation of voids in the porous alumina layer. (After Ref. [285].)

the protrusion when the oxide/metal interface recedes into the aluminum bulk during oxide film formation. After undercutting the void, a new void nucleates on the apex of the protrusion because aluminum becomes rapidly saturated with vacancies. The new void grows, is then detached, and the overall process is repeated.

The vacancy condensation mechanism has been expanded for the description of voids formation in the outer and inner oxide layers [281]. The expanded model was based on the appearance in the oxide/metal interface of a few protrusion apexes after the vacancy detachment instead of that assumed in Macdonald's model.

1.2.5
Crystal Structure of Oxide

Most investigators have agreed that the anodic alumina films formed during anodization are amorphous oxides, or are amorphous with some forms of crystalline γ-Al_2O_3 or γ'-Al_2O_3. The recent observation of anodic films formed, over a wide range of anodizing conditions, does not confirm the presence of crystalline alumina in coatings [200]. The crystallization of γ or γ' alumina can be induced by electron beam irradiation of the sample in the scanning electron microscope [277], with crystallization occurring preferentially in the anion-free inner layer of the cell walls [286–289]. Ono et al. [290] found that the crystallization rate of films formed in various anodizing electrolytes decreases with the increasing content of incorporated anions and H_2O/OH^- in the film. The rate of crystallization increases in the following order:

$$H_2SO_4 < C_2H_2O_4 < H_3PO_4 < H_2CrO_4 \tag{28}$$

Crystallization in the amorphous alumina can occur during sealing and heating of the film [195,291]. A recent study of porous alumina films formed in phosphoric [275,292], oxalic [293] and sulfuric [291] acids, showed that the originally formed anodic porous alumina films are definitely amorphous. On the other hand, anodizing in chromic acid can result in amorphous alumina with traces of γ-Al_2O_3 [291].

1.2.6
Density and Charge of Oxide Film

The density of anodic alumina films varies significantly with anodizing conditions. For constant current anodizing conducted for 30 min in sulfuric acid, the density of oxide film was found to be $2.78\,g\,cm^{-3}$ [233]. The notably greater value of $2.90\,g\,cm^{-3}$ was obtained when the anodizing time was extended to 60 min [233]. Different values of the oxide density were obtained by Ebihara et al. [238] for anodization in 2 M H_2SO_4. For anodizing potentials from 3 to 14 V and in the range of temperature from 10 to 40 °C, anodic alumina density varied between 3.2 and $3.46\,g\,cm^{-3}$. It was also found that increasing the anodizing temperature, and increasing the electrolyte concentration as well as increasing the forming potential leads to a slight decrease in oxide density. According to Gabe [294], the density of oxide film formed by anodization in sulfuric acid can vary from 2.4 to $3.2\,g\,cm^{-3}$. A slight variation in the density of oxide films with anodizing temperature was found for anodizing in oxalic acid at the constant current density regime [237]. The density of oxide varied between 3.07 and 3.48 g cm^{-3} in the range of temperature from 10 to 40 °C, and increased with decreasing anodizing current density. Recently, the density of amorphous porous alumina was estimated at 3.2 g cm^{-3} [235,242,295]. The density of anodic alumina film formed in dehydrated, high-temperature electrolytes with organic solvents was reported as 2.4 g cm^{-3} [296].

Since anodic porous alumina membranes are often used as templates for nanomaterials fabrication, the charge present on the anodic oxide film is a parameter of major interest. The discussion concerning the space charge in anodic alumina oxide, its distribution and kinetics of space charge accumulation in porous alumina films, has been presented in detail [202]. It is generally accepted that the charge of the oxide film is a result of electrochemical processes occurring during anodization and incorporation of anions into the oxide film. The adsorption of species on anodic alumina membranes and solution–membrane interactions are fundamental factors which determine applications of the membrane for micro- and ultrafiltration, and in this respect especially biotechnological separation aspects are extremely important. The space charge in anodic porous alumina also has a negative influence on the dielectric parameters of oxides, as well as on the long-term drift of oxide properties [202]. A negative space charge in anodic porous alumina was reported [202], while in contrast the positively charged alumina film was discussed in the adsorption of SO_4^{2-} ions occurring during the anodization of aluminum in sulfuric acid [208]. The process of spontaneous adsorption of anions from aqueous solutions was studied for commercially available anodic porous membranes [297]. The maximum number of accessible sites for adsorption was found to be much higher than for neutral polycarbonate or nylon membranes. Moreover, at neutral pH a positive charge of anodic alumina membranes formed by anodization was estimated at 4.2 and 8.5 mC m^{-2} for phosphoric and oxalic acids, respectively. The positive charge in porous alumina attributed to the residual electrolyte in the structure was also reported recently [298]. According to Vrublevsky et al. [245], the oxide layer formed in 0.42 M phosphoric acid below 38 V is negatively charged, whilst above 38 V the positive charge of the oxide surface increases with increasing anodizing potential. For anodization conducted in 0.45 M oxalic acid, the anodizing potential at which the surface charge equals zero was estimated at 55 V [247].

1.2.7
Miscellaneous Properties of Anodic Porous Alumina

A growing scientific interest in the fabrication of anodic porous alumina films, and their further application for the synthesis of various nanomaterials, has given rise to many studies of porous alumina self-properties. The thermal characteristics of highly ordered porous alumina structures formed on aluminum have been studied widely [299–306], with thermal analyses being performed in order to collect data on chemical resistance [299,300], structural and optical properties [292,300–303], thermal conductivity and diffusivity along the channel axis [304], self-repair rearrangement of ordered nanopore arrays induced by long-term heat treatment [305], and mechanical properties [306]. The fracture mechanism, Young's modulus, hardness, fracture toughness [306] and fracture behavior in cylindrical ordered porous alumina [307] were also investigated. The electrical properties of porous alumina films [298,308–311], electron energy-loss, associated with Cherenkov radiation during electron beams traveling parallel to the pores of a porous alumina membrane [312], and surface roughness factors of the film [313], were also studied. Contact angle studies on porous alumina were employed to characterize the liquid/surface interactions at the nanoscale and wetting properties of membranes in contact with different solvents and liquids [314]. It was found that aqueous solutions are not suitable to fill completely the nanopores, and the preferred solvents are dimethyl sulfoxide (DMSO) and N,N′-dimethyl formamide (DMF). Small-angle X-ray scattering (SAXS) techniques were employed for the *in-situ* investigation of filling behavior of porous alumina membranes with a pore diameter of 20 nm in contact with perfluoromethylcyclohexane as solvent [315].

The optical characterization of highly ordered porous materials can provide a variety of information on their nanostructural properties, and especially on pore volume fraction, pore shape, pore diameter, anodic layer thickness and the incorporation of additives occurring during anodizing [316–318]. It is widely recognized that porous alumina exhibits a blue photoluminescence (PL) band [319–326], with the emission bands being attributed to optical transition in the singly ionized oxygen vacancies (F^+ centers) or to electrolyte impurities embedded in the porous alumina membranes. The oxygen vacancies are produced in the alumina matrix as a result of enhanced consumption of OH^- in the electrolyte near the anode occurring during the anodization of aluminum [326]. In contrast, the influence of various additives, such as sulfosalicylic acid, Eu^{3+} and Tb^{3+} ions, on the photoluminescence spectra were investigated [322,327–329]. Weak optical radiation, in the visible range of the spectrum emitted during the anodizing of aluminum in acid electrolytes [known as galvanoluminescence (GL) or electroluminescence] was also studied [330–332]. The concentration of impurities and the pre-treatment of aluminum samples, including degreasing, chemical cleaning and electropolishing, were found strongly to affect GL intensity. Furthermore, the intensity of GL depends on anodizing conditions such as current density, temperature, and electrolyte concentration. It was also suggested that, depending on the nature of the electrolyte – whether organic or inorganic – there are two different mechanisms or two different types of luminescence center responsible for GL.

1.3
Kinetics of Self-Organized Anodic Porous Alumina Formation

1.3.1
Anodizing Regimes and Current/Potential-Time Transient

A hexagonal-shaped oxide cell can easily be formed by anodizing aluminum at a constant current density or constant anodizing potential regime. A typical current

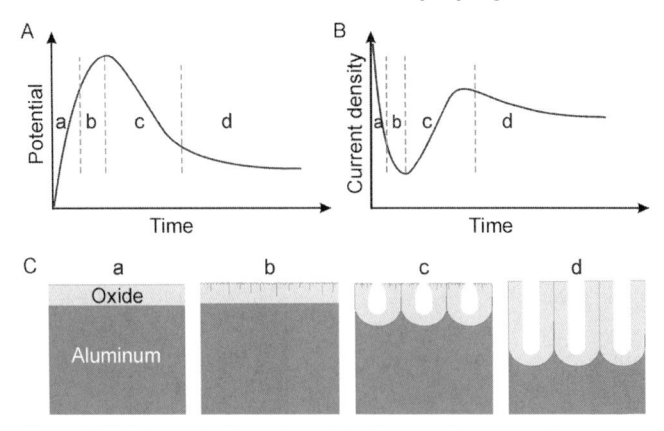

Figure 1.10 Schematic illustration of the kinetics of porous oxide growth in galvanostatic (A) and potentiostatic (B) regimes, together with stages of anodic porous oxide development (C).

density–time and anodizing potential–time transient recorded during the anodization of aluminum in 20% H_2SO_4 at 1 °C are shown in Figure 1.10. When a constant current is applied for porous alumina growth, the potential rises linearly with time until the local maximum is reached, and then decreases gradually to the steady-state-forming potential. During the initial period of anodization (Figure 1.10, stage a), the linear increase in potential is associated with a linear growth of high-resistant oxide film (barrier film) on aluminum. Further anodizing (stage b) results in the propagation of individual paths (pores precursors) through the barrier film. At the maximum of potential (stage c), the breakdown of the tight barrier film occurs and the porous structure begins to be built. Finally, the steady-state growth of porous alumina proceeds (stage d) and a forming potential is almost unchanged. At the start of the process conducted under the constant anodizing potential, current density decreases rapidly with time, and a minimum of current density is quickly achieved. A linear increase then leads to a local maximum. After reaching the maximum, the current density decreases slightly and a steady-state current density of the porous oxide formation is achieved.

The rate of current decrease, the time at which the minimum current is observed, and the steady-state-forming current density depends directly on the anodizing conditions, such as applied anodizing potential, temperature and electrolyte concentration. In general, the minimum of current density decreases with increasing electric field strength, increasing anodizing potential and temperature. The decrease in the minimum value of current is also observed with increasing concentration of acids. The minimum current density occurs earlier, with the higher anodizing potential and lower pH of the electrolyte.

Recently, the constant potential of anodizing instead of constant current regime has been commonly used for the fabrication of closed-packed highly ordered anodic porous alumina films with a desired pore diameter. According to Hoar and Yahalom [333], the relationship between current density and time observed under the

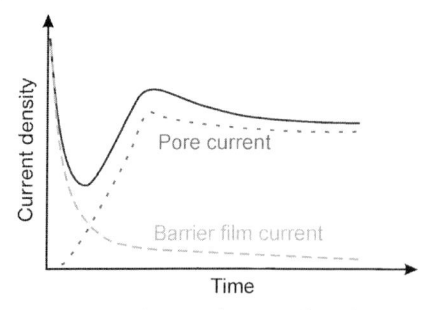

Figure 1.11 Schematic diagram of overlapping processes occurring during the porous oxide growth under constant anodizing potential regime. (After Ref. [333].)

constant anodizing potential is a resultant of two overlapping processes, as shown in Figure 1.11. A first, exponential decrease is connected with the barrier film formation, and the second represents the process of pore formation.

The potential–time transients recorded during anodizing was studied carefully by Kanankala et al. [334]. The theoretical model describing the potential–time behavior at the initial stage of anodizing conducted under the constant current regime in 0.21 M H_2SO_4 or 0.3 M $C_2H_2O_4$ was developed. The range of applied current density for anodizing varied between 20 and 50 mA cm^{-2}. The model was based on the rate of oxide formation and rate of oxide dissolution. The predicted potential–time curves fit perfectly with experimentally recorded data. It was also found that the time when the local current maximum appears on the potential–time curve can be estimated as follows:

$$t_{max} = \frac{75}{i} + 1.5 \qquad (29)$$

where t_{max} is a time (in s) when the maximum is reached and i is a current density in mA cm^{-2}.

The effect of alloy type on the kinetics of porous oxide growth at constant anodizing potential in sulfuric acid was investigated, and anodizing current density–time transients were studied for various types of aluminum alloys at the anodizing potential of 15 and 18 V [335]. The rearrangement of pores, just before the current density reaches a steady-state value, was found not to occur for some alloys. The rearrangement of pores is usually indicated on the current–time curve (Figure 1.10B) as a local maximum presence at stage c. The lack of pore rearrangement was attributed to an accumulation of the alloying elements (e.g., Cu, Fe and Si) at the oxide/metal interface.

The presence of additives in the anodizing electrolyte can cause a slight modification in the kinetics of the process and the current–time transient. It is widely recognized that, during anodizing, a sulfonated triphenylmethane acid dye ("Light Green") concentrates at the pore base and reacts with released Al^{3+} ions [336,337]. As a result, the movement of Al^{3+} ions into the electrolyte (direct ejection) is strongly inhibited and the field-assisted dissolution of oxide is significantly reduced. Consequently, a decreasing local current density at the pore base increases a radius of pore base curvature and interpore distance in the presence of an additive. The influence of "Light Green" as an additive on current density–time behavior was studied in detail for the anodizing conducted in sulfuric acid solutions under the constant potential of 15 V [338]. The decrease in current density with increasing concentration of the additive was also observed. It was found also that increasing the concentration of additive also decreases the slope of current density at the pore initiation stage (stage c in Figure 1.10B).

Recently, the mixing of both galvanostatic and potentiostatic regimes has been reported as a new and efficient method for the high-field anodization of aluminum [213,223,251,293,339]. The sample was pre-anodized at a constant current density and, after a certain period of time and after reaching a specified potential, the potentiostatic mode was switched. Depending on the applied initial current density, the duration of the constant current mode varied from a few seconds to 10 min [251]. The method was used successfully for aluminum anodization in sulfuric, oxalic, and phosphoric acids. The requisite anodizing conditions are collected in Table 1.8.

For oxalic acid pre-anodizing, the constant current density should be lower than 10 mA cm^{-2} in order to avoid any breakdown of the oxide film and active film dissolution. In the case when pre-anodizing is conducted in phosphoric acid, the initial current density cannot exceed 15 mA cm^{-2}, and the steady-state current density during anodizing should be kept at about 70 mA cm^{-2} to avoid any possible overheating of the sample in a high electric field [213]. It was also found that increasing concentration of Al^{3+} allows higher current densities to be achieved during pre-anodizing, and higher anodizing potentials after switching the mode to potentiostatic. In other words, the fresh electrolyte should be aged by additional electrolysis with the aluminum anode [251]. For this reason the additional pre-electrolysis of electrolyte was employed, with the amount of charge passed through sulfuric acid and other electrolytes (oxalic and phosphoric acids) being 30 and 2 A·h, respectively.

Table 1.8 Conditions for the high-field anodizing of aluminum at various electrolytes.

Electrolyte	Concentration (M)	Temperature (°C)	Current density of preanodizing (mA cm^{-2})	Time of constant current mode (s)	Anodizing potential (V)	Reference
H_2SO_4	1.1 (10%)	0.1	160	—	40	[251]
		0	200	420	70	[213]
$H_2C_2O_4$	0.03 (0.25%)	1	12	60	160	[213]
H_3PO_4	0.1 (1%)	0	16	120	235	[213]
H_3BO_3	0.5 (3%)	RT	1	—	1500	[293]
$(NH_4)_2C_4H_4O_6$	0.17 (3%)	—	1	—	25–100	[223]

RT: room temperature.

A completely novel anodizing technique which employed a pulse sequential voltage with a pulse frequency of 100 Hz was proposed by Inada et al. [340]. The technique was used for the strict control of pore diameter in the range of low anodizing potentials (below 3 V) where a linear dependence between pore diameter and forming voltage does not exist. It should be noted that the smallest pores formed by conventional anodizing have a diameter of 7 nm. However, by using the pulse method pores with diameters of 3 and 4 nm can be successfully formed at pulse sequential voltages of 1 and 2 V, respectively.

The anodizing of aluminum resulting in a porous film can be also realized in sulfuric and phosphoric acids under alternating current or potential regimes [341–344]. The observed potential/current–time transients and morphology of films were similar to the conventional anodizing conducted under constant current density or constant potential [343,344]. The percentage of incorporated SO_4^{2-} anions (13%) was the same as found for DC-anodizing [344], but a significant difference was seen to exist between the AC and DC-anodizing. Hydrogen evolution on anodized aluminum occurs during cathodic cycles of AC-anodizing; moreover, for AC-anodizing conducted in sulfuric acid a secondary reaction of SO_4^{2-} reduction proceeds during the cathodic half-cycles, and as a result of this reduction sulfur and/or sulfide is released [341]. In order to reduce the cathodic reduction of sulfate ions, a wide variety of additives (e.g., Fe^{3+}, As^{3+}, Co^{2+}) was studied [341,342]. In contrast, the side cathodic reduction of anions was not observed during AC-anodizing in phosphoric acid [343].

1.3.2
Pores Initiation and Porous Alumina Growth

The phenomena of anodic porous alumina film formation has been studied extensively over several decades, with considerable scientific effort directed towards elucidation of the mechanism of self-organized growth of the porous layer. Thus, several theories have been proposed and developed. Although the anodizing of aluminum was successfully and widely applied for the synthesis of high-ordered nanostructures, it remains unclear as to which physical factors control pore ordering during oxide growth, and especially how the surface features of the aluminum affect the ordering of pores

1.3.2.1 Historical Theories
The early theories of porous film growth take into account a passage through the barrier layer nascent oxygen formed from water in pores [345], peptization of aluminum hydroxide gel on aluminum [346], and current action on the barrier layer resulting in pores [347]. Baumman [348,349] proposed the existence of vapor film over the active layer at the bottom of pores where, at a gas/electrolyte interface, oxygen anions are generated. The growth of oxide kernels occurs simultaneously at the base of pores, and the porous structure is a result of oxide dissolution of previously formed breakthroughs (zig-zags).

Keller et al. [197] extended this theory and proposed the model in which, at the beginning of the anodizing process, formation of the homogeneous barrier film occurs and further dissolution of the oxide in the barrier film is followed by the current which repairs the damage to the oxide layer. The passing current increases the local temperature of the electrolyte such that consequently oxide dissolution is enhanced and current breakdowns form pores in the oxide layer. The anodic structure exhibiting the hexagonally arrangement of cells is derived from pores due to the existing tendency of spherical distribution of potential and current about the certain point (pore). It should be noted that a closed-packed hexagonal arrangement of oxide cells is formed as a result of steric factors.

On the other hand, Akahori's hypothesis of porous oxide growth [350] suggested that, after formation of the barrier layer and pores, evaporation of the electrolyte and melting of aluminum occur at the bottom of pores due to the high local temperature. The oxygen ions are then formed at the end of pore in the gaseous electrolyte. Oxygen ions pass the oxide layer at the pore bottom and react with a liquid aluminum base.

A completely different approach to porous oxide growth was proposed by Murphy and Michelson [351]. According to these authors, the outer part of the barrier film formed on aluminum is transformed into hydroxide and hydrate compounds as a result of interaction with water. Simultaneously, the continuous build-up of dense oxide (inner layer) proceeds at the oxide/metal interface. Hydroxide and hydrate compounds have a tendency to bond or to adsorb water and anions from the electrolyte with creation of a gel-like matrix. Submicrocrystallites of aluminum oxide are embedded in this gel. In fact, oxidation takes place at the border between the barrier oxide film (inner layer) and the outer hydrated oxide layer. This model assumed also the transmission of current by Al^{3+} ions, and a responsibility of oxygen ions for the movement of the barrier layer towards the metal base. In this model pores are formed at defect sites where local differences in solubility or electrical breakdown exist.

An interesting view on the initial stages of aluminum oxide formation during anodization and further growth of the oxide layer was presented by Csokan [352]. At the start of the anodizing process, oxygen atoms or electrolyte anions are adsorbed or chemisorbed onto active sites (defects, faults and grain boundaries) on the aluminum surface. The mono or oligomolecular layer of oxide nuclei is then formed. Perpendicular oxide growth is much slower than at the edges of the nucleus, and as a result of this lateral growth the oxide covers the entire aluminum surface. The local differences in chemical solubility of the oxide film and in structural deformations (different state of energy) are directly responsible for the formation of pores. Csokan's theory explains not only the porous anodic alumina structure but also a fibrous structure formed on aluminum by anodizing. The internal stress influence on the oxide structure was also studied by Csokan [352], who found that structural deformations in the oxide films are produced by the isotropic and anisotropic mechanical stresses.

A different interpretation of fibrous anodic alumina structure was given by Ginsberg *et al.* [353–356], according to who the external part of the alumina tube walls is formed from amorphous alumina oxide, while the internal part consists of a gel containing hydroxide and embedded anions. Moreover, the interior of the tube is filled with the electrolyte and plays an important role in oxygen exchange between electrolyte and metal.

Hoar and Yahalom [333] conducted extensive studies on the pore initiation process during anodizing of aluminum, and suggested that this was a consequence of proton entry into the barrier film when the field strength decreased sufficiently at the certain region. Proton-assisted dissolution of the oxide then occurred.

1.3.2.2 Field-Assisted Mechanism of Porous Film Growth

It is generally accepted that the porous structure of anodic alumina film develops from the barrier-type coating formed on aluminum at the start of anodization. Growth of the barrier film occurs due to the high field ionic conduction and at the constant field strength, defined as a ratio of the potential drop across the barrier film to its thickness [203]. Governed by the constant field strength, the uniform film with a uniform current distribution is developed on the whole surface, as shown in Figure 1.12A.

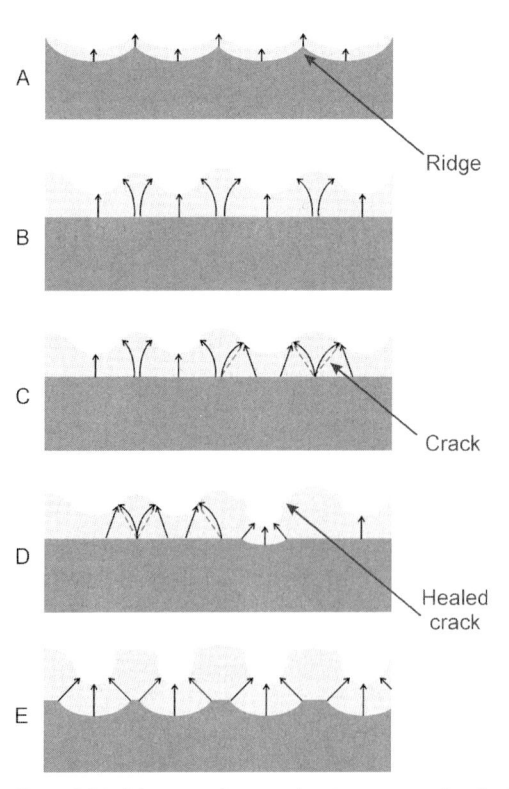

Figure 1.12 Schematic diagram showing current distribution during pore initiation and development of pores on anodized alumina. (After Refs. [200,224].)

The uniform growth results in a smoothing effect on the initial roughness of the aluminum. However, some local variations in field strength can appear on a surface with defects, impurities or preexisting features including subgrain boundaries, ridges and troughs as remains of pre-treatment procedures (e.g., mechanical or electrochemical polishing, etching) [200,208,274]. This non-uniform current distribution leads consequently to the enhanced field-assisted dissolution of oxide and a local thickening of the film (Figure 1.12B). The higher current above metal ridges, accompanied by a local Joule heating, results in the development of a thicker oxide layer [208,255,274]. Simultaneously, the enhanced field-assisted dissolution of oxide tends to flatten the oxide/metal interface. Recently, the effect of local heat transfer on current density was studied for anodizing in sulfuric acid [357]. It was found that increasing the local temperature enhances the local field-assisted oxide dissolution at the pore bases, and consequently increases the local current density. According to Thompson [200,203], the oxide layer grown above the ridges (flaw sites with impurities, scratches) is prone to generate a highly localized stress. Consequently, successive cracking of the film and its rapid healing at the high local current density occur (Figures 1.12C and D). Therefore, with a consumption of aluminum base and enhanced progress in the oxide thickness build-up above the flaw sites, the crack–heal events are more pronounced and the curvature of the film at the oxide/metal interface increases (Figure 1.12E). Shimizu et al. [224] suggested that a tensile stress at surface ridges leads to the formation of cracks which can act as conductive pathways for film growth and where rapid healing effect occurs. The preferential growth of oxide above flaw sites, and thickening of the barrier layer, proceed continuously until the moment when the current is concentrated in the thinner film region at the bottom of the future pore (Figure 1.12E). On the other hand, increasing pore curvature (increasing pore diameter) decreases the effective current density across the barrier layer. As a result, the growth of other pores from other incipient pores is initiated in order to maintain a uniform field strength across the barrier layer. When the curvatures of the oxide film at the oxide/metal interface have increased sufficiently and intersection of the scalloped regions has occurred, the steady-state conditions of pore growth are reached. For steady-state porous oxide growth, there is a dynamic equilibrium between oxide growth at the oxide/metal interface and field-assisted oxide dissolution at the electrolyte/oxide interface [208].

The generation of pores was recently studied on aluminum with a tungsten traces layer incorporated into the anodic pre-film formed in phosphoric acid [358]. During the subsequent anodization, the material was found to flow from the region of the barrier layer towards the cell wall regions. This was attributed to the existence of stresses associated with the film growth and field-assisted plasticity of the film material.

When conducted anodization in chromic acid, Thompson [203] suggested that interaction of electrolyte with the barrier film, resulting in the development of penetrating paths is responsible for the local field strength increase just beneath the tip of the penetration path. When the penetrating paths are more advanced, the local field strength increases and further enhanced field-assisted dissolution of paths occurs until an embryo pore at the oxide/metal interface is developed (Figure 1.13).

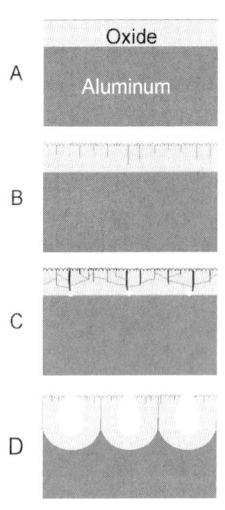

Figure 1.13 Development of penetrating paths and pores during anodizing in chromic acid. (After Ref. [203].)

1.3.2.3 Steady-State Growth of Porous Alumina

Although the anodization of aluminum has been investigated widely, some aspects of the complex process are not yet fully elucidated. It is not clear yet which oxygen-carrying anion species O^{2-} or OH^- ions are involved in the anodic process. The OH^- ions are generated in the anodizing electrolyte from water by simple splitting, or by the cathodic reduction of water and dissolved oxygen through the following reactions:

$$H_2O + 2e^- \rightarrow 2OH^- + H_2 \tag{30}$$

$$O_2 + 2H_2O + 4e^- \rightarrow 4OH^- \tag{31}$$

On the other hand, O^{2-} ions can be formed at the electrolyte/oxide interface from adsorbed OH^- ions by oxygen vacancy annihilation [359], simple splitting of water at the interface, or from water by interaction with adsorbed electrolyte anions in the process shown schematically in Figure 1.14 [208]. In the latter process, OH^- ions may also be produced.

Anodic polarization of aluminum in the acidic electrolyte leads to the amorphous oxide film growth according to the following reactions:

$$2Al^{3+} + 3H_2O \rightarrow Al_2O_3 + 6H^+ + 6e \tag{32}$$

and

$$2Al + 6OH^- \rightarrow Al_2O_3 + 3H_2O + 3e^- \tag{33}$$

$$2Al + 3O^{2-} \rightarrow Al_2O_3 + 6e^- \tag{34}$$

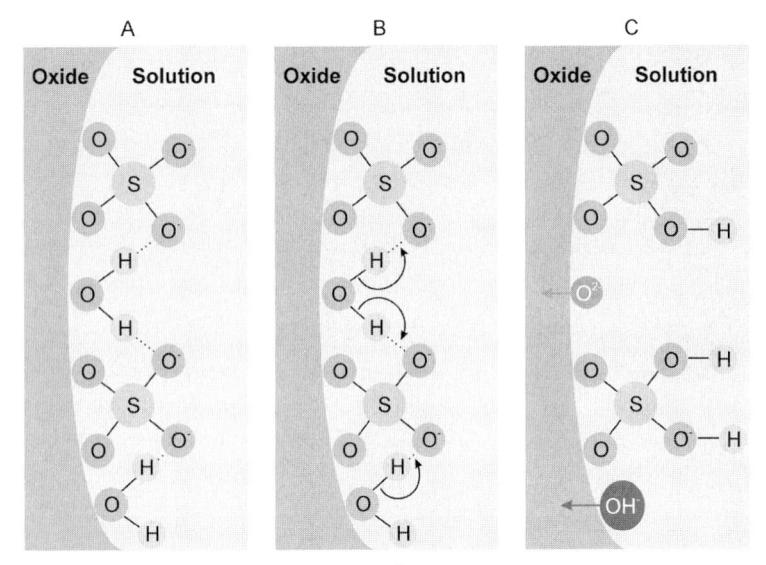

Figure 1.14 Schematic representation of O^{2-} and OH^- ions formation at the oxide/electrolyte interface from water interacting with adsorbed SO_4^{2-} anions. (After Ref. [208].)

The main contribution to the anodic current is made by the given reaction:

$$Al \rightarrow Al^{3+} + 3e^- \tag{35}$$

The evolution of oxygen near the metal/oxide interface was reported for anodizing of aluminum in various electrolytes [204,360–362]. This side reaction can be represented as follows:

$$2H_2O \rightarrow O_2 + 4H^+ + 4e^- \tag{36}$$

$$4OH^- \rightarrow 2H_2O + O_2 + 4e^- \tag{37}$$

or alternatively,

$$2O^{2-} \rightarrow O_2 + 4e^- \tag{38}$$

According to Chu et al. [363], during aluminum anodizing at high anodizing potentials the generation of oxygen is also possible through the following reaction:

$$H_2O_2 \rightarrow O_2 + 2H^+ + 2e^- \tag{39}$$

The formation of oxygen bubbles within anodic alumina, according to Eq. (38) in the vicinity of the metal/film interface, can proceed due to the presence and compositions of impurities or second-phase particles [360,361]. It was also suggested that oxygen evolution is directly connected with growth of the porous alumina film, and the process can be used for testing transition from barrier-type to porous-type coatings [204]. The anodic process is further complicated in the presence of electrolyte anions susceptible to oxidation [342]:

$$2SO_4^{2-} \rightarrow S_2O_8^{2-} + 2e^- \tag{40}$$

It widely recognized that pore formation is attributed to the thermally assisted, field-accelerated dissolution of oxide at the base of pores. On the other hand, the growth of oxide occurs mainly at the oxide/metal interface for typical acidic electrolytes including sulfuric, phosphoric, and oxalic acids. The growth of porous alumina involves the inward migration of oxygen-containing ions (O^{2-} or OH^-) from the electrolyte through the barrier layer, and the simultaneous outward drift of Al^{3+} ions across the oxide layer. It was found that only a part of the Al^{3+} ion flux takes part in the oxide formation at the metal/oxide and oxide/electrolyte interfaces. Depending on the anodizing conditions – and especially on the current efficiency of the oxide growth – oxide formation can also proceed at the oxide/electrolyte interface. For a current efficiency close to 100%, and for phosphates and chromate electrolytes, the majority of Al^{3+} ions reaching the oxide/electrolyte interface are involved in the formation of oxide at the oxide/electrolyte interface [256]. This means that for high-efficiency oxide growth, about 40% of the film material is formed at the oxide/electrolyte interface. This behavior is consistent with the relative transport numbers of Al^{3+} and O^{2-}/OH^- ions observed for various anodizing electrolytes and anodizing conditions. These relative transport numbers for cations and anions are about 0.4 and 0.6, respectively. In contrast, for anodizing in oxalate electrolytes and for processes occurring with a low current efficiency, there is no evidence of any oxide formation at the oxide/electrolyte interface [256]; rather, a direct ejection of Al^{3+} ions to the electrolyte was observed [256]. The remainder of the Al^{3+} ions flux reaching the oxide/electrolyte interface is lost into the electrolyte by field-assisted dissolution [208,274] or by a mechanism involving direct Al^{3+} ion ejection to the solution [359]. A schematic representation of elementary processes involved in porous alumina growth is shown in Figure 1.15.

Additionally, the electrolyte anions migrate inwards to the barrier layer and can be incorporated into the oxide material. The chemical dissolution of oxide in acidic electrolyte results in a thinning of the oxide layer.

For steady-state porous oxide growth, the locally increased field at the electrolyte/oxide interface affects the dissolution of oxide in the pore bases [203]. Increasing field-assisted dissolution of oxide increases the oxide growth rate at the metal/oxide interface due to a dynamic equilibrium between the rate of field-assisted dissolution of oxide and the rate of oxide formation. The field-assisted dissolution of oxide starts from the polarization of Al–O bonds, followed by the removal of Al^{3+} ions from the oxide structure [208]. The removal of Al^{3+} ions occurs more easily in the presence of the field.

The field-assisted oxide dissolution is most likely thermally enhanced through Joule heating effects [208]. According to Li et al. [236], a rise in the temperature of the aluminum anode, calculated for a first 12 s of anodizing, is about 25 °C.

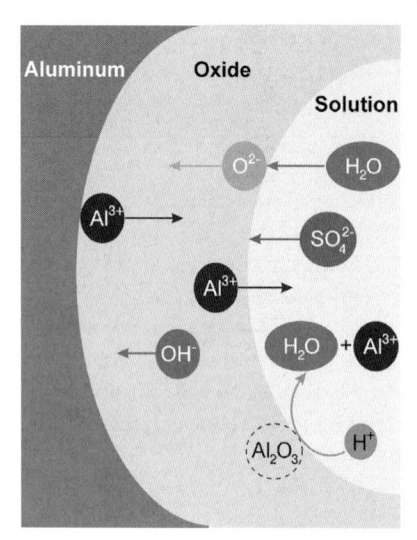

Figure 1.15 Schematic illustration of ions movement and dissolution of oxide in sulfuric acid solution.

1.3.2.4 Growth Models Proposed by Patermarakis and Colleagues

Patermarakis et al. [271–273, 364–378] described, in several publications, a compact theoretical model of porous alumina growth in various electrolytes. Different approaches have been employed to obtain information about the kinetics and mechanism of the oxide growth mainly in sulfate solutions. The main conclusions derived from these models are outlined in the following sections.

A strict kinetic model of anodic porous alumina growth based on the field-assisted dissolution of oxide was presented by Patermarakis et al. [364–368]. The elongated columnar pore with increasing diameter in the direction from the pore bottom to the film's surface was assumed in the model. It was found that, for anodizing conducted in 1.53 M H_2SO_4 under the constant current density regime, oxide dissolution at pores is essentially a field-assisted process, while dissolution of cell walls is governed by first-order kinetics and is thermally activated [364]. The main structural futures of the porous oxide film formed in the unstirred [364,365] and stirred [366] sulfuric acid bath were evaluated. The density of the compact cell walls estimated from the model was about 3.42 g cm^{-3} [364]. For the stirred electrolyte, the concentration of electrolyte in the pores increases linearly on moving from the film's surface to the pore bottoms [366]. The influence of sulfuric acid concentration on applicability of the model in the range between 0.20 and 10.71 M was studied [367]. It was found that the derived kinetic model of porous alumina growth is not valid for the critical concentration of sulfuric existing between 0.51 and 1.53 M [367]. For low acid concentrations, an abnormal oxide growth (pitting and burning behaviors) is suggested. A general formulation of the kinetic model applicable to other galvanostatic anodizations was also proposed [368].

The theoretical model considering mass and charge transport inside pores during the self-organized growth of porous alumina in sulfuric acid was also described

[369]. The proposed model employed both kinetics and transport phenomena equations, and showed that the concentration of $Al_2(SO_4)_3$ at the pore bases is maximal in comparison to other parts of the pore depth; moreover, it increases monotonically with increasing oxide thickness, current density or decreasing temperature [369]. The electrolyte concentration at the pore bases depends on the current density, and various phenomena were observed. The model showed that the sulfuric acid concentration can increase, decrease or reach a local minimum when the analysis of concentration is performed along the pore depth in the direction from the top of the film to the pore bottoms [369]. The transport analysis criterion, which allows prediction of the conditions of regular and abnormal oxide growth, was also performed [370,371]. The most important parameter influencing abnormal oxide growth is electrolyte concentration at the pore base in relation to the current density [370]. The promotion of pitting occurs at low temperature and sulfuric acid concentration or high current density and sulfate additives including $Al_2(SO_4)_3$ [371,372]. For the saturated $Al_2(SO_4)_3$ solution in H_2SO_4, the incorporation of colloidal micelles of aluminum sulfate was observed in the cell walls [373,374]. The kinetics and mechanism of anodizing were also found to be strongly influenced by colloidal micelles. Recently, a comparative study was performed in order to establish the kinetics and mechanism of anodizing in oxalic acid [375]. The mechanism of porous oxide growth in oxalic acid was the same as observed for sulfuric acid, but with an easier and stronger growth of colloidal micelles being reported.

A model for charge transport across the barrier layer was also proposed by Patermarakis *et al.* [376–378]. The prolonged anodization proved that the rate-controlling step of the steady-state growth of porous alumina is charge transport across the barrier layer, with the native oxide being produced mainly in the region adjacent

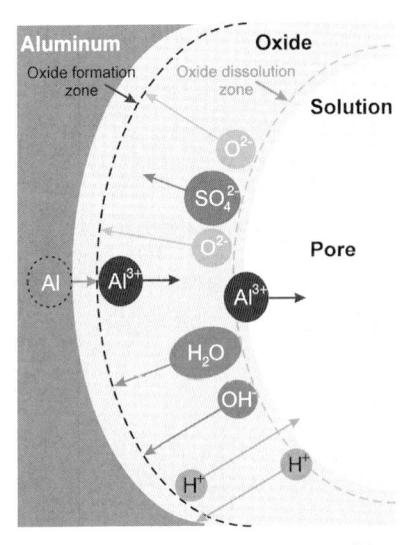

Figure 1.16 Schematic representation of the cross-sectional view of the barrier layer in anodic porous alumina. (After Ref. [376].)

to the metal/oxide interface (Figure 1.16). The zone where field-assisted dissolution of oxide occurs is located inside the oxide film, near the oxide/electrolyte interface.

According to the derived model, the ionization of aluminum as shown in Eq. (35) proceeds through successive one-electron-transfer elementary steps. Charge transport across the barrier layer is realized by the migration of O^{2-}, OH^- and SO_4^{2-} ions. The model conclusions predicted that the O^{2-} ions necessary for aluminum oxidation are derived from the oxide lattice and from OH^- ions adsorbed on the oxide surface. Al^{3+} ions are rather immobilized inside the oxide film, especially at a lower field strength. In the dissolution zone, the Al^{3+} ions are solvated and then move towards the bulk of electrolyte. In the oxide film, SO_4^{2-} ions migrate along a boundary surface of microcrystallites through the successive vacancies. Sulfate anions are much larger than O^{2-} ions, and their movement towards the metal/oxide interface is gradually blocked by greater-sized microcrystallites. The electrolyte anions are unable to reach the oxide formation zone and finally become embedded on microcrystallites in the bulk of the oxide. The model calculation is consistent with experimental data, and proved that there is no significant difference in the incorporation of various type of anion derived from the same electrolyte, such as SO_4^{2-} or HSO_4^-. Other anions, including OH^- and O^{2-}, migrate at similar rates due to the comparable ratio of ion charge to the radius. The migration process is probably realized through vacancies inside crystallites. The OH^- ions can be incorporated on/in crystallites in the oxide bulk, or they can decompose inside the oxide to O^{2-} and H^+. It is most likely that the OH^- ions reach the oxide formation zone and form the oxide film according to Eq. (33). In accordance with the model assumptions [377,378], the process of field-assisted dissolution of oxide and OH^- ions migration in the oxide lattice can be represented as shown in Figure 1.17.

Due to the positive surface charge of oxide in acidic electrolytes, decomposition of water occurs in the double layer. Consequently, the OH^- ion is adsorbed at the oxide/electrolyte interface on the Al^{3+} ion derived from the oxide lattice (Figure 1.17A). Under the high field strength, the H^+ ion is rapidly removed from the double layer to the bulk of solution. The adsorption of OH^- ion onto Al^{3+} weakens the bonds between Al^{3+} and O^{2-} ions in the oxide lattice (Figure 1.17B). The O^{2-} ion can then leave its position in the lattice and migrate further towards the metal/oxide interface. This results in anion void formation in the lattice and further possible occupation of the void by the previously adsorbed OH^- ion (Figure 1.17C and D). Simultaneously, the Al^{3+} ion can move slowly towards the bulk of electrolyte. The process of adsorption and further migration of other OH^- ions is faster than the Al^{3+} ejection to the solution, and can easily be repeated several-fold by the time the Al^{3+} ion leaves its position [377].

1.3.2.5 Other Phenomenological Models of Porous Alumina Growth

Although the field-assisted mechanism of oxide growth is commonly used to describe the self-organized growth of anodic porous alumina, some other models have been developed. The formation of colloidal layer at the electrolyte/metal interface was assumed by Heber [379,380] in his model of pore development and film growth. A chemical interaction between hydroxide, electrolyte and adsorbed water molecules

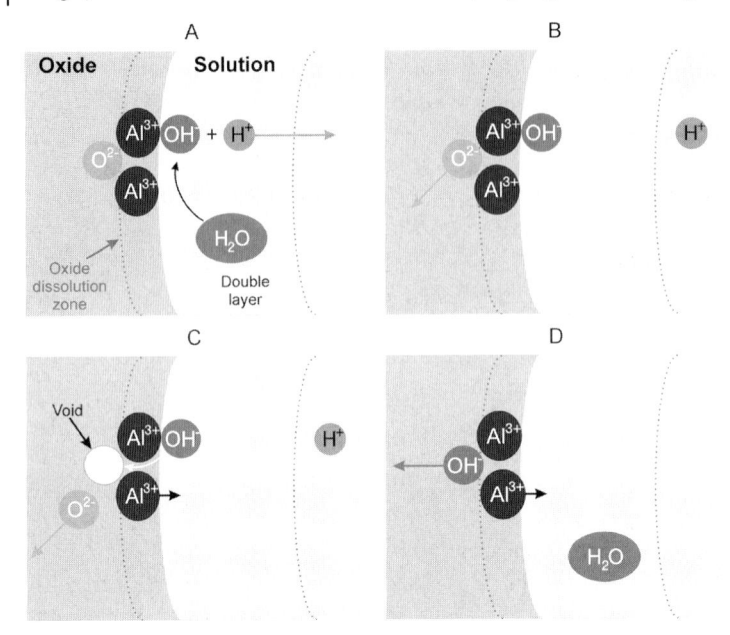

Figure 1.17 The process of field-assisted dissolution of oxide and OH⁻ ions movement in the oxide lattice according to the Patermarakis' model.

within the colloidal layer leads to the formation of droplets and pockets. Pressure inside the pocket is responsible for the pore development and further growth. On the other hand, the formation of a gel-like nascent oxide was assumed in various other studies [227,274,351,381,382]. The gel layers promote the formation of anodic alumina in antimonate, molybdate, silicate and tungstate electrolytes [382]. The gel layer, which is formed above the growing oxide film, eliminates the field-assisted ejection of Al^{3+} ions to the electrolyte. Moreover, it was found that the gel layer could shrink and easily crack on drying.

Stress generation during anodic oxidation of aluminum was studied by Nelson and Oriani [383]. Tensile stress was found to develop rapidly at the metal/oxide interface due to differences in the volume between the metal ions and the oxide. The compressive stress remains constant during the whole process of anodizing. Various possible reactions of aluminum oxidation occurring at the metal/oxide and oxide/electrolyte interfaces were suggested. One proposed model assumed that ions move only through the vacancy exchange mechanism, and that the concentration of vacancies does not change within the oxide film, except at the metal/oxide interface. Compressive stress at the oxide/metal interface, associated with volume expansion during oxide formation, was indicated by Jassensky et al. [384] as a main force responsible for the cellular growth of porous alumina. The generated compressive stress creates repulsive forces between neighboring pores and, as a consequence, induces a close-packed hexagonal arrangement of oxide cells. Since the oxide film formation involves oxidation occurring at the oxide/metal interface, the only direction in which material can expand is by a vertical upwards growth of the cell walls. For the steady-state growth of porous alumina resulting in a high-order of nanopores, a uniform local compressive stress was postulated [385]. The suitable compressive stress provides the correct conditions for steady-state oxide growth, whilst a higher stress might break the oxide walls. A lower compressive stress alters the penetration direction of pores and modifies their growth. An external tensile stress which influences the ordering of pores in self-organized anodization was also studied [231]. Well-ordered structures on anodized aluminum were obtained even on the stressed surface, but a relatively high tensile stress was seen to destroy the arrangement of the pores and, instead of nanopores, huge holes and pits are formed on the highly stressed surface.

The Macdonald's point defect model [283–285] assumed that during film growth, cation vacancies are produced at the oxide/electrolyte interface and are consumed at the metal/oxide interface. In contrast, anion vacancies are formed at the metal/oxide interface but are consumed at the oxide/electrolyte interface. The film formed above metal ridges may contain a high concentration of vacancies (vacancies condensation). According to the model, vacancies at the metal/oxide interface are responsible for the breakdown of the anodic passive film and for large local cation flux through the film. This model predicts that the steady-state thickness of the barrier layer and the logarithm of the steady-state current density should vary linearly with applied anodizing potential.

According to Palibroda et al. [227,234,254,386,387], the steady-state growth of the porous oxide layer is a consequence of the electrical breakdown of the barrier layer by a series of non-destructive avalanches that provide easy pathways for the oxidation of aluminum. The model of oxide growth consists of three steps, the first step being rate-determining in nature [254]. The first step consists of transformation of the existing barrier layer from a compact to porous oxide film. In the second step, aluminum is ionized according to Eq. (35), followed by the reaction of new barrier layer formation [Eq. (32)] in the third step. The proposed model assumed that the barrier layer behaves as a semiconductor, and electrical breakdown of the barrier layer proceeds through non-destructive avalanches [386] which have been previously proposed to explain the electroluminiscence phenomena of anodic alumina oxide [388]. In contrast, Shimizu et al. [389] found that even a few Angstroms difference in barrier layer thickness caused a remarkable variation in local electron tunneling probability. In summary, it can be stated that local differences in oxide thickness promote the electrical breakdown of the barrier layer. The Hoar–Yaholm mechanism of barrier layer protonation [333] and possible proton conductivity in the barrier layer was studied by Palibroda et al. [254,386,387], but no evidence was observed that this mechanism might contribute to the rate-determining step.

Li et al. [236] considered the model of porous alumina growth in which ionization of aluminum according to Eq. (35) proceeds at the metal/oxide interface. The formation of O^{2-} anions occurring by a water-splitting reaction at the oxide/electrolyte interface is a rate-determining step. The increasing electric field at the pore bottom increases local acid-catalyzed corrosion of oxide, and reduces the barrier layer thickness. As a result, the pores are initiated and their further growth proceeds through the self-catalyzed oxide dissolution. An increase in local temperature at the pore bottom (ca. 21 °C), due to Joule heating and acid-catalyzed dissolution of oxide, may lead to a local dehydration of the hydroxide. Consequently, the development of voids and a lateral non-uniform compulsive stress within the middle shell of the cell is observed.

A cellular growth mechanism of self-organized anodic porous alumina was proposed by Zhang *et al.* [390]. The oxide film propagation and growth takes place at a curved metal/oxide interface, which can be counted as an unstable planar front. The dissolution of aluminum and growth of the oxide are controlled by the distribution of the electrical field (instability), which results in a re-stabilized curved metal/oxide interface. Under steady-state conditions, the curved interface then propagates.

1.3.2.6 Other Theoretical Models of Porous Alumina Growth

The model of electronic conduction of anodic oxide based on the hopping conductivity of real amorphous dielectrics of limited thickness, with or without incorporated ions, was presented by Parkhutik and Shershulsky [239]. The hopping transport was modeled as a quasi-Marcovian process. The current density–voltage relationships recorded during aluminum anodizing were fully consistent with model predictions for a dielectric with negative space charge. The slope of the current–voltage relationship was found to be related directly to the transition from the bulk-limiting hopping conduction to surface-limited conduction. Another theoretical model based on electrical field distribution was proposed by Parkhutik *et al.* [202,391]. Basic conclusions derived from the field-assisted model of oxide growth were assumed. The model also considered oxide dissolution by the electrochemical or electric field-enhanced mechanism and 3D electric field distribution in the scalloped barrier layer at pore bottoms. According to this model, the film geometry largely influences the local conduction rate, and is therefore a main parameter responsible for the formation of a porous oxide. The model predicted that steady-state porous oxide growth is a time-independent phenomenon, and that pore formation occurs in a self-consistent manner. The relationship between pore geometry and some anodizing conditions, including anodizing potential, temperature and pH, were successfully established using this model, and the theoretical predictions obtained were consistent with experimental data [239]. Parkhutik *et al.* [392] also studied the effect of combined barrier and porous oxide growth in the mixture of chromic acid with sulfuric acid. The results were attributed to an enhanced formation of insoluble aluminum chromates at high electric field. The effect of extreme dependence of pore growth rate on the electric field was suggested as a feasible explanation for this situation.

A mathematical model of oxide morphology evolution during anodizing was presented by Wu *et al.* [393]. The main assumption of this model was based on the established conduction behavior of anodic alumina oxide and interfacial reactions. Additionally, the effect of current distribution on the near-concentric hemispherical contour between existing surface ridges was taken into account. The enhanced film growth above metal ridges due to a lower local resistance to the metal/oxide interface results in a 2D potential distribution in the oxide film. Although the process of porous oxide formation was described satisfactorily, a lack of explanation for porous oxide growth in a narrow window of experimental conditions was evident. The same modeling school also presented a simulation of the breakdown mechanism of passive oxide films and the growth of pits leading to tunnel formation in anodized alumina during the anodic etching in solutions containing chloride ions [394,395].

In addition to these previously mentioned models, many different approaches to the anodic porous alumina growth have been described in the literature. The morphology of anodic porous alumina membranes was simulated using a radial function of distribution of cells in the triangular network [396], rate equations for competitive processes of alumina formation and etching [334,397], or linear stability analysis showing instability of oxide layer with respect to perturbations with a well-defined wavelength [398]. An electrical bridge model based on the analysis of ion transport in the oxide film and electrical field distribution was also proposed in an attempt to elucidate the self-organized growth of anodic porous alumina [399].

One very promising approach to the anodic porous alumina description seems to be a fractal model of the porous layer formation [400], and the appearance of hexagonally ordered patterns as a result of Turing systems modeling [401,402].

1.3.3
Volume Expansion: The Pilling–Bedworth Ratio (PBR)

The volume expansion of an anodic porous alumina, R, known also as the Pilling–Bedworth ratio (PBR), is defined as the ratio of the volume of aluminum oxide, which is produced by anodizing process, to the consumed aluminum volume:

$$R = \frac{V_{Al_2O_3}}{V_{Al}} = \frac{M_{Al_2O_3} \cdot d_{Al}}{2 \cdot M_{Al} \cdot d_{Al_2O_3}} \tag{41}$$

where: $M_{Al_2O_3}$ is the molecular weight of aluminum oxide, M_{Al} the atomic weight of aluminum, d_{Al} and $d_{Al_2O_3}$ are densities of aluminum (2.7 g cm^{-3}) and porous alumina (3.2 g cm^{-3}), respectively. The theoretical value of the PBR for porous alumina formation with a 100% current efficiency is 1.6. Therefore, the aluminum specimen volume increases significantly during anodizing (see Figure 1.18).

Experimental values of the volume expansion differ slightly from the theoretical predictions due to the lower current efficiency of anodizing, and usually vary between 0.9 and 1.6 [224,384]. Jassensky et al. [384] found that increasing the anodizing potential increases the PBR for anodization conducted in sulfuric acid and the optimal conditions of anodizing; this results in the formation of high-ordered anodic porous alumina corresponding to a moderate expansion of aluminum ($R = 1.22$).

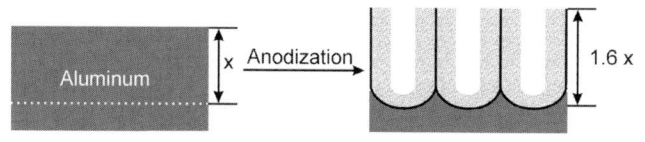

Figure 1.18 Volume expansion observed during anodization of aluminum.

Table 1.9 Anodizing potential influence on the Pilling–Bedworth ratio at 20 °C and galvanostatic regime of anodizing.

Electrolyte	Anodizing potential (V)	$R = f(U)$	Reference(s)
0.45 M $H_2C_2O_4$	$22 < U < 45$	$1.092 + 0.007 \cdot U$	[405]
	$U \leq 55$	$1.144 + 0.0057 \cdot U$	[247,404]
	$U > 55$	$1.308 + 0.003 \cdot U$	
1.1 M H_2SO_4	$13 < U < 24$	$1.1 + 0.0217 \cdot U$	[404]

According to Li et al. [403], the volume expansion factor for optimal anodizing conditions leading to the hexagonally arranged nanopores should be close to 1.4, independent of the electrolyte. For anodizing conducted in 0.34 M oxalic acid at 40 V, the volume expansion factor was found to be about 1.18 [236]. An anodic alumina porosity of 10% guarantees the best ordering of nanopores and volume expansion of about 1.23 [235]. By increasing the PBR above 1.3, a decrease in the size of the ordered domains was observed. The volume expansion factor for the anodization of aluminum in a 0.15 M citric acid solution at 6 and 10 mA cm^{-2} was found to be 1.4 and 1.5, respectively [217].

The influence of anodizing potential on the volume expansion factor was studied by Vrublevsky et al. [247,404,405], for the anodization of aluminum conducted under a constant current density. Measurements were performed using a mechanical profiler with a computer signal processing for oxalic and sulfuric acids, whereupon the PBR was found to be linearly dependent on the anodizing potential (Table 1.9).

In contrast, the concentration of oxalic acid in the range between 0.22 and 0.92 M (2% and 8%) did not affect the volume expansion factor [405]. For anodizing under constant current density, increasing the temperature causes a decrease in anodizing potential and volume expansion factor. Yet, increasing the current density increases the volume expansion factor. This effect can be attributed to the fact that the PBR depends on the electric field strength in the barrier layer. The dependence of the current density logarithm on the inverse volume expansion factor of anodic porous alumina is linear [247,404,405].

1.3.4
Rates of Oxide Formation and Oxide Dissolution

A wide variety of methods were employed to measure the thickness of the oxide layer formed by anodization of aluminum [205]. Recently, optical and microscopic techniques including TEM or SEM have mainly been used to evaluate anodic oxide layer thickness.

For the constant current density anodization, the total thickness of the oxide layer can be calculated from the pore-filling method, using the formula produced by Takashi and Nagayama [406]:

$$h = 10^{-7} \cdot B_U \cdot V_p - \frac{i \cdot t_p \cdot M_{Al}}{n \cdot F \cdot k \cdot d_{Al_2O_3}} (1 - T_{Al^{3+}}) \tag{42}$$

where, B_U is the barrier layer thickness per volt (nm V^{-1}), i is the current density (mA cm^{-2}), M_{Al} is the atomic weight of aluminum, n is the number of electrons associated with oxidation of aluminum, F is Faraday's constant, k is the weight fraction of aluminum in alumina (0.529), $d_{Al_2O_3}$ is the density of porous alumina (3.2 g cm^{-3}), $T_{Al^{3+}}$ is the transport number of Al^{3+} ions (about 0.4), and V_p and t_p are the voltage and time, respectively measured at the point where two straight parts of the voltage–time transient meet.

On the other hand, the thickness of the oxide layer can be calculated from Faraday's law. As the efficiency of anodizing is not usually 100%, the recorded current density cannot be used simply for theoretical estimation of the grown oxide layer, and the current efficiency should be considered as follows:

$$m_{Al_2O_3} = k_{Al_2O_3} \cdot j \cdot t \cdot \eta = \frac{M_{Al_2O_3}}{z \cdot F} \cdot j \cdot t \cdot \eta \tag{43}$$

where, $m_{Al_2O_3}$ is a mass of formed oxide, $k_{Al_2O_3}$ is the electrochemical equivalent for aluminum oxide, j is the passing current (A), t is the time (s), η is the current efficiency, $M_{Al_2O_3}$ is the molecular weight of aluminum oxide (g mol^{-1}), z is the number of electrons associated with oxide formation, and F is Faraday's constant (C mol^{-1}). Taking into account that the oxide mass can be expressed as the product of oxide density ($d_{Al_2O_3}$) and oxide volume ($V_{Al_2O_3}$) or as the product of density, the surface area (S) and oxide height (h):

$$m_{Al_2O_3} = d_{Al_2O_3} \cdot V_{Al_2O_3} = d_{Al_2O_3} \cdot S \cdot h \tag{44}$$

the oxide layer thickness formed at constant current anodizing is:

$$h = \frac{M_{Al_2O_3}}{z \cdot F \cdot d_{Al_2O_3}} \cdot \frac{j}{S} \cdot t \cdot \eta = \frac{M_{Al_2O_3}}{z \cdot F \cdot d_{Al_2O_3}} \cdot i \cdot t \cdot \eta \tag{45}$$

where i is the current density. Consequently, for the constant potential anodizing the thickness of the oxide layer can be expressed by:

$$h = \frac{M_{Al_2O_3}}{z \cdot F \cdot d_{Al_2O_3}} \cdot \eta \int_0^t i(t) \, dt \tag{46}$$

It is generally accepted that the constant current anodizing, the oxide thickness increases linearly with increasing current density according to the following relationship:

$$h = k \cdot i \cdot t \tag{47}$$

where k is a constant value independent of current density and temperature [364]. Therefore, the oxide layer thickness formed under constant potential anodizing can be calculated:

$$h = k \int_0^t i(t)\, dt \tag{48}$$

The value of k was estimated at about $3.09 \times 10^{-6}\,cm^3\,(mA \cdot min)^{-1}$ for the constant current density anodizing carried out in 1.53 M H_2SO_4 [364,367].

The thickness of the oxide layer grown under the constant potential regime in 0.3 M oxalic acid can be easily estimated from SEM cross-sectional views and rates of oxide formation (R_h in nm min^{-1}) at various temperatures according to the given equation [241]:

$$(5\,°C) \quad R_h = 392.30 - 26.92\,U + 0.63\,U^2 \tag{49}$$

$$(15\,°C) \quad R_h = 123.43 - 9.19\,U + 0.23\,U^2 \tag{50}$$

$$(30\,°C) \quad R_h = 51.33 - 3.71\,U + 0.095\,U^2 \tag{51}$$

Sulka et al. [240,407] determined the experimental rate of oxide formation in 2.4 M H_2SO_4 under the constant potential anodizing regime. The growth rate of the oxide layer was calculated for Al samples treated by one-, two-, and three-step anodizing conducted in the overflow cell [407] at 1 °C (Figure 1.19A). No difference was found between the experimental growth rate of oxide layers obtained by the one-, two-, and three-step anodizing procedures. The experimental growth rates of oxide anodized in the simple electrochemical cell with magnetic stirring were also determined for various anodizing temperatures (Figure 1.19B). The effective growth rate of the oxide layer on anodized aluminum increases exponentially with increasing cell potential.

In general, the steady-state growth of anodic alumina is a result of equilibrium between the rate of chemical formation and rate of oxide dissolution. The total amount of dissolved oxide is a sum of electrochemical dissolution (field-assisted process) and chemical etching. Hence, dissolution of the oxide layer should be a function of the hydrogen ion concentration in the anodizing electrolyte, and is especially accelerated by H^+ ion adsorption [199,202,254]. The typical rate of field-assisted dissolution of alumina at room temperature is about 300 nm min^{-1}, compared to a value of 0.1 nm min^{-1} in the absence of the field (chemical dissolution) [203]. For anodizing conducted in 1.5 M H_2SO_4 under 17 V (12.9 mA cm^{-2}) at 21 °C, the rates of field-assisted dissolution and chemical dissolution were calculated to be about 372.5 and 0.084 nm min^{-1}, respectively [244]. According to Nagayama and Tamura [408], the rate of field-assisted dissolution of oxide is about 1040 nm min^{-1} for anodization carried out in 1.1 M H_2SO_4 at 11.9 V and 27 °C.

A strict control of the anodic nanoporous alumina formation requires the gaining of access to the correct information concerning the rate of chemical dissolution of oxide in the acidic electrolyte. The chemical dissolution of oxide, especially in acidic

1.3 Kinetics of Self-Organized Anodic Porous Alumina Formation | 49

A

▲ Three-step anodizing
■ Two-step anodizing
◆ One-step anodizing

$y = 2.746\text{E}{-}03e^{2.421\text{E}{-}01x}$

R_h (mm/min) vs Anodizing potential (V)

B

◆ $T = -8\ °C$
■ $T = 1\ °C$
▲ $T = 10\ °C$

$y = 1.930\text{E}{-}03e^{2.877\text{E}{-}01x}$

$y = 1.237\text{E}{-}03e^{2.879\text{E}{-}01x}$

$y = 6.256\text{E}{-}04e^{2.873\text{E}{-}01x}$

R_h (mm/min) vs Anodizing potential (V)

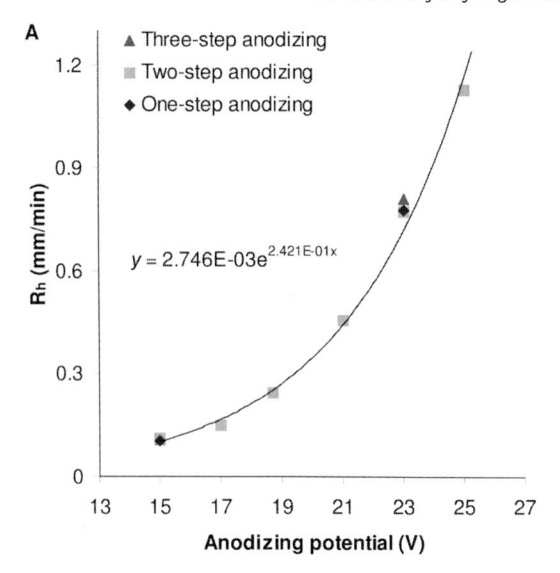

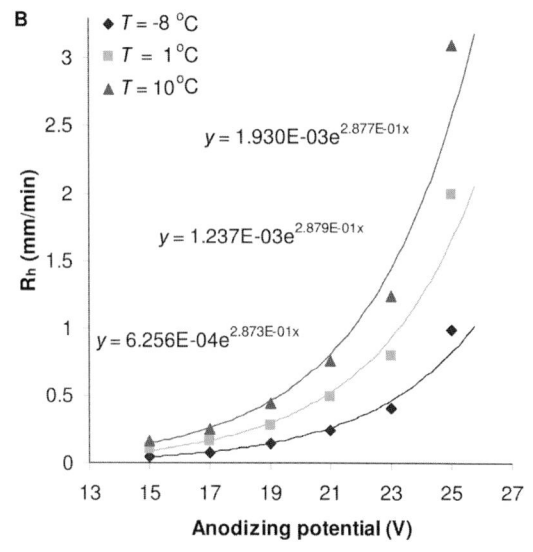

Figure 1.19 Growth rate of the oxide layer (R_h) versus anodizing time. Anodizing was conducted at a constant cell potential in a 2.4 M (20%) sulfuric acid electrolyte in the overflow cell (A) at 1 °C and simple electrochemical cell (B) at various temperatures. (Figure 1.19A reproduced with permission from Ref. [407], © 2002, The Electrochemical Society.)

Table 1.10 The rate of chemical etching of anodic alumina.

Electrolyte	Concentration (M)	Temperature (°C)	Etching rate (nm min^{-1})	Reference
H_2SO_4	1.7 (15%)	20	0.076	[244]
		25	0.114	
		30	0.172	
		50	0.873	
		70	4.434	
	0.1 (1%)	38	0.25	
	0.21 (2%)		0.27	
	1.4–3.1 (12–25%)		0.33	
	7.0 (48%)		0.25	
	15.3 (85%)		0.125	
	2.4 (20%)	20–22	0.05	[252]
	1.53 (13.7%)	20–40	0.052–0.41	[364]
	1.1 (10%)	27	0.074	[364]
	1.1 (10%)	27	0.075	[408]
	2.0 (11.1%)	60	1.6	[206]
$H_2C_2O_4$	0.63	40	0.43	[406]
H_3PO_4	0.45 (4.25%)	20–22	0.02–0.02	[252]

electrolytes, is of major significance for the development of post-treatment procedures that allow the strict control of the pore diameter of nanostructures. Moreover, the rate of chemical etching of oxide in certain media serves as valuable information in the process of selecting optimal anodizing conditions. The chemical dissolution of alumina was normally studied in a sulfuric acid solution due to a strong chemical aggression of the concentrated electrolyte used for anodizing. The concentration of phosphoric or oxalic acid used as an electrolyte for the anodization of aluminum is at least a few-fold lower than the sulfuric acid concentration. Consequently, the expected rate of chemical etching of alumina in phosphoric or oxalic acids solutions is significantly lower. Selected values of the chemical etching rate in various electrolytes are listed in Table 1.10.

The rate of oxide dissolution in 1.53 M oxalic acid is much slower than for the same concentration of sulfuric acid [375]. For example, the rate of chemical etching of oxide at 35 °C in oxalic acid is comparable with the rate of dissolution at 25 °C in sulfuric acid [375]. The incorporated phosphate anions enhanced the chemical dissolution of oxide [268]. Further details on the dissolution rate of the barrier layer are presented in Section 1.4.3.

1.4
Self-Organized and Prepatterned-Guided Growth of Highly Ordered Porous Alumina

The process of alumina template formation by anodization is relatively simple and results in a high density of parallel nanopores. Hence, anodic porous alumina (AAO)

1.4 Self-Organized and Prepatterned-Guided Growth of Highly Ordered Porous Alumina

is a key template material for the fabrication of various nanostructured materials. In general, there are two widely used methods of AAO template synthesis: (i) a self-organized, two-step anodization leading to a quasi-monodomain structure; and (ii) a prepatterned-guided anodization resulting in a perfectly ordered pore lattice.

A flow diagram of the self-organized formation of anodic alumina membranes by the anodization of aluminum with a typical two-step anodizing procedure is presented in Figure 1.20.

The formation of nanopores by self-organized anodizing of aluminum is a multistage process consisting of a pre-treatment, anodizing, and post-treatment steps. The pre-treatment procedure includes annealing of aluminum foil in a non-oxidizing atmosphere, degreasing of samples, and electropolishing. The two-step anodizing procedure is usually based on the initial anodizing at the pre-selected cell potential and subsequent chemical etching of the grown aluminum oxide layer. Following the chemical removal of oxide, a periodic concave triangular pattern formed on the aluminum surface acts as self-assembled masks for the second anodizing. The second anodization is conducted at the same cell potential as used during the first anodizing step. Finally, the synthesized hexagonally arranged nanopore structure can be removed from the base aluminum, and the pores may be opened and widened.

PRETREATMENT → **ANODIZING** → **POST-TREATMENT**

STARTING MATERIAL
Al rolled foil
purity: 99.997 %
thickness: 0.25 mm

↓

THERMAL ANNEALING
T = 400 °C
Ar atmosphere
t = 5 h

↓

DEGREASING
Acetone then ethanol

↓

ELECTROCHEMICAL POLISHING
$HClO_4$ (60 %) : C_2H_5OH = 1:4 (vol.)
T = 10 °C
I = 500 mA/cm^2
t = 1 min

FIRST STEP OF ANODIZATION
20 % H_2SO_4
T = 1 °C
t = 10 min
Potential: 15 - 25 V

↓

OXIDE REMOVAL
6 % H_3PO_4 + 1.8 % H_2CrO_4
T = 60 °C
t = 10 - 15 min

↓

FIRST STEP OF ANODIZATION
20 % H_2SO_4
T = 1 °C
t = 870 - 45 min
Potential: 15 - 25 V

ALUMINUM BASE REMOVAL
Saturated $HgCl_2$
Room temperature
t = 2 - 3 h

↓

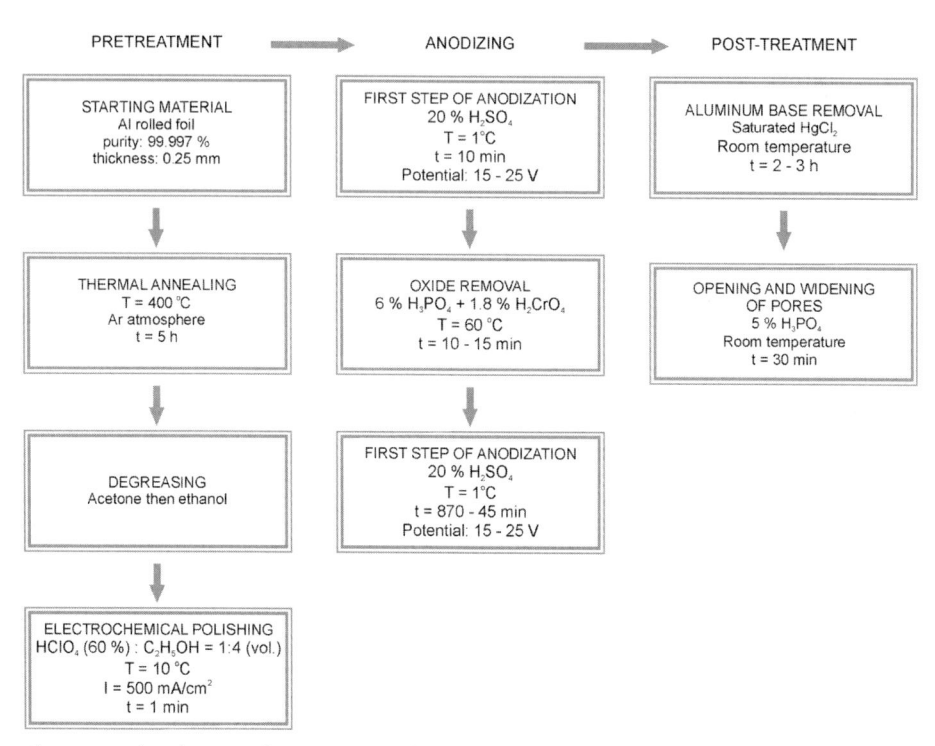

OPENING AND WIDENING
OF PORES
5 % H_3PO_4
Room temperature
t = 30 min

Figure 1.20 Flow diagram of AAO templates formation by two-step anodizing in sulfuric acid.

In contrast, prepatterned-guided anodizations based on the pre-texturing of electropolished aluminum prior to anodizing are used for the synthesis of ideally ordered nanopores. Among these methods, a direct indentation of the aluminum surface with a tip of the scanning probe microscope [410,411], focused-ion beam lithography [412–415], holographic lithography [416] and resist-assisted focused-ion beam lithography [417–419] were used successfully to form the pattern on the aluminum surface. In the direct patterning of aluminum, each sample can be indented individually, although this makes the techniques time-consuming and also limits their applications to the laboratory scale. Therefore, the imprinting lithography with a master stamp (mold) is the most widespread method used for prepatterning of aluminum [420–436]. Stamps with an arranged array of convexes are usually prepared lithographically, and can be used several times for the pre-texturing of aluminum. The mold used for nanoindentation of aluminum can be made from SiC [420–429], Si_3N_4 [430–432], Ni [433–435] and poly(dimethylsiloxane) (PDMS) [436]. The imprinting of aluminum using a master stamp and fabrication of anodic porous alumina from prepatterned aluminum is illustrated in Figure 1.21.

The imprinting of aluminum with a master stamp is usually carried out using an oil press. After imprinting, the generated array of concaves on the aluminum surface is a negative replication of convexes of the master stamp. The typical depth of the concave formed by the molding process is about 20 nm [420]. The different shape and arrangement of convexes in the master mold leads to various nanopore arrays in anodic porous alumina. The indentation of triangular, square, and graphite structure lattices results in hexagonal, square, and triangular cells, respectively (Figure 1.22) [424–427,430,432,434,436]. The porous alumina array with a Moiré pattern was also fabricated using indentation with the mold [431].

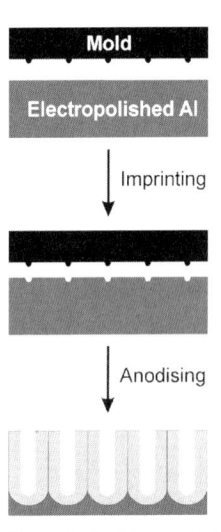

Figure 1.21 Schematic diagram of fabrication of ideally ordered porous alumina using an imprint stamp.

1.4 Self-Organized and Prepatterned-Guided Growth of Highly Ordered Porous Alumina

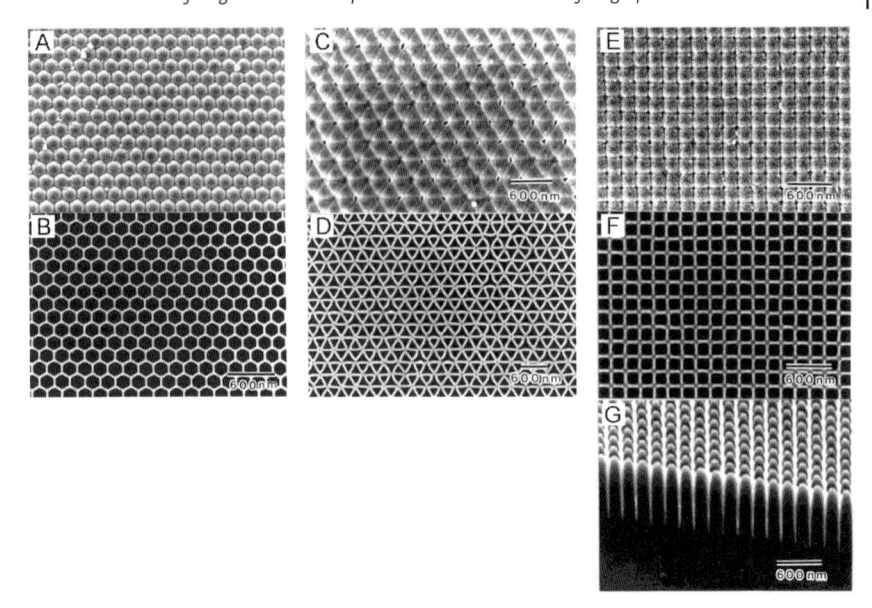

Figure 1.22 SEM images of cells (A, C, E), openings (B, D, F) at the bottom side of anodic alumina and cross-section (G) of anodic alumina formed by indentation of triangular, graphite structure and square lattices. (Reprinted with permission from Ref. [427], © 2001, Wiley-VCH Verlag GmbH.)

Recently, the prepatterned aluminum surface has been obtained using nanosphere lithography (NSL) employing a 2D hexagonal close-packed array of polystyrene [433,437] and Fe_2O_3 spheres [438] formed on the Si or glass supporting substrate. The aluminum layer was sputtered over the self-organized array of spherical monodisperse nanoparticles. After the removal of particles, shallow concaves formed on the Al surface serve as initiation points for pore formation during anodizing (Figure 1.23).

A self-assembly, close-packed 3D ordered lattice of spherical polystyrene particles formed on aluminum was used as a template for the synthesis of 2D/3D composite porous alumina structure on the Al sample [439]. The formation of ordered anodic porous alumina from the aluminum surface covered with a metallic Ta mask was also presented [440].

1.4.1
Aluminum Pre-Treatment

The quality of aluminum substrates and their surface pre-treatment have major influences on the surface nanostructuring by self-organized anodizing. The structure of a pre-existing film on the aluminum surface, which may develop in air, thermally or during chemical and electrochemical treatment, depends on the applied pre-treatment procedure. During the self-organized anodizing of aluminum, the process of pore nucleation is a combination of random nucleation and nucle-

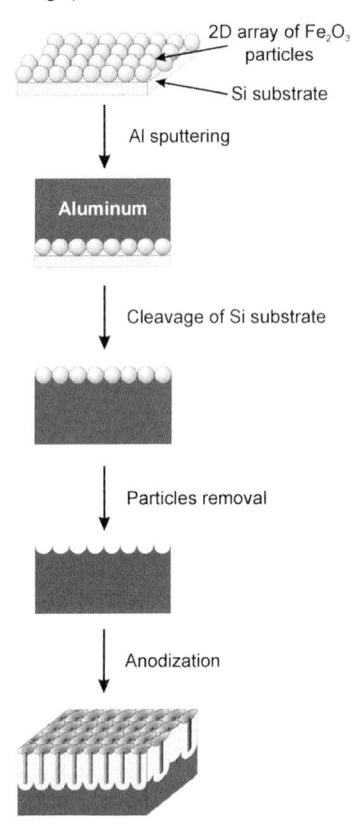

Figure 1.23 Schematic diagram of fabrication of high-ordered anodic porous alumina using a 2D array of spherical monodisperse particles as template. (Reprinted with permission from Ref. [438], © 2006, Wiley-VCH Verlag GmbH & Co. KGaA.)

ation at surface faults. Moreover, the grain boundaries and scratches on the aluminum surface are sites for preferential pore development [440]. The pre-treatment process control must be focused on the reduction of surface faults, or on their controlled and reproducible generation, in the required morphology. Therefore, the most desirable starting material for self-organized nanopore array formation by anodizing is a high-purity, annealed aluminum foil.

The effect of starting material on pore ordering was studied by Terryn et al. [441]. Anodic porous film development in sulfuric and phosphoric acids was studied on rolled and AC-grained aluminum, with a similar pore ordering, perpendicular to the surface, being observed. The anodization of relatively pure aluminum and various aluminum alloys was studied in a sulfuric acid solution at $0\,°C$ [442], and the uniformity of oxide growth was seen to increase with the increasing purity of aluminum.

The annealing of foil reduces stresses in the material and increases the average size of the grain [393,443], which is usually more than $100\,\mu m$ (Figure 1.24). The

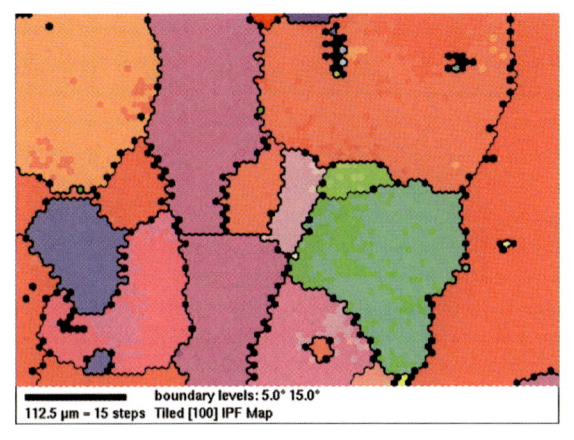

1.4 Self-Organized and Prepatterned-Guided Growth of Highly Ordered Porous Alumina

Figure 1.24 Orientation image microscopy (OIM) top-view image of the annealed and electropolished Al surface with marked grain boundaries. The annealing was conducted in an argon atmosphere at 400 °C for 5 h.

typical annealing of aluminum foil is conducted under an argon or nitrogen atmosphere at 400 or 500 °C for 3–5 h.

The degreasing and polishing of Al samples is then carried out before anodizing. Among various solvents used for degreasing, acetone and ethanol are commonly employed. Due to their moderate or high carcinogenicity, dichloromethane [387], trichloroethylene [218], benzene and methanol [444] have been used only sporadically. The immersion of samples in 5% NaOH at 60 °C for 30 s or 1 min and subsequent neutralization in a 1:1 $HNO_3 + H_2O$ solution for several seconds was also proposed for Al degreasing and cleaning [236,309,335,343]. A mixed solution of $HF + HNO_3 + HCl + H_2O$ (1:10:20:69) was also proposed for degreasing and cleaning of the aluminum surface before anodizing, leading to a highly ordered, self-organized nanopore array [403].

The most important step in the pre-treatment of aluminum before anodizing is the polishing of samples. For aluminum, this can be achieved by means of mechanical, chemical, or electrochemical polishing. Mechanical polishing of aluminum has been used rather sporadically to prepare a smooth surface before anodizing [252,390,442]. However, detailed TEM analyses of mechanically polished aluminum have shown the procedure incapable of producing a microscopically smooth and undeformed surface, even when conducted with great care [445]. Chemical polishing of aluminum to prepare surfaces before anodizing may lead to highly ordered nanostructures, and is not widely used (Table 1.11).

A combined chemical–mechanical polishing of aluminum with a slurry containing 0.05-µm Al_2O_3 particles with hydrogen peroxide, citric acid and phosphoric acid, was reported in detail [447].

It should be noted that electrochemical polishing in a 60% $HClO_4 + C_2H_5OH$ (1:4, v/v) solution at 10 °C and 500 mA cm^{-2} for 1 min is commonly used to prepare smooth Al surfaces before anodizing. Other less-frequently used electropolishing

Table 1.11 Chemical polishing of aluminum before anodizing.

Electrolyte	Temperature (°C)	Time (min)	Remarks	Reference
70% HNO_3 + 85% H_3PO_4 (15:85)	85	2	Followed by immersing in stirred 1 M NaOH, 10 min, RT	[278]
3.5% H_3PO_4 + 45 g L^{-1} CrO_3	80	10	Followed by electrochemical polishing	[299]
70 mL H_3PO_4 + 25 mL H_2SO_4 + 5 mL HNO_3	85	1.5	After initial mechanical polishing	[446]
91% H_3PO_4 + 98% H_2SO_4 + 70% HNO_3 (75:11:11, v/v) + 0.8 g L^{-1} $FeSO_4 \cdot 7H_2O$	95–100	45–75	Gentle stirring of solution	[333]
70% H_3PO_4 + 20% H_2SO_4 + 10% HNO_3 + 5 g L^{-1} $Cu(NO_3)_2$	–	–	–	[341]
25% NaOH + 20% $NaNO_3$	80	3	–	[220]

^{e}RT: room temperature.

mixtures, together with conditions of electrochemical polishing, are listed in Table 1.12.

An AFM study showed that the typical surface roughness of an electropolished sample is about 20–30 nm on a lateral length scale of 10 µm [443]. The examination of various procedures of pre-treatment of aluminum surfaces, including mechanical polishing or electropolishing, in a standard perchloric acid–ethanol mixture followed by chemical etching was studied in detail [450]. The influence of the electropolishing of aluminum on pore ordering occurring during anodizing was also studied [211,443,451]. Typical SEM images of the aluminum surface after electropolishing in a mixture of perchloric acid and ethanol are presented in Figure 1.25.

Similar stripes and mound patterns formed on the surface during the electrochemical polishing of aluminum were reported elsewhere [185,452–457]. The Al (1 1 0) surface was found to exhibit a regular striped array, whereas on the Al (1 1 1) and Al (1 0 0) surfaces hexagonally ordered patterns were observed. These patterns act as self-assembled masks for nanopore growth in the anodizing process. However, a cellular structure was also observed sporadically on the electropolished aluminum surface [449,450].

Apart from high-purity aluminum foil, the most frequently used starting materials for the anodizing of aluminum have included sputtered or evaporated aluminum on various substrates, including soda-lime glass covered with a tin-doped indium oxide (ITO) [213,363,458–468], Si [298,310–312,321,322,468–485], Ti [486], InP [487] and GaAs [488]. The transfer of nanopore order from anodic porous alumina to semiconductor materials (e.g., Si) is of special interest from the point of view of

1.4 Self-Organized and Prepatterned-Guided Growth of Highly Ordered Porous Alumina

Table 1.12 Non-typical conditions of electrochemical polishing of aluminum.

Electrolyte	Temperature (°C)	Current density or potential	Time (min)	Reference(s)
20% $HClO_4$ + 80% C_2H_5OH	0–5	15 V	2	[259]
20% $HClO_4$ + 80% C_2H_5OH	10	20 V	2	[256]
20% $HClO_4$ + 80% C_2H_5OH	10	100 mA cm^{-2}	5	[224,262–265]
$HClO_4$ + C_2H_5OH (1:4, v/v)	7	20 V	1.0–1.5	[241]
	30	160 mA cm^{-2}	3.5	[222]
51.7% $HClO_4$ + 98% C_2H_5OH + Glycerol (2:7:1, v/v)	5	17 V	4	[448]
$HClO_4$ + CH_3COOH (1:4, v/v)	10–15	100 mA cm^{-2}	–	[268]
$HClO_4$ + C_2H_5OH + 2-butoxyethanol + H_2O (6.2:70:10:13.8, v/v)	10	500 mA cm^{-2}	1	[236]
$HClO_4$ + C_2H_5OH + 2-butoxyethanol + H_2O	10	40 V or 170 mA cm^{-2}	4.5	[305]
H_3PO_4 + C_2H_5OH + H_2O (40:38:25, v/v)	40	5 mA cm^{-2}	2	[393]
H_3PO_4 + $C_4H_{11}OH$ (80:20, v/v)	60–65	10–40 V 30–50 mA cm^{-2}	10–15	[449]
H_3PO_4 + H_2SO_4 + H_2O (4:4:2, v/v)	–	–	–	[384,443]
H_3PO_4 + H_2SO_4 + H_2O (7:2:1, v/v) + 35 g dm^{-3} CrO_3	40	135 mA cm^{-2}	10	[299]

integrating AAO with silicon circuit industrial processes, and the potential application of such structures as biosensors, bioreactors, and magnetic recording media. The main drawback of those methods which employ sputtered or evaporated aluminum is a limited thickness of the aluminum layer. Consequently, the anodizing

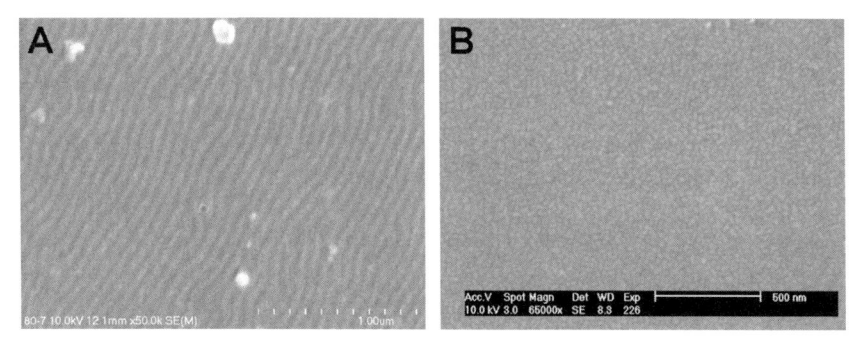

Figure 1.25 SEM top-view images of electropolished aluminum foil with typical stripes (A) and mounds (B) patterns. The electropolishing was conducted in a 60% $HClO_4$ + C_2H_5OH (1:4, v/v) solution at 10 °C and 500 mA cm^{-2} for 1 min.

time required for steady-state growth of pores, and resulting in an ordered arrangement of nanopores, might be not reached before the entire aluminum layer is completely converted into aluminum oxide.

The self-organized anodizing of aluminum was carried out sporadically on aluminum alloys [489,490] or a curved cylindrical sample [491]. In the latter case, cylindrically and pentagonally shaped 3D alumina nanotemplates with hexagonally arranged nanopores were successfully fabricated.

1.4.2
Self-Organized Anodizing of Aluminum

The self-organized growth of ordered pores on anodized aluminum occurs within a relatively narrow window of experimental conditions. Generally, a mild anodizing process leading to porous alumina formation is conducted at low temperatures and employs mainly sulfuric, oxalic, and phosphoric acids as an electrolyte. For each electrolyte, there is a certain range of potential which can be applied for anodizing without burning or breakdown of the oxide film (Table 1.13). Moreover, there is a certain value of the anodizing potential (self-ordering regime) at which, the best arrangement of nanopores is observed. When anodizing is conducted under different values of anodizing potential, the degree of pore ordering is reduced drastically. The temperature of electrolyte and alcohol addition have no influence on the self-ordering regime for the certain anodizing electrolyte. In mild, self-organized anodization, the rate of oxide growth is low due to the low current density, and the highest or most moderate rate of oxide growth is observed in sulfuric acid [448]. More recently, hard anodizing (also known as a high-field anodizing) has been successfully applied for self-organized oxide formation [213,251,499]. In this process the range of anodizing potential and self-ordering values of potentials differ widely (Table 1.13).

The passing current density used during hard anodization is much higher than in mild anodization, and the rate of oxide growth increases about 2500- to 3500-fold [499]. For mild anodizing conducted in phosphoric acid, the rate of oxide growth is about 0.05 to 0.2 $\mu m\,min^{-1}$ over a range of anodizing potential from 80 to 195 V, whereas for hard anodizing at 195 V the rate varies between 4 and 10 $\mu m\,min^{-1}$, depending on the applied electric field [339,448]. Thus, in order to avoid any breakdown of the oxide film during hard anodization, aluminum is usually pre-anodized at a constant current density or constant potential for a few minutes. It should be noted that the high-field applied during hard anodizing promotes much evolution of heat, and any excessive heat should be effectively removed from the sample in order to prevent burning.

Following the discovery by Masuda and Satoh [500] of a two-step anodization, the one-step, self-organized anodizing procedure (which resulted in a worse-ordered arrangement of nanopores) is no longer used. Today, a two-step anodizing procedure is commonly used for the formation of high-ordered arrangements of nanopores by anodizing aluminum in sulfuric, oxalic, or phosphoric acid. Following an initial anodization at the pre-selected cell potential, chemical etching of the grown aluminum oxide layer is carried out in a mixed solution containing

Table 1.13 Mild and hard anodization conditions of self-organized anodizing of aluminum and self-ordering regimes.

	Mild anodization				Hard anodization			
Electrolyte concentration (M)	Temperature (°C)	Range of potential (V)	Self-ordering regime (V)	Reference(s)	Electrolyte concentration (M)	Temperature (°C)	Range of potential (V)	Self-ordering regime (V)	Reference(s)
0.3 H_2SO_4	10	10–25	25	[403,492,493]	1.8 H_2SO_4	0	40–70	70	[213,251]
2.4 H_2SO_4	1			[216,226,240]					
6.0 H_2SO_4	20		18	[494]					
0.3 $H_2C_2O_4$	1–5	30–100	40	[403,493,495]	0.03–0.06 $H_2C_2O_4$	3	100–160	160	[213]
	15–20			[216,241,496]	0.3 $H_2C_2O_4$	1	110–150	120–150	[499]
0.2–0.3 H_3PO_4	0–5	160–195	195	[216,443,497]	0.1 H_3PO_4	0	195–235	235	[213]
H_3PO_4-CH$_3$OH-H$_2$O (1:10:89, v/v)	−4			[498]	H_3PO_4-C_2H_5OH-H_2O (v/v 1:20:79)	−10 to 0	195	195	[339]

H_3PO_4 (6 wt.%) and H_2CrO_4 (1.8 wt.%) at a temperature of 60–80 °C (e.g., Refs. [226,231,240,248,318,407,409,493,498]). A slightly modified composition of solution consisting of 0.4 or 0.5 M H_3PO_4 and 0.2 M H_2CrO_4 was also proposed for the oxide removal [235,496,501]. The time required for the chemical etching of oxide depends heavily on the thickness of the oxide film grown during the first anodization, and can vary from a few minutes to several hours. It should be noted that the rate of oxide growth during anodization depends significantly on the anodizing electrolyte, with the highest rate being observed for sulfuric acid anodizing. A completely different method for the removal of oxide formed during the first-step of anodization was proposed by Schneider et al. [502]. This method employs a voltage detachment procedure, whereby a reverse voltage is applied to the anodized sample (the value of the reverse voltage is the same as was used for the anodization). Such a voltage detachment procedure allows the production of a fully flexible anodic porous alumina membrane, as opposed to oxide removal by chemical etching. The enhanced mechanical flexibility of the membrane is an undeniable advantage of the voltage detachment procedure over the chemical etching. One application of the electrochemical voltage pulse technique for the effective detachment of oxide layer at various electrolytes was investigated systematically [503]. The procedure results in a freestanding detached AAO membrane with open pores. The proposed mechanism of the pulse detachment involved oxide removal independently of the applied procedure, and resulted in the formation of a periodic concave pattern on the aluminum surface, which acted as a mask for the second anodizing. Following oxide removal, a second anodization is carried out at the same anodizing potential as used for the first anodization. Selected procedures of two-step anodization of aluminum resulting in a high-order arrangement of nanopores are detailed in Table 1.14.

A three-step anodizing procedure, consisting of two full cycles of pre-anodization and subsequent oxide removal, was verified as a method of self-organized nanopore arrangement in oxalic [236,390] and sulfuric acid [407]. The duration of each anodizing step is usually different. For example, times for the first, second, and third anodizations in three-step anodizing in oxalic acid were fixed at 10 min, 690 min and 3 min, respectively [236]. It was fount that, for the three-step anodization, the order of self-assembled nanopores is comparable with the arrangement of pores obtained by the two-step anodizing process [390,407]. Brändli et al. [512] investigated the effect of multi-step anodization in oxalic acid on a long-range uniformity in pore diameter, with the anodization and oxide dissolution cycles being repeated even three or four times; however, a rather disordered nanopore structure was obtained. It was also found that increasing number of anodization–etching cycles led to an increase in the uniformity of pore size and pore diameter.

1.4.2.1 Structural Features of Self-Organized AAO

The structural features of anodic porous alumina formed by anodization under a potentiostatic regime depend on the electrolyte and the applied anodizing potential. The interpore distance of the AAO lattice formed by the self-organized anodization depends heavily on the anodizing potential. Figure 1.26 includes SEM images of the pore arrangements with the same magnification, taken from the bottom of pores

1.4 Self-Organized and Prepatterned-Guided Growth of Highly Ordered Porous Alumina | 61

Table 1.14 Selected procedures of two-step anodizing of aluminum conducted in sulfuric, oxalic, and phosphoric acids.

Electrolyte	Temperature (°C)	Potential (V)	Time of first anodization (min)	Mixture type*	Oxide removal Temperature (°C)	Time (min)	Second anodization Time (min)	Oxide thickness (μm)	References(s)
0.3 M H_2SO_4	10	25	1320	A	60	–	5940	~200	[493]
1.1 M H_2SO_4	1	19	1440	A	60	600	5700	200	[504]
2.4 M H_2SO_4	1	15–25	10	A	60	10–15	870–45	90	[240,248]
							1240–125	140	[407]
6–8 M H_2SO_4	0–20	18	60	A	–	–	5	0.6	[494]
0.2 M $H_2C_2O_4$	18	40	1800	A	60	–	50	–	[505]
0.3 M $H_2C_2O_4$	0	40	20	B	60	5	120	10	[506]
	0–25		600	C	65	720	60	–	[496]
	1		1440	A	60	600	5700	200	[504]
	10		120	A	60	240	2–10	0.2–1.3	[507]
	15		900	A	65	–	10–600	–	[390]
	17		600	A	60	840	5	0.7	[500]
	20		40	A	65	10–20	120	–	[508]
	24		300	A	60	600	5	1.0	[509]
0.5 M $H_2C_2O_4$	5–30	20–60	120–740	A	65	60–240	740	–	[241]
	5	40	1320	A	60	–	5940	~170	[493]
0.15–0.5 M $H_2C_2O_4$	5	10–60	1200	D	–	–	30	1.2–1.5	[495]
0.15 M H_3PO_4	2	195	1440	A	60	600	1200	110	[510]
0.5 M H_3PO_4	0	140	120	A	60	240	5–10	–	[511]
H_3PO_4-CH_3OH-H_2O (1:10:89, V/v)	–4	195	1200	A	80	–	1200	~100	[498]

*Mixture types: A: phosphoric acid (6 wt%) and chromic acid (1.8 wt%); B: 0.4 M phosphoric acid and 0.2 M chromic acid; C: 0.5 M phosphoric acid and 0.2 M chromic acid; D: phosphoric acid.

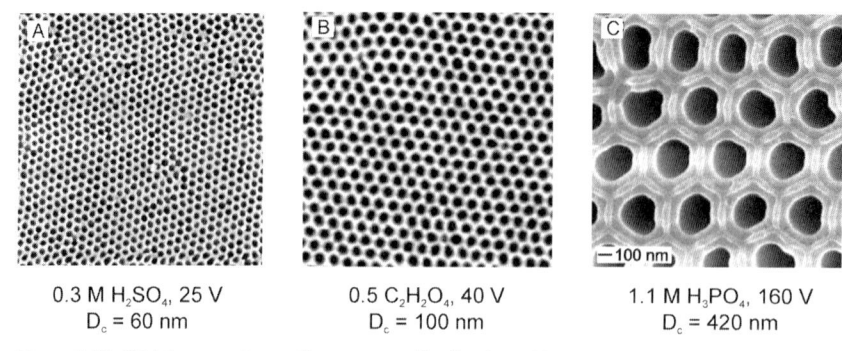

0.3 M H$_2$SO$_4$, 25 V   0.5 C$_2$H$_2$O$_4$, 40 V   1.1 M H$_3$PO$_4$, 160 V
D$_c$ = 60 nm   D$_c$ = 100 nm   D$_c$ = 420 nm

Figure 1.26 SEM bottom-views of porous anodic alumina with hexagonally arranged nanopore structures after opening pore bottoms. Nanostructures were formed by self-organized anodization in different electrolytes at 10 °C (A), 5 °C (B) and 3 °C (C). (Reprinted with permission from Ref. [493], 1999, AVS The Science & Technology Society.)

after their opening, and the interpore distance values of nanostructures. The high-ordered anodic porous alumina films were obtained by the two-step anodizing in sulfuric, oxalic, and phosphoric acid solutions under potentials of 25, 40, and 160 V, respectively. Pores formed in sulfuric, oxalic and phosphoric acid were opened by immersing in 5 wt.% phosphoric acid at 30 °C for 30 min, 35 °C for 30 min, and 45 °C for 30 min, respectively.

SEM bottom-views of the AAO layers synthesized in the same electrolyte, but at various anodizing potentials, are presented in Figure 1.27. The increasing anodizing potential is seen to increase the interpore distance.

The typical range of potentials used for anodizing or hard anodizing (dotted arrows) and corresponding interpore distances for various electrolytes are presented in Figure 1.28. The diagonal of the graph denotes the theoretical dependence of interpore distance on anodizing potential according to Eq. (5).

Irrespective of the type of electrolyte used, the linear relationship between the anodizing potential and interpore distance of the formed nanostructures is clearly visible. Each electrolyte is related with a specific range of anodizing potential and corresponding interpore distance of the anodic porous alumina film. The interpore distance of the formed nanostructure can be controlled over the whole range of nanometric scale by choosing an appropriate anodizing electrolyte and corresponding anodizing potential. It should be pointed out that the type of electrolyte used for anodizing is directly related with the operating conditions, including the anodizing potential, the concentration of electrolyte, and the temperature. On the other hand, the structural features of anodic porous alumina are dependent upon conditions used for anodizing. Selected operating conditions commonly used for anodizing of aluminum are presented in Tables 1.15, 1.16 and 1.17 for sulfuric, oxalic, and phosphoric acid anodizing, respectively.

In addition, the characteristic features of fabricated nanostructures including pore diameter, interpore distance, porosity and pore density of the nanostructure, are also collected in Tables 1.15–1.17. The operating conditions of anodization and

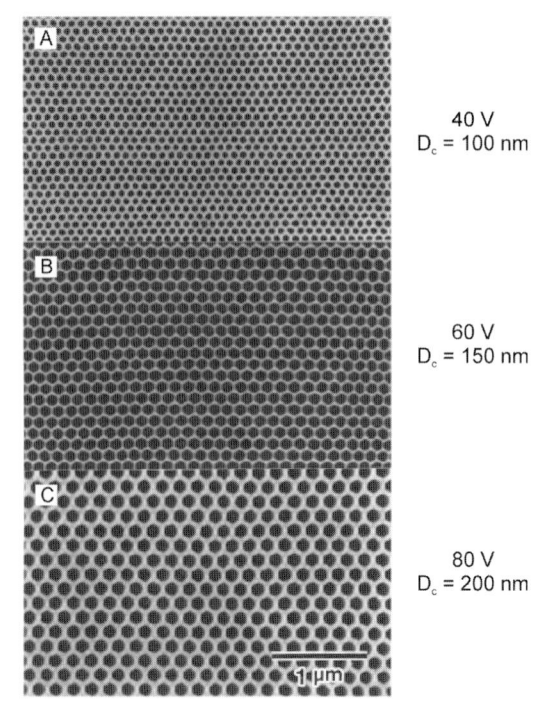

Figure 1.27 SEM images of perfectly ordered porous anodic alumina formed from pre-textured Al at different anodizing potential in 0.3 M oxalic acid at 17 °C (A and B) and 3 °C (C). (Reprinted with permission from Ref. [420], © 1997, American Institute of Physics.)

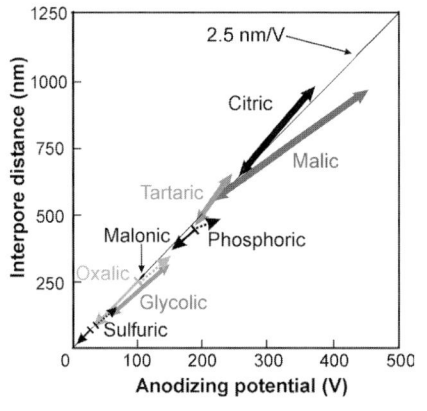

Figure 1.28 Anodizing potential influence on the interpore distance for anodic porous alumina formed in various electrolytes. Dotted arrows for phosphoric, oxalic and sulfuric acids corresponds to hard anodizing conditions. (Data for glycolic, malonic, tartaric, malic and citric acids were taken from Refs. [211,213].)

Table 1.15 Structural features of nanostructures obtained by anodizing in H_2SO_4 at various anodizing conditions.

Concentration (M)	Temperature (°C)	Potential (V)	D_p (nm)	D_c (nm)	α (%)	n (1 cm^{-2})	Remarks	Reference(s)
0.18	10	25	24	66.3	12.0	2.63×10^{10}	2-step	[235,493]
0.3	10	25	n.a.	60	n.a.	3.2×10^{10}	1-step, volume expansion factor: 1.40	[403]
0.5–4.0	10–40	3–18	12–18	18.1–47.9	41.0–13.0	3.2–0.5×10^{11}	1-step	[238]
1.1	5	25	36	64	28.7	2.8×10^{10}	2-step	[502]
1.0–1.5	20	2–20	9.4–16.0	17.8–50.3	29.3–10.5	n.a.	1-step	[249,513]
1.53	18	15	13.5	40.5	10.1	8.1×10^{10}	1-step, rate of oxide growth: ~330 nm/min	[448]
1.7	n.a.	10–15	10–20	n.a.	n.a.	n.a.	1-step, poor periodicity	[514]
1.8	0.1–5	15–25	10–20	44.8–65.0	4.5–8.6	5.8–2.7×10^{10}	1 step	[251]
1.8	0.1	40–70	30–50	90–130	10.0–13.5	1.4–0.7×10^{10}	1-step, hard anodizing,	[213,251]
2.4	−8 to 10	15–25	13.4–27.0	39.7–68.7	10.3–20.5	7.3–2.4×10^{10}	2-step	[240,248,407,409]
6.0–8.0	0 or 20	18	30	45	40.3	5.7×10^{10}	2-step	[494]

D_p: pore diameter;
D_c: interpore distance;
α: porosity;
n: pore density;
n.a.: data not available.

Table 1.16 Structural features of nanostructures formed by anodizing in $H_2C_2O_4$ at various anodizing conditions.

Concentration (M)	Temperature (°C)	Potential (V)	D_P (nm)	D_c (nm)	α (%)	n (1 cm^{-2})	Remarks	Reference(s)
0.03–0.06	3	100–160	50–100	220–440	4.7	$0.24–0.06 \times 10^{10}$	1-step, hard anodising	[213]
0.15	−1	70	90.4	n.a.	26.9	0.42×10^{10}	1-step	[253,515]
	16		87.4		21.9	0.36×10^{10}		
0.15	5	10–60	11–39	40.6–141	6.6–6.9	$7.0–0.6 \times 10^{10}$	2-step	[495]
0.5		40	27	95	8.1	1.3×10^{10}		
0.2	18	40	40	103	13.7	1.2×10^{10}	1-step, rate of oxide growth: ~71 nm min^{-1}	[448]
0.2	18	40	43.8	104.2	16.0	1.1×10^{10}	2-step	[505]
0.3	1	110–150	49–59	220–300	3.3–3.4	$1.3–1.9 \times 10^{10}$	Hard anodization	[499]
0.3	1	40	31	105	8.0	1.05×10^{10}	2-step, rate of oxide growth: 28.6 nm min^{-1} ordered domains: 1–5 μm^2	[235,493]
0.3	15	20–60	n.a.	49.8–159.8	n.a.	$4.6–0.45 \times 10^{10}$	2-step, rate of oxide growth for 40 V: $−147.75 + 125.53t$ [nm min^{-1}]	[241]
		40	24	109	4.4	0.97×10^{10}		
0.3	15	40	63	100	36.0	1.15×10^{10}	3-step, ordered domains: 0.5–2 μm^2	[390]
0.3	20	40	50	92	26.8	1.36×10^{10}	2-step	[509]
0.3	30	2–40	11.5–34.2	20.3–97.3	31.4–12.5	n.a.	1-step	[249,513]
0.34	15	40	60	104	31.6	1.12×10^{10}	3-step, ordered domains: ~4 μm^2	[236]
0.45	5	40–50	36–57	105–114	10.7–22.7	$1.05–0.89 \times 10^{10}$	2-step	[502]
0.45–0.9	10–40	3–20	14.5–21.1	20.5–54.5	45.7–13.7	$3.1–0.4 \times 10^{11}$	1-step	[237]
		20–40	21.1–37.7	54.5–110.7	13.7–10.6	$0.4–0.1 \times 10^{11}$		
0.63	20–40	4–36	7.3–18.5	n.a.	32.0–10.4	$19.1–0.97 \times 10^{20}$	1-step	[406]

D_P: pore diameter;
D_c: interpore distance;
α: porosity;
n: pore density;
n.a.: data not available.

Table 1.17 Structural features of nanostructures formed by anodizing in H_3PO_4 at various anodizing conditions.

Concentration (M)	Temperature (°C)	Potential (V)	D_p (nm)	D_c (nm)	α (%)	n (1 cm^{-2})	Remarks	Reference(s)
0.04–0.4	−1 to 16	160	170–200	n.a.	14.0–26.0	5.0×10^{12}	1-step	[515]
0.1–0.25	0	195–235	130–n.a.	420–480	8.7–n.a.	6.5–n.a. $\times 10^8$	1-step, hard anodizing	[213]
0.1	3	195	158.4	501	9.0	4.6×10^8	1-step	[235, 493]
0.3–1.1	0	195	n.a.	500	n.a.	4.6×10^8	1-step	[497]
0.4	25	80	80	208	13.4	3.1×10^9	1-step, rate of oxide growth: ~55.6 nm min^{-1}	[448]
0.4	25	5–40	20.8–74.7	23.5–105.4	80.0–50.5	n.a.	1-step	[249]
0.42	23.9	20–120	33	73–274		18.9–0.9 $\times 10^9$	1-step, pore diameter constant	[197]
0.42	25	87–117.5	64–79	236–333	n.a.	n.a.	1-step	[232]
0.53	n.a.	40–120	60–200	n.a.	n.a.	n.a.	1-step, rate of oxide growth: 1–2 nm min^{-1}, poor periodicity	[514]
1.1	3	160	n.a.	420	n.a.	6.5×10^8	1-step, volume expansion factor: 1.45	[403]
H_3PO_4-CH_3OH-H_2O (v/v 1:10:89)	−4	195	200	460	17.1	5.5×10^8	2-step, ordered domains: 1–5 μm^2	[498]
H_3PO_4-C_2H_5OH-H_2O (v/v 1:20:79)	−10 or −5	195 (at current density 150–400 mA cm^{-2})	80–140	380–320	4.0–17.4	8.0–11.3 $\times 10^8$	2-step, hard anodizing, rate of oxide growth: 4–10 μm min^{-1}, anodizing ratio: 0.62 nm V^{-1}	[339]
0.42 M H_3PO_4 + 0.5–1% Hydracrylic triprotic carboxylic acid + 0.05–0.1% Cerium salt	15–20	120–130	100–200	n.a.	9.0–43.3	1.1–1.4 $\times 10^{13}$	1-step	[219, 516]

D_p: pore diameter;
D_c: interpore distance;
α: porosity;
n: pore density;
n.a.: data not available.

corresponding structural features of anodic porous alumina films for infrequently used electrolytes, such as chromic, glycolic, malic, tartaric, citric and malonic acids or mixed sulfuric and oxalic acids, are listed in Table 1.18.

An analysis of the data contained with Tables 1.15–1.18 shows that the highest density of nanopores can be obtained by anodizing in sulfuric acid. Anodization carried out in other electrolytes results in a lower pore density of the anodic porous alumina film. It should be noted that the pore diameter is less strictly related to the anodizing potential than the interpore distance, and is seen to differ slightly even for the same anodizing conditions. Consequently, there is no consistency among the porosity of nanostructures obtained under the same anodizing conditions. This can be attributed to various concentrations of the electrolyte, or to various hydrodynamic conditions of anodizing. It is generally accepted that stirring of the electrolyte during anodization is an unquestionable necessity for in order to form ordered hexagonal nanostructures [221,236,366,443]. Indeed, without stirring the electrolyte the temperature at pore bottoms increases significantly [236]. Subsequently, in the event of weak heat abstraction, breakdown of the oxide film or its anodic dissolution can easily occur, especially during anodization at high potential or high electric field (hard anodization). Furthermore, the electrolyte composition at pore bottoms differs from the bulk electrolyte [366,443]. In general, with an increasing speed of electrolyte stirring and decreasing concentration of the acidic solution used for anodizing, a shift of the self-ordering regime occurs and higher values of anodizing potentials can be applied (compare mild and hard anodizations in Table 1.13) [494]. On the other hand, an increasing concentration of acid, especially in combination with a high anodizing temperature or a long process duration, greatly enhances the chemical etching of oxide; the result of this is that a significant widening of pores occurs.

The effect of electrolyte temperature on interpore distance, pore diameter, wall thickness and barrier layer thickness of anodic porous alumina formed by two-step anodizing in 2.4 M H_2SO_4 is illustrated graphically in Figure 1.29. These data were collected in the range of anodizing potential from 15 to 25 V for two different geometries of the electrochemical cell. A simple double-wall cell with magnetic stirring [240,248] and overflow cell with a pump system [407,409] were used for anodizing aluminum over a range of temperatures from −8 to 10 °C. For the overflow cell, anodizations were conducted only at a temperature of 1 °C. With increasing temperature of the electrolyte, however, the range of anodizing potential narrows such that, at 10 °C, the self-ordering of nanopores is observed only between 17 and 25 V.

An analysis of the data presented in Figure 1.29 shows that the interpore distance and barrier layer thickness do not vary significantly with the temperature of anodization. Moreover, the different geometry of the set-up, resulting in the different hydrodynamics conditions, does not have any considerable influence on the interpore distance and barrier layer thickness. These experimental data are in a close agreement with the theoretical predictions calculated from Eqs. (5) and (15). For the pore diameter and wall thickness, such good consistency between the experimental values is not observed, and the data differ depending on the anodizing temperature. In general, increasing the temperature of the electrolyte leads to an increase in pore

Table 1.18 Structural features of nanostructures formed by anodizing in a non-commonly used electrolytes at various anodizing conditions.

Electrolyte	Temperature (°C)	Potential (V)	D_p (nm)	D_c (nm)	α (%)	n (1 cm^{-2})	Remarks	Reference(s)
0.3 M H_2SO_4 + 0.3 M $H_2C_2O_4$	3	20–48	n.a	50–98	n.a.	$4.6–1.2 \times 10^{10}$	1-step, best ordering at 36 V	[222]
0.026–0.044 M H_3PO_4 + 0.023–0.11 M $H_2C_2O_4$ + 0.023–0.14 M H_3PO_2 + 3.4×10^{-3} M Na_2WO_4	35	120	250–500	350–650	n.a.	$3.0–7.0 \times 10^8$	1-step, various compositions of electrolytes were studied, rate of oxide growth: 2.33 μm min^{-1}, barrier layer thickness 400–500 nm	[220]
0.3 M Chromic acid	40	5–40	17.1–44.8	25.6–109.4	44.6–17.0	n.a.	1-step	[249]
0.3 M Chromic acid	60	45	23–50	74–17	n.a.	$2.3–2.8 \times 10^{10}$	1-step, stepwise increase in potential, random pores	[218]
0.44 M Chromic acid	n.a.	40	70–100	n.a.	n.a.	n.a.	1-step	[514]
1.3 M (10%) Glycolic acid	10	50–150	35–n.a.	150–320	10.1–n.a.	5.1×10^9	1-step or 2-step	[213]
0.13 M (2%) Tartaric acid	1–5	235–240	n.a.	630–650	n.a.	$2.9–2.7 \times 10^8$	1-step	[213]
3 M Tartaric acid	5	195	n.a.	500	n.a.	4.6×10^8	1-step	[211]
0.15–0.3 M (2–4%) Malic acid	10	220–450	n.a.	550–950	n.a.	$3.8–1.3 \times 10^8$	1-step	[213]
0.1–0.2 M (2–4%) Citric acid	20	270–370	n.a.	650–980	n.a.	$2.7–1.2 \times 10^8$	1-step	[213]
0.125–0.15 M Citric acid	21	260–450	130–250	650–1100	n.a.	n.a.	1-step, Al on Si, anodizing ratio: 1.1 nm V^{-1}	[217]
2 M Citric acid	20	240	180	600	10.0	3.2×10^8	1-step, rate of oxide growth: 0.12 μm min^{-1}	[216]
0.1 M Malonic acid	20	150	n.a.	n.a.	10.0	n.a.	1-step	[230]
5.0 M Malonic acid	5	120	n.a.	300	n.a.	1.3×10^9	1-step	[211]
γ-butyrolactone + ethylene glycol (9:1) + 0.9% H_2O	RT^a	250–500	n.a	275–550	n.a.	n.a.	1-step, anodizing ratio: 0.8 nm V^{-1}, random order	[295, 517]
0.3 M $Na_2B_4O_7$	60	60	up to 235	n.a.	32.0	2.5×10^{14}	1-step, pH = 9.2, random pores, rate of oxide growth: 60 nm min^{-1}	[201]

D_p: pore diameter;
D_c: interpore distance;
α: porosity;
n: pore density;
n.a.: data not available;
RT: room temperature.

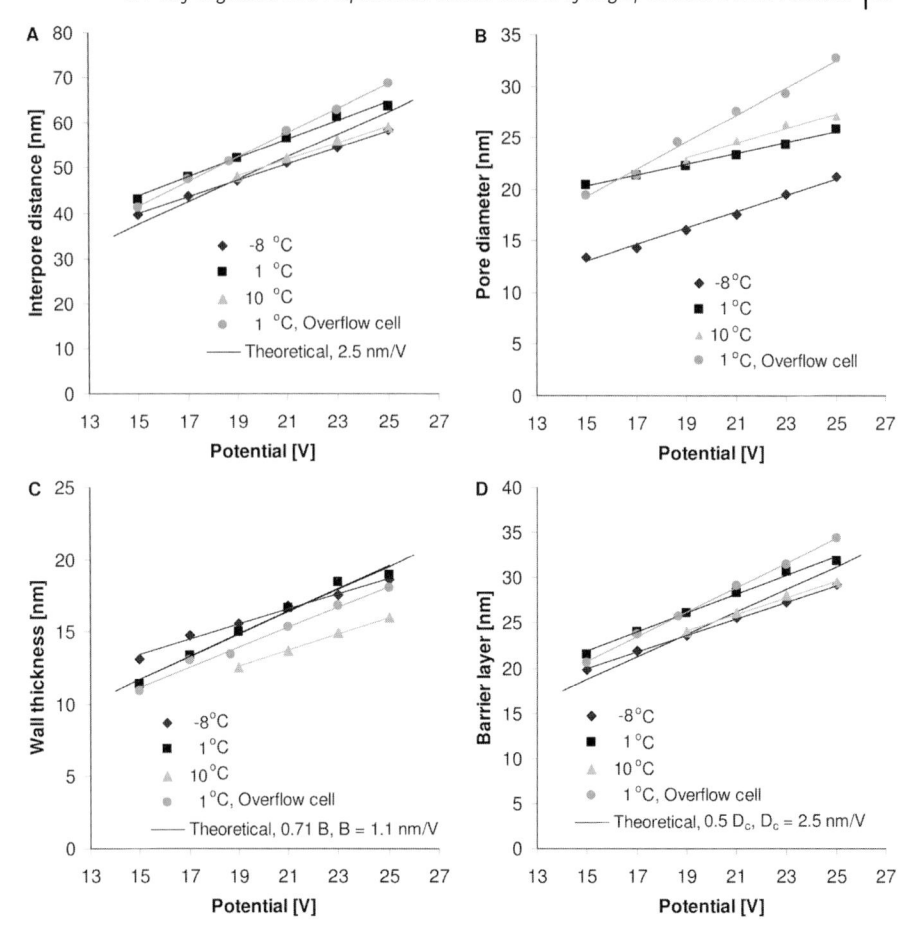

Figure 1.29 Average interpore distance (A) pore diameter (B), wall thickness (C) and barrier layer thickness (D) versus anodizing potential at various temperatures. The anodization in the overflow cell and double-wall cell with magnetic stirrer was carried out in 2.4 M H_2SO_4.

diameter and a decrease in wall thickness. Due to the slow chemical dissolution of oxide occurring at −8 °C, the smallest pore sizes were obtained. Consequently, the thinnest cell walls are observed at the highest studied temperature. Since porosity depends on the pore diameter according to Eq. (19), the observed changes in pore diameter at different anodizing temperatures should also be indicated in the calculated porosity of the nanostructure.

The influence of temperature on the porosity and pore density of self-organized anodic porous alumina formed in various electrochemical cells and temperatures are presented in Figure 1.30.

Indeed, the data in Figure 1.30 indicate that the porosity of anodic alumina varies between 10 and 20% in the whole range of the anodizing potential, and for all

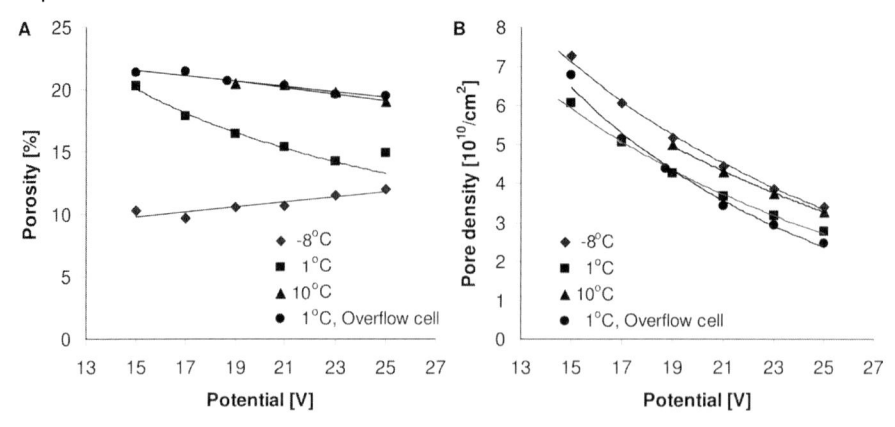

Figure 1.30 Porosity (A) and pore density of the nanostructure (B) versus anodizing potential at various temperatures. The anodization in the overflow cell and double-wall cell with magnetic stirrer was carried out in 2.4 M H_2SO_4.

studied temperatures. With increasing anodizing potential, both behaviors – an increase and decrease in porosity – were observed. As might be expected from Eq. (25), the pore density decreases with increasing anodizing potential for all temperatures.

The current density recorded during steady-state growth of anodic porous alumina in 2.4 M sulfuric acid under a constant potential regime shows an exponential dependency on the anodizing potential (Figure 1.31A).

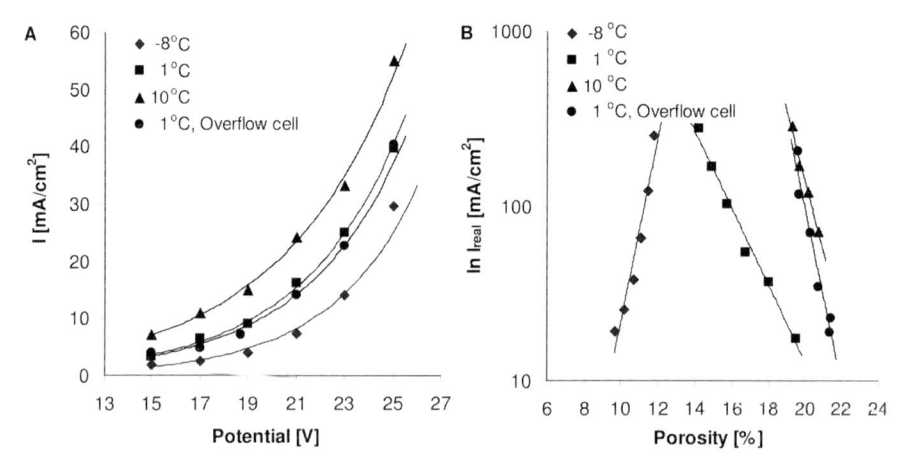

Figure 1.31 Dependence of steady-state current density on anodizing potential (A) and ln I_{real} on porosity of nanostructure (B) at various temperatures. The anodization in the overflow cell and double-wall cell with magnetic stirrer was carried out in 2.4 M H_2SO_4.

1.4 Self-Organized and Prepatterned-Guided Growth of Highly Ordered Porous Alumina

Although the increasing temperature increases the current density of the porous oxide growth, a different hydrodynamics flux in electrochemical cells does not have any significant influence on the steady-state current density. It is widely recognized that the current density accompanied by steady-state growth of the porous oxide film is related exponentially with the electric field strength, defined as a ratio of forming potential to barrier layer thickness. Therefore, the recorded steady-state current density at 1 °C is similar for different electrochemical cells. Due to the porosity of the anodic film, the real current density at the bottom of the pores is much higher than recorded. The real current density, I_{real}, as the ratio of average current density for the certain anodizing potential to the porosity of nanostructure, shows a linear dependency on the porosity of nanostructure (Figure 1.31B).

The structural features of anodic porous alumina formed in 2.4 M H_2SO_4 (especially pore diameter) depend on the electrolyte temperature as, with increasing temperature, an enhanced oxide dissolution is observed (Figure 1.32).

As a result, a considerable widening of pores occurs at the higher anodizing potential. The 3D SEM top-view image of anodic porous alumina formed at 25 V shows that, at this particular magnification, nanostructures do not exhibit defects in the nanopores arrangement. The distances between neighboring pores and pore diameter are fully uniform over the analyzed surface area for the temperature of

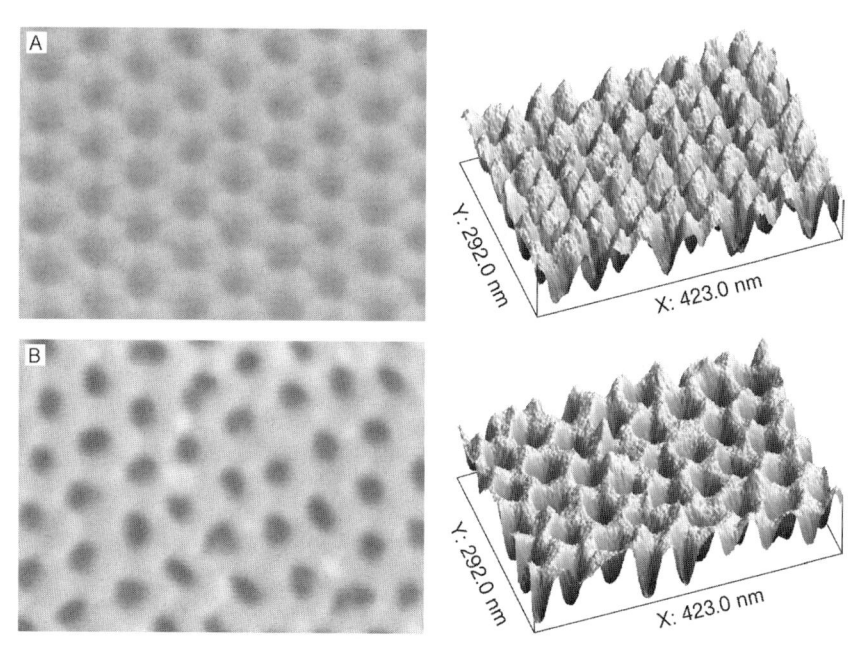

Figure 1.32 SEM top-view and 3D images for the anodic porous alumina surface formed at −8 °C (A) and 10 °C (B). The two-step anodization in the double-wall cell with magnetic stirrer was carried out in 2.4 M H_2SO_4 under 25 V. (Reprinted with permission from Ref. [240], © 2007, Elsevier Ltd.).

−8 °C. At a temperature of 10 °C, however, a variation is indicated in pore diameter over the surface area (Figure 1.32B). Moreover, the pores are reasonably wider than those observed at −8 °C, a finding which is especially apparent in the 3D SEM image, where the surface of the sample anodized at 10 °C is smoother due to the enhanced chemical etching of oxide. Uniformity of the pore diameter and interpore distance were also analyzed with, for all studied temperatures, 1000 independent measurements of pore diameter and interpore distance being carried out from SEM top-views images of anodic porous alumina formed under various cell potentials. The collected values of the pore diameter and interpore distance, for each of the anodizing potentials, were classified into seven groups and distribution diagrams were constructed for all studied temperatures. The typical pore size and interpore distance distribution diagrams obtained for an anodizing potential of 19 V at various temperatures are shown in Figure 1.33.

In order to obtain a better understanding of the influence of temperature on uniformity of pore diameter and interpore distance, the breadths of the distribution diagram (being a maximum difference between extreme values), were calculated from the diagrams and are collected in Table 1.19. Most importantly, it should be noted that for anodizations carried out at 10 °C under 15 and 17 V, the pore growth was not exhibited on the surface. Rather, a fiber-like oxide or porous random oxide structure was formed on the top of anodized aluminum surface.

In general, the breadth of distribution diagrams at −8 and 1 °C decreases with increasing anodizing potential, over the range of potentials between 15 and 23 V. A further increase in anodizing potential to 25 V causes a slight increase in the breadth of the distribution diagrams. At an anodizing temperature of 10 °C, there is no a direct relationship between anodizing potential and the breadth of the diagram. This can be explained in terms of the enhanced chemical dissolution of oxide at 10 °C and, consequently, a significant change in pore diameter. Thus, the uniformity of pore diameter is heavily affected by the chemical attack of acid.

It is generally accepted that whilst the duration of anodizing influences the pore diameter, the cell size stays unchanged [223]. With increasing anodizing time an increase in pore sizes was observed elsewhere (e.g., Refs. [201,223,233,255,408]). The pore diameter increases with anodizing time due not only to the chemical dissolution of oxide but also to the coalescence and formation of stable pore walls [223]. The anodizing duration also alternates the regularity of the self-organized anodic porous alumina structure. An improvement in nanopore arrangement and an increase in the average size of the domains exhibiting an ideally arranged pore structure, were observed with increasing anodizing time for sulfuric [216,407,492], oxalic [216,518], phosphoric [216,497], and malonic acid [211] solutions. Thus, an extended anodizing time improves the regularity of nanopore arrangement and produces the ideally arranged triangular lattice. Moreover, the extended anodizing not only rearranges the cells but also reduces the number of defects and dislocations in the nanostructure. In contrast, Nielsch *et al.* [235] found that on extending the anodizing time from 24 to 48 h, the size of ordered domain decreased for the anodization conducted in 0.1 M H_3PO_4.

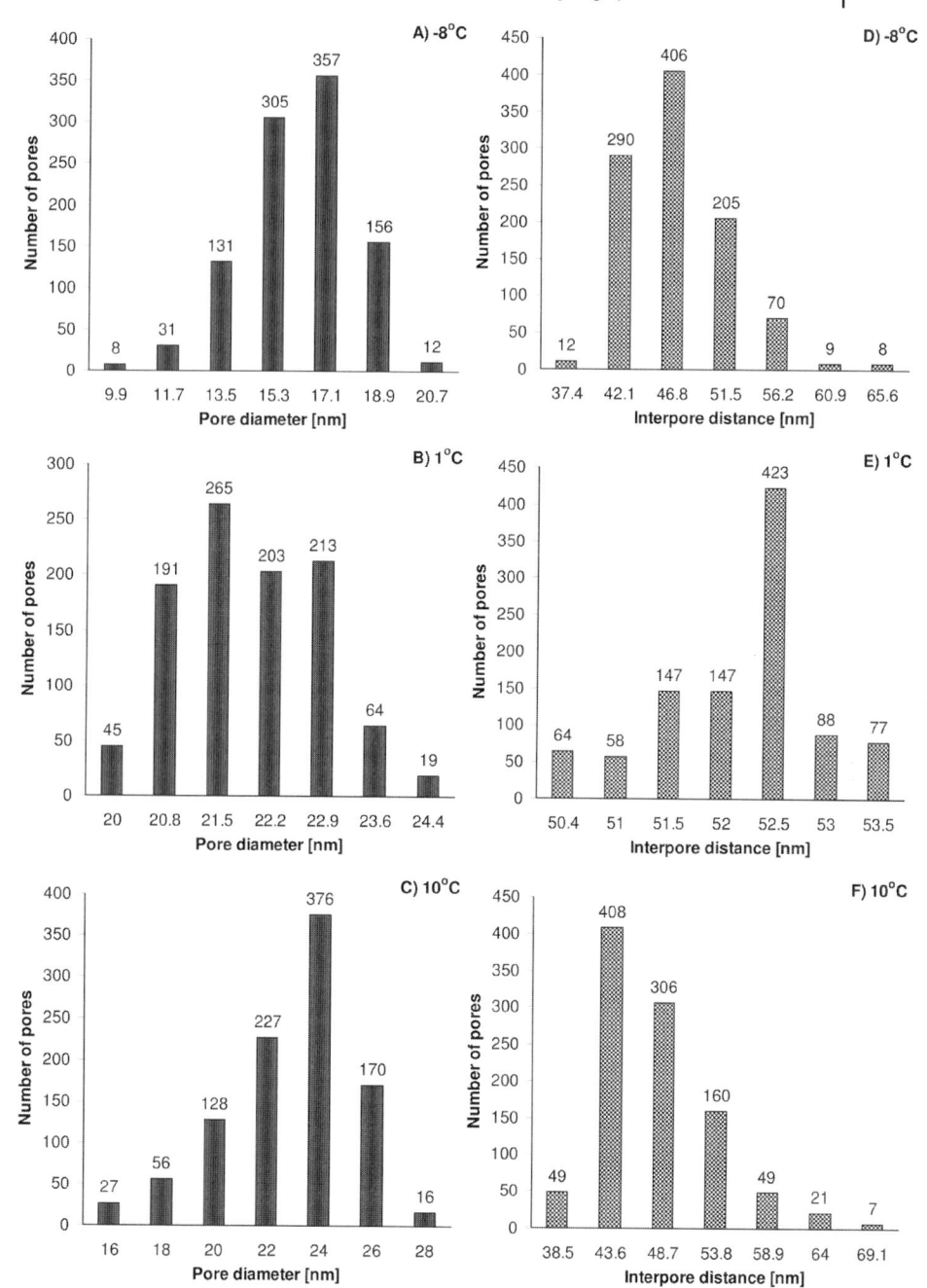

1.4 Self-Organized and Prepatterned-Guided Growth of Highly Ordered Porous Alumina | 73

Figure 1.33 Pore size and interpore distance distribution diagrams of anodic porous alumina formed under 19 V at various temperatures: −8 °C (A and D), 1 °C (B and E) and 10 °C (C and F). The two-step anodization in the double-wall cell with magnetic stirrer was carried out in 2.4 M H_2SO_4.

Table 1.19 Distribution diagrams breadth at various temperatures and anodizing potentials.

	Breadth of the distribution diagram (nm)					
	Pore diameter			Interpore distance		
Potential (V)	−8 °C	1 °C	10 °C	−8 °C	1 °C	10 °C
15	14.0	5.7	–	35.0	3.7	–
17	14.0	6.3	–	34.6	2.4	–
19	12.6	5.1	13.0	32.9	3.6	35.7
21	10.4	4.9	18.2	32.9	3.1	31.4
23	6.3	4.2	19.6	30.1	2.8	33.6
25	11.5	4.9	15.3	37.0	2.7	34.6

The two-step anodization in double-wall cell with magnetic stirrer was carried out in 2.4 M H_2SO_4. (Reprinted with permission from Ref. [240]; © 2007 Elsevier Ltd.).

1.4.2.2 Order Degree and Defects in Nanopore Arrangement

The broad range of various applications of the high-ordered, self-organized anodic porous alumina requires close control of the nanopore arrangement over a large surface area. From the nanoengineer's perspective, one of the most desired material is a close-packed nanostructure with a perfect arrangement of uniform pores. Therefore, the avoidance of defects formation in self-organized anodic porous alumina and a subsequent strict analysis of ordering degree in the nanostructure is extremely important. The two-step anodization of aluminum, conducted under precisely selected operating conditions including the specific self-ordering regime of anodizing potential, can lead to perfectly arranged nanostructures. In the case when operating conditions are fixed differently from the self-ordering regime, the hexagonal arrangement of nanopores is disturbed and some defects appears. An analysis of the order degree of nanopore arrangement can be performed from images of the anodized aluminum surface, or from images showing the pore bottoms. It should be noted that the order degree determined from the surface of anodized aluminum is usually slightly worse than the order at the pore bottoms. This difference is attributed to the rearrangement of pores occurring at the metal/oxide interface during the anodic porous alumina growth.

SEM images taken from the anodized surface and pore bottoms, together with 2D fast Fourier transforms (FFTs) and average profiles along the intensity of the 2D FFT image, are shown in Figure 1.34. The two-step, self-organized anodizing of aluminum was performed in 2.4 M H_2SO_4 under 21 V at 1 °C.

The regular triangular nanopore array with a limited number of defects in the pore arrangement is clearly visible in SEM images, and proven by 2D FFT images. The Fourier transform of the lattice provides useful information about the periodicity of the structure in the inverse scale. For a perfectly triangular lattice, a FFT pattern should consist of six distinct spots on the edges of a hexagon. When the order of the lattice is worse, ring-shaped or even disc-shaped forms can appear in the FFT image. The non-perfect regularity of the nanostructure shown in Figure 1.34 is

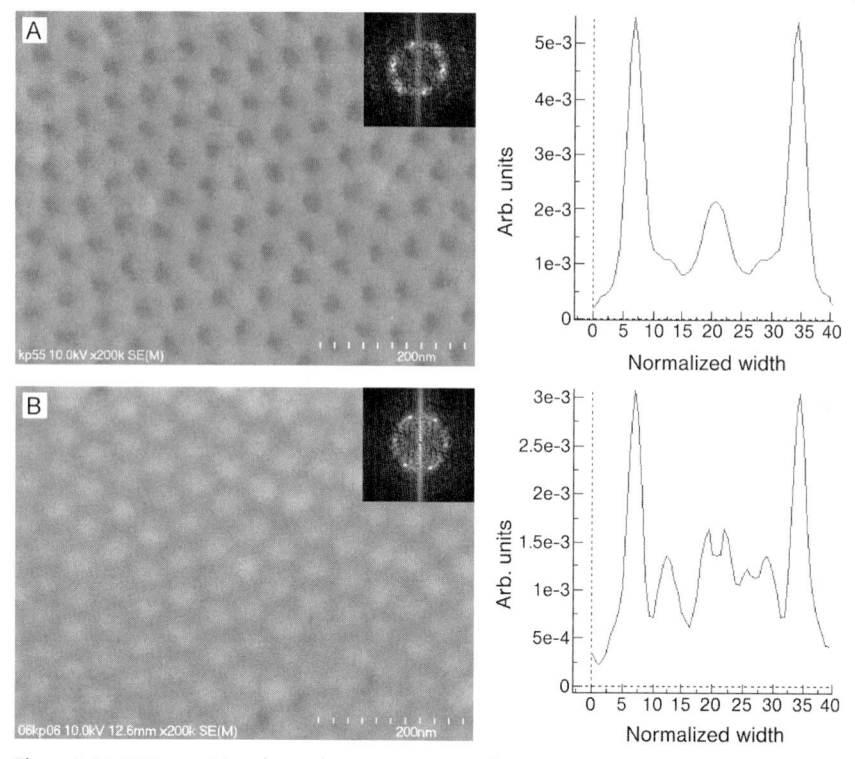

Figure 1.34 SEM top (A) and pore bottom (B) views of anodic porous alumina layer with 2D FFT images and profile analyses of the FFT intensity. The two-step anodization in the double-wall cell with magnetic stirrer was carried out in 2.4 M H_2SO_4 under 21 V at 1 °C.

indicated in the ring-shaped pattern of the 2D FFT. The higher order degree at the pore bottoms appears as distinctly marked spots in the 2D FFT image (Figure 1.34B), and narrower main peaks in the average profile along the intensity of the FFT image.

The cross-section of anodic porous alumina layer formed by two-step, self-organized anodizing in 2.4 M H_2SO_4 at 21 V shows a net of parallel straight channels (Figure 1.35A). In order to observe a channel structure of holes grown in the anodized layer, a He^+ ion beam was transmitted vertically through the anodic layer with open pores, and dissipation of the beam after membrane transmission was analyzed. In the dissipation diagram (Figure 1.35B), a relatively narrow peak proves that channels formed during anodizing are both straight and parallel.

A two-step, self-organizing anodization of aluminum in sulfuric acid can result in the perfect triangular arrangement of nanopores. Figure 1.36A shows the SEM-top view of anodic porous alumina formed in sulfuric acid at 25 V and −8 °C.

The perfectly arranged triangular lattice of nanopores is indicated in the SEM image as a net of black circles. The dimensions of pores and distances between neighboring pores are fully uniform. The 3D SEM image of the anodized surface

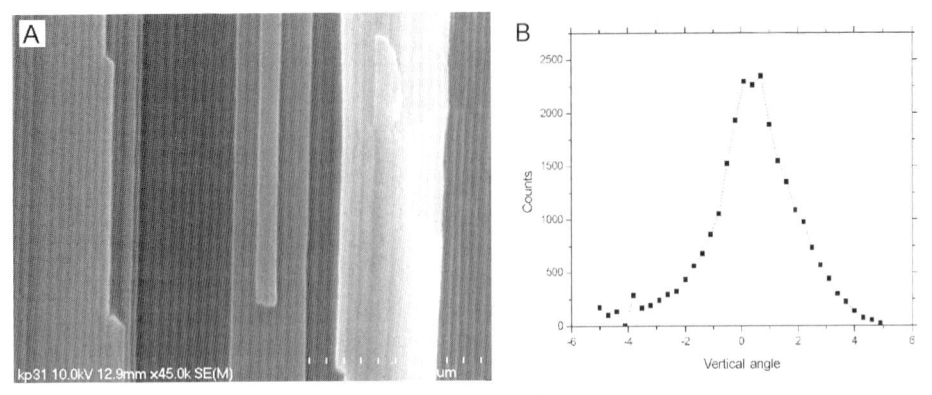

Figure 1.35 Cross-sectional view of anodic porous alumina formed at 21 V (A) and dissipation diagram recorded after transmission of the He$^+$ ion beam through the channels of AAO (B). The two-step anodization in the double-wall cell with magnetic stirrer was carried out in 2.4 M H$_2$SO$_4$ at 1 °C.

(Figure 1.36B) does not exhibit any defects in the arrangement of pores, while the high-degree of pore order is proved by the 2D FFT pattern (Figure 1.36C) consisting of six distinct spots.

It is widely recognized that the ideally ordered array of nanopores grown by self-organization processes can be observed only within perfectly ordered domains, over a surface area of about 0.5 to 5 µm^2 [236,403,493,498]. Defects in the hexagonal arrangement of nanopores appear at boundaries between ordered domains. Figure 1.37A shows the SEM top-view image of anodized aluminum taken from the surface area of 1 µm^2. The anodic porous alumina film was formed by two-step anodizing in 2.4 M H$_2$SO$_4$ under the constant potential of 25 V at −8 °C. A close-packed hexagonal arrangement of nanopores is still visible, but some defected pores have appeared that are not six-fold coordinated by neighboring pores. The 3D SEM image (Figure 1.37B) reveals clearly existing centers of defects as the protruding hills much higher than others. The order degree in the nanostructure was reduced slightly, and six distinct spots in the FFT image (Figure 1.37C) were broadened out.

The effect of anodizing temperature on the pore order degree of anodic porous alumina formed in sulfuric acid was studied at cell potentials between 15 and 25 V. A typical SEM top-view and 2D FFT images are shown in Figure 1.38A for a sample anodized at 23 V and 1 °C.

In order to better characterize the order degree of self-organized nanostructures, quantitative analyses of the FFT intensity profile were performed for various temperatures and anodizing potentials. Figure 1.38B presents the intensity profile taken from the FFT image along the direction marked with a solid line. From other two directions, marked in the FFT image, the intensity profiles were also taken and the regularity ratio, $H/W_{1/2}$ (defined as a ratio of the maximum intensity of the peak to the wide of the peak at half-maximum; Figure 1.38B), was calculated for various

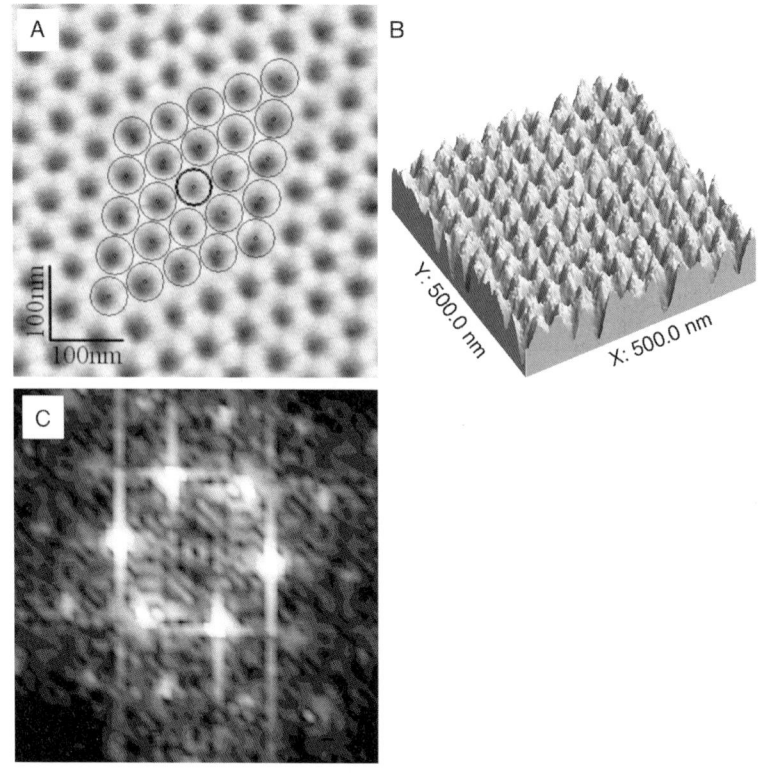

Figure 1.36 SEM top-view (A), 3D topography representation (B), and 2D FFT (C) image of a perfect triangular nanohole array. The porous alumina layer was formed at 25 V by two-step anodizing in 2.4 M H_2SO_4 at −8 °C. The analyzed surface area was 0.25 µm². (Reprinted from Ref. [226].)

anodizing potentials and temperatures. The average regularity ratio versus anodizing potential is shown in Figure 1.39.

The results show that the hexagonal arrangement of nanopores obtained by two-step, self-organized anodization of aluminum in sulfuric acid can be considerably improved by increasing the anodizing potential. The average regularity ratio increases with anodizing potential such that a maximum is reached at 25 V, independently of the anodizing temperature. These results show that the formation of nanostructure with the best nanopore order occurs at the anodizing potential of 25 V.

The second method of quantitative analysis of pore arrangement in anodic porous alumina formed by self-organized anodization is based on a defect maps known as Delaunay's triangulations. Defects maps constructed from SEM top-view images for the arrangement of over 1000 pores formed in 2.4 M sulfuric acid at various potentials and temperatures are shown in Figure 1.40. Each pore that is not six-fold

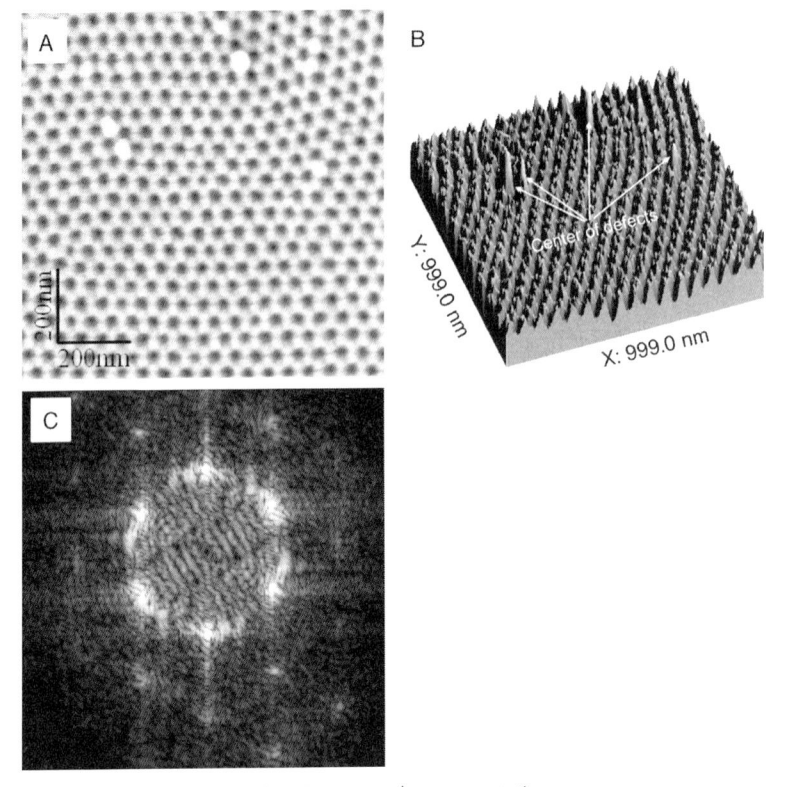

Figure 1.37 SEM top-view (A), 3D topography representation (B), and 2D FFT (C) image of anodic porous alumina with a regular nanohole array. The porous alumina layer was formed at 25 V by two-step anodizing in 2.4 M H_2SO_4 at $-8\,°C$. The analyzed surface area was $1\,\mu m^2$. (Reprinted from Ref. [226].)

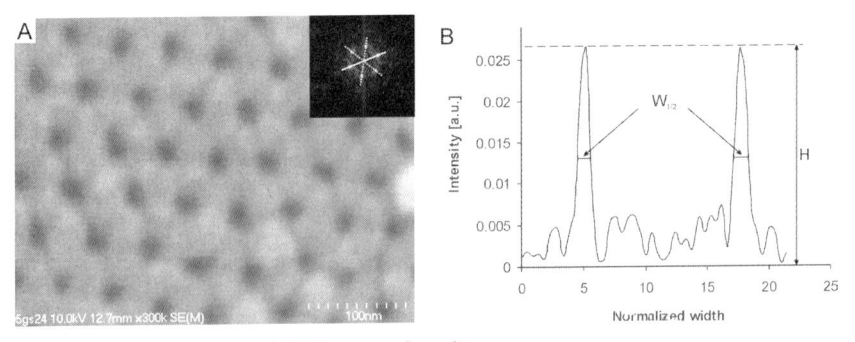

Figure 1.38 SEM top-view and FFT images of anodic porous alumina (A) and profile analysis of the FFT intensity along the marked solid line (B). The porous alumina layer was formed at 23 V by two-step anodizing in 2.4 M H_2SO_4 at $1\,°C$. (Reprinted with permission from Ref. [240], © 2007 Elsevier Ltd.).

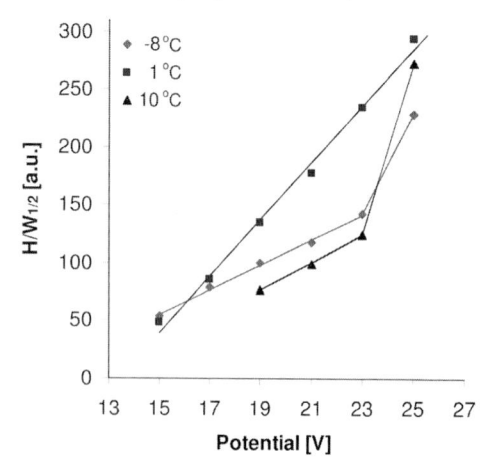

Figure 1.39 Anodizing potential influence on the average regularity ratio ($H/W_{1/2}$) at various temperatures. (Reprinted with permission from Ref. [240], © 2007, Elsevier Ltd.).

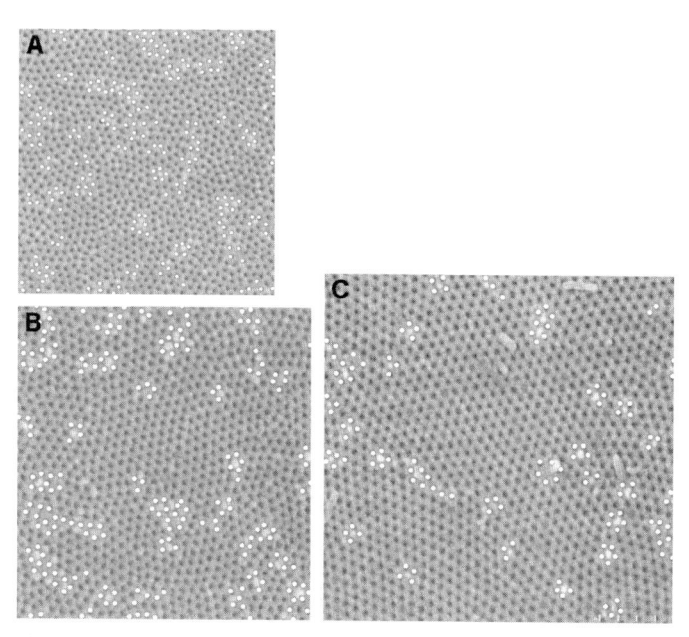

Figure 1.40 Defect maps of the arrangement of over 1000 pores constructed from SEM images taken from a surface anodized at various cell potentials: 17 V (A), 21 V (B) and 25 V (C). The porous alumina layer was formed at 23 V by two-step anodizing in 2.4 M H_2SO_4 at 1 °C. (Reprinted with permission from Ref. [248], © 2006, Elsevier B.V.).

Table 1.20 Percentage of defects in the arrangement of nanopores at various temperatures.

Anodizing potential (V)	Surface area (μm)	−8 °C		1 °C		10 °C	
		Number of pores	Percentage of defects (%)	Number of pores	Percentage of defects (%)	Number of pores	Percentage of defects (%)
15	1.24 × 1.24	1230	30.81	1054	20.87	–	–
17	1.32 × 1.32	1046	29.70	1072	20.43	–	–
19	1.48 × 1.48	1077	30.18	1040	21.15	1100	30.73
21	1.57 × 1.57	1111	30.06	1004	20.92	1038	29.67
23	1.76 × 1.76	1193	30.60	1134	20.99	1249	30.50
25	1.76 × 1.88 or 1.82 × 1.82	1268	9.54	1137	11.70	1191	10.75

Reprinted from Ref. [226].

coordinated by neighboring pores is marked with a white dot. The percentage of defected pores was calculated and collected in Table 1.20.

For anodizing potentials from 15 to 23 V, the percentage of defected pores is constant, independent of the anodizing temperature. At a temperature of 1 °C, a 20% defect level is observed, while at −8 or 10 °C the percentage of defects increases to about 30%. A perfectly ordered anodic porous alumina with a number of defects equals to about 10% can be obtained by a self-organized anodizing of aluminum carried out at 25 V.

The analysis of nanostructure regularity and generated defects showed that the best nanostructure resulting in a highly ordered and uniform nanopore arrangement with a limited number of defects can be formed by a self-organized anodizing of aluminum conducted in sulfuric acid at 25 V and at 1 °C.

The effect of electrolyte concentration and anodizing time on the regularity of nanostructures formed by self-organized anodization of aluminum were also investigated for oxalic acid [241,385,495,496]. The regularity ratio analysis performed at various temperatures showed that the best nanopore arrangement with a largest size of ordered domains can be observed for the nanostructure formed under 40 V in 0.5 M oxalic acid and 5 °C [495], or in 0.3 M oxalic acid at 15–17 °C [241,496]. In contrast, Ba et al. [385] suggested that the best ordered anodic porous alumina could be formed by anodizing under 40 V in 0.6 M oxalic acid at 0 °C. The influence of anodizing time on the size of the ordered domains grown during the anodization of aluminum in 0.3 M oxalic acid at 40 V and 0 °C was studied [236,519]. It was found that the ordered domain size, equal to about 2.8 μm², could be obtained after 10 h of anodization, and the ordered domain area was doubled after 7 h of anodizing [519]. The linear relationship between the size of domain, D (μm²), and time of anodization, t (h), was found to be at 0 °C:

$$[519] \quad D = 0.550 \cdot t \tag{52}$$

$$[236] \quad D = 0.195 \cdot t \tag{53}$$

and at 15 °C [236]:

$$D = 0.404 \cdot t \tag{54}$$

The influence of temperature of aluminum annealing before anodization on the ordering degree of nanopores formed by two-step anodizing in 0.3 M oxalic acid was also investigated [520]. It was found that, with an increasing annealing temperature, the size of ordered domains increases gradually.

The structural uniformity of nanopore arrangement was studied from SEM images using adaptive thresholding with an analysis of wave propagation [521]. The percentage of hexagonality was calculated on the basis of Veronoi diagrams. The average hexagonality percentage for a commercially available membrane with a pore size of 20 nm (Anapore, Whatman) was found to range between 37.5 and 41.3%. However, for the laboratory-prepared membrane, synthesized by the self-organized one- or two-step anodizing in oxalic acid, the determined hexagonality ranged from 47.2 to 54.5%. The arrangement of nanopores formed by anodizing of aluminum was studied by Sui and Sangier using AFM [522]. The hexagonal array percentage over a surface area of 0.25 µm^2 was found to be about 66% and 100% for anodizing in sulfuric and oxalic acids, respectively.

1.4.3
Post-Treatment of Anodic Porous Alumina

Recently, anodic porous alumina films with a regular arrangement of nanopores have been widely used as templates for the fabrication of other nanomaterials. Therefore, AAO films fabricated by self-organized anodization are usually subjected to an optional further post-treatment procedure which included removal of the aluminum base, removal or thinning of the barrier layer, and re-anodizing.

1.4.3.1 Removal of the Aluminum Base

In order to detach the aluminum oxide from the remaining aluminum base, an electrochemical detachment method employing a reversal and pulse voltage techniques can be used [502,503]. The separation of aluminum oxide from the remaining aluminum substrate might also be performed by using electrochemical etching in 20% HCl, with an operating potential between 1 and 5 V [302]. However, the most widespread method is based on a wet chemical removal of aluminum. For this procedure, the unoxidized aluminum substrate is usually dissolved by immersing an anodized sample for a few hours in a saturated $HgCl_2$ solution (e.g., Refs. [220,407,409]). Other solutions that are infrequently used for aluminum base removal are listed, together with operating conditions, in Table 1.21.

The effect of different $CuSO_4$/HCl contents and solution temperature on the rate of chemical etching of aluminum substrate was investigated with great care [507]. Subsequently, it was found that the time necessary for effective dissolution of a

Table 1.21 Non-typical solutions used for the chemical removal of aluminum base after anodizing.

Solution	Temperature (°C)	Time (min)	Reference(s)
$Br_2 + CH_3OH$	40	10	[252,295]
$CuCl_2$ (saturated)	–	–	[523]
$CuCl_2$ (saturated) + 8% HCl	RT	–	[305,324]
0.08 M $CuCl_2$ + 8% HCl	20	60	[524]
0.1 M $CuCl_2$ + 20% HCl	5–RT	10	[253,299,491,501,515]
$CuSO_4$ (saturated)	–	–	[303]
$CuSO_4$ (saturated) + 38% HCl	0–20	2–5	[318,507]

RT: room temperature.

0.2 mm-thick aluminum base is less than 2 min within a range of HCl concentrations from 25% to 65%. The temperature of solution has a negligible influence on the time of removal. Indeed, following removal of the aluminum substrate, the spherical pore bottoms can be observed (see Figure 1.34B).

1.4.3.2 Removal of the Barrier Layer

The AAO template preparation consists of pore opening after separation of the oxide from the Al substrate, or pore widening before the subsequent deposition of metals and semiconductors into pores. Removal of the barrier layer in a separated AAO membrane, formed by self-organized anodization, is usually carried out by chemical etching of the oxide. The nanopore bottoms are opened by immersing in a H_3PO_4 solution, with the opening time depending directly on the barrier layer thickness, and consequently on anodizing conditions. Some selected procedures of barrier layer removal are listed in Table 1.22.

If the opening time is prolonged, the widening of pores can occur simultaneously [498]. The diameter of the opened pores can be adjusted by changing the time of the chemical etching treatment in a phosphoric acid solution. Pore opening and widening were studied in detail by Xu et al. [318]. Samples before widening treatment were anodized in 0.3 M $H_2C_2O_4$ at 40 V and 12 °C. The dissolution rate of the barrier layer in 0.5 M H_3PO_4 was found to be approximately 1.3 nm min^{-1}, and to decrease inside the columnar pore channel with increasing pore depth. The horizontal and vertical dissolution rates were also distinguished. The horizontal rate of oxide dissolution is mainly responsible for barrier layer removal, whereas the vertical rate plays a key role in the widening of pores and nanopore thickness reduction. A different etching technique using a reactive ion-beam (mainly Ar$^+$) was also proposed for removal of the barrier layer in anodic porous alumina [385,524,528]. Most importantly, it should be noted that dry etching by ion-beams requires the use of a sophisticated apparatus.

In order to obtained an array of ordered nanopores with a desired pore diameter, it is important first to carry out a widening treatment (Table 1.23).

Table 1.22 Selected procedures of pore opening after two-step anodizing of aluminum in sulfuric, oxalic and phosphoric acids.

	Anodization conditions			Opening conditions				
Electrolyte	Temperature (°C)	Potential (V)		Electrolyte	Temperature (°C)	Time (min)	D_p (nm)	Reference(s)
0.3 M H_2SO_4	10	25		0.5 M H_3PO_4	30	30	–	[403,493]
1.1 M H_2SO_4	5	25			30	23	33	[502]
0.2 M $H_2C_2O_4$	0	40			25	45	43	[506]
0.3 M $H_2C_2O_4$	0	60			RT	120	80	[501]
	1	40			35	30	–	[403,493]
	17	40			30	60	50	[500]
0.45 M $H_2C_2O_4$	5	40–50			30	75–85	40–72	[502]
1.1 M H_3PO_4	3	160			45	30	–	[403,493]
H_3PO_4-CH_3OH-H_2O (1:10:89, v/v)	−4	195		1.1 M H_3PO_4	20	300	350	[498,525,526]
H_3PO_4-C_2H_5OH-H_2O (1:20:79, v/v)	−10 to 0	195[a]		0.5 M H_3PO_4	45	30	–	[339]

RT: room temperature.
[a] Hard anodizing.

Table 1.23 Selected procedures of pore widening after a two-step anodizing of aluminum in sulfuric, oxalic and phosphoric acid solutions.

	Anodization conditions				Widening conditions			
Electrolyte	Temperature (°C)	Potential (V)	D_P (nm)	Electrolyte	Temperature (°C)	Time (min)	D_P (nm)	Reference(s)
1.1 M H_2SO_4	10	20	–	0.1 M H_3PO_4	35	20	–	[522]
6.0 M H_2SO_4	20	18	–	0.5 H_3PO_4	30	12	30	[494]
0.04 M $H_2C_2O_4$	5	85	90	0.5 H_3PO_4	20	30	120	[498,525,526]
0.2 M $H_2C_2O_4$	18	40	–	0.1 H_3PO_4	–	40	44	[505]
0.3 M $H_2C_2O_4$	–	40	31	0.1 H_3PO_4	30	30	46	[527]
						50	60	
	15	40	24	0.1 H_3PO_4	30	50	74	[241]
	15	40	–	0.1 H_3PO_4	35	50	52–60	[522]
	20	40	25	0.63 M H_3PO_4	20	85	75	[305]
	24	40	40	0.5 M H_3PO_4	25	60	80	[509]
0.63 M $H_2C_2O_4$	20	30	14.5	0.63 M $H_2C_2O_4$	40	30	20	[406]
						60	26	
						90	32	
						120	38	
0.3 M H_3PO_4	0	195	–	1.1 M H_3PO_4	–	240	320	[497]
0.45 M H_3PO_4	20–22	140	100	2.4 M H_2SO_4	20	60	103	[252]
						80	109	
						300	115	

A relationship between pore diameter (D_p, in nm) and widening time (t_w, in min) for samples anodized in 0.3 M oxalic acid under 40 V at 15 °C was reported by Hwang et al. [241]:

$$D_p = 24.703 - 0.116 \cdot t_w + 0.0221 \cdot t_w^2 \tag{55}$$

The widening of pores was conducted in 0.1 M H_3PO_4 at a temperature of 30 °C. In contrast, Choi et al. [529] suggested that the rate of pore widening in 1 M H_3PO_4 at 30 °C was equal to 1.83 nm min^{-1}. The diameter of the widened pores can easily be determined from SEM or AFM images [527]. Recently, capacitance–voltage measurements on thin alumina films have been proposed for characterization of the pore-widening process [530]. Pore widening in 1.2 M NaOH, leading to a simultaneous separation of AAO nanowires and nanotubes from the grown nanostructure, was also reported [523]. When ultrasonic waves were also used to assist the pore-widening process [531], the widening time was found to be shortened to a significant degree.

1.4.3.3 Structure and Thinning of the Barrier Layer

The structure of the barrier layer in anodic porous alumina, as well as its dielectric characterization modified by oxide dissolution or ion implantation, was studied using *in-situ* spectroscopic ellipsometry [532,533], Brillouin spectroscopy [534], electrochemical impedance spectroscopy (EIS) [448,533,535,536], and re-anodizing techniques [245–247,267,535,537,538].

It is generally accepted that the barrier layer in anodic porous alumina consists of two or three sublayers, each with a different susceptibility to chemical dissolution in acidic solution. Furthermore, the number and thicknesses of the sublayers, as well as the percentage share of the barrier layer thickness, depends on the anodizing potential. An anodic porous alumina was first formed under various anodizing potentials in 0.4 M H_3PO_4 and 0.45 M $H_2C_2O_4$ at 20 °C, and then subjected to dissolution by immersion in sulfuric acid [246,247,532,537,538]. The dissolution rate of the barrier layer in 2 M H_2SO_4 or H_3PO_4 at 50 °C was investigated using the re-anodizing technique (Figs. 1.41 and 1.42).

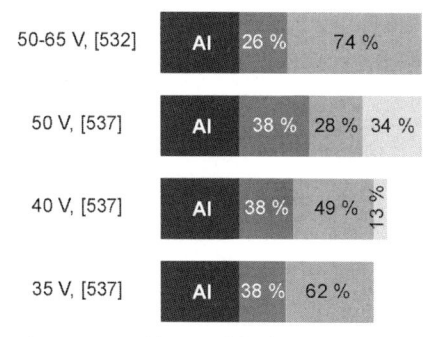

50-65 V, [532]	Al	26 %	74 %	
50 V, [537]	Al	38 %	28 %	34 %
40 V, [537]	Al	38 %	49 %	13 %
35 V, [537]	Al	38 %	62 %	

Figure 1.41 Sublayers of the barrier layer with different dissolution rates for anodic porous alumina formed by anodizing of aluminum in 0.4 M H_3PO_4 at 20 °C. (After Refs. [532,537].)

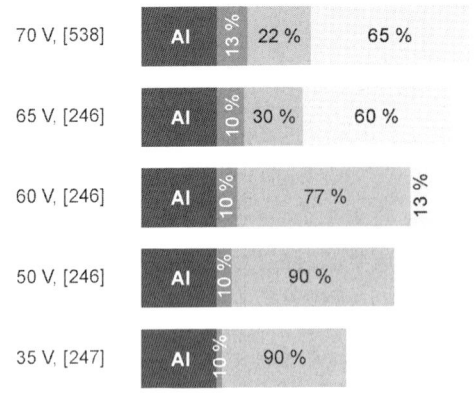

70 V, [538]	Al	13%	22%	65%	
65 V, [246]	Al	10%	30%	60%	
60 V, [246]	Al	10%		77%	13%
50 V, [246]	Al	10%		90%	
35 V, [247]	Al	10%		90%	

Figure 1.42 Sublayers of the barrier layer with different dissolution rates for anodic porous alumina formed by anodizing of aluminum in 0.44 M oxalic acid at 20 °C. (After Refs. [246,247,538].)

Dissolution of the barrier layer formed in a 0.1 M ammonium adipate solution at the anodizing potential between 5 and 80 V and 20 °C was studied using the re-anodizing technique [267]. The rate of dissolution was investigated in 2 M sulfuric acid at 60 °C, whereupon the results obtained suggested the dissolution rate to be similar for anodic films formed at the anodizing potential between 20 and 80 V. Although the barrier layer was seen to be composed of one layer, for lower anodizing potentials two sublayers were apparent, each with different dissolution rates.

A close control of the nucleation sequence of electrodeposits within all pores of the anodic porous alumina is required in order to create a significant thinning of the barrier layer. Such effective thinning provides the necessary conditions for electrons to tunnel through the barrier layer, with subsequent uniform pore filling by electrodeposition. Thinning of the barrier layer in anodized aluminum always occurs together with pore widening. According to Stein et al. [532], the barrier layer in anodic porous alumina formed in 0.4 M H_3PO_4 under the constant current density of 2 mA cm^{-2} at 20 °C consists of two sublayers each with a different dissolution rate (see Figure 1.41). The interface barrier layer, which is close to the metal/oxide interface, does not dissolved in 2 M H_3PO_4, whereas the layer at the pore bottoms is in contact with the electrolyte and dissolves easily. The rate of barrier layer dissolution in a mixture of 0.4 M phosphoric and 0.2 M chromic acids at various temperatures was investigated in detail [533]. The barrier layer in the anodic porous alumina film was formed by anodization in 0.34 M oxalic acid under 30 V at 0 °C. Subsequently, the dissolution rate of the 36 nm-thick barrier layer was estimated as approximately 0.3 and 1.0 nm min^{-1} at temperatures of 38 and 60 °C, respectively. Under the same anodizing conditions, the dissolution rate of porous oxide formed by anodization was estimated at 6 nm min^{-1} for 38 °C.

O'Sullivan and Wood [208,232] suggested that decreasing the anodizing potential would result in a substantial barrier layer thinning, due to a field-assisted dissolution of the oxide. On the basis of this assumption, an extraordinary electrochemical method of homogeneous thinning of the barrier layer was derived [539]. A progressive reduction of the anodizing potential led to a thinning of the barrier layer, and finally to its perforation. The step-wise reduction of the anodizing potential from 160 to 0.1 V, with decrements of 0.3 V or 5% of the existing potential, was applied to thin the barrier layer formed by the anodization g of aluminum in 0.4 M H_3PO_4 at 25 °C [539]. A similar, exponentially decaying, potential was applied for thinning of the barrier layer in anodic porous alumina formed in 0.1 M phosphoric acid at 195 V [529]. The process of thinning involves two different electrolytes used for the different ranges of decaying potential. For the thinning process from 195 to 80 V, 0.1 M phosphoric acid was used, and each potential decrement of 2 V was applied for 180 s. In the range of decaying potential from 80 to 1 V, a 0.3 M oxalic acid solution was used for barrier layer thinning. The exponential step-wise reduction began from a step of 2 V and finished at the last step with the potential difference of 0.01 V. Each step of the potential reduction was applied for 30 s. This thinning method was successfully applied during fabrication of the ordered Ni nanowire array by electrodeposition of metal into nanopores of the AAO template. The thinning of the barrier layer in AAO formed in 0.3 M oxalic acid at 40 V and 2 °C was performed by means of two subsequent additional anodizations conducted for 15 min under constant current densities of 290 and 135 mA cm^{-2}, respectively [540]. Prior to the electrochemical barrier layer thinning, a chemical widening of the pores in 0.3 M oxalic acid was carried out for 2 h at 30 °C.

1.4.3.4 Re-Anodization of Anodic Porous Alumina

The re-anodizing technique, which is known also as a pore-filling method, is widely used to characterize porous oxide films formed by anodization. The re-anodization of anodic alumina is usually carried out under the constant current density in a neutral electrolyte, for example a mixture of 0.5 M H_3BO_3 + 0.05 M $Na_2B_4O_7$ [245–247,406,537,538]. The dissolution rate of the barrier layer in acidic solutions, as well as the anodizing potential influence on the thickness of the barrier layer formed during anodization in ammonium adipate [267], oxalic [246,247,406,538] and phosphoric acid [245,537] solutions, were investigated using the re-anodization technique. The re-anodizing method allows one to calculate the anodic oxide thickness from Eq. (42). Girginov et al. [533] showed EIS to be a suitable technique for the estimation of barrier layer thickness modified by re-anodization. Moreover, the electrical conductivity of the complex anodic oxide films formed during re-anodizing of anodic porous alumina was investigated in detail [541]. The modification of barrier layer properties by the re-anodization of anodic porous alumina in oxalic acid with fluoride additives was also studied [536], with a significant reduction in electrical resistance being observed during the re-anodizing.

The re-anodizing methods can be used for determination of anodic alumina porosity, α, from the following equation:

$$\alpha = \frac{T_{Al^{3+}} \cdot (m_2/m_1)}{1-(1-T_{Al^{3+}})(m_2/m_1)} = \frac{T_{Al^{3+}} \cdot (m_2/m_1)}{1-T_{O^{2-}} \cdot (m_2/m_1)} \tag{56}$$

where: $T_{Al^{3+}}$ and $T_{O^{2-}}$ are transport numbers of Al^{3+} and O^{2-}, m_1 is the slope of the voltage–time curve recorded during re-anodizing, and m_2 is the slope of the voltage–time curve recorded during the anodizing of aluminum. Since the transport numbers of Al^{3+} and O^{2-} were estimated at 0.4 and 0.6 for various electrolytes used for anodizing, the porosity of anodic porous alumina can easily be calculated. The re-anodization was also successfully employed for determination of the porosity of anodic porous alumina films formed in sulfuric [216,249], oxalic [216,249,406,541], phosphoric [216,249,336], and chromic acids [249].

The kinetics of growth of the complex anodic alumina films during re-anodization was studied by Girginov et al. [542–545]. These authors found that, during re-anodization, the current flowed mainly through the bottom of pores, with distributions of the current density at the pore bottoms corresponding directly to the hemispherical curvature of the pore bottom.

1.5
AAO Template-Assisted Fabrication of Nanostructures

The self-organized anodization of aluminum, which results in a hexagonal pattern of nanopores with an extended long-range perfect order, appears to be a very promising powerful and inexpensive method used for the synthesis of nanostructured materials. AAO templates are widely used for the transfer of nanopore arrangements to other materials. In general, the use of anodic porous alumina for nanomaterials fabrication can be carried out in two ways: (i) the anodic film with a remaining aluminum substrate can be used directly for the electrodeposition of metals; or (ii) it can be separated from the aluminum base and further processed into a freestanding membrane with open pores at the top and bottom of the membrane. The second approach to employing a freestanding AAO template represents a key method for fabricating highly ordered nanostructures. Liang et al. [546] have reported details of the most widespread methods of nanoarrays fabrication employing anodic porous alumina (Figure 1.43).

An enormous variety of nanoscale materials including metal and semiconductors nanopores, rings, particles, nanorods and nanowires, metal oxides, metallic, Si and diamond nanopore arrays, carbon and polymer nanotubes have been successfully fabricated on a basis of ordered anodic porous alumina. However, unfortunately only selected nanostructured materials fabricated on the basis of the AAO template can be discussed here.

Most nanomaterials synthesized with the assistance of anodic porous alumina may be classified into the following groups:

- Metal nanodots, nanowires, nanorods and nanotubes
- Metal oxide nanodots, nanowires and nanotubes
- Semiconductor nanodots, nanowires, nanopillars and nanopore arrays
- Polymer, organic and inorganic nanowires and nanotubes
- Carbon nanotubes

1.5 AAO Template-Assisted Fabrication of Nanostructures

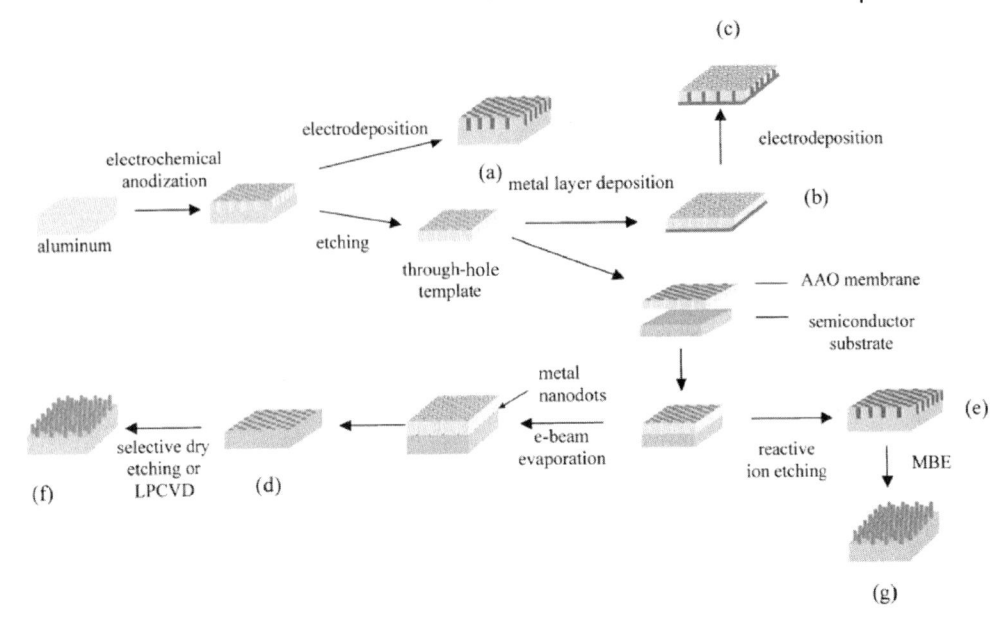

Figure 1.43 Schematic diagram of the fabrication of nanostructured materials with utilization of anodic porous alumina. Step a: metallic nanowire array electrodeposited in AAO; step b: AAO template with metal deposited on its surface; step c: metal nanowire array electrodeposited within the AAO template; step d: metallic nanodot array deposited on semiconductor substrate; step e: semiconductor substrate with nanopore array; step f: semiconductor freestanding nanopillar array; step g: heterostructure quantum dot array by molecular beam epitaxy (MBE). (Reprinted with permission from Ref. [546], © 2002, Institute of Electrical and Electronics Engineers, Inc.).

- Photonic crystals
- Other nanomaterials (metallic and diamond membranes, biomaterial, etc.).

1.5.1
Metal Nanodots, Nanowires, Nanorods, and Nanotubes

Metallic nanodot arrays are often fabricated by the evaporation of metals into the AAO template (step d in Figure 1.43). The advantage of the evaporation method over electrochemical deposition is an easier control of nanodot growth. Various nanodot arrays were obtained including Ni [4], Cu [547] Au and Au–Ag [548] nanodots on a Si substrate. Recently, Park et al. [549] reported Au, Al, Ag, Pb, Cu, Sn and Zn nanodot arrays on the silicon substrate obtained by metal evaporation through the AAO template. A second approach to nanodot or nanoparticle synthesis is a metal deposition via a chemical route. For example, Ag [550] and Pd [551] nanodots were fabricated successfully using the chemical method.

Arrays of metallic nanowires and nanorods are mainly synthesized by the electrodeposition of materials into the nanochannels of anodic porous alumina (step a in

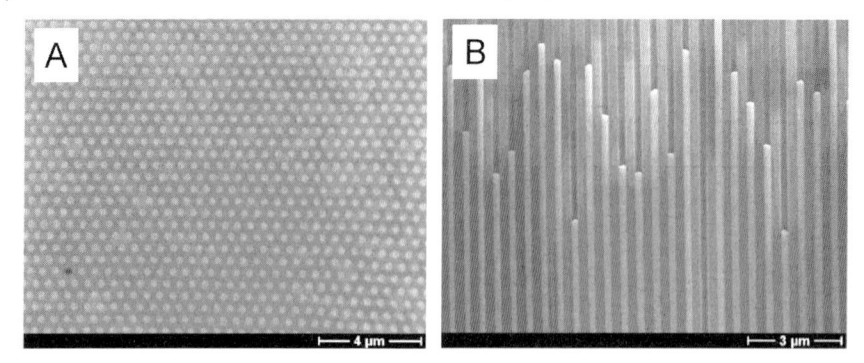

Figure 1.44 SEM images of silver wires embedded in a monodomain anodic porous alumina template. Top view (A) and cross-sectional view (B). (Reprinted in part with permission from Ref. [529], © 2004, American Chemical Society.)

Figure 1.43) or in the AAO template with open pores (steps b and c in Figure 1.43). Arrays of Ag [529,552,553], Au [554–556], Co [557–559], Cu [560], Ni [540,556,561–566], Pb [567], and Pd [568] nanowires and nanorods were obtained successfully by electrodeposition. A highly ordered array of silver nanowires electrodeposited in the AAO template is presented in Figure 1.44.

A single Ni nanowire and complex nanowire pattern were also obtained by electrodeposition with a supported AAO template [569]. Recently, the electrochemical deposition of multilayered Co/Cu [570] and alloyed Co–Cr [571], Co–Pt [572,573], Fe–Pt [573] and Fe–Pd [574] nanowires into anodic porous alumina was also reported, while the evaporation technique was used to fabricate arrays of Au nanowires [575]. Ag [576] and Ni doped with P [577] nanowires were obtained by the electroless deposition of metals into the AAO template.

Metallic nanotubes with outer diameters ranging from 50 to 100 nm were successfully prepared by the electrodeposition of Fe, Co and Ni inside nanopores of the through-hole AAO template (steps b and c in Figure 1.43) [578]. The immobilization of Ag nanoparticles, followed by the electrochemical deposition of Au and Ni, was applied for the synthesis of Au and multi-segmented Au–Ni nanotubes [579]. Thin-walled Co [510,580] or Pd [581–583] nanotubes coated with a polymer layer on their inner walls were formed in anodic porous alumina. These composite magnetic nanotubes were obtained within the AAO template by thermal decomposition of the appropriate precursor; the same process was also used to synthesize Au nanotubes inside the AAO template [584]. The spontaneous coalescence of Au nanoparticles on the pore walls of a silane-treated AAO template was used for the fabrication of Au nanotubes [585]. The electroless deposition as a method of Co, Ni and Cu nanotube synthesis was reported by Wang et al. [586]. Aluminum nanotubes were formed by physical vapor deposition on the AAO template, followed by atmospheric pressure injection of the evaporated layer into pores of the template [587].

1.5.2
Metal Oxide Nanodots, Nanowires, and Nanotubes

By using the AAO template, metal oxide nanodots and nanowires are usually formed in nanopores of the template by electrochemical or electroless deposition of metals (steps a–c in Figure 1.43), followed by their oxidation. This method was employed successfully for the fabrication of ZnO nanowires [588]. The electrodeposition technique was used for a direct deposition of Cu_2O nanowires into the channels of porous alumina template [589]. Metal evaporation through the AAO template with open pores (step d in Figure 1.43) was used for the preparation of TiO_2 nanodots [546]. Recently, the complex structure of In_2O_3 rods in dots was synthesized by the deposition of individual nanorods within the AAO template on a Si substrate using a radiofrequency magnetron sputtering system [590]. A highly ordered array of tantalum oxide nanodots with a narrow size distribution was fabricated by the anodization of Al on TaN [591] and Ta(Nb) [592] substrates. During anodizing, the aluminum layer is converted into porous alumina oxide, and the growth of tantalum oxide beneath the anodic porous alumina occurs simultaneously. Alternatively, ZnO nanowires can be formed by a catalyzed epitaxial crystal growth [593], and the AAO template was used to form ordered Au dots on GaN. The regular array of straight ZnO nanowires was grown on Au nucleation sites. TiO_2 [594] and WO_3 [595] porous films were fabricated using a two-step replication process. The first step of the method was based on the fabrication of negative-type AAO with poly(methacrylate) (PMMA). In the second step, the replicated PMMA-negative with the evaporated thin Au layer was used either for the subsequent electrodeposition of WO_3 [595] or for a sol–gel synthesis of TiO_2 [594]. Sander *et al.* [596] used atomic layer deposition (ALD) to create an array of ordered TiO_2 nanotubes inside the channels of anodic porous alumina template (Figure 1.45). Recently, ferroelectric $PbZrTiO_3$ nanotube arrays were also fabricated by using a sol–gel method of the AAO template infiltration [597].

1.5.3
Semiconductor Nanodots, Nanowires, Nanopillars, and Nanopore Arrays

Recently, research interest has focused on the synthesis of semiconductor nanostructures, due mainly to their fascinating potential technological applications. The hexagonal pattern of the AAO template can be replicated directly in semiconductors by using dry etching methods, such as plasma etching, ion milling, and reactive ion etching. This positive transfer of the nanopore arrangement (step e in Figure 1.43) results in antidot (pore) arrays in the semiconductor material. Alternatively, the negative transfer of AAO nanopores, employing molecular beam epitaxy (MBE) and metal–organic chemical vapor deposition (MOCVD) (steps e and g in Figure 1.43) or vapor–liquid–solid (VLS) growth and low-pressure chemical vapor deposition (LPCVD) (steps d and f in Figure 1.43), produces quantum dots or nanopillars. Quantum dot arrays can also be fabricated by the extended anodization of aluminum on a semiconductor substrate [598]. The process is continued until the moment when a whole aluminum layer is consumed, after which a barrier layer of anodic

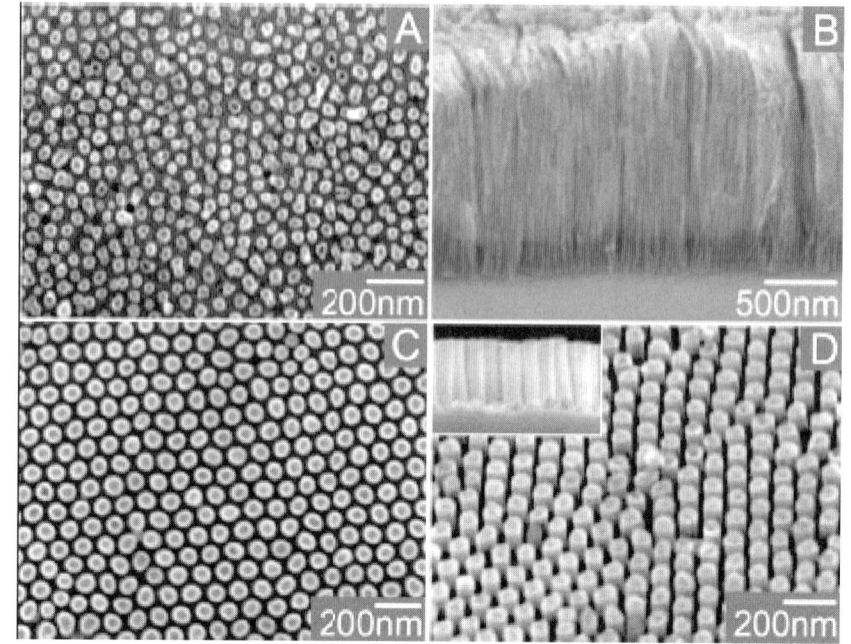

Figure 1.45 SEM images of TiO$_2$ nanotube arrays on silicon substrate. Top view (A) and side view (B) of nanotubes with outer diameters of 40 nm and average intertube spacing of 60 nm. Top (C) and oblique (D) of hexagonally ordered array of nanotubes with outer diameters of 80 nm and average intertube spacing of 100 nm. The inset in (D) shows a side view of the nanotubes on the substrate. (Reprinted with permission from Ref. [596], © 2004, Wiley-VCH Verlag GmbH & Co. KGaA.)

porous alumina is dissolved and oxidation of the semiconductor substrate occurs. The great variety of quantum dot arrays, including InAs [598], GaN [599], CdTe [600,601], InGaN [602], SiO$_2$ [603] and nc-Si:H [604] dots on various semiconductor substrates (Si, GaAs or GaN), were fabricated with AAO template assistance. Wang et al. [602] reported that a strict control of the growth time can result in nanorings and nanoarrows of regular patterns formed by MOCVD (Figure 1.46).

Anodic porous alumina is widely used as a template for the synthesis of semiconductor nanowires and nanopillars. Si, InP, GaAs and GaN nanopillars were fabricated using the reactive ion or reactive beam etching process (step e in Figure 1.43) [546,605,606]. Uniformity of the pore diameter in GaAs and InP nanohole arrays was fairy acceptable for 2D photonic bandgap crystals. The hexagonally arranged nanowire array of GaN grown in the VLS process (step f in Figure 1.43) was also reported [4]. The AAO template-assisted electrochemical deposition was used for the preparation of PbS [607], CdS [608–610] and Se [611] nanowires. Semiconductor InN nanowires were fabricated by the electrodeposition of indium into pores of anodic porous alumina, followed by a direct reaction of indium with ammonia at 550–700 °C [612]. The hydrothermal method with the AAO membrane was used for the synthesis of In$_2$S$_3$ nanofibers [613].

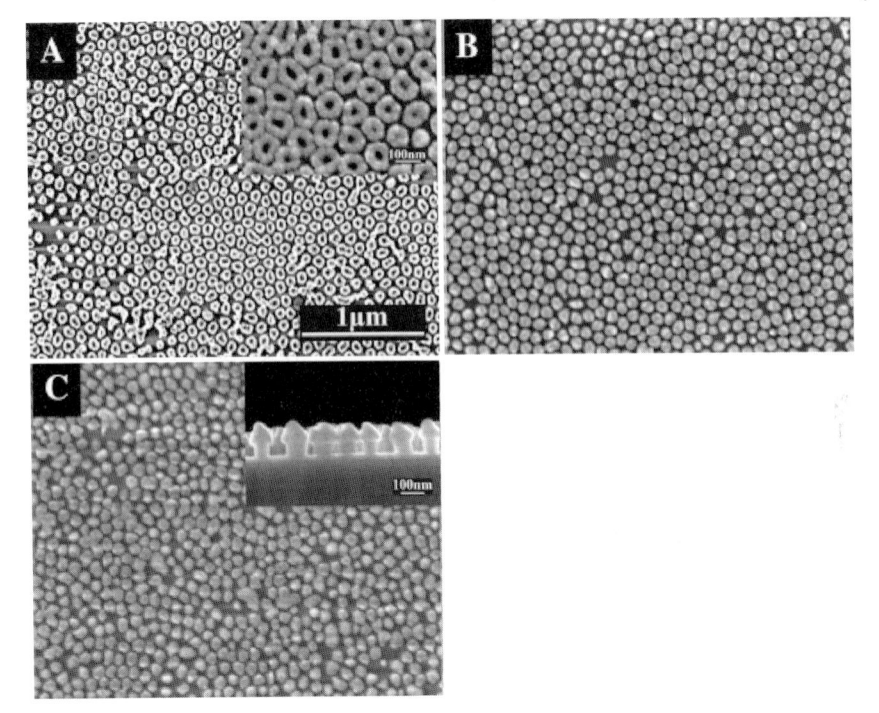

Figure 1.46 SEM images of InGaN nanostructures. Nanorings (A) (inset shows the larger magnitude); nanodots (B); and nanoarrows (C) (inset shows the cross-sectional view). (Reprinted with permission from Ref. [602], © 2006, American Chemical Society.)

1.5.4
Polymer, Organic and Inorganic Nanowires and Nanotubes

Ordered porous alumina membranes are considered to be one of the most suitable host materials for the fabrication of polymer nanowires and nanotubes. High-conductivity polypyrrole (PPy) nanowires with a pore diameter of 220 nm and length of 20 µm were prepared by using anodic porous alumina as the template (Figure 1.47). The high-density array of uniform nanowires was electrosynthesized in a medium of 75% isopropyl alcohol + 20% boron trifluoride diethyl etherate + 5% poly(ethylene glycol) [614].

A melt-assisted template method employing anodic porous alumina was used for the synthesis of semi-conducting polythiophene mesowires [615]. The AAO template was also used for the fabrication of organic–inorganic hybrid polymer nanowires consisting of poly(N-vinylcarvazole), rhodamine 6G and TiO_2 nanoparticles [616]. A wide variety of polymer nanotubes, such as perylene [617], polyaniline [618], polystyrene [582,619], polytetrafluoroethylene (PTFE) [619,583], polyvinylidene fluoride (PVDF) [582,583], and epoxy resin [620] were successfully prepared by a simple wetting of the AAO templates. Poly(p-xylene) (PPX) nanowires were synthesized by CVD polymerization [621] and poly(2-hydroxyethyl methacrylate) (PHEMA) by atom

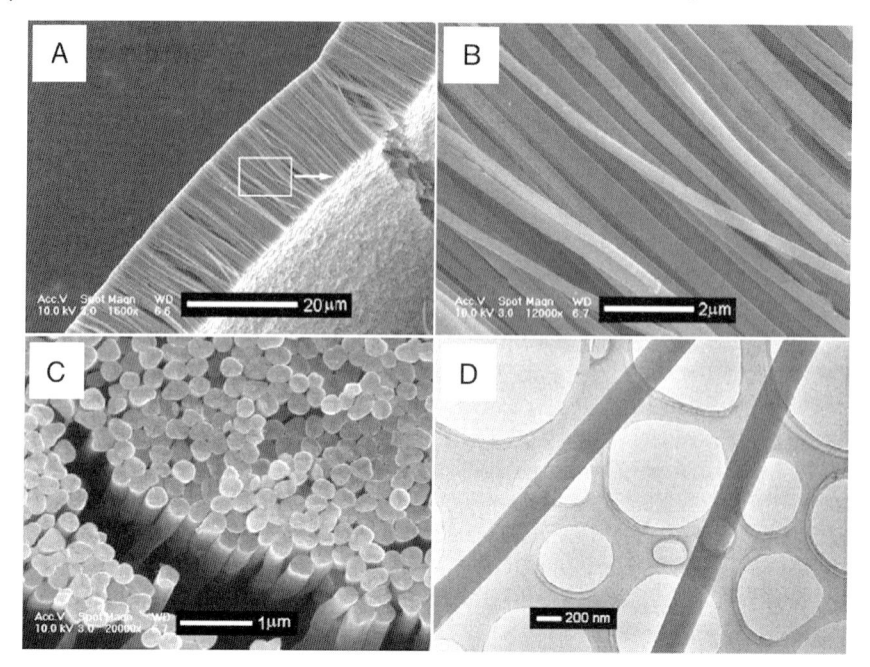

Figure 1.47 SEM images (A)–(C) and TEM image (D) of the PPy nanowires. (Reprinted with permission from Ref. [614], © 2006, IOP Publishing Ltd.).

transfer radical polymerization [622], with the assistance of AAO templates. The highly ordered array of PMMA nanopores, fabricated on the basis of the AAO template [594,595,623,624] is frequently used for the fabrication of various nanoporous membranes by the two-step replication process.

Recently, the preparation of crystalline organic nanorods of 2,7-di-t-butylpyrene (DTBP) [625] and organometallic nanotubes synthesized from a di-Rh N-heterocyclic carbene complex [626] were also reported.

By using highly ordered anodic porous alumina as templates, inorganic nanowires such as AgI [627], CuS, Ag_2S, CuSe and Ag_2Se [628] were prepared successfully by the electrochemical or electroless deposition processes. A high aspect ratio Prussian blue nanotube [629] with an outer diameter of 60 nm, and nickel sulfide [630] nanotubes with a diameter of approximately 200 nm, were each grown inside the pores of the AAO template.

1.5.5
Carbon Nanotubes

In recent years, carbon nanotubes have attracted much attention due to their remarkable electronic, optical, and mechanical properties. AAO templates have often been utilized for carbon nanotube growth [621,631–638], while cylindrical carbon nanotubes with various diameters [631–637] and triangular [638] cross-sections have been synthesized using the CVD approach. An array of parallel and multiwalled carbon nanotubes with uniform diameter and periodic arrangement was achieved by using the AAO template [4,634,635]. The production of carbon nanotubes with deposited Pt nanoparticles on the interior walls of the AAO template has also been reported [637].

1.5.6
Photonic Crystals

Two-dimensional photonic crystals are periodic dielectric materials with an ability to control the propagation of light. The ideally ordered array of nanopores in anodic porous alumina can be directly used as a 2D photonic crystal in the near-infrared (IR) and visible wavelength regions. Numerous publications have been dedicated to 2D photonic band-gap crystals made from ideally ordered anodic porous alumina [639–646]. A monodomain porous alumina with a perfect nanopore order and high aspect ratio, for photonic crystals applications, is normally fabricated by anodizing of the pre-patterned aluminum foil.

1.5.7
Other Nanomaterials (Metallic and Diamond Membranes, Biomaterials)

The growing interest in nanoimprinting and the development of regular patterns has resulted in a considerable increase in the number of attempts to transfer the perfect order of nanopores from the AAO template to other materials with a greater mechanical stability. A variety of nanomaterials with a highly ordered arrangement of nanopores has been fabricated with aid from the AAO template, including a negative array of highly ordered Ni nanorods [647], and negative membranes from Ni [648], Au [593,649,650], Pt [651] and diamond [652,653]. A freestanding gold nanotube membrane replicated by Fan *et al.* [593] from a monodomain AAO template is illustrated in Figure 1.48.

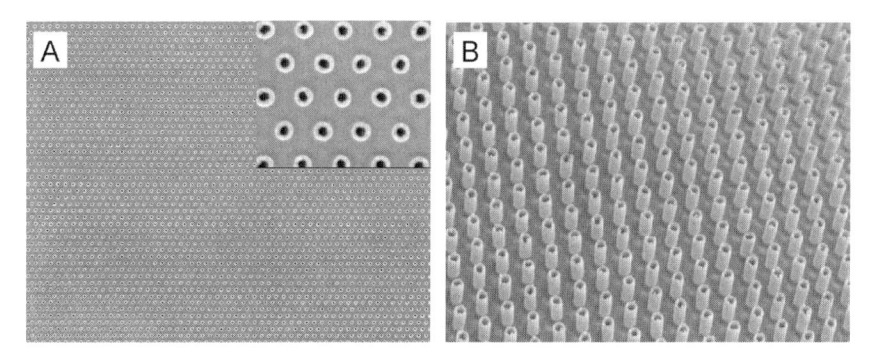

Figure 1.48 (A) SEM image of a gold nanotube membrane viewed from the front side (inset: enlarged top view). The inner diameter of the nanotube is about 130 nm. (B) Inclined view of the gold nanotube membrane showing the perfect alignment of the nanotubes. (Reprinted with permission from Ref. [593], © 2005, IOP Publishing Ltd.).

Recently, a number of nanostructured biomaterials have been synthesized on the base of the AAO template [622,654–657]. Protein binding to an AAO membrane covered with a polymer layer and functionalized by a nitrotriacetate–Cu^{2+} complex was studied [622]. Such modified membranes reveal an unusually high capacity for rapid protein binding, and can be applied to the purification of His-tag proteins. A flow-through-type DNA array, prepared by affixing a single-stranded probe DNA to the sidewall of holes in the perfectly ordered anodic porous alumina film with the evaporated Pt layer, was recently described [654]. The formation of a highly ordered DNA array on the Au array formed in anodic porous alumina was also studied [655,656]. Controlled DNA nanopatterning was carried out via the fabrication of Au disk arrays on the AAO template, with a interpore distance of 63 nm and a pore diameter of 40 nm. DNA molecules were fixed to the ordered Au disc array by the formation of Au–S bonds between the Au surface and thiolated groups introduced at the end of the DNA sequences. The DNA nanopatterns (Figure 1.49) were formed upon immersion of the Au disk array into DNA solution [656].

The potential applications of such materials lie in the field of highly functional biodevices for analysis in hereditary diseases, in DNA computers, molecular motors, heterogeneous sensors, transducers, and reactors. An electrochemical biosensor for the determination of blood glucose was constructed by the condensation polymerization of dimethyldichlorosilane at the surface of a porous alumina membrane [658]. Anodic porous alumina films can also be used as implants with enhanced bone-bonding performance, by filling the nanopores with a bioactive material that supports normal osteoblastic activity [659].

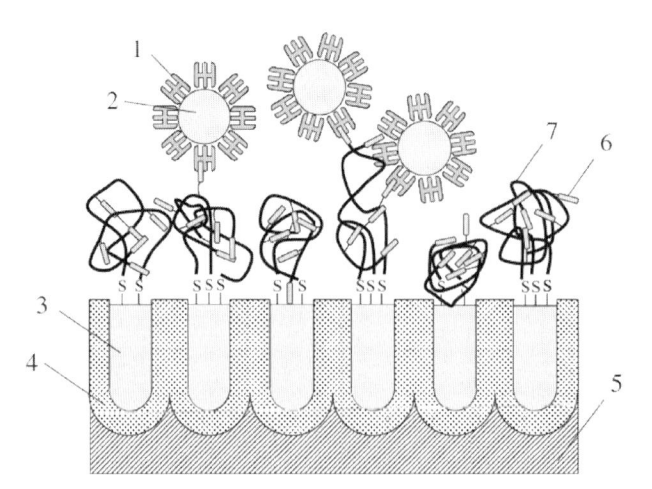

Figure 1.49 Schematic of Au-particle-DNA array prepared using Au disk array formed from highly ordered anodic porous alumina. (1) Streptavidin; (2) Au particles; (3) deposited Au; (4) anodic porous alumina; (5) aluminum; (6) biotin; (7) DNA. (Reprinted with permission from Ref. [656], © 2005, The Japan Society of Applied Physics.)

Acknowledgments

Some parts of these investigations were supported by the Ministry of Science and Information Society Technologies, Poland (grant 3 T08D 001 27). The author kindly acknowledges the Laboratory of Field Emission Scanning Electron Microscopy and Microanalysis at the Institute of Geological Sciences, Jagiellonian University (Poland), where the SEM imaging was performed.

References

1. Parades, J.I., Martínez-Alonso, A. and Tascón, J.M.D. (2003) *Micropor. Mesopor. Mater.*, **65**, 93–126.
2. Rosei, F. (2004) *J. Phys. Condens. Matter*, **16**, S1373–S1436.
3. DiNardo, N.J. (1994) in *Materials Science and Technology*, (eds R.W. Cahn, P. Haasen and E.J. Kramer), Vol. 2B, *Characterization of Materials, Part II*, E. Lifshin (Vol. Ed.), VCH Verlagsgesellschaft mbH, Weinheim, Germany, pp. 2–158.
4. Chik, H. and Xu, J.M. (2004) *Mater. Sci. Eng. R*, **43**, 103–138.
5. Noda, S. (2001) *MRS Bull.*, August, 618–620.
6. Polman, A., Wiltzius, P. and Editors, G. (2001) *MRS Bull.*, August, 608–610.
7. Kamp, M., Happ, T., Mahnkopf, S., Duan, G., Anand, S. and Forchel, A. (2004) *Physica E*, **21**, 802–808.
8. Matthias, S., Müller, F., Jamois, C., Wehrspohn, R.B. and Gösele, U. (2004) *Adv. Mater.*, **16**, 2166–2170.
9. Létant, S.E., van Buuren, T.W. and Terminello, L.J. (2004) *Nano Lett.*, **4**, 1705–1707.
10. Cui, Y., Wei, Q., Park, H. and Lieber, C.M. (2001) *Science*, **293**, 1289–1292.
11. Vaseashta, A. and Dimova-Malinovska, D. (2005) *Sci. Technol. Adv. Mater.*, **6**, 312–318.
12. Alexson, D., Chen, H., Cho, M., Dutta, M., Li, Y., Shi, P., Raichura, A., Ramadurai, D., Parikh, S., Stroscio, M.A. and Vasudev, M. (2005) *J. Phys.: Condens. Matter*, **17**, R637–R656.
13. Petroff, P.M., Lorke, A. and Imamoglu, A. (2001) *Physics Today*, May, 46–52.
14. Hazani, M., Hennrich, F., Kappes, M., Naaman, R., Peled, D., Sidorov, V. and Shvarts, D. (2004) *Chem. Phys. Lett.*, **391**, 389–392.
15. Pedano, M.L. and Rivas, G.A. (2004) *Electrochem. Commun.*, **6**, 10–16.
16. Gooding, J.J. (2005) *Electrochim. Acta*, **50**, 3049–3060.
17. Ben-Ali, S., Cook, D.A., Evans, S.A.G., Thienpont, A., Bartlett, P.N. and Kuhn, A. (2003) *Electrochem. Commun.*, **5**, 747–751.
18. Ben-Ali, S., Cook, D.A., Bartlett, P.N. and Kuhn, A. (2005) *J. Electroanal. Chem.*, **579**, 181–187.
19. Mai, L.H., Hoa, P.T.M., Binh, N.T., Ha, N.T.T. and An, D.K. (2000) *Sensors Actuators B*, **66**, 63–65.
20. Islam, T., Mistry, K.K., Sengupta, K. and Saha, H. (2004) *Sensors Mater.*, **16**, 345–356.
21. Fert, A. and Piraux, L. (1999) *J. Magn. Magn. Mater.*, **200**, 338–358.
22. Oepen, H.P. and Kirschner, J. (1999) *Curr. Opin. Solid State Mater. Sci.*, **4**, 217–221.
23. Ross, C.A. (2001) *Annu. Rev. Mater. Res.*, **31**, 203–235.
24. Aranda, P. and García, J.M. (2002) *J. Magn. Magn. Mater.*, **249**, 214–219.
25. Hasegawa, R. (2002) *J. Magn. Magn. Mater.*, **249**, 346–350.
26. Juang, J-Y. and Bogy, D.B. (2005) *Microsyst. Technol.*, **11**, 950–957.
27. Chou, S.Y. (1997) *Proceed. IEEE*, **85**, 652–671.
28. Guo, L., Leobandung, E. and Chou, S.Y. (1997) *Science*, **275**, 649–651.

29 Oh, S.-W., Rhee, H.W., Lee, Ch., Kim, Y. Ch., Kim, J.K. and Yu, J-W. (2005) *Curr. Appl. Phys.*, **5**, 55–58.
30 Walter, E.C., Favier, F. and Penner, R.M. (2002) *Anal. Chem.*, **74**, 1546–1553.
31 Yun, M., Myung, N.V., Vasquez, R.P., Lee, Ch., Menke, E. and Penner, R.M. (2004) *Nano Lett.*, **4**, 419–422.
32 Routkevitch, D., Tager, A.A., Haruyama, J., Almawlawi, D., Moskovits, M. and Xu, J.M. (1996) *IEEE Trans. Electron Devices*, **43**, 1646–1658.
33 Samuelson, L. (2003) *Materials Today*, October, 22–31.
34 Forshaw, M., Stadler, R., Crawley, D. and Nikolić, K. (2004) *Nanotechnology*, **15**, S220–S223.
35 Dreselhaus, M.S., Lin, Y.M., Rabin, O., Jorio, A., Souza Filho, A.G., Pimenta, M. A., Saito, R., Samsonidze, Ge.G. and Dresselhaus, G. (2003) *Mater. Sci. Eng. C*, **23**, 129–140.
36 Dresselhaus, M.S., Lin, Y.M., Rabin, O., Black, M.R. and Dresselhaus, G. (2004) in *Springer Handbook of Nanotechnology*, (ed. B. Bhushan), Springer-Verlag, Heidelberg, Germany, pp. 99–145.
37 Martín, J.I., Nogués, J., Liu, K., Vicent, J. L. and Schuller, I.K. (2003) *J. Magn. Magn. Mater.*, **256**, 449–501.
38 Schwarzacher, W., Kasyutich, O.I., Evans, P.R., Darbyshire, M.G., Yi, G., Fedosyuk, V.M., Rousseaux, F., Cambril, E. and Decanini, D. (1999) *J. Magn. Magn. Mater.*, **198–199**, 185–190.
39 Maria, J., Jeon, S. and Rogers, J.A. (2004) *J. Photochem. Photobiol. A*, **166**, 149–154.
40 Lin, B.J. (2006) *Microelectron. Eng.*, **83**, 604–613.
41 Chou, S.Y., Krauss, P.R. and Kong, L. (1996) *J. Appl. Phys.*, **79**, 6101–6106.
42 Duvail, J.L., Dubois, S., Piraux, L., Vaurès, A., Fert, A., Adam, D., Champagne, M., Rousseaux, F. and Decanini, D. (1998) *J. Appl. Phys.*, **84**, 6359–6365.
43 Dumpich, G., Krome, T.P. and Hausmanns, B. (2002) *J. Magn. Magn. Mater.*, **248**, 241–247.
44 Papaioannou, E., Simeonidis, K., Valassiades, O., Vouroutis, N., Angelakeris, M., Ppolopoulos, P., Kostic, I. and Flevaris, N.K. (2004) *J. Magn. Magn. Mater.*, **272–276**, e1323–e1325.
45 Chen, A., Chua, S.J., Chen, P., Chen, X.Y. and Jian, L.K. (2006) *Nanotechnology*, **17**, 3903–3908.
46 Lodder, J.C. (2004) *J. Magn. Magn. Mater.*, **272–276**, 1692–1697.
47 Ji, Q., Chen, Y., Ji, L., Jiang, X. and Leung, K-N. (2006) *Microelectron. Eng.*, **83**, 796–799.
48 Miramond, C., Fermon, C., Rousseaux, F., Decanini, D. and Carcenac, F. (1997) *J. Magn. Magn. Mater.*, **165**, 500–503.
49 Turberfield, A.J. (2001) *MRS Bulletin*, August, 632–636.
50 Ross, C.A., Hwang, M., Shima, M., Smith, H.I., Farhoud, M., Savas, T.A., Schwarzacher, W., Parrochon, J., Escoffier, W., Neal Bertram, H., Humphrey, F.B. and Redjdal, M. (2002) *J. Magn. Magn. Mater.*, **249**, 200–207.
51 Rosa, W.O., de Araújo, A.E.P., Gobbi, A. L., Knobel, M. and Cescato, L. (2005) *J. Magn. Magn. Mater.*, **294**, e63–e67.
52 Chou, S.Y., Krauss, P.R. and Renstrom, P. R. (1996) *J. Vac. Sci. Technol. B*, **14**, 4129–4133.
53 Chou, S. (2003) *Technol Rev.*, February, 42–44.
54 Harnagea, C., Alexe, M., Schilling, J., Choi, J., Wehrspohn, R.B., Hesse, D. and Gösele, U. (2003) *Appl. Phys. Lett.*, **83**, 1827–1829.
55 Grujicic, D. and Pesic, B. (2005) *J. Magn. Magn. Mater.*, **288**, 196–204.
56 Li, W., Hsiao, G.S., Harris, D., Nyffenegger, R.M., Virtanen, J.A. and Penner, R.M. (1996) *J. Phys. Chem.*, **100**, 20103–20113.
57 Nyffenegger, R.M. and Penner, R.M. (1997) *Chem. Rev.*, **97**, 1195–1230.
58 Lee, S., Kim, J., Shin, S., Lee, H-J., Koo, S. and Lee, H. (2004) *Mater. Sci. Eng. C*, **24**, 3–9.
59 Kolb, D.M. and Simeone, F.C. (2005) *Electrochim. Acta*, **50**, 2989–2996.

60 Raimundo, D.S., Stelet, A.B., Fernandez, F.J.R. and Salcedo, W.J. (2005) *Microelectron. J.*, **36**, 207–211.
61 Jones, B.A., Searle, J.A. and O'Grady, K. (2005) *J. Magn. Magn. Mater.*, **290–291**, 131–133.
62 Komiyama, H., Yamaguchi, Y. and Noda, S. (2004) *Chem. Eng. Sci.*, **59**, 5085–5090.
63 Zhukov, A.A., Ghanem, M.A., Goncharov, A.V., Bartlett, P.N. and de Groot, P.A.J. (2004) *J. Magn. Magn. Mater.*, **272–276**, e1369–e1371.
64 Li, Y., Cai, W., Duan, G., Sun, F., Cao, B. and Lu, F. (2005) *Mater. Lett.*, **59**, 276–279.
65 Ma, W., Hesse, D. and Gösele, U. (2006) *Nanotechnology*, **17**, 2536–2541.
66 Teichert, C., Bean, J.C. and Lagally, M.G. (1998) *Appl. Phys. A*, **67**, 675–685.
67 Oster, J., Huth, M., Wiehl, L. and Adrian, H. (2004) *J. Magn. Magn. Mater.*, **272–276**, 1588–1589.
68 Schmidt, O.G., Rastelli, A., Kar, G.S., Songmuang, R., Kiravittaya, S., Stoffel, M., Denker, U., Stufler, S., Zrenner, A., Grützmacher, D., Nguyen, B-Y. and Wennekers, P. (2004) *Physica E*, **25**, 280–287.
69 Adair, J.H., Li, T., Kido, T., Havey, K., Moon, J., Mecholsky, J., Morrone, A., Talham, D.R., Ludwig, M.H. and Wang, L. (1998) *Mater. Sci. Eng. R*, **23**, 139–242.
70 Tartaj, P., Morales, M.P., González-Carreño, T., Veintemillas-Verdaguer, S. and Serna, C.J. (2005) *J. Magn. Magn. Mater.*, **290–291**, 28–34.
71 Vázquez, M., Luna, C., Morales, M.P., Sanz, R., Serna, C.J. and Mijangos, C. (2004) *Physica B*, **354**, 71–79.
72 Yakutik, I.M. and Shevchenko, G.P. (2004) *Surf. Sci.*, **566–568**, 414–418.
73 Kim, K.D., Han, D.N. and Kim, H.T. (2004) *Chem. Eng. J.*, **104**, 55–61.
74 Lee, G-J., Shin, S-I., Kim, Y-Ch. and Oh, S-G. (2004) *Mater. Chem. Phys.*, **84**, 197–204.
75 Chen, S., Feng, J., Guo, X., Hong, J. and Ding, W. (2005) *Mater. Lett.*, **59**, 985–988.
76 Zhou, H. and Li, Z. (2005) *Mater. Chem. Phys.*, **89**, 326–331.
77 Liu, Y-Ch. and Lin, L-H. (2004) *Electrochem. Commun.*, **6**, 1163–1168.
78 Zhang, J., Han, B., Liu, M., Liu, D., Dong, Z., Liu, J., Li, D., Wang, J., Dong, B., Zhao, H. and Rong, L. (2003) *J. Phys. Chem. B*, **107**, 3679–3683.
79 Murray, B.J., Li, Q., Newberg, J.T., Menke, E.J., Hemminger, J.C. and Penner, R.M. (2005) *Nano Lett.*, **5**, 2319–2324.
80 Völkel, B., Kaltenpoth, G., handrea, M., Sahre, M., Nottbohm, Ch.T., Küller, A., Paul, A., Kautek, W., Eck, W. and Gölzhäuser, A. (2005) *Surf. Sci.*, **597**, 32–41.
81 Wang, C., Chen, M., Zhu, G. and Lin, Z. (2001) *J. Colloid Interfac. Sci.*, **243**, 362–364.
82 Groves, J.T., Ulman, N. and Boxer, S.G. (1997) *Science*, **275**, 651–653.
83 Castellana, E.T. and Cremer, P.S. (2006) *Surf. Sci. Rep.*, **61**, 429–444.
84 Whang, D., Jin, S. and Lieber, C.M. (2003) *Nano Lett.*, **3**, 951–954.
85 Rao, C.N.R., Govindaraj, A., Gundiah, G. and Vivekchand, S.R.C. (2004) *Chem. Eng. Sci.*, **59**, 4665–4671.
86 Kim, Y., Kim, Ch. and Yi, J. (2004) *Mater. Res. Bull.*, **39**, 2103–2112.
87 Thiruchitrambalam, M., Palkar, V.R. and Gopinathan, V. (2004) *Mater. Lett.*, **58**, 3063–3066.
88 Pavasupree, S., Suzuki, Y., Pivsa-Art, S. and Yoshikawa, S. (2005) *Sci. Technol. Adv. Mater.*, **6**, 224–229.
89 Choi, H., Sofranko, A.C. and Dionysiou, D.D. (2006) *Adv. Funct. Mater.*, **16**, 1067–1074.
90 Yang, P., Deng, T., Zhao, D., Feng, P., Pine, D., Chmelka, B.F., Whitesides, G.M. and Stucky, G.D. (1998) *Science*, **282**, 2244–2246.
91 Karakassides, M.A., Gournis, D., Bourlinos, A.B., Trikalitis, P.N. and Bakas, T. (2003) *J. Mater. Chem.*, **13**, 871–876.
92 Zach, M.P. and Penner, R.M. (2000) *Adv. Mater.*, **12**, 878–883.
93 Liu, H., Favier, F., Ng, K., Zach, M.P. and Penner, R.M. (2001) *Electrochim. Acta*, **47**, 671–677.

94 Penner, R.M. (2002) *J. Phys. Chem. B*, **106**, 3339–3353.
95 Zach, M.P., Ng, K.H. and Penner, R.M. (2000) *Science*, **290**, 2120–2123.
96 Zach, M.P., Inazu, K., Ng, K.H., Hemminger, J.C. and Penner, R.M. (2002) *Chem. Mater.*, **14**, 3206–3216.
97 Walter, E.C., Zach, M.P., Favier, F., Murray, B., Inazu, K., Hemminger, J.C. and Penner, R.M. (2002) in *Physical Chemistry of Interface Nanomaterials*, (eds J.Z. Zhang and Z.L. Wang), Proceedings of the SPIE, Vol. 4807, pp. 83–92.
98 Walter, E.C., Ng, K., Zach, M.P., Penner, R.M. and Favier, F. (2002) *Microelectron. Eng.*, **61–62**, 555–561.
99 Walter, E.C., Murray, B.J., Favier, F., Kaltenpoth, G., Grunze, M. and Penner, R.M. (2002) *J. Phys. Chem.*, **106**, 11407–11411.
100 Walter, E.C., Zach, M.P., Favier, F., Murray, B.J., Inazu, K., Hemminger, J.C. and Penner, R.M. (2003) *ChemPhysChem.*, **4**, 131–138.
101 Walter, E.C., Murray, B.J., Favier, F. and Penner, R.M. (2003) *Adv. Mater.*, **15**, 396–399.
102 Atashbar, M.Z., Banerji, D., Singamaneni, S. and Bliznyuk, V. (2004) *Nanotechnology*, **15**, 374–378.
103 Li, Q., Olson, J.B. and Penner, R.M. (2004) *Chem. Mater.*, **16**, 3402–3405.
104 Li, Q., Walter, E.C., van der Veer, W.E., Murray, B.J., Newberg, J.T., Bohannan, E.W., Switzer, J.A., Hemminger, J.C. and Penner, R.M. (2005) *J. Phys. Chem. B*, **109**, 3169–3182.
105 Menke, E.J., Li, Q. and Penner, R.M. (2004) *Nano Lett.*, **4**, 2009–2014.
106 Petit, C., Legrand, J., Russier, V. and Pilcni, M.P. (2002) *J. Appl. Phys.*, **91**, 1502–1508.
107 Mazur, M. (2004) *Electrochem. Commun.*, **6**, 400–403.
108 Shi, S., Sun, J., Zhang, J. and Cao, Y. (2005) *Physica B*, **362**, 231–235.
109 Liu, F-M. and Green, M. (2004) *J. Mater. Chem.*, **14**, 1526–1532.
110 Dryfe, R.A.W., Walter, E.C. and Penner, R.M. (2004) *ChemPhysChem.*, **5**, 1879–1884.
111 Yanagimoto, H., Deki, S., Akamatsu, K. and Gotoh, K. (2005) *Thin Solid Films*, **491**, 18–22.
112 Sun, L., Searson, P.C. and Chien, C.L. (1999) *Appl. Phys. Lett.*, **74**, 2803–2805.
113 Chien, C.L., Sun, L., Tanase, M., Bauer, L.A., Hultgren, A., Silevitch, D.M., Meyer, G.J., Searson, P.C. and Reich, D.H. (2002) *J. Magn. Magn. Mater.*, **249**, 146–155.
114 Ziegler, K.J., Polyakov, B., Kulkarni, J.S., Crowley, T.A., Ryan, K.M., Morris, M.A., Erts, D. and Holmes, J.D. (2004) *J. Mater. Chem.*, **14**, 585–589.
115 Kazakova, O., Erst, D., Crowley, T.A., Kulkarni, J.S. and Holmes, J.D. (2005) *J. Magn. Magn. Mater.*, **286**, 171–176.
116 Aylett, B.J., Earwaker, L.G., Forcey, K., Giaddui, T. and Harding, I.S. (1996) *J. Organometal. Chem.*, **521**, 33–37.
117 Rastei, M.V., Meckenstock, R., Devaux, E., Ebbesen, Th. and Bucher, J.P. (2005) *J. Magn. Magn. Mater.*, **286**, 10–13.
118 Matthias, S., Schilling, J., Nielsch, K., Müller, F., Wehrspohn, R.B. and Gösele, U. (2002) *Adv. Mater.*, **14**, 1618–1621.
119 Zhao, L., Yosef, M., Steinhart, M., Göring, P., Hofmeister, H., Gösele, U. and Schlecht, S. (2006) *Angew. Chem. Int. Ed.*, **45**, 311–315.
120 Martin, C.R. (1994) *Science*, **266**, 1961–1966.
121 Szklarczyk, M., Strawski, M., Donten, M.L. and Donten, M. (2004) *Electrochem. Commun.*, **6**, 880–886.
122 Ounadjela, K., Ferré, R., Louail, L., George, J.M., Maurice, J.L., Piraux, L. and Dubois, S. (1997) *J. Appl. Phys.*, **81**, 5455–5457.
123 Piraux, L., Dubois, S., Ferain, E., Legras, R., Ounadjela, K., George, J.M., Mayrice, J.L. and Fert, A. (1997) *J. Magn. Magn. Mater.*, **165**, 352–355.
124 Scarani, V., Doudin, B. and Ansermet, J-P. (1999) *J. Magn. Magn. Mater.*, **205**, 241–248.
125 Kazadi Mukenga Bantu, A., Rivas, J., Zaragoza, G., López-Uintela, M.A. and

Blanco, B.C. (2001) *J. Non-Crystall. Solids*, **287**, 5–9.

126 Kazadi Mukenga Bantu, A., Rivas, J., Zaragoza, G., López-Uintela, M.A. and Blanco, B.C. (2001) *J. Appl. Phys.*, **89**, 3393–3397.

127 Valizadeh, S., George, J.M., Leisner, P. and Hultman, L. (2001) *Electrochim. Acta*, **47**, 865–874.

128 Ge, S., Li, Ch., Ma, X., Li, W., Xi, L. and Li, C.X. (2001) *J. Appl. Phys.*, **90**, 509–511.

129 Ge, S., Ma, X., Li, Ch. and Li, W. (2001) *J. Magn. Magn. Mater.*, **226–230**, 1867–1869.

130 Encinas, A., Demand, M., George, J-M. and Piraux, L. (2002) *IEEE Trans. Magn.*, **38**, 2574–2576.

131 Vila, L., George, J.M., Faini, G., Popa, A., Ebels, U., Ounadjela, K. and Piraux, L. (2002) *IEEE Trans. Magn.*, **38**, 2577–2579.

132 Rivas, J., Kazadi Mukenga Bantu, A., Zaragoza, G., Blanco, M.C. and López-Quintela, M.A. (2002) *J. Magn. Magn. Mater.*, **249**, 220–227.

133 Demand, M., Encinas-Oropesa, A., Kenane, S., Ebels, U., Huynen, I. and Piraux, L. (2002) *J. Magn. Magn. Mater.*, **249**, 228–233.

134 Motoyama, M., Fukunaka, Y., Sakka, T., Ogata, Y.H. and Kikuchi, S. (2005) *J. Electroanal. Chem.*, **584**, 84–91.

135 Konishi, Y., motoyama, M., Matsushima, H., Fukunaka, Y., Ishii, R. and Ito, Y. (2003) *J. Electroanal. Chem.*, **559**, 149–153.

136 Dubois, S., Michel, A., Eymery, J.P., Duvail, J.L. and Piraux, L. (1999) *J. Mater. Res.*, **14**, 665–671.

137 Han, G.C., Zong, B.Y. and Wu, Y.H. (2002) *IEEE Trans. Magn.*, **38**, 2562–2564.

138 Blondel, A., Meier, J.P., Doudin, B. and Ansermet, J-Ph. (1994) *Appl. Phys. Lett.*, **65**, 3019–3021.

139 Piraux, L., George, J.M., Despres, J.F., Leroy, C., Ferain, E., Legras, R., Ounadjela, K. and Fert, A. (1994) *Appl. Phys. Lett.*, **65**, 2484–2486.

140 Dubois, S., Beuken, J.M., Piraux, L., Duvail, J.L., Fert, A., George, J.M. and Maurice, J.L. (1997) *J. Magn. Magn. Mater.*, **165**, 30–33.

141 Piraux, L., Dubois, S., Duvail, J.D., Ounadjela, K. and Fert, A. (1997) *J. Magn. Magn. Mater.*, **175**, 127–136.

142 Maurice, J-L., Imhoff, D., Etienne, P., Dubois, O., Piraux, L., George, J-M., Galtier, P. and Fret, A. (1998) *J. Magn. Magn. Mater.*, **184**, 1–18.

143 Dubois, S., Marchal, C., Beuken, J.M., Piraux, L., Fert, A., George, J.M. and Maurice, J.L. (1997) *Appl. Phys. Lett.*, **70**, 396–398.

144 Valizadeh, S., George, J.M., Leisner, P. and Hultman, L. (2002) *Thin Solid Films*, **402**, 262–271.

145 Xue, S.H. and Wang, Z.D. (2006) *Mater. Sci. Eng. B*, **135**, 74–77.

146 Piraux, L., Dubois, S., Duvail, J.L., Radulescu, A., Demoustier-Champagne, S., Ferain, E. and Legras, R. (1999) *J. Mater. Res.*, **14**, 3042–3050.

147 Curiale, J., Sánchez, R.D., Troiani, H.E., Pastoriza, H., levy, P. and Leyva, A.G. (2004) *Physics B*, **354**, 98–103.

148 Park, I-W., Yoon, M., Kim, Y.M., Kim, Y., Kim, J.H., Kim, S. and Volkov, V. (2004) *J. Magn. Magn. Mater.*, **272–276**, 1413–1414.

149 Huang, E., Rockford, L., Russell, T.P. and Hawker, C.J. (1998) *Nature*, **395**, 757–758.

150 Otsuka, H., Nagasaki, Y. and Kataoka, K. (2001) *Mater. Today*, **4** (May/June), 30–36.

151 Hamley, I.W. (2003) *Nanotechnology*, **14**, R39–R54.

152 Yalçın, O., Yıldız, F., Özdemir, M., Aktaş, B., Köseoğlu, Y., Bal, M. and Tuominen, M.T. (2004) *J. Magn. Magn. Mater.*, **272–274**, 1684–1685.

153 Jeoung, E., Galow, T.H., Schotter, J., Bal, M., Ursache, A., Tuominen, M.T., Stafford, C.M., Russell, T.P. and Rotello, V.M. (2001) *Langmuir*, **17**, 6396–6398.

154 Wu, X.C. and Tao, Y.R. (2004) *J. Cryst. Growth*, **242**, 309–312.

155 Wang, L., Zhang, X., Zhao, S., Zhou, G., Zhou, Y. and Qi, J. (2005) *Mater. Res. Soc. Symp. Proc.*, **879E**, Z3.21.1–Z3.21.6.

156 Huang, M.H., Mao, S., Feick, H., Yan, H., Wu, Y., Kind, H., Weber, E., Russo, R. and Yang, P. (2001) *Science*, **292**, 1897–1899.

157 Xu, C.X., Sun, X.W., Dong, Z.L. and Yu, M.B. (2004) *J. Cryst. Growth*, **270**, 498–504.

158 Stelzner, T., Andrä, G., Wendler, E., Weschl, W., Scholz, R., Gösele, U. and Christiansen, S.H. (2006) *Nanotechnology*, **17**, 2895–2898.

159 Chou, S.Y. and Zhuang, L. (1999) *J. Vac. Sci. Technol. B*, **17**, 3197–3202.

160 Chou, S.Y., Keimel, C. and Gu, J. (2002) *Nature*, **417**, 835–837.

161 Szafraniak, I., Hesse, D. and Alexe, M. (2005) *Solid State Phenomena*, **106**, 117–122.

162 Yasuda, H., Ohnaka, I., Fujimoto, S., Sugiyama, A., Hayashi, Y., Yamamoto, M., Tsuchiyama, A., Nakano, T., Uesugi, K. and Kishio, K. (2004) *Mater. Lett.*, **58**, 911–915.

163 Hassel, A.W., Bello-Rodriguez, B., Milenkovic, S. and Schneider, A. (2005) *Electrochim. Acta*, **50**, 3033–3039.

164 Yasuda, H., Ohnaka, I., Fujimoto, S., Takezawa, N., Tsuchiyama, A., Nakano, T. and Uesugi, K. (2006) *Scripta Mater.*, **54**, 527–532.

165 Grüning, U., Lehmann, V., Ottow, S. and Busch, K. (1996) *Appl. Phys. Lett.*, **68**, 747–749.

166 Jessensky, O., Müller, F. and Gösele, U. (1997) *Thin Solid Films*, **297**, 224–228.

167 Birner, A., Grüning, U., Ottow, S., Schneider, A., Müller, F., Lehmann, V., Föll, H. and Gösele, U. (1998) *Phys. Stat. Sol. A*, **165**, 111–117.

168 Motohashi, A. (2000) *Jpn. J. Appl. Phys.*, **39**, 363–367.

169 Bisi, O., Ossicini, S. and Pavesi, L. (2000) *Surf. Sci. Rep.*, **38**, 1–126.

170 Hamm, D., Sasano, J., Sakka, T. and Ogata, Y.H. (2002) *J. Electrochem. Soc.*, **149**, C331–C337.

171 Chazalviel, J.-N., Ozanam, F., Gabouze, N., Fellah, S. and Wehrspohn, R.B. (2002) *J. Electrochem. Soc.*, **149**, C511–C520.

172 Föll, H., Christophersen, M., Carstensen, J. and Hasse, G. (2002) *Mater. Sci. Eng. R*, **39**, 94–141.

173 Seo, M. and Yamaya, T. (2005) *Electrochim. Acta*, **51**, 787–794.

174 Hasegawa, H. and Sato, T. (2005) *Electrochim. Acta*, **50**, 3015–3027.

175 Choi, J., Wehrspohn, R.B., Lee, J. and Gösele, U. (2004) *Electrochim. Acta*, **49**, 2645–2652.

176 Bourdet, P., Vacandio, F., Argème, L., Rossi, S. and Massiani, Y. (2005) *Thin Solid Films*, **483**, 205–210.

177 Macak, J.M., Sirotna, K. and Schmuki, P. (2005) *Electrochim. Acta*, **50**, 3679–3684.

178 Prida, V.M., Hernández-Vélez, M., Cervera, M., Pirota, K., Sanz, R., Navas, D., Asenjo, A., Aranda, P., Ruiz-Hitzky, E., Batallán, F., Vázquez, M., Hernando, B., Menéndez, A., Bordel, N. and Pereiro, R. (2005) *J. Magn. Magn. Mater.*, **294**, e69–e72.

179 Raja, K.S., Misra, M. and Paramguru, K. (2005) *Electrochim. Acta*, **51**, 154–165.

180 Raja, K.S., Misra, M. and Paramguru, K. (2005) *Mater. Lett.*, **59**, 2137–2141.

181 Tsuchiya, H., Macak, J.M., Ghicov, A., Taveira, L. and Schmuki, P. (2005) *Corros. Sci.*, **47**, 3324–3335.

182 Zhao, J., Wang, X., Chen, R. and Li, L. (2005) *Solid State Commun.*, **134**, 705–710.

183 Bayoumi, F.M. and Ateya, B.G. (2006) *Electrochem. Commun.*, **8**, 38–44.

184 Ricker, R.E., Miller, A.E., Yue, D-F., Banerjee, G. and Bandyopadhyay, S. (1996) *J. Electron. Mater.*, **25**, 1585–1592.

185 Yuzhakov, V.V., Chang, H-Ch. and Miller, A.E. (1997) *Phys. Rev. B*, **56**, 12608–12624.

186 Tsuchiya, H. and Schmuki, P. (2004) *Electrochem. Commun.*, **6**, 1131–1134.

187 Sieber, I., Hildebrand, H., Friedrich, A. and Schmuki, P. (2005) *Electrochem. Commun.*, **7**, 97–100.

188 Karlinsey, R.L. (2005) *Electrochem. Commun.*, **7**, 1190–1194.

189 Heidelberg, A., Rozenkranz, C., Schultze, J.W., Schäpers, Th. and Staikov, G. (2005) *Surf. Sci.*, **597**, 173–180.

190 Habazaki, H., Ogasawara, T., Konno, H., Shimizu, K., Asami, K., Saito, K., Nagata, S., Skeldon, P. and Thompson, G.E. (2005) *Electrochim. Acta*, **50**, 5334–5339.

191 Tsuchiya, H. and Schmuki, P. (2005) *Electrochem. Commun.*, **7**, 49–52.

192 Shin, H-C., Dong, J. and Liu, M. (2004) *Adv. Mater.*, **16**, 237–240.

193 Cochran, W.C. and Keller, F. (1963) in *The Finishing of Aluminum* (ed. G.H. Kissin), New York Reinhold Publishing Corporation, pp. 104–126.

194 Lowenheim, F.A. (1978) in *Electroplating*, McGraw-Hill Book Company New York, pp. 452–478.

195 Wernick, S., Pinner, R. and Sheasby, P.G. (1987) in *The Surface Treatment and Finishing of Aluminium and its Alloys*, ASM International, Finishing Publication Ltd., 5th edition, pp. 289–368.

196 Jelinek, T.W. (1997) in *Oberflächenbehandlung von Aluminium*, Eugen G. Leuze Verlag –D –88348 Saulgau/Württ, pp. 187–219 (in German).

197 Keller, F., Hunter, M.S. and Robinson, D.L. (1953) *J. Electrochem. Soc.*, **100**, 411–419.

198 Young, L. (1961) in *Anodic Oxide Films*, Academic Press, London and New York, pp. 193–221.

199 Despić, A.R. (1985) in *J. Electroanal. Chem.*, **191**, 417–423.

200 Thompson, G.E. and Wood, G.C. (1983) in *Treatise on Materials Science and Technology*, (ed. J.C. Scully), Academic Press New York, Vol. 23, pp. 205–329.

201 Pakes A., Thompson, G.E., Skeldon, P., Morgan, P.C. and Shimizu, K. (1999) *Trans. IMF*, **77**, 171–177.

202 Despić, A. and Parkhutik, V.P. (1989) in *Modern Aspects of Electrochemistry*, (eds J.O'M. Bockris, R.E. White and B.E. Conway), Plenum Press, New York and London, Vol. 20, pp. 401–503.

203 Thompson G.E. (1997) *Thin Solid Films*, **297**, 192–201.

204 Zhu, X.F., Li, D.D., Song, Y. and Xiao, Y.H. (2005) *Mater. Lett.*, **59**, 3160–3163.

205 Diggle, J.W., Downie, T.C. and Goulding, C.W. (1969) *Chem. Rev.*, **69**, 365–405.

206 Takahashi, H., Fujimoto, K. and Nagayama, M. (1988) *J. Electrochem. Soc.*, **135**, 1349–1353.

207 Morks, M.F., Hamdy, A.S., Fahim, N.F. and Shoeib, M.A. (2006) *Surf. Coat. Technol.*, **200**, 5071–5076.

208 O'Sullivan, J.P. and Wood, G.C. (1970) *Proc. Roy. Soc. Lond. A*, **317**, 511–543.

209 Surganov, V., Morgen, P., Nielsen, J.C., Gorokh, G. and Mozalev, A. (1987) *Electrochim. Acta*, **32**, 1125–1127.

210 Surganov, V., Gorokh, G., Poznyak, A. and Mozalev, A. (1988) *Zhur. Prikl. Khimii (Russ. J. Appl. Chem.)*, **61**, 2011–2014 (in Russian).

211 Ono, S., Saito, M. and Asoh, H. (2005) *Electrochim. Acta*, **51**, 827–833.

212 Surganov, V. and Gorokh, G. (1993) *Mater. Lett.*, **17**, 121–124.

213 Chu, S.Z., Wada, K., Inoue, S., Isogai, M., Katsuta, Y. and Yasumori, A. (2006) *J. Electrochem. Soc.*, **153**, B384–B391.

214 Mozalev, A., Surganov, A. and Magaino, S. (1999) *Electrochim. Acta*, **44**, 3891–3898.

215 Surganov, V. and Gorokh, G. (1993) *Zhur. Prikl. Khimii (Russ. J. Appl. Chem.)*, **66**, 683–685 (in Russian).

216 Ono, S., Saito, M., Ishiguro, M. and Asoh, H. (2004) *J. Electrochem. Soc.*, **151**, B473–B478.

217 Mozalev, A., Mozaleva, I., Sakairi, M. and Takahashi, H. (2005) *Electrochim. Acta*, **50**, 5065–5075.

218 Zhou, X., Thompson, G.E. and Potts, G. (2000) *Trans. IMF*, **78**, 210–214.

219 Wang, H. and Wang, H.W. (2006) *Mater. Chem. Phys.*, **97**, 213–218.

220 Jia, Y., Zhou, H., Luo, P., Luo, S., Chen, J. and Kuang, Y. (2006) *Surf. Coat. Technol.*, **201**, 513–518.

221 Mozalev, A., Poznyak, A., Mozaleva, I. and Hassel, A.W. (2001) *Electrochem. Commun.*, **3**, 299–305.

222 Shingubara, S., Morimoto, K., Sakaue, H. and Takahagi, T. (2004) *Electrochem. Solid-State Lett.*, **7**, E15–E17.

223 Choo, Y.H. and Devereux, O.F. (1975) *J. Electrochem. Soc.*, **122**, 1645–1653.

224 Shimizu, K., Kobayashi, K., Thompson, G.E. and Wood, G.C. (1992) *Phil. Mag. A*, **66**, 643–652.

225 Arrowsmith, D.J. and Moth, D.A. (1986) *Trans. IMF*, **64**, 91–93.

226 Sulka, G.D. and Jaskuła, M. (2006) *J. Nanosci. Nanotechnol.*, **6**, 3803–3811.

227 Palibroda, E., Farcas, T. and Lupsan, A. (1995) *Mater. Sci. Eng. B*, **32**, 1–5.

228 Martin, T. and Hebert, K.R. (2001) *J. Electrochem. Soc.*, **148**, B101–B109.

229 Abdel-Gaber, A.M., Abd-El-Nabey, B.A., Sidahmed, I.M., El-Zayady, A.M. and Saadawy, M. (2006) *Mater. Chem. Phys.*, **98**, 291–297.

230 Ono, S., Saito, M. and Asoh, H. (2004) *Electrochem. Solid-State Lett.*, **7**, B21–B24.

231 Sulka, G.D., Stroobants, S., Moshchalkov, V., Borghs, G. and Celis, J-P. (2004) *J. Electrochem. Soc.*, **151**, B260–B264.

232 Wood, G.C., O'Sullivan, J.P. and Vaszko, B. (1968) *J. Electrochem. Soc.*, **115**, 618–620.

233 Paolini, G., Masaero, M., Sacchi, F. and Paganelli, M. (1965) *J. Electrochem. Soc.*, **112**, 32–38.

234 Palibroda, E. (1984) *Surf. Technol.*, **23**, 341–351 (in French).

235 Nielsch, K., Choi, J., Schwirn, K., Wehrspohn, R.B. and Gösele, U. (2002) *Nano Lett.*, **2**, 677–680.

236 Li, F., Zhang, L. and Metzger, R.M. (1998) *Chem. Mater.*, **10**, 2470–2480.

237 Ebihara, K., Takahashi, H. and Nagayama, M. (1983) *J. Met. Finish. Soc. Japan (Kinzoku Hyomen Gijutsu)*, **34**, 548–553 (in Japanese).

238 Ebihara, K., Takahashi, H. and Nagayama, M. (1982) *J. Met. Finish. Soc. Japan (Kinzoku Hyomen Gijutsu)*, **33**, 156–164 (in Japanese).

239 Parkhutik, V.P. and Shershulsky, V.I. (1992) *J. Phys. D: Appl. Phys.*, **25**, 1258–1263.

240 Sulka, G.D. and Parkoła, K.G. (2007) *Electrochim. Acta*, **52**, 1880–1888.

241 Hwang, S-K., Jeong, S-H., Hwang, H-Y., Lee, O-J. and Lee, K-H. (2002) *Korean J. Chem. Eng.*, **19**, 467–473.

242 Marchal, D. and Demé, B. (2003) *J. Appl. Cryst.*, **36**, 713–717.

243 Hunter, M.S. and Fowle, P.E. (1954) *J. Electrochem. Soc.*, **101**, 481–485.

244 Hunter, M.S. and Fowle, P.E. (1954) *J. Electrochem. Soc.*, **101**, 514–519.

245 Vrublevsky, I., Parkoun, V. and Schreckenbach, J. (2005) *Appl. Surf. Sci.*, **242**, 333–338.

246 Vrublevsky, I., Parkoun, V., Sokol, V. and Schreckenbach, J. (2005) *Appl. Surf. Sci.*, **252**, 227–233.

247 Vrublevsky, I., Parkoun, V., Schreckenbach, J. and Marx, G. (2004) *Appl. Surf. Sci.*, **227**, 282–292.

248 Sulka, G.D. and Parkoła, K. (2006) *Thin Solid Films*, **515**, 338–345.

249 Ono, S. and Masuko, N. (2003) *Surf. Coat. Technol.*, **169–170**, 139–142.

250 Ono, S., Asoh, H., Saito, M. and Ishiguro, M. (2003) *Electrochemistry*, **71**, 105–107 (in Japanese).

251 Chu, S-Z., Wada, K., Inoue, S., Isogai, M. and Yasumori, A. (2005) *Adv. Mater.*, **17**, 115–2119.

252 Nakamura, S., Saito, M., Huang, Li-F., Miyagi, M. and Wada, K. (1992) *Jpn. J. Appl. Phys.*, **31**, 3589–3593.

253 Bocchetta, P., Sunseri, C., Bottino, A., Capannelli, G., Chiavarotti, G., Piazza, S. and Di Quarto, F. (2002) *J. Appl. Electrochem.*, **32**, 977–985.

254 Palibroda, E. (1995) *Electrochim. Acta*, **40**, 1051–1055.

255 Wood, G.C. and O'Sullivan, J.P. (1970) *Electrochim. Acta*, **15**, 1865–1876.

256 Xu, Y., Thompson, G.E., Wood, G.C. and Bethune, B. (1987) *Corros. Sci.*, **27**, 83–102.

257 Wood, G.C., Skeldon, P., Thompson, G.E. and Shimizu, K. (1996) *J. Electrochem. Soc.*, **143**, 74–83.

258 Habazaki, H., Shimizu, K., Skeldon, P., Thompson, G.E., Wood, G.C. and Zhou, X. (1997) *Corros. Sci.*, **39**, 731–737.

259 Thompson, G.E., Skeldon, P., Shimizu, K. and Wood, G.C. (1995) *Phil. Trans. R. Soc. Lond. A*, **350**, 143–168.

260 Shimizu, K., Brown, G.M., Kobayashi, K., Skeldon, P., Thompson, G.E. and Wood, G.C. (1999) *Corros. Sci.*, **41**, 1835–1847.

261 Shimizu, K., Habazaki, H., Skeldon, P., Thompson, G.E. and Wood, G.C. (1999) *Surf. Interface Anal.*, **27**, 1046–1049.

262 Shimizu, K., Brown, G.M., Habazaki, H., Kobayashi, K., Skeldon, P., Thompson, G.E. and Wood, G.C. (1999) *Surf. Interface Anal.*, **27**, 24–28.

263 Shimizu, K., Brown, G.M., Habazaki, H., Kobayashi, K., Skeldon, P., Thompson, G.E. and Wood, G.C. (1999) *Surf. Interface Anal.*, **27**, 153–156.

264 Shimizu, K., Habazaki, H., Skeldon, P., Thompson, G.E. and Wood, G.C. (1999) *Surf. Interface Anal.*, **27**, 998–1002.

265 Shimizu, K., Brown, G.M., Habazaki, H., Kobayashi, K., Skeldon, P., Thompson, G.E. and Wood, G.C. (1999) *Electrochim. Acta*, **44**, 2297–2306.

266 Shimizu, K., Habazaki, H., Skeldon, P., Thompson, G.E. and Wood, G.C. (2000) *Electrochim. Acta*, **45**, 1805–1809.

267 Ono, S., Wada, Ch. and Asoh, H. (2005) *Electrochim. Acta*, **50**, 5103–5110.

268 Takahashi, H., Fujimoto, K., Konno, H. and Nagayama, M. (1984) *J. Electrochem. Soc.*, **131**, 1856–1861.

269 Habazaki, H., Shimizu, K., Skeldon, P., Thompson, G.E. and Zhou, X. (1997) *Corros. Sci.*, **39**, 719–730.

270 Ono, S. and Masuko, N. (1992) *Corros. Sci.*, **33**, 503–507.

271 Patermarakis, G. and Pavlidou, C. (1994) *J. Catal.*, **147**, 140–155.

272 Patermarakis, G., Moussoutzanis, K. and Nikolopoulos, N. (1999) *J. Solid State Electrochem.*, **3**, 193–204.

273 Patermarakis, G., Chandrinos, J. and Moussoutzanis, K. (2001) *J. Electroanal. Chem.*, **510**, 59–66.

274 Thompson, G.E., Furneaux, R.C., Wood, G.C., Richardson, J.A. and Goode, J.S. (1978) *Nature*, **272**, 433–435.

275 Ono, S. and Masuko, N. (1992) *Corros. Sci.*, **33**, 841–850.

276 Patermarakis, G. and Kerassovitou, P. (1992) *Electrochim. Acta*, **3**, 125–137.

277 Palibroda, E. and Marginean, P. (1994) *Thin Solid Films*, **240**, 73–75.

278 Alwitt, R.S., Dyer, C.K. and Noble, B. (1982) *J. Electrochem. Soc.*, **129**, 711–717.

279 Ono, S., Ichinose, H. and Masuko, N. (1991) *J. Electrochem. Soc.*, **138**, 3705–3710.

280 Ono, S., Ichinose, H. and Masuko, N. (1992) *J. Electrochem. Soc.*, **139**, L80–L81.

281 Mei, Y.F., Wu, X.L., Shao, X.F., Huang, G.S. and Siu, G.G. (2003) *Phys. Lett. A*, **309**, 109–113.

282 Pu, L., Chen, Z-Q., Tan, Ch., Yang, Z., Zou, J-P., Bao, X-M., Feng, D., Shi, Y. and Zheng, Y-D. (2002) *Chin. Phys. Lett.*, **19**, 391–394.

283 Macdonald, D.D. (1992) *J. Electrochem. Soc.*, **139**, 3434–3449.

284 Macdonald, D.D., Biaggio, S.R. and Song, H. (1992) *J. Electrochem. Soc.*, **139**, 170–177.

285 Macdonald, D.D. (1993) *J. Electrochem. Soc.*, **140**, L27–L30.

286 Thompson, G.E., Furneaux, R.C. and Wood, G.C. (1978) *Corros. Sci.*, **18**, 481–498.

287 Thompson, G.E., Xu, Y., Skeldon, P., Shimizu, K., Han, S.H. and Wood, G.C. (1987) *Phil. Mag. B*, **55**, 651–667.

288 Shimizu, K., Thompson, G.E. and Wood, G.C. (1981) *Thin Solid Films*, **81**, 39–44.

289 Thompson, G.E., Wood, G.C. and Williams, J.Q. (1986) *Chemtronics*, **1**, 125–129.

290 Ono, S., Ichinose, H., Kawaguchi, T. and Masuko, N. (1990) *Corros. Sci.*, **31**, 249–254.

291 Parkhutik, V.P., Belov, V.T. and Chernyckh, M.A. (1990) *Electrochim. Acta*, **35**, 961–966.

292 Brown, I.W.M., Bowden, M.E., Kemmitt, T. and MacKenzie, K.J.D. (2006) *Curr. Appl. Phys.*, **6**, 557–561.

293 Yakovleva, N.M., Yakovlev, A.N. and Chupakhina, E.A. (2000) *Thin Solid Films*, **366**, 37–42.

294 Gabe, D.R. (2000) *Trans. IMF*, **78**, 207–209.

295 Liu, Y., Alwitt, R.S. and Shimizu, K. (2000) *J. Electrochem. Soc.*, **147**, 1382–1387.

296 Lu, Q., Skeldon, P., Thompson, G.E., Habazaki, H. and Shimizu, K. (2005) *Thin Solid Films*, **471**, 118–122.

297 Hernández, A., Martínez, F., Martín, A. and Prádanos, P. (1995) *J. Coll. Interface Sci.*, **173**, 284–296.

298 Theodoropoulou, M., Karahaliou, P.K., Georgia, S.N., Krontiras, C.A., Pisanias, M.N., Kokonou, M. and Nassiopoulou, A.G. (2005) *Ionics*, **11**, 236–239.

299 Lee, C.W., Kang, H.S., Chang, Y.H. and Hahm, Y.M. (2000) *Korean J. Chem. Eng.*, **17**, 266–272.

300 Yang, S.G., Li, T., Huang, L.S., Tang, T., Zhang, J.R., Gu, B.X., Du, Y.W., Shi, S.Z. and Lu, Y.N. (2003) *Phys. Lett. A*, **318**, 440–444.

301 Hoang, V.V. and Oh, S.K. (2004) *Physica B*, **352**, 73–85.

302 Mata-Zamora, M.E. and Saniger, J.M. (2005) *Rev. Mex. Fis.*, **51**, 502–509.

303 Xu, W.L., Zheng, M.J., Wu, S. and Shen, W.Z. (2004) *Appl. Phys. Lett.*, **85**, 4364–4366.

304 Borca-Tasciuc, D-A. and Chen, G. (2005) *J. Appl. Phys.*, **97**, 084303/1-9.

305 Chen, C.C., Chen, J.H. and Chao, Ch.G. (2005) *Jpn. J. Appl. Phys.*, **44**, 1529–1533.

306 Xia, Z., Riester, L., Sheldon, B.W., Curtin, W.A., Liang, J., Yin, A. and Xu, J.M. (2004) *Rev. Adv. Mater. Sci.*, **6**, 131–139.

307 Chen, C.H., Takita, K., Honda, S. and Awaji, H. (2005) *J. Eur. Ceram. Soc.*, **25**, 385–391.

308 Kato, S., Nigo, S., Uno, Y., Onisi, T. and Kido, G. (2006) *J. Phys.: Conference Series*, **38**, 148–151.

309 Fernandes, J.C.S., Picciochi, R., Da Cunha Belo, M., Moura e Silva, T., Ferreira, M.G.S. and Fonseca, I.T.E. (2004) *Electrochim. Acta*, **49**, 4701–4707.

310 Karahaliou, P.K., Theodoropulou, M., Krontiras, C.A., Xanthopoulos, N., Georga, S.N. and Pisanias, M.N. (2004) *J. Appl. Phys.*, **95**, 2776–2780.

311 Theodoropoulou, M., Karahaliou, P.K., Georgia, S.N., Krontiras, C.A., Pisanias, M.N., Kokonou, M. and Nassiopoulou, A.G. (2005) *J. Phys.: Conference Series*, **10**, 222–225.

312 Zabala, N., Pattantyus-Abraham, A.G., Rivacoba, A., García de Abajo, F.J. and Wolf, M.O. (2003) *Phys. Rev. B*, **68**, 245407/1-13.

313 Oh, H-J., Park, G-S., Kim, J-G., Jeong, Y. and Chi, Ch-S. (2003) *Mater. Chem. Phys.*, **82**, 331–334.

314 Redón, R., Vázquez-Olmos, A., Mata-Zamora, M.E., Ordóñez-Medrano, A., Rivera-Torres, F. and Saniger, J.M. (2006) *Rev. Adv. Mater. Sci.*, **11**, 79–87.

315 Alvine, K.J., Shpyrko, O.G., Pershan, P.S., Shin, K. and Russell, T.P. (2006) *Phys. Rev. Lett.*, **97**, 175503/1-4.

316 Thompson, D.W., Snyder, P.G., Castro, L. and Yan, L. (2005) *J. Appl. Phys.*, **97**, 113511/1-9.

317 Zhang, D-X., Zhang, H-J., Lin, X-F. and He, Y-L. (2006) *Spectroscopy Spectral Anal.*, **26**, 411–414. (in Chinese).

318 Xu, W.L., Chen, H., Zheng, M.J., Ding, G.Q. and Shen, W.Z. (2006) *Optital Mater.*, **28**, 1160–1165.

319 Hohlbein, J., Rehn, U. and Wehrspohn, R.B. (2004) *Phys. Stat. Sol. A*, **201**, 803–807.

320 Chen, J., Cai, W-L. and Mou, J-M. (2001) *J. Inorgan. Mater.*, **16**, 677–682. (in Chinese).

321 Kokonou, M., Nassiopoulou, A.G. and Travlos, A. (2003) *Mater. Sci. Eng. B*, **101**, 65–70.

322 Yang, Y., Chen, H-L. and Bao, X-M. (2003) *Acta Chim. Sinica*, **61**, 320–324 (in Chinese).

323 Huang, G.S., Wu, X.L., Siu, G.G. and Chu, P.K. (2006) *Solid State Commun.*, **137**, 621–624.

324 Chen, J.H., Huang, C.P., Chao, C.G. and Chen, T.M. (2006) *Appl. Phys. A*, **84**, 297–300.

325 Du, Y., Cai, W.L., Mo, C.M., Chen, J., Zhang, L.D. and Zhu, X.G. (1999) *Appl. Phys. Lett.*, **74**, 2951–2953.

326 Huang, G.S., Wu, X.L., Kong, F., Cheng, Y.C., Siu, G.G. and Chu, P.K. (2006) *Appl. Phys. Lett.*, **89**, 073114/1–3.

327 Yang, Y. and Gao, Q. (2004) *Phys. Lett. A*, **333**, 328–333.

328 Shi, Y-L., Zhang, X-G. and Li, H-L. (2001) *Spectroscopy Lett.*, **34**, 419–426.

329 de Azevedo, W.M., de Carvalho, D.D., Khoury, H.J., de Vasconcelos, E.A. and da Silva, E.F. Jr. (2004) *Mater. Sci. Eng. B*, **112**, 171–174.

330 Stojadinovic, S., Zekovic, Lj., Belca, I., Kasalica, B. and Nikolic, D. (2004) *Electrochem. Commun.*, **6**, 708–712.

331 Kasalica, B., Stojadinovic, S., Zekovic, Lj., Belca, I. and Nikolic, D. (2005) *Electrochem. Commun.*, **7**, 735–739.

332 Stojadinovic, S., Belca, I., Kasalica, B., Zekovic, Lj. and Tadic, M. (2006) *Electrochem. Commun.*, **8**, 1621–1624.

333 Hoar, T.P. and Yahalom, J. (1963) *J. Electrochem. Soc.*, **110**, 614–621.

334 Kanakala, R., Singaraju, P.V., Venkat, R. and Das, B. (2005) *J. Electrochem. Soc.*, **152**, J1–J5.

335 Tsangaraki-Kaplanoglou, I., Theohari, S., Dimogerontakis, Th., Wang, Y-M., Kuo, H-H. and Kia, S. (2006) *Surf. Coat. Technol.*, **200**, 2634–2641.

336 Dimogerontakis, Th. and Kaplanoglou, I. (2001) *Thin Solid Films*, **385**, 182–189.

337 Dimogerontakis, Th. and Kaplanoglou, I. (2002) *Thin Solid Films*, **402**, 121–125.

338 Dimogerontakis, T. and Tsangaraki-Kaplanoglou, I. (2003) *Plating Surf. Finish.*, May, 80–82.

339 Li, Y., Zheng, M., Ma, L. and Shen, W. (2006) *Nanotechnology*, **17**, 5101–5105.

340 Inada, T., Uno, N., Kato, T. and Iwamoto, Y. (2005) *J. Mater. Res.*, **20**, 114–120.

341 Kneeshaw, J.A. and Gabe, D.R. (1984) *Trans. IMF*, **62**, 59–63.

342 Gabe, D.R. (1987) *Trans. IMF*, **65**, 152–154.

343 Li, L. (2000) *Solar Energ. Mater. Solar Cells*, **64**, 279–289.

344 De Graeve, I., Terryn, H. and Thompson, G.E. (2006) *Electrochim. Acta*, **52**, 1127–1134.

345 Setoh, S. and Miyata, A. (1932) *Sci. Pap. Inst. Phys.Chem. Res. Tokyo*, **17**, 189–236.

346 Wernick, S. (1934) *J. Electrodepos. Tech. Soc.*, **9**, 153–176.

347 Rummel, T. (1936) *Z. Physik*, **99**, 518–551 (in German).

348 Baumann, W. (1936) *Z. Physik*, **102**, 59–66 (in German).

349 Baumann, W. (1939) *Z. Physik*, **111**, 707–736 (in German).

350 Akahori, H. (1961) *J. Electron Microscopy (Tokyo)*, **10**, 175–185.

351 Murphy, J.F. and Michaelson, C.E. (1962) in Proceedings Conference on Anodizing, University of Notthingham, UK, 12–14 September 1961, Aluminium Development Association, London pp. 83–95.

352 Csokan, P. (1980) in *Advances in Corrosion Science and Technology*, (eds M.G. Fontana and R.W. Staehle), Plenum Press New York and London Vol. 7, pp. 239–356.

353 Ginsberg H. and Wefers, K. (1962) *Metall.*, **16**, 173–175 (in German).

354 Ginsberg, H. and Wefers, K. (1963) *Metall.*, **17**, 202–209 (in German).

355 Ginsberg, H. and Wefers, K. (1961) *Aluminium (Dusseldorf)*, **37**, 19–28 (in German).

356 Ginsberg, H. and Kaden, W. (1963) *Aluminium (Dusseldorf)*, **39**, 33–41 (in German).

357 De Graeve, I., Terryn, H. and Thompson, G.E. (2002) *J. Appl. Electrochem.*, **32**, 73–83.

358 Garcia-Vergara, S.J., Skeldon, P., Thompson, G.E. and Habazaki, H. (2006) *Electrochim. Acta*, **52**, 681–687.

359 Siejka, J. and Ortega, C. (1977) *J. Electrochem. Soc.*, **124**, 883–891.

360 Crossland, A.C., Habazaki, H., Shimizu, K., Skeldon, P., Thompson, G.E., Wood, G.C., Zhou, X. and Smith, C.J.E. (1999) *Corros. Sci.*, **41**, 1945–1954.

361 Shimizu, K., Habazaki, H. and Skeldon, P. (2002) *Electrochim. Acta*, **47**, 1225–1228.

362 Zhuravlyova, E., Iglesias-Rubianes, L., Pakes, A., Skeldon, P., Thompson, G.E., Zhou, X., Quance, T., Graham, M.J., Habazaki, H. and Shimizu, K. (2002) *Corros. Sci.*, **44**, 2153–2159.

363 Chu, S.Z., Wada, K., Inoue, S. and Todoroki, S. (2002) *J. Electrochem. Soc.*, **149**, B321–B327.

364 Patermarakis, G., Lenas, P., Karavassilis, Ch. and Papayiannis, G. (1991) *Electrochim. Acta*, **36**, 709–725.

365 Patermarakis, G. (1996) *Electrochim. Acta*, **41**, 2601–2611.

366 Patermarakis, G. and Papandreadis, N. (1993) *Electrochim. Acta*, **38**, 2351–2361.

367 Patermarakis, G. and Tzouvelekis, D. (1994) *Electrochim. Acta*, **39**, 2419–2429.

368 Patermarakis, G. and Moussoutzanis, K. (1995) *J. Electrochem. Soc.*, **142**, 737–743.

369 Patermarakis, G. (1998) *J. Electroanal. Chem.*, **447**, 25–41.

370 Patermarakis, G. and Moussoutzanis, K. (2001) *Corros. Sci.*, **43**, 1433–1464.

371 Patermarakis, G. and Moussoutzanis, K. (2003) *Chem. Eng. Commun.*, **190**, 1018–1040.

372 Patermarakis, G. and Moussoutzanis, K. (2002) *Corros. Sci.*, **44**, 1737–1753.

373 Patermarakis, G. and Moussoutzanis, K. (2002) *J. Solid State Electrochem.*, **6**, 475–484.

374 Patermarakis, G. and Moussoutzanis, K. (2005) *J. Solid State Electrochem.*, **9**, 205–233.

375 Patermarakis, G. and Masavetas, K. (2006) *J. Electroanal. Chem.*, **588**, 179–189.

376 Patermarakis, G. and Moussoutzanis, K. (1995) *Electrochim. Acta*, **40**, 699–708.

377 Patermarakis, G., Moussoutzanis, K. and Chandrinos, J. (2001) *J. Solid State Electrochem.*, **6**, 39–54.

378 Patermarakis, G., Moussoutzanis, K. and Chandrinos, J. (2001) *J. Solid State Electrochem.*, **6**, 71–72.

379 Heber, K. (1978) *Electrochim. Acta*, **23**, 127–133.

380 Heber, K. (1978) *Electrochim. Acta*, **23**, 135–139.

381 Wada, K., Shimohira, T., Yamada, M. and Baba, N. (1986) *J. Mater. Sci.*, **21**, 3810–3816.

382 Morlidge, J.R., Skeldon, P., Thompson, G.E., Habazaki, H., Shimizu, K. and Wood, G.C. (1999) *Electrochim. Acta*, **44**, 2423–2435.

383 Nelson, J.C. and Oriani, R.A. (1993) *Corros. Sci.*, **34**, 307–326.

384 Jessensky, O., Müller, F. and Gösele, U. (1998) *Appl. Phys. Lett.*, **72**, 1173–1175.

385 Ba, L. and Li, W.S. (2000) *J. Phys. D: Appl. Phys.*, **33**, 2527–2531.

386 Palibroda, E. (1984) *Surf. Technol.*, **23**, 353–365 (in French).

387 Palibroda, E., Lupsan, A., Pruneanu, S. and Savos, M. (1995) *Thin Solid Films*, **256**, 101–105.

388 Shimizu, K. and Tajima, S. (1979) *Electrochim. Acta*, **24**, 309–311.

389 Shimizu, K., Brown, G.M., Kobayashi, K., Thompson, G.E. and Wood, G.C. (1993) *Corros. Sci.*, **34**, 1853–1857.

390 Zhang, L., Cho, H.S., Li, F., Metzger, R.M. and Doyle, W.D. (1998) *J. Mater. Sci. Lett.*, **17**, 291–294.

391 Makushok, Yu.E., Parkhutik, V.P., Martinez-Duart, J.M. and Albella, J.M. (1994) *J. Phys. D: Appl. Phys.*, **27**, 661–669.

392 Parkhutik, V.P., Albella, J.M., Makushok, Yu.E., Montero, I., Martinez-Duart, J.M. and Shershulskii, V.I. (1990) *Electrochim. Acta*, **35**, 955–960.

393 Wu, H., Zhang, X. and Hebert, K.R. (2000) *J. Electrochem. Soc.*, **147**, 2126–2132.

394 Hebert, K.R. (2001) *J. Electrochem. Soc.*, **148**, B236–B242.

395 Huang, R. and Hebert, K.R. (2004) *J. Electroanal. Chem.*, **565**, 103–114.

396 Randon, J., Mardilovich, P.P., Govyadinov, A.N. and Paterson, R. (1995) *J. Colloid Interface Sci.*, **169**, 335–341.

397 Singaraju, P., Venkat, R., Kanakala, R. and Das, B. (2006) *Eur. Phys. J. Appl. Phys.*, **35**, 107–111.

398 Singh, G.K., Golovin, A.A., Aranson, I.S. and Vinokur, V.M. (2005) *Europhys. Lett.*, **70**, 836–842.

399 Pan, H., Lin, J., Feng, Y. and Gao, H. (2004) *IEEE Trans. Nanotechnol.*, **3**, 462–467.
400 Aroutiounian, V.M. and Ghulinyan, M.Zh. (2000) *Modern Phys. Letters B*, **14**, 39–46.
401 Leppänen, T., Karttunen, M., Barrio, R.A. and Kaski, K. (2004) *Brazil. J. Phys.*, **34**, 368–372.
402 Boissonade, J., Dulos, E. and De Kepper, P. (1995) *Chemical Waves and Patterns* (eds R. Kapral and K. Showalter), Kluwer Academic Publishers, pp. 221–268.
403 Li, A-P., Müller, F., Birner, A., Nielsch, K. and Gösele, U. (1998) *J. Appl. Phys.*, **84**, 6023–6026.
404 Vrublevsky, I., Parkoun, V., Sokol, V., Schreckenbach, J. and Marx, G. (2004) *Appl. Surf. Sci.*, **222**, 215–225.
405 Vrublevsky, I., Parkoun, V., Schreckenbach, J. and Marx, G. (2003) *Appl. Surf. Sci.*, **220**, 51–59.
406 Takahashi, H. and Nagayama, M. (1978) *Corros. Sci.*, **18**, 911–925.
407 Sulka, G.D., Stroobants, S., Moshchalkov, V., Borghs, G. and Celis, J-P. (2002) *J. Electrochem. Soc.*, **149**, D97–D103.
408 Nagayama, M. and Tamura, K. (1967) *Electrochim. Acta*, **12**, 1097–1107.
409 Sulka, G.D., Stroobants, S., Moshchalkov, V., Borghs, G. and Celis, J-P. (2002) *Bulletin du Cercle d'Etudes des Métaux*, **17**, P1/1-8.
410 Masuda, H., Kanezawa, K. and Nishio, K. (2002) *Chem. Lett.*, 1218–1219.
411 Shingubara, S., Murakami, Y., Morimoto, K. and Takahagi, T. (2003) *Surf. Sci.*, **532–535**, 317–323.
412 Iwasaki, T. and Den, T. (2002) Patent US 6 476 409 B2.
413 Aiba, T., Nojiri, H., Motoi, T., Den, T. and Iwasaki, T. (2003) Patent US 6 541 386 B2.
414 Liu, C.Y., Datta, A. and Wang, Y.L. (2001) *Appl. Phys. Lett.*, **78**, 120–122.
415 Peng, C.Y., Liu, C.Y., Liu, N.W., Wang, H.H., Datta, A. and Wang, Y.L. (2005) *J. Vac. Sci. Technol. B*, **23**, 559–562.
416 Sun, Z. and Kim, H.K. (2002) *Appl. Phys. Lett.*, **81**, 3458–3460.
417 Liu, N.W., Datta, A., Liu, C.Y. and Wang, Y.L. (2003) *Appl. Phys. Lett.*, **82**, 1281–1283.
418 Cojocaru, C.S., Padovani, J.M., Wade, T., Mandoli, C., Jaskierowicz, G., Wegrowe, J.E., Fontcuberta i Morral, A. and Pribat, D. (2005) *Nano Lett.*, **5**, 675–680.
419 Li, A-P., Müller, F., Birner, A., Nielsch, K. and Gösele, U. (1999) *Adv. Mater.*, **11**, 483–486.
420 Masuda, H., Yamada, H., Satoh, M., Asoh, H., Nakao, M. and Tamamura, T. (1997) *Appl. Phys. Lett.*, **71**, 2770–2772.
421 Asoh, H., Nishio, K., Nakao, M., Yokoo, A., Tammamura, T. and Masuda, H. (2001) *J. Vac. Sci. Technol. B*, **19**, 569–572.
422 Masuda, H., Yotsuya, M., Asano, M., Nishio, K., Nakao, M., Yakoo, A. and Tamamura, T. (2001) *Appl. Phys. Lett.*, **78**, 826–828.
423 Asoh, H., Nishio, K., Nakao, M., Tammamura, T. and Masuda, H. (2001) *J. Electrochem. Soc.*, **148**, B152–B156.
424 Asoh, H., Ono, S., Hirose, T., Nakao, M. and Masuda, H. (2003) *Electrochim. Acta*, **48**, 3171–3174.
425 Asoh, H., Ono, S., Hirose, T., Takatori, I. and Masuda, H. (2004) *Jpn. J. Appl. Phys.*, **43** (9A), 6342–6346.
426 Masuda, H. (2001) *Electrochemistry*, **69**, 879–883 (in Japanese).
427 Masuda, H., Asoh, H., Watanabe, M., Nishio, K., Nakao, M. and Tamamura, T. (2001) *Adv. Mater.*, **13**, 189–192.
428 Masuda, H., Abe, A., Nakao, M., Yokoo, A., Tamamura, T. and Nishio, K. (2003) *Adv. Mater.*, **15**, 161–164.
429 Matsumoto, F., Harada, M., Nishio, K. and Masuda, H. (2005) *Adv. Mater.*, **17**, 1609–1612.
430 Choi, J., Nielsch, K., Reiche, M., Wehrspohn, R.B. and Gösele, U. (2003) *J. Vac. Sci. Technol. B*, **21**, 763–766.
431 Choi, J., Wehrspohn, R.B. and Gösele, U. (2003) *Adv. Mater.*, **15**, 1531–1534.
432 Choi, J., Wehrspohn, R.B. and Gösele, U. (2005) *Electrochim. Acta*, **50**, 2591–2595.
433 Matsui, Y., Nishio, K. and Masuda, H. (2005) *Jpn. J. Appl. Phys.*, **44**, 7726–7728.

434 Lee, W., Ji, R., Ross, C.A., Gösele, U. and Nielsch, K. (2006) *Small*, **2**, 978–982.

435 Yasui, K., Nishio, K., Nunokawa, H. and Masuda, H. (2005) *J. Vac. Sci. Technol. B*, **23**, L9–L12.

436 Nishio, K., Fukushima, T. and Masuda, H. (2006) *Electrochem. Solid-State Lett.*, **9**, B39–B41.

437 Masuda, H., Matsui, Y., Yotsuya, M., Matsumoto, F. and Nishio, K. (2004) *Chem. Lett.*, **33**, 584–585.

438 Matsui, Y., Nishio, K. and Masuda, H. (2006) *Small*, **2**, 522–525.

439 Asoh, H. and Ono, S. (2005) *Appl. Phys. Lett.*, **87**, 103102/1-3.

440 Zhao, X., Jiang, P., Xie, S., Feng, J., Gao, Y., Wang, J., Liu, D., Song, L., Liu, L., Dou, X., Luo, S., Zhang, Z., Xiang, Y., Zhou, W. and Wang, G. (2006) *Nanotechnology*, **17**, 35–39.

441 Terryn, H., Vereecken, J. and Landuyt, J. (1990) *Trans. IMF*, **68**, 33–37.

442 Fratila-Apachitei, L.E., Terryn, H., Skeldon, P., Thompson, G.E., Duszczyk, J. and Katgerman, L. (2004) *Electrochim. Acta*, **49**, 1127–1140.

443 Jessensky, O., Müller, F. and Gösele, U. (1998) *J. Electrochem. Soc.*, **145**, 3735–3740.

444 Nagayama, M., Tamura, K. and Takahashi, H. (1970) *Corros. Sci.*, **10**, 617–627.

445 Da Silva, M.F., Shimizu, K., Kobayashi, K., Skeldon, P., Thompson, G.E. and Wood, G.C. (1995) *Corros. Sci.*, **37**, 1511–1514.

446 van den Brand, J., Snijders, P.C., Sloof, W.G., Terryn, H. and de Witt, J.W.H. (2004) *J. Phys. Chem. B***108**, 6017–6024.

447 Wang, Y-L., Tseng, W-T. and Chang, S-Ch. (2005) *Thin Solid Films*, **474**, 36–43.

448 Jagminienė, A., Valinčius, G., Riaukaitė, A. and Jagminas, A. (2005) *J. Cryst. Growth*, **274**, 622–631.

449 Holló, M.Gy. (1960) *Acta Metall.*, **8**, 265–268.

450 Shimizu, K., Kobayashi, K., Skeldon, P., Thompson, G.E. and Wood, G.C. (1997) *Corros. Sci.*, **39**, 701–718.

451 Wu, M.T., Leu, I.C. and Hon, M.H. (2002) *J. Vac. Sci. Technol. B*, **20**, 776–782.

452 Ricker, R.E., Miller, A.E., Yue, D-F., Banerjee, G. and Bandyopadhyay, S. (1996) *J. Electron. Mater.*, **25**, 1585–1592.

453 Bandyopadhyay, S., Miller, A.E., Chang, H.C., Banerjee, G., Yuzhakov, V., Yue, D-F., Ricker, R.E., Jones, S., Eastman, J.A., Baugher, E. and Chandrasekhar, M. (1996) *Nanotechnology*, **7**, 360–371.

454 Konovalov, V.V., Zangari, G. and Metzger, R.M. (1999) *Chem. Mater.*, **11**, 1949–1951.

455 Yuzhakov, V.V., Takhistov, P.V., Miller, A.E. and Chang, H-Ch. (1999) *Chaos*, **9**, 62–77.

456 Kong, L-B., Huang, Y., Guo, Y. and Li, H-L. (2005) *Mater. Lett.*, **59**, 1656–1659.

457 Liu, H.W., Guo, H.M., Wang, Y.L., Wang, Y.T., Shen, C.M. and Wei, L. (2003) *Appl. Surf. Sci.*, **219**, 282–289.

458 Miney, P.G., Colavita, P.E., Schiza, M.V., Priore, R.J., Haibach, F.G. and Myrick, M.L. (2003) *Electrochem. Solid-State Lett.*, **6**, B42–B45.

459 Inoue, S., Chu, S-Z., Wada, K., Li, D. and Haneda, H. (2003) *Sci. Technol. Adv. Mater.*, **4**, 269–276.

460 Chu, S.Z., Wada, K., Inoue, S. and Todoroki, S. (2003) *Surf. Coat. Technol.*, **169–170**, 190–194.

461 Chu, S.Z., Wada, K., Inoue, S. and Todoroki, S. (2003) *Electrochim. Acta*, **48**, 3147–3153.

462 Chu, S-Z., Wada, K., Inoue, S., Hishita, S-ichi. and Kurashima, K. (2003) *J. Phys. Chem. B*, **107**, 10180–10184.

463 Chu, S-Z., Inoue, S., Hishita, S-ichi. and Kurashima, K. (2004) *J. Phys. Chem. B*, **108**, 5582–5587.

464 Chu, S.Z., Inoue, S., Wada, K. and Hishita, S. (2004) *J. Electrochem. Soc.*, **151**, C38–C44.

465 Chu, S.Z., Inoue, S., Wada, K. and Kurashima, K. (2005) *Electrochim. Acta*, **51**, 820–826.

466 Chu, S.Z., Inoue, S., Wada, K., Kanke, Y. and Kurashima, K. (2005) *J. Electrochem. Soc.*, **152**, C42–C47.

467 Kemell, M., Färm, E., Leskelä, M. and Ritala, M. (2006) *Phys. Stat. Sol. A*, **203**, 1453–1458.
468 Rabin, O., Herz, P.R., Lin, Y-M., Akinwande, A.I., Cronin, S.B. and Dresselhaus, M.S. (2003) *Adv. Funct. Mater.*, **13**, 631–638.
469 Crouse, D., Lo, Yu-Hwa Miller, A.D. and Crouse, M. (2000) *Appl. Phys. Lett.*, **76**, 49–51.
470 Asoh, H., Matsuo, M., Yoshihama, M. and Ono, S. (2003) *Appl. Phys. Lett.*, **83**, 4408–4410.
471 Choi, J., Sauer, G., Göring, P., Nielsch, K., Wehrspohn, R.B. and Gösele, U. (2003) *J. Mater. Chem.*, **13**, 1100–1103.
472 Toh, C-S., Kayes, B.M., Nemanick, E.J. and Lewis, N.S. (2004) *Nano Lett.*, **4**, 767–770.
473 Pu, L., Shi, Y., Zhu, J.M., Bao, X.M., Zhang, R. and Zheng, Y.D. (2004) *Chem. Commun.*, 942–943.
474 Le Paven-Thivet, C., Fusil, S., Aubert, P., Malibert, C., Zozime, A. and Houdy, Ph. (2004) *Thin Solid Films*, **446**, 147–154.
475 Myung, N.V., Lim, J., Fleurial, J-P., Yun, M., West, W. and Choi, D. (2004) *Nanotechnology*, **15**, 833–838.
476 Wu, M.T., Leu, I.C., Yen, J.H. and Hon, M.H. (2004) *Electrochem. Solid-State Lett.*, **7**, C61–C63.
477 Shiraki, H., Kimura, Y., Ishii, H., Ono, S., Itaya, K. and Niwano, M. (2004) *Appl. Surf. Sci.*, **237**, 369–373.
478 Chen, Z. and Zhang, H. (2005) *J. Electrochem. Soc.*, **152**, D227–D231.
479 Kokonou, M., Nassiopoulou, A.G., Giannakopoulos, K.P. and Boukos, N. (2005) *J. Phys.: Conference Series*, **10**, 159–162.
480 Mei, Y.F., Wu, X.L., Qiu, T., Shao, X.F., Siu, G.G. and Chu, P.K. (2005) *Thin Solid Films*, **492**, 66–70.
481 Kimura, Y., Shiraki, H., Ishibashi, K-I., Ishii, H., Itaya, K. and Niwano, M. (2006) *J. Electrochem. Soc.*, **153**, C296–C300.
482 Yang, Y., Chen, H., Mei, Y., Chen, J., Wu, X. and Bao, X. (2002) *Solid State Commun.*, **123**, 279–282.

483 Xu, C-L., Li, H., Zhao, G-Y. and Li, H-L. (2006) *Mater. Lett.*, **60**, 2335–2338.
484 Xu, C-L., Li, H., Xue, T. and Li, H-L. (2006) *Scripta Mater.*, **54**, 1605–1609.
485 Mei, Y.F., Wu, X.L., Shao, X.F., Siu, G.G. and Bao, X.M. (2003) *Europhys. Lett.*, **62**, 595–599.
486 Briggs, E.P., Walpole, A.R., Wilshaw, P.R., Karlsson, M. and Pålsgård, E. (2004) *J. Mater. Sci. Mater. Med.*, **15**, 1021–1029.
487 Yasui, K., Sakamoto, Y., Nishio, K. and Masuda, H. (2005) *Chem. Lett.*, **34**, 342–343.
488 Zhou, H.Y., Qu, S.C., Wang, Z.G., Liang, L.Y., Cheng, B.C., Liu, J.P. and Peng, W.Q. (2006) *Mater. Sci. Semicond. Process.*, **9**, 337–340.
489 Yakovlev, N.M., Yakovlev, A.N. and Denisov, A.I. (2003) *Investigated in Russia*, **6**, 673–582 (in Russian).
490 Garcia-Vergara, S.J., El Khazmi, K., Skeldon, P. and Thompson, G.E. (2006) *Corros. Sci.*, **48**, 2937–2946.
491 Yoo, B-Y., Hendricks, R.K., Ozkan, M. and Myung, N.V. (2006) *Electrochim. Acta*, **51**, 3543–3550.
492 Masuda, H., Hesegwa, F. and Ono, S. (1997) *J. Electrochem. Soc.*, **144**, L127–L130.
493 Li, A-P., Müller, F., Birner, A., Nielsch, K. and Gösele, U. (1999) *J. Vac. Sci. Technol. A*, **17**, 1428–1431.
494 Masuda, H., Nagae, M., Morikawa, T. and Nishio, K. (2006) *Jpn. J. Appl. Phys.*, **45**, L406–L408.
495 Shingubara, S., Okino, O., Sayama, Y., Sakaue, H. and Takahagi, T. (1997) *Jpn. J. Appl. Phys.*, **36**, 7791–7795.
496 Almasi Kashi, M. and Ramazani, A. (2005) *J. Phys. D: Appl. Phys.*, **38**, 2396–2399.
497 Masuda, H., Yada, K. and Osaka, A. (1998) *Jpn. J. Appl. Phys.*, **37**, L1340–L1342.
498 Li, A.P., Müller, F. and Gösele, U. (2000) *Electrochem. Solid State Lett.*, **3**, 131–134.
499 Lee, W., Ji, R., Gösele, U. and Nielsch, K. (2006) *Nature Mat.*, **5**, 741–747.

500 Masuda, H. and Satoh, M. (1996) *Jpn. J. Appl. Phys.*, **35**, L126–L129.
501 Zhao, Y., Chen, M., Zhang, Y., Xu, T. and Liu, W. (2005) *Mater. Lett.*, **59**, 40–43.
502 Schneider, J.J., Engstler, J., Budna, K.P., Teichert, Ch. and Franzka, S. (2005) *Eur. J. Inorg. Chem.*, 2352–2359.
503 Yuan, J.H., Chen, W., Hui, R.J., Hu, Y.L. and Xia, X.H. (2006) *Electrochim. Acta*, **51**, 4589–4595.
504 Zhou, W.Y., Li, Y.B., Liu, Z.Q., Tang, D.S., Zou, X.P. and Wang, G. (2001) *Chin. Phys.*, **10**, 218–222.
505 Suh, J.S. and Lee, J.S. (1999) *Appl. Phys. Lett.*, **75**, 2047–2049.
506 Yan, J., Rama Rao, G.V., Barela, M., Brevnov, D.A., Jiang, Y., Xu, H., López, G.P. and Atanassov, P.B. (2003) *Adv. Mater.*, **15**, 2015–2018.
507 Ding, G.Q., Zheng, M.J., Xu, W.L. and Shen, W.Z. (2005) *Nanotechnology*, **16**, 1285–1289.
508 Hou, K., Tu, J.P. and Zhang, X.B. (2002) *Chin. Chem. Lett.*, **13**, 689–692.
509 Ono, T., Konoma, Ch., Miyashita, H., Kanaori, Y. and Esashi, M. (2003) *Jpn. J. Appl. Phys. Part 1*, **42** (6B), 3867–3870.
510 Nielsch, K., Castaño, F.J., Ross, C.A. and Krishnan, R. (2005) *J. Appl. Phys.*, **98**, 034318/1-6.
511 Ding, G.Q., Shen, W.Z., Zheng, M.J. and Zhou, Z.B. (2006) *Nanotechnology*, **17**, 2590–2594.
512 Brändli, Ch., Jaramillo, T.F., Ivanovskaya, A. and McFarland, E.W. (2001) *Electrochim. Acta*, **47**, 553–557.
513 Ono, S., Takeda, K. and Masuko, N. (2000–2001) *ATB Metall.*, **40/41**, 398–403.
514 Sadasivan, V., Richter, C.P., Menon, L. and Williams, P.F. (2005) *AIChE J.*, **51**, 649–655.
515 Bocchetta, P., Sunseri, C., Chiavarotti, G. and Quarto, F.D. (2003) *Electrochim. Acta*, **48**, 3175–3183.
516 Wang, H. and Wang, H-W. (2004) *Trans. Nonferrous Met. Soc. China*, **14**, 166–169.
517 Shimizu, K., Alwitt, R.S. and Liu, Y. (2000) *J. Electrochim. Soc.*, **147**, 1388–1392.
518 Masuda, H. and Fukuda, K. (1995) *Science*, **268**, 1466–1468.
519 Ghorbani, M., Nasirpouri, F., Iraji zad, A. and Saedi, A. (2006) *Mater. Design*, **27**, 983–988.
520 Yu, W.H., Fei, G.T., Chen, X.M., Xue, F.H. and Xu, X.J. (2006) *Phys. Lett. A*, **350**, 392–395.
521 da Fontoura Costa, L., Riveros, G., Gómez, H., Cortes, A., Gilles, M., Dalchiele, E.A. and Marotti, R.E. (2005) *Cond-mat/0504573v*, 11–12.
522 Sui, Y.-Ch. and Saniger, J.M. (2001) *Mater. Lett.*, **48**, 127–136.
523 Xu, X.J., Fei, G.T., Zhu, L.Q. and Wang, X.W. (2006) *Mater. Lett.*, **60**, 2331–2334.
524 Rehn, L.E., Kestel, B.J., Baldo, P.M., Hiller, J., McCormick, A.W. and Birchter, R.C. (2003) *Nuc. Instrum. Meth. Phys. Res. B*, **206**, 490–494.
525 Wehrspohn, R.B., Li, A.P., Nielsch, K., Müller, F., Erfurth, W. and Gösele U. (2000) in *Oxide Films* (eds K.R. Hebert R.S. Lillard and B.R. MacDougall), PV-2000-4, Toronto, Canada, Spring, pp. 271–282.
526 Wehrspohn, R.B., Nielsch, K., Birner, A., Schilling, J., Müller, F., Li, A-P. and Gösele, U. (2001) in *Pits and Pores II: Formation Properties and Significance for Advanced Materials* (eds P. Schumki D.J. Lockwood, Y.H. Ogata and H.S. Isaacs), PV 2000-25, Phoenix, Arizona, Fall 2000, pp. 168–179.
527 Choi, D.H., Lee, P.S., Hwang, W., Lee, K.H. and Park, H.C. (2006) *Curr. Appl. Phys.*, **6S1**, e125–e129.
528 Xu, T., Zangari, G. and Metzger, R.M. (2002) *Nano Lett.*, **2**, 37–41.
529 Choi, J., Sauer, G., Nielsch, K., Wehrspohn, R.B. and Gösele, U. (2003) *Chem. Mater.*, **15**, 776–779.
530 Das, B. and Garman, Ch. (2006) *Microelectron. J.*, **37**, 695–699.
531 Tang, M., He, J., Zhou, J. and He, P. (2006) *Mater. Lett.*, **60**, 2098–2100.
532 Stein, N., Rommelfangen, M., Hody, V., Johann, L. and Lecuire, J.M. (2002) *Electrochim. Acta*, **47**, 1811–1817.

533 Brevnov, D.A., Rama Rao, G.V., López, G.P. and Atanassov, P.B. (2004) *Electrochim. Acta*, **49**, 2487–2494.

534 Lefeuvre, O., Pang, W., Zinin, P., Comins, J.D., Every, A.G., Briggs, G.A.D., Zeller, B.D. and Thompson, G.E. (1999) *Thin Solid Films*, **350**, 53–58.

535 Girginov, A., Popova, A., Kanazirski, I. and Zahariev, A. (2006) *Thin Solid Films*, **515**, 1548–1551.

536 Jagminas, A., Kurtinaitienė, M., Angelucci, R. and Valinčius, G. (2006) *Appl. Surf. Sci.*, **252**, 2360–2367.

537 Vrublevsky, I., Parkoun, V., Schreckenbach, J. and Goedel, W.A. (2006) *Appl. Surf. Sci.*, **252**, 5100–5108.

538 Vrublevsky, I., Parkoun, V., Sokol, V. and Schreckenbach, J. (2004) *Appl. Surf. Sci.*, **236**, 270–277.

539 Furneaux, R.C., Rigby, W.R. and Davidson, A.P. (1989) *Nature*, **337**, 147–149.

540 Nielsch, K., Müller, F., Li, A.P. and Gösele, U. (2000) *Adv. Mater.*, **12**, 582–586.

541 Girginov, A.A., Zahariev, A.S. and Klein, E. (2002) *J. Mater. Sci.: Mater. Electron.*, **13**, 543–548.

542 Zahariev, A. and Girginov, A. (2003) *Bull. Mater. Sci.*, **26**, 349–353.

543 Girginov, A., Kanazirski, I., Zahariev, A. and Popova, A. (2004) *Bull. Electrochem.*, **20**, 103–106.

544 Girginov, A., Zahariev, A., Kanazirski, I. and Tzvetkoff, T. (2004) *Bull. Electrochem.*, **20**, 405–407.

545 Girginov, A.A., Zahariev, A.S. and Machakova, M.S. (2002) *Mater. Chem. Phys.*, **76**, 274–278.

546 Liang, J., Chik, H. and Xu, J. (2002) *IEEE J. Selected Topics Quant. Electron.*, **8**, 998–1008.

547 Shimizu, T., Nagayanagi, M., Ishida, T., Sakata, O., Oku, T., Sakaue, H., Takahagi, T. and Shingubara, S. (2006) *Electrochem. Solid-State Lett.*, **9**, J13–J16.

548 Masuda, H., Yasui, K. and Nishio, K. (2000) *Adv. Mater.*, **12**, 1031–1033.

549 Park, S.K., Noh, J.S., Chin, W.B. and Sung, D.D. (2007) *Curr. Appl. Phys.*, **7**, 180–185.

550 Zhao, H., Jiang, Z., Zhang, Z., Zhai, R. and Bao, X. (2006) *Chinese J. Catalysis*, **27**, 381–385 (in Chinese).

551 Kordás, K., Tóth, G., Levoska, J., Huuhtanen, M., Keiski, R., Härkönen, M., George, T.F. and Vähäkangas, J. (2006) *Nanotechnology*, **17**, 1459–1463.

552 Pang, Y.T., Meng, G.W., Fang, Q. and Zhang, L.D. (2003) *Nanotechnology*, **14**, 20–24.

553 Xu, X.J., Fei, G.T., Wang, X.W., Jin, Z., Yu, W.H. and Zhang, L.D. (2007) *Mater. Lett.*, **61**, 19–22.

554 Sander, M.S. and Tan, L-S. (2003) *Adv. Funct. Mater.*, **13**, 393–397.

555 Pan, S.L., Zeng, D.D., Zhang, H.L. and Li, H.L. (2000) *Appl. Phys. A*, **70**, 637–640.

556 Evans, P., Hendren, W.R., Atkinson, R., Wurtz, G.A., Dickson, W., Zayats, A.V. and Pollard, R.J. (2006) *Nanotechnology*, **17**, 5746–5753.

557 Chaure, N.B., Stamenov, P., Rhen, F.M.F. and Coey, J.M.D. (2005) *J. Magn. Magn. Mater.*, **290–291**, 1210–1213.

558 Yasui, K., Morikawa, T., Nishio, K. and Masuda, H. (2005) *Jpn. J. Appl. Phys.*, **44**, L469–L471.

559 Xu, J. and Xu, Y. (2006) *Mater. Lett.*, **60**, 2069–2072.

560 Pang, Y.T., Meng, G.W., Zhang, Y., Fang, Q. and Zhang, L.D. (2003) *Appl. Phys. A*, **76**, 533–536.

561 Nielsch, K., Wehrespohn, R.B., Barthel, J., Kirschner, J., Gösele, U., Fisher, S.F. and Kronmüller, H. (2001) *Appl. Phys. Lett.*, **79**, 1360–1362.

562 Nielsch, K., Wehrspohn, R.B., Fischer, S.F., Kronmüller, H., Barthel, J., Kirschner, J., Schweinböck, T., Weiss, D. and Gösele, U. (2002) *Mat. Res. Soc. Symp. Proc.*, **705**, Y9.3.1–Y9.3.6.

563 Nielsch, K., Wehrspohn, R.B., Barthel, J., Kirschner, J., Fischer, S.F., Kronmüller, H., Schweinböck, T., Weiss, D. and Gösele, U. (2002) *J. Magn. Magn. Mater.*, **249**, 234–240.

564 Nielsch, K., Hertel, R., Wehrspohn, R.B., Barthel, J., Kirschner, J., Gösele, U.,

Fischer, S.F. and Kronmüller, H. (2002) *IEEE Trans. Magn.*, **38**, 2571–2573.

565 Sauer, G., Brehm, G., Schneider, S., Graener, H., Seifert, G., Nielsch, K., Choi, J., Göring, P., Gösele, U., Miclea, P. and Wehrspohn, R.B. (2006) *Appl. Phys. Lett.*, **88**, 023106/1-3.

566 Kumar, A., Fähler, S., Schlörb, H., Leistner, K. and Schultz, L. (2006) *Phys. Rev. B*, **73**, 064421/1-5.

567 Pang, Y.-T., Meng, G.-W., Zhang, L.-D., Qin, Y., Gao, X-Y., Zhao, A-W. and Fang, Q. (2002) *Adv. Funct. Mater.*, **12**, 719–722.

568 Kim, K., Kim, M. and Cho, S.M. (2006) *Mater. Chem. Phys.*, **96**, 278–282.

569 Vlad, A., Mátéfi-Tempfli, M., Faniel, S., Bayot, V., Melinte, S., Piraux, L. and Mátéfi-Tempfli, S. (2006) *Nanotechnology*, **17**, 4873–4876.

570 Tang, X-T., Wang, G-C. and Shima, M. (2006) *J. Appl. Phys.*, **99**, 033906/1-7.

571 Chaure, N.B. and Coey, J.M.D. (2006) *J. Magn. Magn. Mater.*, **303**, 232–236.

572 Gao, T.R., Yin, L.F., Tian, C.S., Lu, M., Sang, H. and Zhou, S.M. (2006) *J. Magn. Magn. Mater.*, **300**, 471–478.

573 Dahmane, Y., Cagnon, L., Voiron, J., Pairis, S., Bacia, M., Ortega, L., Benbrahim, N. and Kadri, A. (2006) *J. Appl. Phys. D: Appl. Phys.*, **39**, 4523–4528.

574 Fei, X.L., Tang, S.L., Wang, R.L., Su, H.L. and Du, Y.W. (2007) *Solid State Commun.*, **141**, 25–28.

575 Losic, D., Shapter, J.G., Mitchell, J.G. and Voelcker, N.H. (2005) *Nanotechnology*, **16**, 2275–2281.

576 Gu, X., Nie, Ch., Lai, Y. and Lin, Ch. (2006) *Mater. Chem. Phys.*, **96**, 217–222.

577 Ren, X., Huang, X-M. and Zhang, H-H. (2006) *Acta Phys-Chim. Sin.*, **22**, 102–105 (in Chinese).

578 Cao, H., Wang, L., Qiu, Y., Wu, Q., Wang, G., Zhang, L. and Liu, X. (2006) *ChemPhysChem.*, **7**, 1500–1504.

579 Lee, W., Scholz, R., Nielsch, K. and Gösele, U. (2005) *Angew. Chem. Int. Ed.*, **44**, 6050–6054.

580 Nielsch, K., Castaño, F.J., Matthias, S., Lee, W. and Ross, C.A. (2005) *Adv. Eng. Mater.*, **7**, 217–221.

581 Steinhart, M., Jia, Z., Schaper, A.K., Wehrspohn, R.B., Gösele, U. and Wendorff, J.H. (2003) *Adv. Mater.*, **15**, 706–709.

582 Steinhart, M., Wendorff, J.H. and Wehrspohn, R.B. (2003) *Chem. Phys. Chem.*, **4**, 1171–1176.

583 Steinhart, M., Wehrspohn, R.B., Gösele, U. and Wendorff, J.H. (2004) *Agnew. Chem. Int. Ed.*, **43**, 1334–1344.

584 Lee, M., Hong, S. and Kim, D. (2006) *Appl. Phys. Lett.*, **89**, 043120/1-3.

585 Lahav, M., Sehayek, T., Vaskevich, A. and Rubinstein, I. (2003) *Agnew. Chem. Int. Ed.*, **42**, 5576–5579.

586 Wang, W., Li, N., Li, X., Geng, W. and Qiu, S. (2006) *Mater. Res. Bull.*, **41**, 1417–1423.

587 Sung, D.D., Choo, M.S., Noh, J.S., Chin, W.B. and Yang, W.S. (2006) *Bull. Korean Chem. Soc.*, **27**, 1159–1163.

588 Li, Y., Meng, G.W., Zhang, L.D. and Phillipp, F. (2000) *Appl. Phys. Lett.*, **76**, 2011–2013.

589 Ko, E., Choi, J., Okamoto, K., Tak, Y. and Lee, J. (2006) *ChemPhysChem.*, **7**, 1505–1509.

590 Ding, G.Q., Shen, W.Z., Zheng, M.J. and Zhou, Z.B. (2006) *Appl. Phys. Lett.*, **89**, 063113/1-3.

591 Wu, C-T., Ko, F-H. and Hwang, H-Y. (2006) *Microelectron. Eng.*, **83**, 1567–1570.

592 Park, I.H., Lee, J.W. and Chung, C.W. (2006) *Integrated Ferroelectrics*, **78**, 245–253.

593 Fan, H.J., Lee, W., Scholz, R., Dadgar, A., Krost, A., Nielsch, K. and Zacharias, M. (2005) *Nanotechnology*, **16**, 913–917.

594 Masuda, H., Nishio, K. and Baba, N. (1992) *Jpn. J. Appl. Phys.*, **31**, L1775–L1777.

595 Nishio, K., Iwata, K. and Masuda, H. (2003) *Electrochem. Solid-State Lett.*, **6**, H21–H23.

596 Sander, M.S., Côte, M.J., Gu, W., Kile, B.M. and Tripp, C.P. (2004) *Adv. Mater.*, **16**, 2052–2057.

597 Min, H-S. and Lee, J-K. (2006) *Ferroelectics*, **336**, 231–235.
598 Alonso-González, P., Martín-González, M.S., Martín-Sánchez, J., González, Y. and González, L. (2006) *J. Cryst. Growth*, **294**, 168–173.
599 Wang, Y.D., Zang, K.Y. and Chua, S.J. (2006) *J. Appl. Phys.*, **100**, 054306/1-4.
600 Jung, M., Lee, H.S., Park, H.L., Lim, H-J. and Mho, S-I. (2006) *Curr. Appl. Phys.*, **6**, 1016–1019.
601 Jung, M., Lee, H.S., Park, H.L. and Mho, S-I. (2006) *Curr. Appl. Phys.*, **6S1**, e187–e191.
602 Wang, Y., Zang, K., Chua, S., Sander, M.S., Tripathy, S. and Fonstad, C.G. (2006) *J. Phys. Chem. B*, **110**, 11081–11087.
603 Konkonou, M., Nassiopoulou, A.G., Giannakopolous, K.P., Travlos, A., Stoica, T. and Kennou, S. (2006) *Nanotechnology*, **17**, 2146–2151.
604 Ding, G.Q., Zheng, M.J., Xu, W.L. and Shen, W.Z. (2006) *Thin Solid Films*, **508**, 182–185.
605 Nakao, M., Oku, S., Tamamura, T., Yasui, K. and Masuda, H. (1999) *Jpn. J. Appl. Phys.*, **38**, 1052–1055.
606 Sai, H., Fujii, H., Arafune, K., Ohshita, Y., Yamaguchi, M., Kanamori, Y. and Yugami, H. (2006) *Appl. Phys. Lett.*, **88**, 201116/, 1–3.
607 Wu, C., Shi, J-B., Chen, C-J. and Lin, J-Y. (2006) *Mater. Lett.*, **60**, 3618–3621.
608 Aguilera, A., Jayaraman, V., Sanagapalli, S., Singh, R.S., Jayaraman, V., Sampson, K. and Singh, V.P. (2006) *Solar Energy Mater. Solar Cells*, **90**, 713–726.
609 Yang, W., Wu, Z., Lu, Z., Yang, X. and Song, L. (2006) *Microelectron. Eng.*, **83**, 1971–1974.
610 Varfolomeev, A., Zaretsky, D., Pokalyakin, V., Tereshin, S., Pramanik, S. and Bandyopadhyay, S. (2006) *Appl. Phys. Lett.*, **88**, 113114/1-3.
611 Zhang, X.Y., Xu, L.H., Dai, J.Y., Cai, Y. and Wang, N. (2006) *Mater. Res. Bull.*, **41**, 1729–1734.
612 Zhang, J., Xu, B., Jiang, F., Yang, Y. and Li, J. (2005) *Phys. Lett. A*, **337**, 121–126.
613 Zhu, X., Ma, J., Wang, Y., Tao, J., Zhou, J., Zhao, Z., Xie, L. and Tian, H. (2006) *Mater. Res. Bull.*, **41**, 1584–1588.
614 Yan, H., Zhang, L., Shen, J., Chen, Z., Shi, G. and Zhang, B. (2006) *Nanotechnology*, **17**, 3446–3450.
615 O'Brien, G.A., Quinn, A.J., Iacopino, D., Pauget, N. and Redmond, G. (2006) *J. Mater. Chem.*, **16**, 3237–3241.
616 Shin, H-W., Cho, S.Y., Choi, K-H., Oh, S-L. and Kim, Y-R. (2006) *Appl. Phys. Lett.*, **88**, 263112/1-3.
617 Zhao, L., Yang, W., Ma, Y., Yao, J., Li, Y. and Liu, H. (2003) *Chem. Commun.*, 2442–2443.
618 Yang, S.M., Chen, K.H. and Yang, Y.F. (2005) *Synth. Metals*, **152**, 65–68.
619 Steinhart, M., Wendorff, J.H., Greiner, A., Wehrspohn, R.B., Nielsch, K., Schilling, J., Choi, J. and Gösele, U. (2002) *Science*, **296**, 1997.
620 Niu, Z-W., Li, D. and Yang, Z-Z. (2003) *Chin. J. Polymer Sci.*, **21**, 381–384.
621 Schneider, J.J. and Engstler, J. (2006) *Eur. J. Inorg. Chem.*, 1723–1736.
622 Sun, L., Dai, J., Baker, G.L. and Bruening, M.L. (2006) *Chem. Mater.*, **18**, 4033–4039.
623 Nishio, K., Nakao, M., Yokoo, A. and Masuda, H. (2003) *Jpn. J. Appl. Phys. Part 2*, **42** (1A/B), L83–L85.
624 Yanagishita, T., Nishio, K. and Masuda, H. (2005) *Adv. Mater.*, **17**, 2241–2243.
625 Al-Kaysi, R.O. and Bardeen, Ch.J. (2006) *Chem. Commun.*, 1224–1226.
626 Ravindran, S., Andavan, G.T.S., Tsai, C., Ozkan, C.S. and Hollis, T.K. (2006) *Chem. Commun.*, 1616–1618.
627 Wang, Y., Ye, Ch., Wang, G., Zhang, L., Liu, Y. and Zhao, Z. (2003) *Appl. Phys. Lett.*, **82**, 4253–4255.
628 Piao, Y., Lim, H., Chang, J.Y., Lee, W-Y. and Kim, H. (2005) *Electrochim. Acta*, **50**, 2997–3013.
629 Johansson, A., Widenkvist, E., Lu, J., Boman, M. and Jansson, U. (2005) *Nano Lett.*, **5**, 1603–1606.
630 Wang, W., Wang, S-Y., Gao, Y-L., Wang, K-Y. and Liu, M. (2006) *Mater. Sci. Eng. B*, **133**, 167–171.

631 Kyotani, T., Tsai, L-fu. and Tomita, A. (1996) *Chem. Mater.*, **8**, 2109–2113.

632 Zhang, X.Y., Zhang, L.D., Zheng, M.J., Li, G.H. and Zhao, L.X. (2001) *J. Crystal Growth*, **223**, 306–310.

633 Sui, Y.C., Cui, B.Z., Guardían, R., Acosta, D.R., Martiínez, L. and Perez, R. (2002) *Carbon*, **40**, 1011–1016.

634 Chen, Q-L., Xue, K-H., Shen, W., Tao, F-F., Yin, S-Y. and Xu, W. (2004) *Electrochim. Acta*, **49**, 4157–4161.

635 Yin, A., Tzolov, M., Cardimona, D.A. and Xu, J. (2006) *IEEE Trans. Nanotechnol.*, **5**, 564–567.

636 Wen, S., Jung, M., Joo, O-S. and Mho, S-i. (2006) *Curr. Appl. Phys.*, **6**, 1012–1015.

637 Yu, K., Ruan, G., Ben, Y. and Zou, J.J. (2007) *Mater. Lett.*, **61**, 97–100.

638 Yanagishita, T., Sasaki, M., Nishio, K. and Masuda, H. (2004) *Adv. Mater.*, **16**, 429–432.

639 Masuda, H., Ohya, M., Asoh, H., Nakao, M., Nohtomi, M. and Tamamura, T. (1999) *Jpn. J. Appl. Phys.*, **38**, L1403–L1405.

640 Masuda, H., Ohya, M., Nishio, K., Asoh, H., Nakao, M., Nohtomi, M., Yakoo, A. and Tamamura, T. (2000) *Jpn. J. Appl. Phys.*, **39**, L1039–L1041.

641 Masuda, H., Ohya, M., Asoh, H. and Nishio, K. (2001) *Jpn. J. Appl. Phys.*, **40**, L1217–L1219.

642 Masuda, H., Yamada, M., Matsumoto, F., Yakoyama, S., Mashiko, S., Nakao, M. and Nishio, K. (2006) *Adv. Mater.*, **18**, 213–216.

643 Wehrspohn, R.B. and Schilling, J. (2001) *MRS Bullet.*, August, 623–626.

644 Mikulskas, I., Juodkazis, S., Tomašiunas, R. and Dumas, J.G. (2001) *Adv. Mater.*, **13**, 1574–1577.

645 Choi, J., Schilling, j., Nielsch, K., Hillebrand, R., Reiche, M., Wehrspohn, R.B. and Gösele, U. (2002) *Mat. Res. Soc. Symp. Proc.*, **722**, L5.2.1–L.5.2.6.

646 Choi, J., Luo, Y., Wehrspohn, R.B., Hillebrand, R., Schilling, J. and Gösele, U. (2003) *J. Appl. Phys.*, **94**, 4757–4762.

647 Yanagishita, T., Nishio, K. and Masuda, H. (2006) *Jpn. J. Appl. Phys.*, **45**, L804–L806.

648 Vázquez, M., Hernández-Vélez, M., asenjo, A., Navas, D., Pirota, K., Prida, V., Sánchez, O. and Badonedo, J.L. (2006) *Physica B*, **384**, 36–40.

649 Masuda, H., Hogi, H., Nishio, K. and Matsumoto, F. (2004) *Chem. Lett.*, **33**, 812–813.

650 Lee, W., Alexe, M., Nielsch, K. and Gösele, U. (2005) *Chem. Mater.*, **17**, 3325–3327.

651 Masuda, H., Matsumoto, F. and Nishio, K. (2004) *Electrochemistry*, **72**, 389–394.

652 Masuda, H., Watanabe, M., Yasui, K., Tryk, D., Rao, T. and Fujishima, A. (2000) *Adv. Mater.*, **12**, 444–447.

653 Masuda, H., Yasui, K., Watanabe, H.M., Nishio, K., Nakao, M., Tamamura, T., Rao, T.N. and Fujishima, A. (2001) *Electrochem. Solid State Lett.*, **4**, G101–G103.

654 Matsumoto, F., Nishio, K. and Masuda, H. (2004) *Adv. Mater.*, **16**, 2105–2108.

655 Matsumoto, F., Nishio, K., Miyasaka, T. and Masuda, H. (2004) *Jpn. J. Appl. Phys.*, **43** (5A), L640–L643.

656 Matsumoto, F., Kamiyama, M., Nishio, K. and Masuda, H. (2005) *Jpn. J. Appl. Phys.*, **44**, L355–L358.

657 Grasso, V., Lambertini, V., Ghisellini, P., Valerio, F., Stura, E., Perlo, P. and Nicolini, C. (2006) *Nanotechnology*, **17**, 795–798.

658 Myler, S., Collyer, S.D., Bridge, K.A. and Higson, S.P.J. (2002) *Biosens. Bioelectron.*, **17**, 35–43.

659 Walpole, A.R., Briggs, E.P., Karlsson, M., Pålsgård, E. and Wilshaw, P.R. (2003) *Mat-wiss. u. Werkstofftech.*, **34**, 1064–1068.

2
Nanostructured Materials Synthesized Using Electrochemical Techniques

Cristiane P. Oliveira, Renato G. Freitas, Luiz H.C. Mattoso, and Ernesto C. Pereira

2.1
Introduction

The visualization and/or handling of atom clusters, or even individual atoms, has been achieved during the past 25 years. One of the most practical instruments for this purpose is the tunneling microscope, which was invented during the 1980s. On such a small scale, material's properties are not the same as those of a macroscopic-sized substance. The first occasion on which a scientist proposed atom handling was in 1959, when Professor Richard Feynman [1], during a lecture at the California Institute Technology, suggested that in the near future engineers would be able to take atoms and place them exactly where they wanted to, without – of course – infringing the laws of Nature. Today, Professor Feynman's lecture, which was entitled *There is Plenty of Room at the Bottom*, is considered a milestone of the current technological and scientific era.

We can define nanoscience as the study of the phenomena and handling of structures on the atomic, molecular and macromolecular scale, where at least one dimension is significantly less than the others. And nanotechnology includes the design, characterization, and production of devices and systems on a nanometer scale. This definition of nanotechnology is wide-ranging, it is not a specific technology, all of the techniques based on physics, chemistry, biology, materials science and engineering – and/or using computers – leading to the development of materials and tools in such a way that the common point among them is the reduced dimension in which they operate. Among these applications are: increased storage and processing capacity for computers; the creation of new mechanisms for drug delivery; and the production of materials which are both lighter and more resistant than metals and plastics. Additional benefits of nanotechnology developments include energy saving, environmental protection, and the more efficient use of increasingly scarce raw materials.

Today, these new investigations into material fabrication are focused mainly on nanostructured devices or, at least, microdevices. The most-often used technique for

Nanostructured Materials in Electrochemistry. Edited by Ali Eftekhari
Copyright © 2008 WILEY-VCH Verlag GmbH & Co. KGaA, Weinheim
ISBN: 978-3-527-31876-6

this purpose is lithography which, despite its associated high costs, is used extensively for the large-scale production of these materials. In fact, it is mainly due to high costs that many research groups worldwide are seeking alternative, new routes of preparation. In this regard, electrochemical tools are one such approach which has emerged and which benefits from both low cost and accuracy. The low cost is related to the use of near-room temperature conditions and aqueous solutions (in most cases), while high accuracy is related to an ability to control the potential or current density on an interfacial region extending over only a few nanometers. In fact, the electric potentials used in an electrochemical procedure range between tenths of volts and hundreds of volts. Otherwise, the electric field, considering the small size of the electrical double layer, ranges from about 10^5 to 10^7 V cm^{-1}. Furthermore, the simplest relationship between potential and current density which flows through the electrode/solution interface is exponential in nature. Today, opinion suggests that the main problem relating to electrochemical tools is the nature of the chemical species in solution which can be incorporated during material growth on electrode surface. Yet, the grain size and shape, crystal structure, doping level – and even the ions' oxidation state in compounds – can at present be controlled.

During the past 15 years, several groups investigating electrochemistry have moved towards a materials science approach, having previously developed electrochemical methods for synthesizing electronic materials such as semiconductors [2–9], metal oxides [10–12], metal nitrides [13,14], porous silicon [15,16], metal, alloy and multilayers, and a variety of layered composites [17–19]. As a consequence of these successes, a new route has emerged for the preparation of nanostructured materials.

Based on these developments, the aim of this chapter is to review the materials prepared using electrochemical methods, and for this purpose it is divided into two principal parts, namely anodic and cathodic approaches. In fact, by using both methods it is possible to prepare nanostructured materials such as metals (and their alloys), oxides, semiconductors, composites and compounds with different shapes, including layered, dots, wells and wires, all of which have become very important in the development of nanostructured materials.

The electrochemical processes employed in nanotechnology can be subdivided as anodic, cathodic, and open-circuit processes [20]. In general, the latter approach is treated as a chemical process, despite involving both cathodic and anodic reactions which follow the electrochemical laws. The first section of the chapter includes details about the first two processes, in addition to an outline of anodic methods. Electropolishing and anodization are among the most important anodic processes; the former approach deals with the material anodic dissolution which leads not only to a leveling and brightening of the dissolving surface but is also used for precision shaping and surface structuring [20], which may in fact be considered a form of electrochemical machining. Otherwise, anodization consists of the growth of an oxide film by anodic polarization. Within this chapter, an example of an electropolishing application will be presented followed by the formation of nanostructured oxide films by anodization. The most well-known oxide of the anodic film type – porous anodic alumina (PAA) – will be described in a separate subsection restricted to the presentation of nanostructures fabricated by this means.

The second section of this chapter is dedicated to the cathodic methods which represent major means of preparing nanostructured systems, including one-dimensional (1D) or even two-dimensional (2D) structures such as nanowires, nanotubes and monolayers. Further discussion will center on template procedures to prepare nanowires and their different compositions, shapes, and magnetic properties. Regarding the formation of the well-ordered 2D structure, discussion will also focus on the mechanical, magnetic and electrocatalytic properties of metallic multilayers and superlattices prepared using electroplating. Finally, underpotential deposition (UPD) formation and deposition at electrochemical interfaces using tips as preparation tools will be outlined.

It should be noted that organic materials such as nanostructured polymers, although, in most cases, are prepared using the template approach, will not be discussed in this chapter, mainly because these materials constitute a chapter in their own right.

2.2
Anodic Synthesis

2.2.1
Electropolishing and Anodization

Electrochemical machining involves the selective anodic dissolution of a metal (or alloy) through a mask in order to achieve a given shape and surface finish [20]. Aluminum electropolishing, otherwise, occurs by selective dissolution, under specific conditions without the use of any form of mask. One common procedure before anodizing the aluminum is the use of electropolishing to level the aluminum surface. During this process, an unexpected effect is the formation of patterns of in hexagonal nanoarrays. Bandyopadhyay et al. [21] investigated such pattern formation during aluminum electropolishing, and found out that the process may lead to spatial ordering depending on the duration and applied voltage. Figure 2.1 shows micrographs of the stripes (Figure 2.1a) and "egg-carton" (Figure 2.1b) patterns formed on the aluminum surface after electropolishing under two different conditions. The period of the patterns is about 100 nm, and the height of a crest above a trough is about 3 nm. The authors described how to use these structures as self-assembled masks for producing highly periodic arrays of quantum wires and dots, and this routine is depicted schematically in Figure 2.2. Here, a thin aluminum film is evaporated onto a given substrate, after which electropolishing is carried out to produce the desired surface morphology. After electropolishing, any troughs are selectively etched away in a suitable solution, and finally the metal dots are obtained in a periodic pattern on the substrate surface.

Such structures have application in nanoelectronic computing architectures using single effects in quantum dots [22]. This simple process, which is used widely in the anodizing industry, was subsequently promoted to a "high technology" for the mass fabrication of dense, 2D quasi-periodic arrays of metallic, semi-conducting, or super-conducting quantum dots.

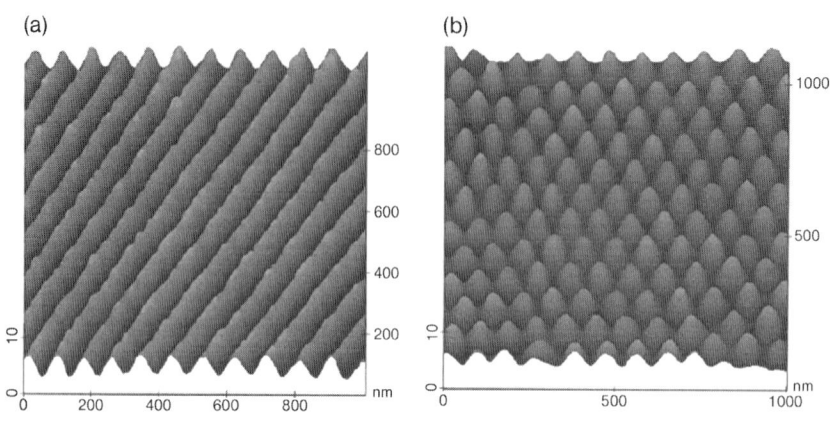

Figure 2.1 Atomic force micrograph of stripes and "egg-carton" patterns formed on an aluminum surface after electropolishing. Stripes form after electropolishing at 50 V for 10 s; egg-cartons form after electropolishing at 60 V for 30 s. (Reprinted from Ref. [21].)

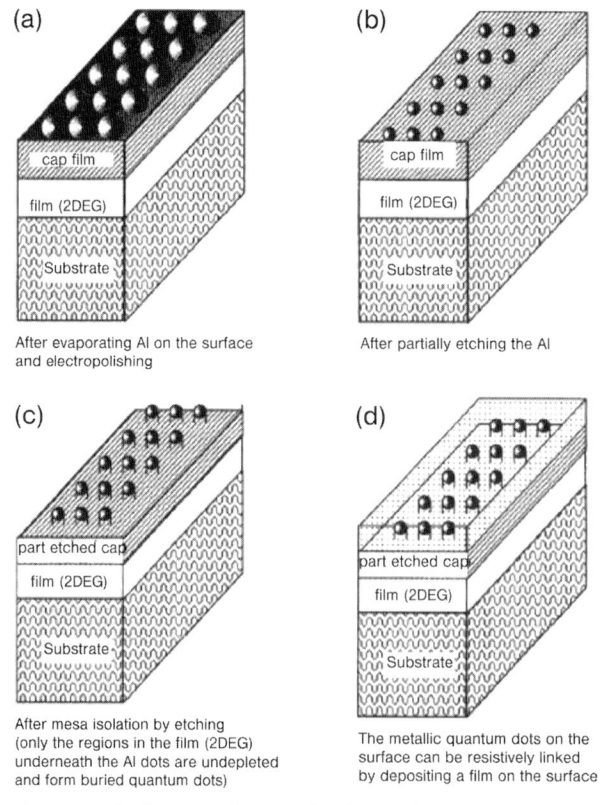

(a) After evaporating Al on the surface and electropolishing

(b) After partially etching the Al

(c) After mesa isolation by etching (only the regions in the film (2DEG) underneath the Al dots are undepleted and form buried quantum dots)

(d) The metallic quantum dots on the surface can be resistively linked by depositing a film on the surface

Figure 2.2 The four steps for using the electropolished aluminum as a self-assembled mask to create a periodic array of electrically linked quantum dots on a surface. (Reprinted from Ref. [21].)

The interest in anodic polarization of valve metals dates back about 60 years, due to the practical importance of anodic oxides, such as capacitor and surface protection applications. These anodic oxides are characterized by a delay in the dissolution reaction of the metal. Recently, different research groups [23–28] have recommenced investigations into valve metal oxidation, mainly because some anodic oxides have unique and excellent properties for optical, electronic, photochemical and biological applications. Among these anodic oxides, titanium oxide is a promising material for photoelectrocatalysis [29] and implants [30]. The latter application is due to the superior mechanical strength and biocompatibility, and has led to the commercial production of implants fabricated from pure titanium or titanium-based alloys [31]. Titanium anodization is used to prepare an interfacial layer (a porous layer) on the metal surface, which has the function of osseointegrating with bones [32–35]. The formation of this layer is important because the native titanium oxide coatings (the oxide formed naturally in air) do not have any ability to form a strong bond with bony tissue; consequently, the surfaces of titanium implants must be covered with a thick oxide layer in order to promote osseointregration.

Aluminum is the most frequently used metal for the fabrication of oxides with ordered pore arrays, although some issues regarding the possibility of preparing these nanostructured arrays during anodization of other valve metals such as Ti, Ta, Nb, V, Hf or W have also been proposed. During the past few years, many different research groups have begun to study the electrochemical anodization of other valve metals [36,37] and bilayers formed by two valve metals [38,39] aimed at this purpose. For example, Choi et al. [36] investigated the preparation of a porous titania (titanium oxide) in different electrolytes. In most reports [35–37,39], the authors used an anodizing solution containing fluoride ions. However, Martelli et al. [40], during the investigation of the effects of the preparation parameters (i.e., electrolyte concentration, anodization temperature and current density) on the surface morphology of the titanium oxide films, reported the preparation of porous titanium oxide without any addition of fluoride ions to the anodization bath. A scanning electron microscopy (SEM) image of porous titanium oxide prepared by galvanostatic anodization is shown in Figure 2.3.

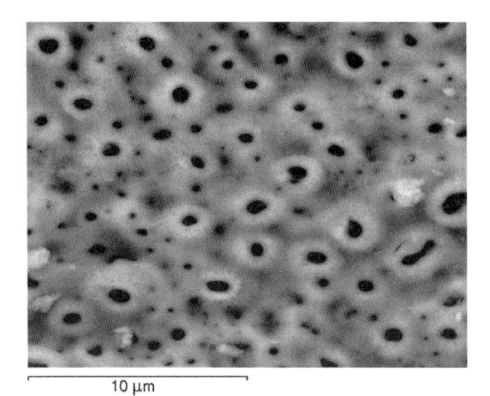

Figure 2.3 SEM image of porous titanium oxide prepared in 1 mol dm^{-3} H$_3$PO$_4$, 15 mA·cm^{-2}, 15°C.

Figure 2.4 SEM image of a titanium oxide cross-section. The nanotubular structure was prepared by anodization in 0.5 mol dm^{-3} H$_3$PO$_4$ + 0.14 mol dm^{-3} HF solution at 20 V. (Reprinted from Ref. [35].)

These authors [40] observed that a large porosity, with pore size ranging from 100 to 700 nm, was obtained during anodization using a low current density and temperature (5°C).

Recently, Raja et al. [35] reported details about the self-ordered growth of nanostructured titanium oxide by anodization. In this report, the authors' intention was to clarify the role of fluoride ions during the formation of titanium oxide nanotubes. A SEM image of the cross-section of this structure formed at 20 V in 0.5 mol dm^{-3} H$_3$PO$_4$ + 0.14 mol dm^{-3} HF solution, is shown in Figure 2.4, which clearly shows the presence of the scalloped barrier film.

Besides the porous titanium oxide (Figure 2.5a) presenting a random pore distribution, similar to that reported by Martelli et al. [40], Choi et al. [36] reported the preparation of monodomain porous titanium oxide (Figure 2.5b) by the nanoimprint of titanium and successive anodization below the breakdown potential. This report [36] was the first to investigate the possibility of preparing ordered structures from different valve metals using nanoimprinting methods.

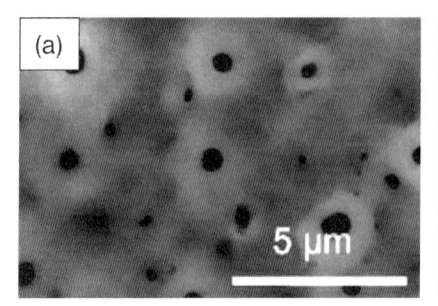

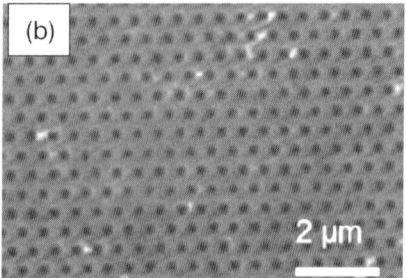

Figure 2.5 (a) SEM image of titanium anodized in 1 mol dm^{-3} phosphoric acid at 210 V for 210 min at 60°C. (b) SEM image of anodization of the nanoimprinted titanium, having a lattice constant of 500 nm at 10 V in ethanolic HF for 240 min. (Reprinted from Ref. [36].)

Cheng et al. [41] reported a surface processing to improve the biocompatibility of titanium. This involved a cathodic pretreatment in acidic solution (1 mol dm^{-3} H$_2$SO$_4$ solution) followed by anodization in alkaline media (5 mol dm^{-3} NaOH solution). In this report [41], the authors demonstrated the presence of nanoparticles of TiH$_2$ phase on the titanium, which was critical to the formation of a multinanoporous TiO$_2$ layer and, consequently, to the improvement in the bioactive performance over that layer. The authors observed that a multinanoporous layer and thicker titanium oxide layer was formed during the anodization performed after the cathodic pretreatment. These features are considered to be responsible for the improvement of the implant biocompatibility [42].

Recently, the growth of nanoscale hydroxyapatite phase using chemically treated titanium oxide nanotubes prepared by anodization [33] was reported. For example, Oh et al. [33] studied the effects of a vertically aligned titanium oxide nanotubes on in-vitro hydroxyapatite by investigating the formation kinetics and morphology, their aim being the use of such nanotubes for bone growth. These authors used a chemical treatment of the titanium oxide nanotubes produced in NaOH solution to explore the possibility of enhancing its bioactivity. Such chemical treatment led to the formation of sodium titanate, which presents an extremely nanofiber-like structure on top of the TiO$_2$ nanotubes. The material formed was exposed to a simulated body solution in order to investigate hydroxyapatite growth [33]. A micrograph of the titanium surface with nanotubes after the anodization process is shown in Figure 2.6. The formation of hydroxyapatite is significantly accelerated compared to the pure TiO$_2$ nanotube surface. In the latter case, a 7-day period was needed for the formation of a detectable quantity of hydroxyapatite, compared to 1 day for the TiO$_2$-NaTiO$_3$ nanofibers.

Recently, photoreactive materials have received much attention due to their inherent photochemical properties. Titanium oxide is a particularly versatile semiconductor with a variety of applications in photocatalysis, as gas sensors, photovoltaic cells, optical coatings, structural ceramics, and biocompatible materials [43,44]. It has also been confirmed that the titania morphology modifies the physico-chemical properties [45]. Both, sol–gel and hydrothermal synthesis have been developed to produce various nanostructured titania powders with high photoactivity [46–48].

Figure 2.6 SEM image of titanium oxide nanotube after 30 min of anodization. (Reprinted from Ref. [33].)

These films showed a limit in their photoelectrocatalytic activity due to a low electron transfer rate between the TiO_2 film and its supporting material. An alternative route for preparing such materials is to use electrochemical methods. Thus, titanium oxide prepared electrochemically as a photoelectrode presents several advantages over other titanium oxide, due to an efficient charge transfer between the oxide and the metal [49]. Moreover, the thickness and morphology of these materials can be controlled by adjusting the electrolyte composition and the applied current density or potential, under galvanostatic or potentiostatic conditions, respectively [40]. Xie et al. [49] prepared two types of electrochemical titanium oxide films: (i) the microstructured TiO_2 formed in H_2SO_4-H_2O_2-H_3PO_4-HF solution after 2 h of anodization; and (ii) the nanostructured TiO_2 thin films prepared in H_3PO_4-HF solution for 30 min followed by a calcination process at high temperature. The photocatalytic performance of the nanostructured films used for bisphenol degradation was more efficient than that of the microstructured, thick-film electrode.

Today, superimposed metallic layers of valve metals (a valve metal substrate covered with a thin, continuous, deposited layer of a different metal) are the subject of intense investigation in the field of anodization. The anodization of superimposed metallic layers represents an alternative approach to the study of ionic transport processes in the anodic oxide growth [50–53]. It should be pointed out that important information on ionic transport in amorphous and crystalline oxides was obtained using the Al–Mo, Al–Ta, Al–Nb, and Al–Zr couples as tracers for studying cation migration during the formation of barrier-type alumina [54–56]. In addition, other groups using the same approach [37–39,54,57] showed that these systems are very complex due to the ionic current transport resistivity through the anodic oxides of the deposited metallic layer and substrate metal. New superimposed metallic layers were investigated, in which both valve metals are sputter-deposited on a dielectric substrate, using mainly aluminum as the overlying metal over hafnium [58], niobium [59] or tantalum [60]. Recently, much technological interest has been expressed in these new types of superimposed metallic layers, also referred to as "double-layered anodic oxides", for both micro- and nanoelectronics, such as in planarized aluminum interconnection [60], precision thin-film resistors [38], thin-film capacitors [61], and nanostructured field-emission cathodes [62]. It has also been reported that a nanostructured anodic oxide film can be formed from the underlying metal through PAA [38]. Such a film is composed of column arrays penetrating the pores in the initially formed upper alumina. When niobium is the underlying metal, structures called "goblets" are formed in the alumina pore [39]. A schematic diagram of the general step for the formation process of nanostructured tantalum anodic oxide under porous alumina is shown in Figure 2.7.

Mozalev et al. [39] reported a systematic investigation on the anodization of Ta–Al and Nb–Al bilayers under conditions where a porous anodic alumina layer is formed. The formation of these structures was carried out under galvanostatic condition until the anodizing voltage reached a steady-state value. At this stage, the anodization was switched to potentiostatic mode. The structures formed using this procedure are presented in Figures 2.8 and 2.9.

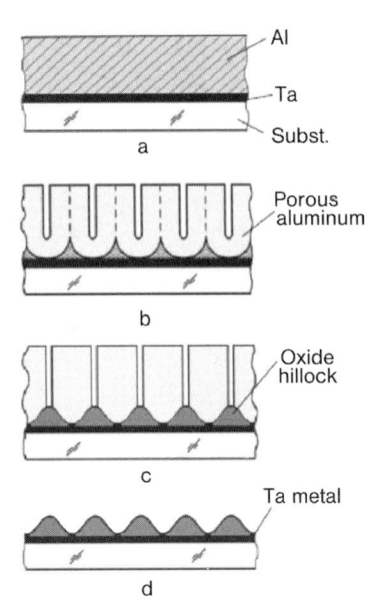

Figure 2.7 Diagrammatic representation of the nanostructured tantalum anodic oxide under porous alumina formation. (a) Sputter-deposition of an Al on Ta system; (b) porous anodizing of the aluminum layer; (c) anodizing of the tantalum underlayer through the alumina pores; (d) selective removal of the overlying porous alumina. (Reprinted from Ref. [39].)

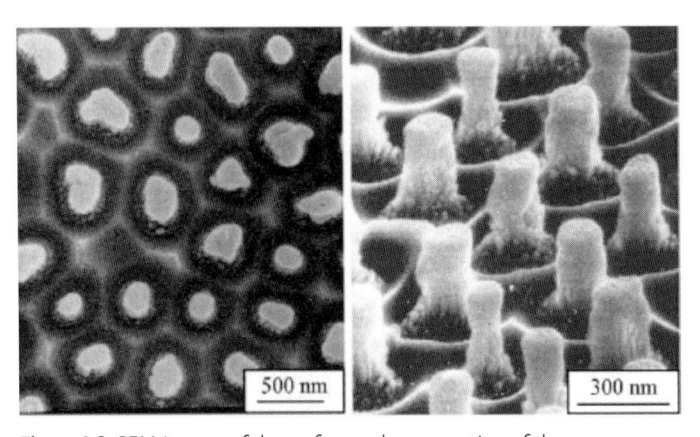

Figure 2.8 SEM images of the surface and cross-section of the nanostructured tantalum oxide formed from Ta–Al bilayers anodized in 0.8 mol dm^{-3} tartaric acid at 200 V. (Reprinted from Ref. [39].)

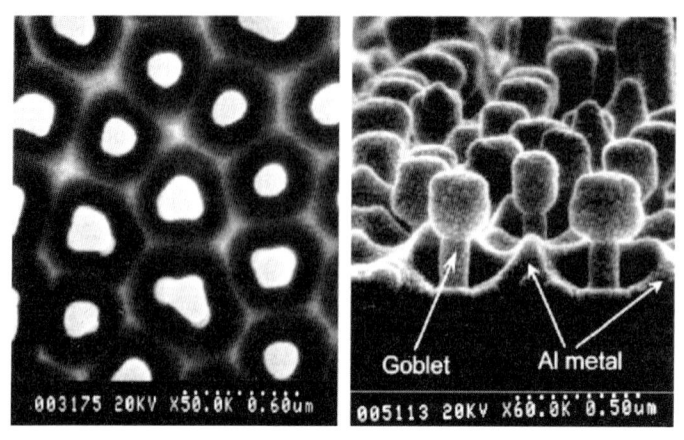

Figure 2.9 SEM images of the surface and cross-section of the Nb–Al bilayer which were anodized in 0.8 mol dm^{-3} tartaric acid at 200 V. After anodization, samples were dipped in selective etching solution at 60 °C for 60 min to remove overlying alumina. (Reprinted from Ref. [39].)

The same authors [39] proposed that, during the anodization, the transport processes involve the outward migration of Ta^{5+} and the inward migration of O^{2-} ions. Moreover, at the tantala (tantalum oxide)/alumina (aluminum oxide) interface, the tantalum oxide is presumed to develop through injection of O^{2-} ions which are released from dissociating Al–O bonds in the alumina barrier layer due to the high electric field. Thus, the underlying tantalum is oxidized by oxygen ions transported through/from the barrier layer of the initially formed porous alumina – that is, without direct contact of tantalum with the electrolyte. With regards to the exotic shape of the nanostructures formed during anodization of the Nb–Al bilayers (goblet-like shape), these authors suggested that this might be explained by the sharp difference in resistivity between the outer and inner parts of the alumina cells, and a relatively large difference in the resistivity in niobia–alumina couple compared with tantala–alumina couple [39].

In a previous report [38], the authors described the anodization of a Ta–Al bilayer to produce nanoscale metal-oxide coatings, and investigated their potential use as an alternative material for microfilm resistor fabrication. The interest in this system is related to the fact that the major limitation of thin-film hybrid technology is in the fabrication of accurate and large-value microfilm resistors [63,64]. As can be observed in Figure 2.8, the coatings comprise regularly arranged nanoscale dielectric bulges, surrounded by unanodized metallic tantalum gaps, which are self-integrated in an ultra-thin continuous conductive grid. This nanoscale topography resulted in large-value sheet resistance, which varied from 700 Ω to 75 kΩ, as well as a low-temperature resistance coefficient (10^{-5} K^{-1}). As noted by Liang *et al.* [65], much research effort has been concentrated on individual nanostructures, and almost none of the groups investigated the collective behavior of the ensemble of these materials. In this sense, such as electrochemical anodizing technique

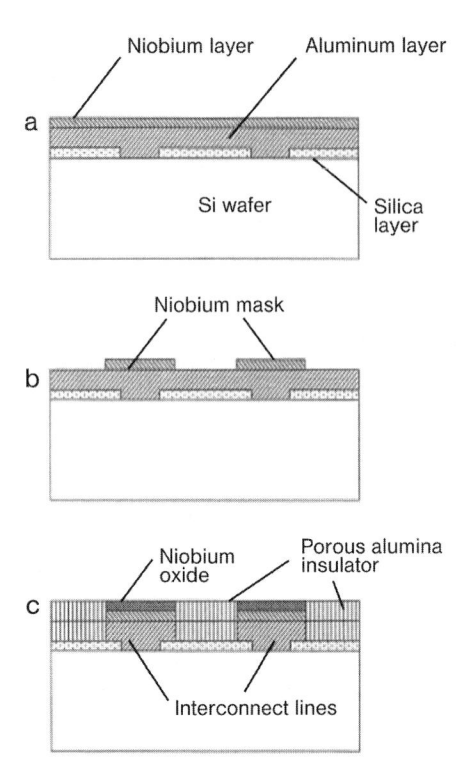

Figure 2.10 Schematic view of the main technological process step of a device preparation. (a) Deposition of a bilayer system of aluminum–niobium onto the silicon substrate; (b) formation of a photoresist mask, plasma etching of the upper niobium layer, removing the photoresist mask; (c) electrochemical anodization of the aluminum layer and formation of a porous alumina film, chemical etching of the porous alumina film. (Reprinted from Ref. [67].)

represents an alternative routine for fabrication, which is fully compatible with the existing integrated circuit (IC) technologies, which can also be used for integration of the nanostructured metal-oxide films with other IC components on a single substrate [66].

In order to enhance the reliability of the silicon IC devices, Lazarouk et al. [67] used an electrochemical anodization technology to prepare a low-constant dielectric material based on porous alumina. This approach could minimize the interconnection problems that are the most severe limiting factors in ICs as their dimensions decrease. The route proposed allows the formation of an intralayer alumina insulator used in the multilevel aluminum metallization. A schematic representation of the main technological process steps is shown in Figure 2.10, while SEM images of the porous alumina surface with built-in aluminum lines after anodization are shown in Figure 2.11. An enlarged view of the multilevel aluminum interconnection with the porous alumina insulator is also shown in Figure 2.11. In order to control the porosity and dielectric constant, the anodization was carried out under

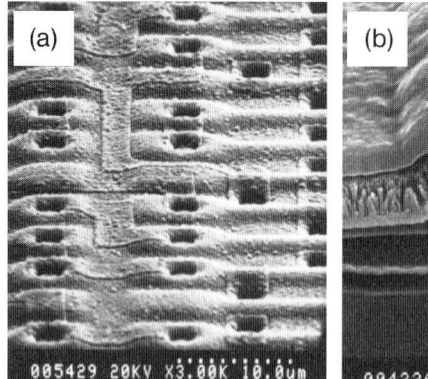

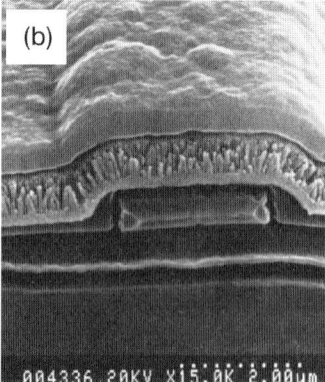

Figure 2.11 (a) SEM image of the porous alumina surface with built-in aluminum lines. (b) Cross-sectional SEM image of the multilevel aluminum interconnections with the porous alumina insulator. (Reprinted from Ref. [67].)

special conditions using a low forming voltage (V), and in a H_2SO_4 and H_3PO_4 mixed solution. The use of the mixed solution during PAA preparation decreases the oxide porosity. This is important because films with high porosity cannot be used as reliable intralayer insulators, since this parameter depends on the chemical and thermal stability of the insulator layer, with a high porosity would lead to a decrease in thermal conductivity. These authors concluded that the reliability of built-in aluminum interconnections with low-ε porous alumina fulfill all the requirements of advanced IC technologies.

2.2.2
Porous Anodic Alumina

Aluminum oxide growth has been studied since the 1930s [68]. Due to the chemical stability of its passive layer, the anodic polarization of aluminum was investigated intensively in order to obtain protective and decorative materials [68]. Subsequently, anodic aluminum oxide was used traditionally as a corrosion-resistant material. In 1953, the porous structure of aluminum oxide was described by Keller and coworkers as a hexagonally close-packed duplex structure consisting of porous and barrier layers [69].

In 1995, Masuda *et al.* [70] presented the so called *two-step anodization method* for the fabrication of honeycomb alumina. This technique allowed the preparation of a highly ordered porous structure with a hexagonal distribution, and since then the main emphasis in this system has been to control the geometry and regularity of pore distribution. Designs of new pore shapes have rarely been explored due to the high regularity attained with the two-step anodization method. The complexity of shape control on the nanometer scale also leads to difficulties in obtaining such structures. As it has been well established [69,70], the pore shape in PAA has,

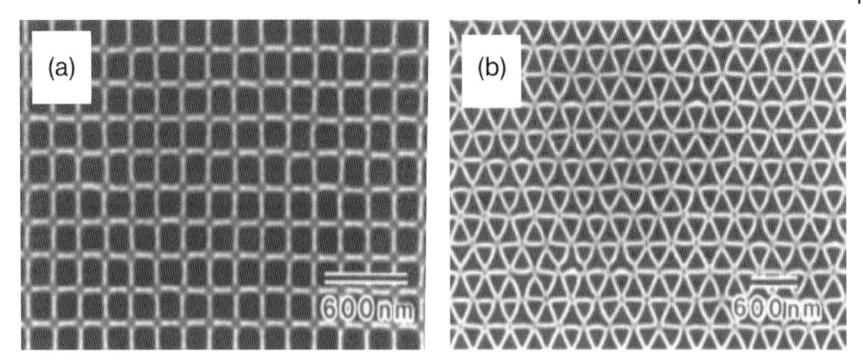

Figure 2.12 SEM images of the pore opening at the bottom of the anodic alumina formed by indentation of square (a) and triangular (b) patterns. Anodization was performed using the following parameters: (a) 0.05 mol dm^{-3} H$_2$C$_2$O$_4$ at 17°C and 80 V for 6 min; (b) 0.5 mol dm^{-3} H$_3$PO$_4$ solution at 17°C and 80 V for 20 min. (Reprinted from Ref. [72].)

spontaneously, a hexagonal array. However, based on the Voronoi tessellation concepts [71], Masuda and coworkers [72] reported the fabrication of architectures with square and triangular pores. These authors combined nanoindentation techniques with anodization, the procedure was based on the stamping of the pore initiation sites in square or triangular lattices before anodization. This procedure allows one to obtain shapes printed on the Al substrates, based on the nanotiling through the close-packed cells. Some micrographs of the anodic alumina formed using this procedure are shown in Figure 2.12.

In another report [73], pore growth with square cells formed during the initial stage of the anodization was investigated. These authors noted that the square lattice was also square at steady state, and concluded that the imprinting pattern – which is different from natural hexagonal cells – controls the growth of anodic porous alumina and so defines the final shape of the pores. These indentation techniques opened new possibilities for the exploration of different cell shapes, although certain unsolved questions remain, for instance, with regards to the true nanowire cross-section prepared inside these types of membrane. According to the findings of Choi et al. [74], the pore shape at the bottom of the porous structure is not controlled by the prepatterned stamp on the Al surface. Rather, the bottom region shape is defined only by the current flow and the electrolyte – that is, the bottom preserves the hexagonal pore structure. These authors compared the surface and the bottom view, and observed that the pore shape changed from rectangles into circles during their growth. Moreover, the same group reported a guided self-organization [75], in which prepatterns on the Al substrate formed by stamping not only acted as seeds for pores but also guided the formation of new pores between the prepatterns. This procedure enabled the preparation of alumina pore arrays with interpore distances smaller than the lattice constant of the stamp print. By using the guided self-organization process [75], these authors prepared a triangular pore array which was generated

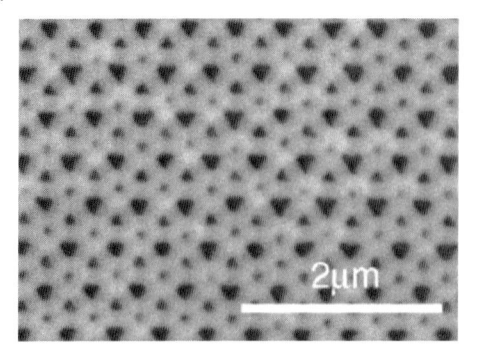

Figure 2.13 SEM image of a porous anodic alumina membrane with triangular pores generated from the lithographic patterns with a square shape. (Reprinted from Ref. [75].)

from the lithographic patterns with a square shape. An SEM image of triangular pores generated during the growth of pores by guided self-organization, as reported by Choi et al. [75], is shown in Figure 2.13.

Due to the high regularity achieved with two-step anodization, as reported by Masuda et al. [70], and the nanoidentation techniques combined with anodization [76,77], little has added with regard to the preparation of new pore architectures. One interesting configuration in PAA are the Y-branched pores reported by Zou et al. [78], which show a ramification along the porous structure (normal to substrate). These authors obtained this pore configuration by using a two-step potentiostatic anodization on aluminum film evaporated onto a Si substrate. According to Zou et al. [78], the Al/Si interface is responsible for the formation of branched cells, since the Y-branched pores are derived from stagnant cells that grew slowly (or even ceased to grow) during the anodization, due to an increase in the electrical resistance at interface. Previously [79], an electrochemical route for the preparation of individual alumina nanotubes had been reported; these alumina nanotubes were also prepared from an aluminum film evaporated onto Si. However, in this case, these authors considered that the semiconductor substrate was principally responsible due to the detachment of cells.

By changing the position where electrical contact is made [79] during the synthesis of individual alumina nanotubes (ANTs), is possible to obtain different structures. Electrical contact can be made at the bottom of the Si substrate [known as normal stepwise anodization (NSA configuration], or laterally on the Si wall [known as lateral stepwise anodization (LSA) configuration] (Figure 2.14d). Here, the current paths are different, and this leads to different nanotube geometrical parameters using the two methods. It should be emphasized here that the orientation of the sample is not important. The transmission electron microscopy (TEM) images in Figure 2.14 show a general view of the ANTs, which are attached to the PAA mother film. The largest ANT observed was 650 nm in length, and had outer and inner diameters of 35 and 12 nm, respectively (Figure 2.14a). Some bundles of ANTs were also present (Figure 2.14b). The ANTs fabricated by the two methods had the same structure and differed only in size, with the NSA tubes being smaller than the LSA

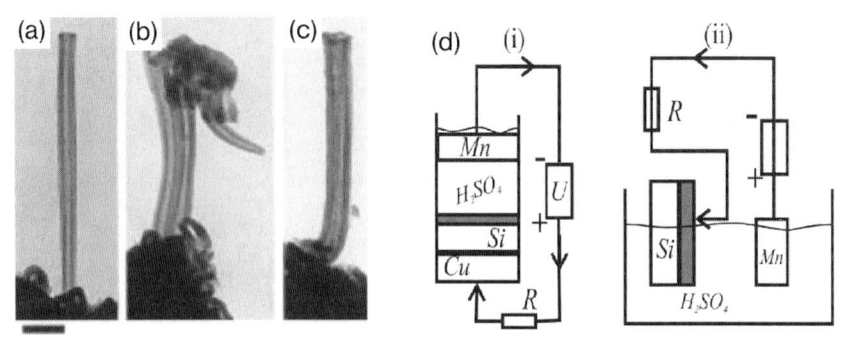

Figure 2.14 TEM images of: (a) alumina nanotubes (ANTs); (b) normal stepwise anodization (NSA) tubes; (c) lateral stepwise anodization (LSA) tubes. (d) A sketch of the preparation methods of ANTs: (i) normal stepwise anodization (NSA); (ii) lateral stepwise anodization. Scale bar = 100 nm. (Reprinted from Ref. [79].)

tubes. The main factor determining tube size may be whether the current path during growth passed through the interface region, or not.

Alumina whiskers and fibers are widely used in advanced materials such as metal matrix composites, because of their high temperature resistance and high modulus values [80,81]. Generally, these materials are prepared using conventional ceramic methods. Additionally, high-performance alumina fibers can be used as catalyst supports, as radar-transparent structures, and antenna windows [82]. Recently, Al_2O_3 fibers [83] and nanopillars [84] have been produced, although high-temperature calcination was needed to obtain both structures. In spite of reports about electrochemical methods being used to prepare branched alumina nanotubes (bANTs) [78] and individual ANTs [79], few additional investigations have been made concerning the fabrication of Al_2O_3 fibers arrays using these procedures. Pang [85] reported the large-scale synthesis of ordered Al_2O_3 nanowire arrays embedded in the PAA nanochannels by electrodeposition at room temperature.

Through-hole PAAs with ordered nanochannels were prepared using two-step anodization method [85], where upon it was observed that bulk quantities of Al_2O_3 nanowires with an equal height and a highly ordered tip array were formed. Tian et al. [86] also reported the preparation of alumina membranes with parallel Y-branched pores using a procedure different to that of Zou et al. [78]. Tian et al. performed a potentiostatic anodization using a high-purity aluminum foil in three steps where, before the third anodization step, the oxide formed during the intermediate step was not removed. However, during the last anodization step the applied voltage was reduced by a factor of $1/\sqrt{2}$. This procedure was based on results obtained by Parkhutik and Shershusky [87] and Almawlawi et al. [88], which suggested that the pore diameter in PAA is proportional to the applied potential. The potential decrease then resulted in ramificated pores. An important result here is that the branching process has occurred at the same depth. A micrograph of a PAA membrane, showing a close image of a branching point of an individual Y-branched

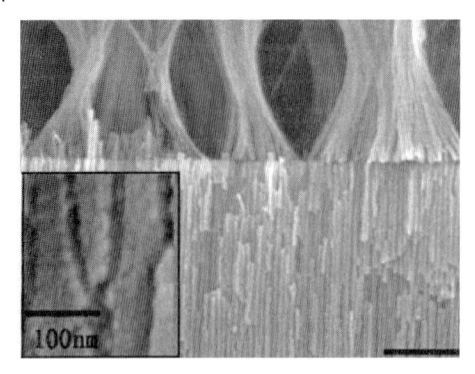

Figure 2.15 SEM side view image of the alumina nanowire array standing a porous alumina membrane on the surface. Scale bar = 1 mm. The inset shows a close-up of the branching point of an individual Y-shaped cell. (Reprinted from Ref. [86].)

pore prepared using this route, is shown in Figure 2.15. Furthermore, the Y-branched nanoporous alumina membrane was etched during an adequate period of time, after which the alumina nanowire arrays stood on the surface of the remaining alumina membrane. This kind of structure was also prepared using the consecutive anodization method, as proposed by Li et al. [89]. In both reports [78,86], the fundamental aspects involved in the formation of the pore ramification were not well understood. However, from a different point of view these membranes may represent an important potential in the preparation of heterojunctions.

Both, alumina nanotubes and branchy alumina can provide new routes to the synthesis of nanostructures with an increasingly complex design. These structures make up the fascinating class of nanoscale cables, heterojunctions, and jacks to the fabrication of nanotube-based electronic devices and circuits.

Two overlapped porous layer configurations in PAA were obtained and reported by the present authors' group [90,91] (see Figure 2.16). These structures were

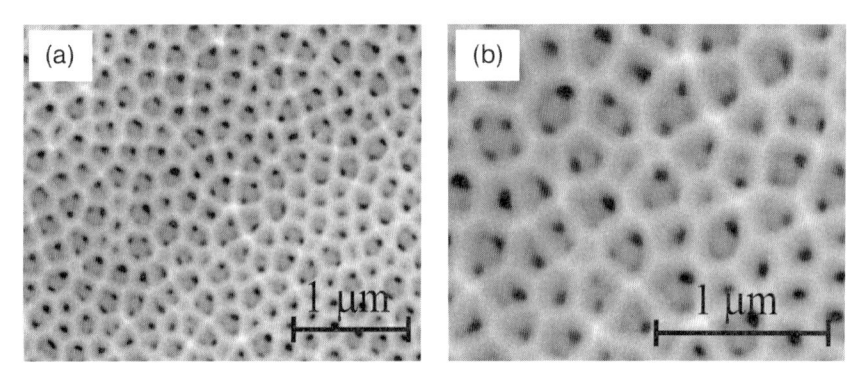

Figure 2.16 SEM images of the porous anodic alumina (PAA) membranes prepared in two-step galvanostatic anodizations that present two porous layers overlapped at (a) low and (b) high magnification.

Table 2.1 Mild anodization (MA) versus hard anodization (HA) in 0.3 M $C_2H_2O_4$ (1°C).

	MA	HA
Voltage (V)	40	110–150
Current density (mA·cm^{-2})	5	30–250
Film growth rate ($\mu m\,h^{-1}$)	2.0–6.0 (linear)	50-70 (non-linear)
Porosity (%)	10	3.3–3.4
Interpore distance (nm)	100	220–300
Pore diameter (nm)	40	49–59
Pore density (pores cm^{-1})	1.0×10^{10}	$1.3–1.9 \times 10^9$

prepared under galvanostatic conditions, using a variation of the two-step anodization method. Unfortunately, it was not possible to define this array in a side view (by SEM), since the superior porous layer was very thin compared to the inferior layer.

Recently, Lee et al. [92] reported a new oxalic acid-based anodization process for long-range ordered alumina membranes. This method can be considered as a new version of the traditional hard anodization (HA) used by industry to fabricate mechanically robust and very thick alumina films. These authors reported a new self-ordering regime with interpore distance from 200 to 300 nm which, to date, have been achieved by using mild anodization (MA). Hence, this represents a second self-ordering regime for oxalic acid anodization, but in this case the range of applied potential used is between 120 and 150 V (three- to fourfold higher than that in MA conditions in oxalic acid). MA is characterized by slow oxide growth rates (2–6 $\mu m\,h^{-1}$) which requires several days of processing in order to obtain thicker, highly ordered Al_2O_3 films. Unfortunately, this feature limits the use of Masuda's approach [70] on an industrial scale. The time-consuming factor is not the only limitation, since the spatial domain, with a high ordering degree, is also reduced. The main features of MA and HA are summarized in Table 2.1.

Figure 2.17 shows a side view of the porous alumina prepared under MA and HA in oxalic acid. In addition, on the basis of the fact that HA yields anodic alumina with one-third lower porosity than MA (i.e., $P_{HA} < 3\%$ for HA and $P_{MA} < 10\%$ for MA), these authors modulated the pore diameter by combining both anodization processes (Figure 2.18). First, the Al sheet is anodized under MA in 0.1 M H_3PO_4 at 110 V using 4 wt.% H_3PO_4 (10°C) for 15 min, and this results in a segment of oxide nanopores with higher porosity. Subsequently, HA was carried out at 137 V using 0.015 M $H_2C_2O_4$ (0.5°C) for 2 min, which resulted in a lower-porosity nanoporous alumina segment. Highly uniform periodic modulations in pore diameters have been achieved by repeating two consecutive anodization processes of MA and HA. In addition, the length of each segment of oxide nanopores can freely be controlled by varying the time of the anodization steps.

The versatility of the porous anodic alumina is one of its most important features. By taking advantage of the highly ordered structure of the PAA, it is possible to design complex nanoscale-structured devices with unique physical properties due to the size effects. The role of the porous anodic alumina within the nanostructure

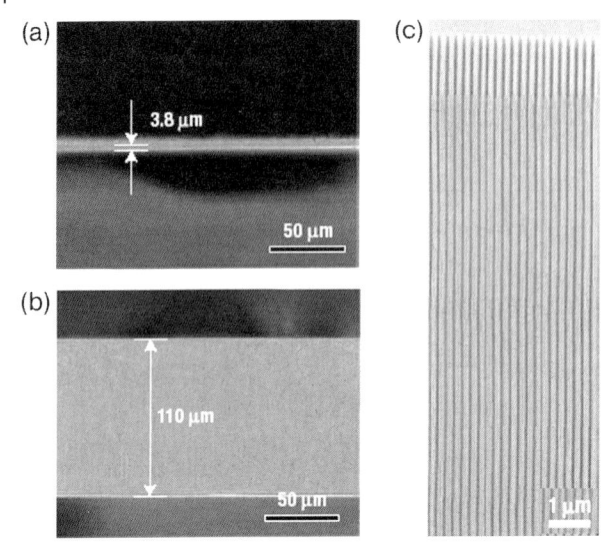

Figure 2.17 SEM images of the corresponding anodic porous alumina (AAO) specimens formed by (a) mild anodization (MA) for 2 h and (b) hard anodization (HA) for 2 h. (c) A close-up view of image (b). The thicknesses of the respective samples are indicated in the cross-sectional SEM images. (Reprinted from Ref. [92].)

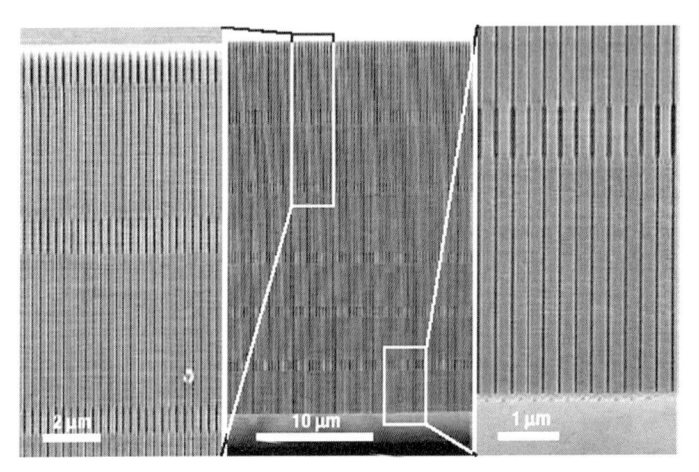

Figure 2.18 Long-range ordered alumina membranes with modulated pore diameters. SEM images showing the cross-sectional view of the corresponding sample with modulated pore diameters. Magnified cross-sectional images of the top part of the membrane are shown on the left and right. (Reprinted from Ref. [92].)

fabrication field can be divided basically in two wide classes: (i) highly ordered porous structure working as a template to the preparation of a nanodevice (which is the most traditional application of this system); and (ii) PAA acting as a nanodevice itself. In the latter case, the highly ordered architecture combined with the physical and mechanical properties are the key point. Some representative examples of the first class will be presented briefly in the next subsection, while details of nanostructures prepared by cathodic methods using PAA as a template are presented in Section 3. (For further details about this topic, see Chapter 2.) Details of the second above-described class are presented in Section 2.2.2.2.

2.2.2.1 Porous Anodic Alumina as Template

Although porous alumina is present at least in one step of all procedures for preparing nanomaterials, its main application in nanotechnology is in template synthesis. The success of porous alumina as a template is due to the fact that a large variety of materials (metallic, semiconductors and insulator nanowires, carbon nanotubes, etc.) can be consistently fabricated with accuracy and ease. For example, nanodots (Au, Ni, Co, Fe, Si and GaAs), nanopores (Si, GaAs and GaN), nanowires (ZnO) and nanotubes (Si, carbon) which can be prepared using highly ordered porous alumina by evaporation or using etching masks. The porous alumina membrane determines the uniformity, diameter size, and length of the nanostructures. It is similar to the design of nanodevices inside a beehive.

Recently, Hillebrenner *et al.* [93] showed interest in a universal delivery vehicle which used PAA as a template to synthesize silica nano test tubes. Currently, much interest exists in the preparation of nanodevices to deliver biomolecules and macromolecules to specific sites inside living systems. An alternative is that of nanoparticles, where drug release is accomplished by chemical or enzymatic degradation of the particle wall [94,95]. These groups are currently exploring an alternative release system which utilizes template-synthesized nano test tubes. It is suggested that, if these nano test tubes were filled with a medicine, and the open end then sealed with a chemically labile cap, they could act as delivery vehicles. A schematic route used to "build" nano test tubes is shown in Figure 2.19, while SEM and TEM images of nano test tubes prepared in this way are shown in Figure 2.20.

By using a free-standing PAA as an evaporation mask, Ding and coworkers [96] were able to fabricate nanocrystalline Si:H (nc-Si:H) artificial quantum nanodot arrays on Si. The thickness and pore diameter of porous alumina play an important role in this fabrication process. The nc-Si:H quantum nanodot obtained has a tunable diameter ranging from 50 to 90 nm, as shown in Figure 2.21. The uniqueness of this approach is the fabrication of nanodevice arrays with true quantum size effects, combined with artificial quantum dot fabrication through porous alumina masks with Si natural quantum dots inside nc-Si:H dots. This alternative procedure allows the preparation of semiconductor devices arrays, where the Si natural quantum dots will play a key role in the quantum size effects, while the spacing of the uniform artificial quantum dots serves as a good electrical insulator [96].

Recently, many research groups [97–99] have been investigating the deposition of carbon nanotubes with controlled sizes, using chemical vapor synthesis (CVD) in

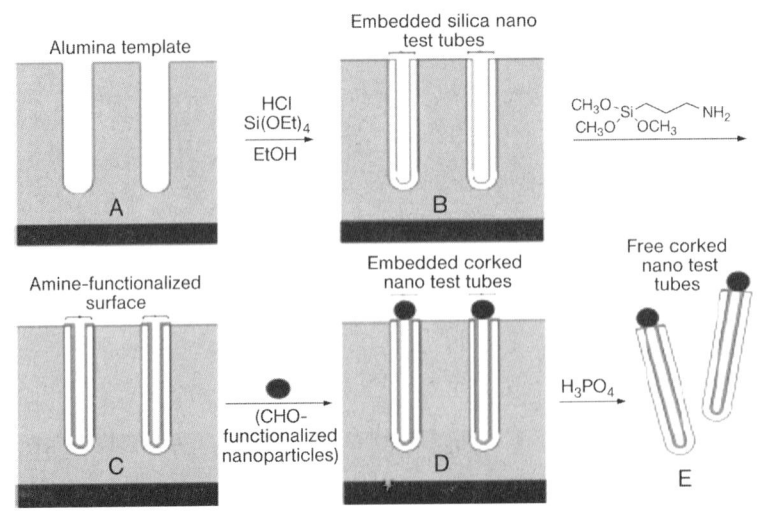

Figure 2.19 Scheme of the test tube synthesis and corking processes. (Reprinted from Ref. [93].)

a porous alumina template. Yanagishita et al. [100] prepared carbon nanotubes using porous alumina with triangular pores as template; these templates were prepared according to the procedure reported by Masuda et al. [72]. Carbon nanotubes were deposited in PAA by CVD. Usually, cobalt particles, which may serve as a catalyst for

Figure 2.20 SEM images (upper) and TEM image (lower) of the nano test tubes prepared by template synthesis. Scale bar = 500 nm. (Reprinted from Ref. [93].)

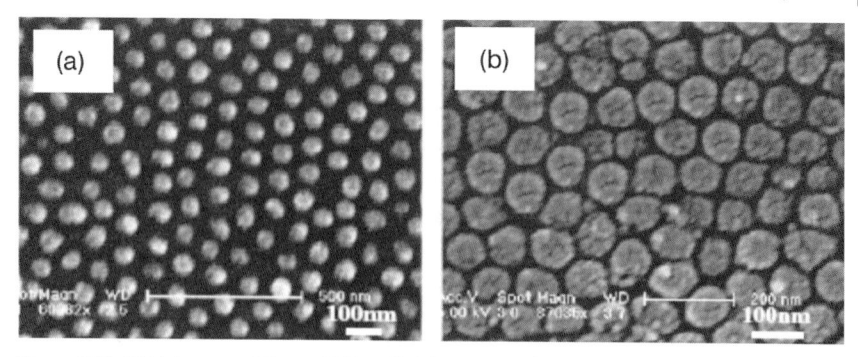

Figure 2.21 SEM images of the nc-Si:H artificial quantum dot arrays with different dot diameters: (a) 50 nm; (b) 90 nm. (Reprinted from Ref. [96].)

the synthesis of carbon nanotubes, are electrodeposited at the bottom of the porous alumina template. An SEM image of the ordered arrays of carbon nanotubes with triangular cross-sections, prepared using an alumina template, is shown in Figure 2.22.

2.2.2.2 Porous Anodic Alumina to Create Nanodevices

Photonic Crystals A recent application of PAA is as photonic crystals [101]. When electromagnetic radiation with a wavelength comparable to the periodicity of the system passes through such an array, the dispersion relation is modified according to its geometry and composition. This is analogous to the effect of a crystalline structure on the behavior of electrons [102]. These artificially engineered periodic dielectric structures, which are referred to as "photonic crystals" [103], are periodic dielectric materials that allow control of the light path. In this system, the light behaves like the electron inside the semiconductor, highlighting forbidden regions. It follows, therefore, that new designs of nanostructured materials may lead to

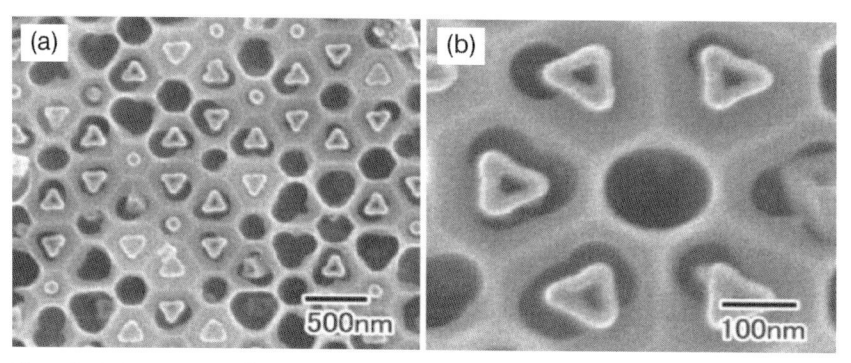

Figure 2.22 SEM images of carbon nanotubes with a triangular cross-section at (a) low magnification and (b) high magnification. (Reprinted from Ref. [100].)

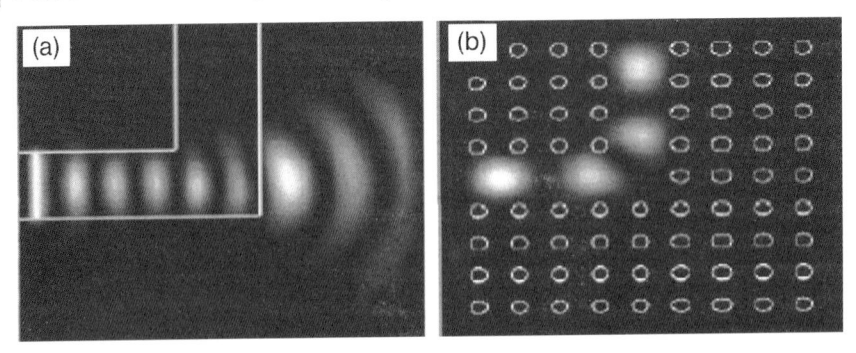

Figure 2.23 Simulation of the optical path performed in a conventional material (a) and a photonic crystal (b). (Reprinted from Ref. [104]; courtesy Prof. Dr. H.K. Kim.)

advances in optical manipulation, such as waveguides, which are devices that present a curvature of the optical light path. A simulation of an optical path, as reported by Kim [104] and performed in a conventional material and in a photonic crystal, is shown in Figure 2.23. Here, the change of dispersion relation in the photonic crystal compared to that in a conventional material can be clearly observed.

On the basis that applications as photonic crystals require a perfect PAA membrane with only one domain, the self-ordered growth and prepattern guided anodization compete on terms of both product quality and cost. Although self-ordering is a cheap and simple route, and has achieved good results [70], more sophisticated technologies, which involve a prepattern of the substrate (e.g., imprint by a pyramid stamp [74]; imprint by a dot-like stamp [76]; electron beam (e-beam) lithography [105]; two-step indentation by commercial gratings [106]), are currently under investigation. According to Mikulskas et al. [106], the key characteristic of these methods is the fabrication of a precise die having appropriate dimensions. Here, it is important to emphasize that, in the self-ordered as well as in prepattern guided anodization, the applied voltage also plays a key role.

In 1999, Li et al. [102] reported the fabrication of parallel, regular nanopore arrays with a high aspect ratio in PAA, which is important for the 2D behavior of photonic crystals. These authors developed a lateral microstructuring technique for micromachining bars of PAA. The preparation process of ordered PAA involved a two-step anodization, without any previous imprint procedure. Subsequently, a lithographic technique was applied to obtain the microstructured porous array. The route to porous array preparation, and the microstructure obtained according to this procedure, are shown schematically in Figure 2.24. The highlights of these studies are the high aspect ratio obtained (ca. 1400), and the accuracy of the design which combines self-ordered growth with lithographic techniques.

Masuda et al. [107] also reported the preparation of a porous array with a 2D photonic band gap in the visible wavelength region. The fabrication was based on Al pretextured by imprinting with a mold, followed by anodization. The 2D photonic crystal was fabricated using PAA with a highly ordered pore array, and an aspect ratio

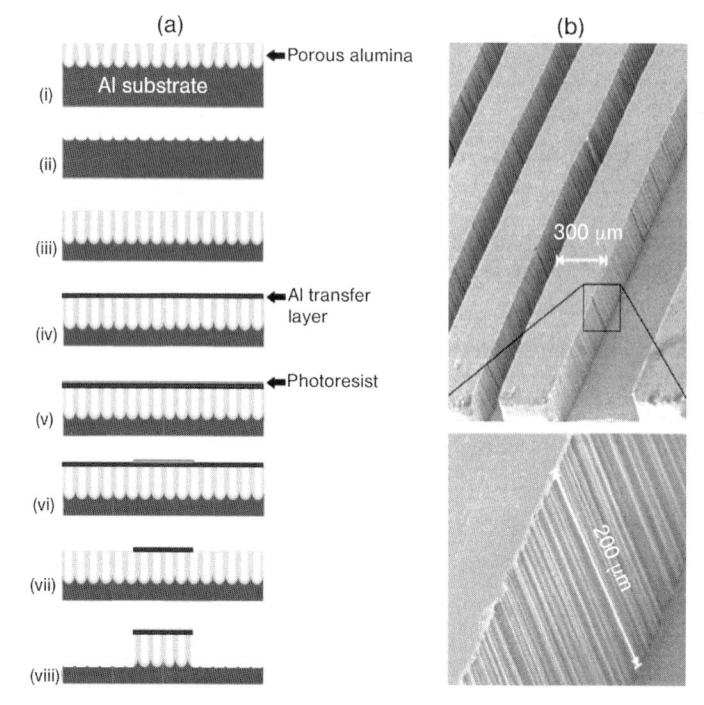

Figure 2.24 (a) Schematic diagram of the fabrication and microstructuring processes on a nanopore array in anodic alumina. (i) Porous alumina obtained in the first anodization; (ii) Al substrate after the chemical removal of the porous alumina formed during the first anodization; (iii) a nanopore array formed in the second anodization; (iv) evaporation of an aluminum transfer layer; (v) coating of photoresist; (vi) patterning of resist; (vii) patterning of aluminum layer; and (viii) etching of vertical microstructures. (b) Typical microstructure depth profile and an enlarged view of a side wall. (Reprinted from Ref. [102].)

in excess of 200. The transmission properties of the obtained ordered pore array in the alumina matrix exhibited a stop band in the spectrum which corresponded to the band gap in 2D photonic crystals.

Mikulskas et al. [106] proposed an alternative substrate prepatterning, using an optical grating. The texture of the grating was transferred to an aluminum surface, covering it completely. The porous array obtained, which has a perfectly ordered structure, is shown in Figure 2.25. Optical transmission measurements confirmed the existence of a photonic band gap in the visible region, thus indicating that the perspectives are good for this fabrication technique, which can be used to create large structures. These results are also important for conventional spectroscopic applications and 2D photonic crystal development. Furthermore, this alternative method provides the possibility of preparing photonic crystals based on porous alumina by using holographic optical lithography, which is commonly used in the production of diffraction gratings.

Choi et al. [108] investigated both the near solution-hydrated (outer) oxide and internal (inner) oxide in detail. The inner layer is a pure alumina with

Figure 2.25 SEM image of the porous anodic alumina (left) and cleaved edge of the porous array (right) prepared according to procedure reported by Mikulskas et al. (Reprinted from Ref. [106].)

thickness ~50 nm, while the outer layer has incorporated anions. Consequently, these layers present different dielectric constants. In addition, the outer layer consists of an outermost part and an intermediate part into which the anions are concentrated. Hence, the outer layer has a non-homogeneous effective dielectric constant, depending on the concentration of impurities at each site. As shown in Figure 2.23, the optical path in this type of material is determined by the arrays formed according to the lattice defects, while the periodic potential modulation will be equivalent to the dielectric constant and defect modulation. These authors reported a detailed discussion on photonic crystals, based on PAA. They also described the preparation of a perfect 2D photonic crystal on an area of 4 cm^2 using imprint methods, followed by conventional potentiostatic anodization.

Electrochemical Double Layer Capacitors Jung et al. [109] used carbon nanotube-modified PAA as electrode materials for electrochemical double layer capacitors (EDLC). These authors reported the preparation of carbon nanotubes embedded in free-standing porous alumina, together with the results of its EDLC properties investigation. The aluminum substrate was seen to be the current collector for the electrochemical cell. The homogeneous carbon nanotubes, grown in highly ordered porous alumina, present good characteristics for EDLC with enhanced specific capacitances. These systems were also investigated as field emitter arrays due to the improved diameter uniformity and inner distance of the tube regularity [110,111].

The use of PAA membranes as separators for Li rechargeable batteries was investigated by Mozalev et al. [112]. Here, an attempt was made to apply the alumina membranes for storing a non-aqueous battery electrolyte, and for the repeatable deposition–dissolution of lithium metal through the pores on Ni and Al substrates. These authors noted that the ionic resistance of the membrane prepared by one-step galvanostatic anodization was lower than those of commercially available polyethylene separators, and this proved attractive for propylene carbonate-based battery

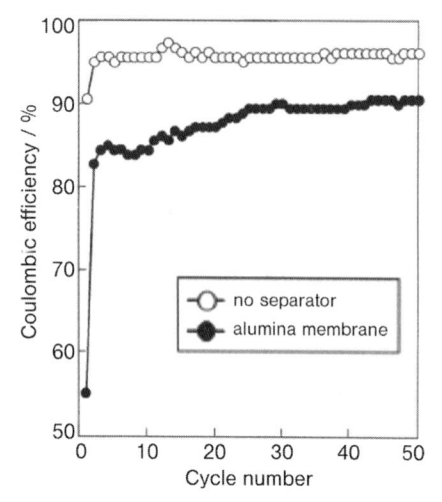

Figure 2.26 Coulombic efficiencies for cycling of Li on the Al substrate directly and through the one-step membrane attached. The cell was charged at: $1\,mA\,cm^{-2}$ for 6 min, 3 min rest time. Discharge at: $1\,mA\,cm^{-2}$, cut-off potential 2 V versus Li/Li^+, 3 min rest time. (Reprinted from Ref. [112].)

electrolytes. Coulombic efficiencies for the cycling of Li through the alumina membrane on an aluminum substrate was improved with cycle number, reaching a constant value of 89% after 25 cycles and showing an upward tendency beyond the 40th lithium deposition (Figure 2.26).

Light-Emitting Diodes An organic light-emitting diode (OLED) is a light-emitting diode (LED) in which the emissive layer comprises a thin film of an organic (O) compound. It has been reported [113] that a thin insulating film on the cathode surface largely enhances the electroluminescence intensity and lifetime. Kukhta et al. [114] proposed a new cathode for organic light-emitting based on PAA films. The model of the organic electroluminescent cell based on the porous alumina is shown in Figure 2.27. In order to improve the efficiency of the electroluminescent structure,

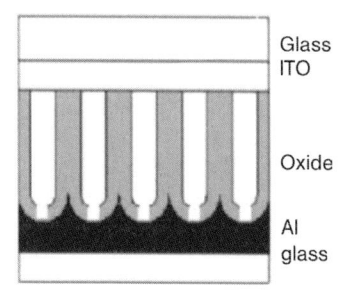

Figure 2.27 Schematic model of the organic electroluminescent cell. (Reprinted from Ref. [114].)

which consists of aluminum, PAA doped with organic phosphor, and a tin-doped indium oxide (ITO) layer, the authors removed the barrier layer by a slow reduction of the anodization voltage to zero. A consequent increase in the alumina porosity was observed which was proportional to the increase in electroluminescence intensity.

Anodic films can serve as effective catalysts or catalyst supports within microreactors due to their thickness being in the micron range, their elevated specific surface area, and their strong adhesion to the substrate. Ganley et al. [115] described the process conditions for preparing PAA as a catalyst support in monolithic microreactors. The process involved the galvanostatic anodization of an 1100 aluminum alloy in oxalic acid, and the deposition of a transition metal (which served as the catalyst). These authors subsequently demonstrated the performance of aluminum–alumina microreactors with optimized PAA films for the decomposition of anhydrous ammonia using supported metal catalysts, thereby opening the possibility of carrying out heterogeneous catalyzed reactions on a small scale.

Medical Applications Porous alumina can also be used as implants in medical applications. One of the main goals of bony tissue engineering is to design better materials to control osteoblast behavior [116]. It has been shown previously that osteoblast responses are extremely sensitive to surface roughness and porosity [117], and in this sense nanoporous designs have proved favorable for osteoblast proliferation. Karlsson [118] described a new procedure to prepare a composite material for use as a bone implant, which consists of porous alumina formed on titanium substrate. Subsequent *in-vitro* studies showed that the porous alumina had a high adhesion to the titanium substrate, and supported the osteoblast and its proliferation.

Popat *et al.* [116] also reported an investigation of the osteoblast response seeded on PAA, by investigating the parameters of short-term adhesion and proliferation, long-term functionality, and matrix production. These authors compared results obtained with PAA to those achieved with amorphous alumina, commercially available ANOPORE™, and glass. PAA showed an improved short-term osteoblast adhesion and proliferation compared to the other surfaces, while the nanoporous alumina support achieved the greatest osteoblast adhesion overall. In addition, cells cultured on porous alumina presented a higher protein content, as indicate by the bi-cinchoninic acid (BCA) assay.

As mentioned earlier [93], porous alumina represents a very important alternative approach for the preparation of nanodevices with biomedical functions, in this case, as a template for the preparation of nano test tubes. Similarly, Gong *et al.* [119] reported the use of nanoporous alumina capsules, with highly ordered pores having diameters ranging from 25 to 55 nm, for controlled drug delivery. Anodization of the aluminum tubes resulted in a tubular porous alumina membrane, the process was based on the studies of Itoh *et al.* [120]. The capsules were filled with fluorescent molecules of different molecular weight, and plugged with silicone. Subsequent controlled-release experiments showed that drug transport was sensitive to the pore diameter, and that it was also possible to control molecular transport by the appropriate selection of the membrane pore size.

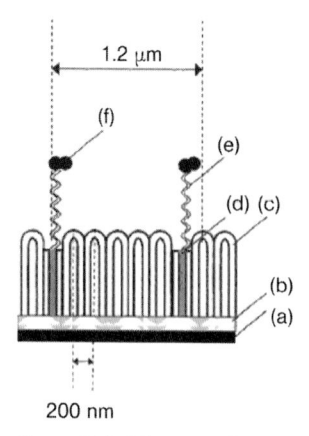

Figure 2.28 Schematic of DNA array prepared using ideally ordered anodic porous alumina. (a) Ni; (b) Pt–Pd layer, which functions as an electrode during electrodeposition of Au inside these opened pores; (c) PAA; (d) Au; (e) DNA; (f) fluorescent dye. (Reprinted from Ref. [122].)

Moreover, these biocapsules prevented the diffusion of molecules larger than a certain cut-off size.

In addition to the development of devices for drug delivery, the area of biomedicine has made many important advances in the construction of highly ordered structures of biological molecules such as proteins, DNA, and antibodies on the nanometer scale. For example, Matsumoto et al. [121] reported the fabrication of highly ordered nanometer-sized arrays of DNA on regularly sized Au disks inside porous alumina. It appeared that these authors encountered some problems during the fabrication of biological molecule nanopatterns using PAA, since the limits of resolution of the optical microscope used was insufficient to observe such patterns. More recently [122], the same group presented a DNA nanoarray on Au disks with a controllable interval suitable for observation by optical microscopy. A representation of the DNA nanoarray produced using ideally ordered porous alumina is shown in Figure 2.28. The Au disk fabrication was made possible by selectively opening the pores and filling them with Au. In order to accomplish this, the group explored the differences in thicknesses between the barrier layers at the imprinted and unimprinted sites of the convex dots on the Al substrate and found that, at for every six imprinted sites there was one unimprinted site. Moreover, the barrier layer which developed in the latter site was thinner than in the former site. An SEM image, clearly showing the period between the Au disks functionalized with DNA, is shown in Figure 2.29. Most importantly, this approach shows how a simple insight can overcome some limitations, in this case, the resolution limit of the optical microscope.

In summary, within this section several studies have been presented on the advances of precise tailoring at the nanometer scale using porous alumina, an approach which can be considered as one of the key aspects in the development of nanotechnology. Indeed, the template approach using PAA, although pioneered independently

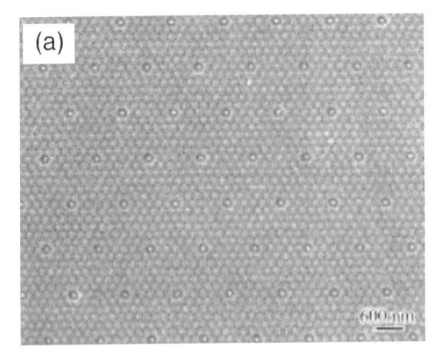

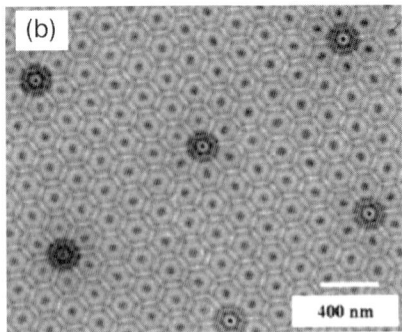

Figure 2.29 (a) SEM image of the fabricated Au disk array. Disk diameter = 50 nm (i.e., the diameter of the alumina pores). Disk period = 1.2 mm. (b) SEM image, taken from the barrier layer side of a porous alumina substrate with selectively opened pores (large dark circles) which were filled with Au to form the Au disk. The small dark points are the pores with unremoved barrier layer. Note that the opened pores are perfectly separated from each other by six cells. (Reprinted from Ref. [122].)

by several groups from different fields [21,70,96,122–124], is currently being used with great success by an increasingly number of research groups worldwide. More significantly, perhaps, a large number of prototypes of nanostructured devices have been produced, thus demonstrating the enhancement of performance that can be achieved compared with conventional devices.

2.3
Cathodic Synthesis

2.3.1
Nanowires

Research into nanostructured systems, particularly of one-dimensional (1D) structures, has included many investigations into the physical and chemical principles of these materials. Today, the properties and effects of nanostructures, such as high storage and information transmission rate, blue shift of absorption edge of nanoparticles, conductance quantization, enhancement of mechanical properties [125–128], are better understood on the basis of theoretical studies and calculations performed on low-dimensional systems [129], notably those related to the next generation of nanoelectronic devices [130]. In spite of the difficulties encountered in obtaining an accurate control of the fabrication parameters, 1D nanostructures are ideal systems compared to dots and 2D systems for studying transport phenomena at the nanometric scale [131].

These facts, together with the current tendency to fabricate functional, nanostructured devices, are clearly the new "trends" in nanotechnology. This applies in particular to 1D systems such as nanowires (NW) and nanotubes, and these are discussed in the following paragraphs.

Nanostructured materials and devices are interesting not only because they may lead to new applications, but also because they exhibit novel quantum phenomena. An interesting example is conductance quantization, which occurs in a semiconductor or metal wire connected between two macroscopic electrodes when the following two conditions are fulfilled. First, the wire must be shorter than the electron mean free path, so that electrons are transported ballistically along the wire. Second, the wire diameter must be comparable to the electron wavelength in order to allow electrons to form standing waves (quantum modes) in the transverse direction of the wire. These systems are also known as "quantum wires" or "quantum point contacts". This phenomenon was first observed in semiconductor devices, where the electron mean free path is of the order of many microns, and the electron wavelength is ~40 nm, much larger than the atomic scale [132,133]. The relatively large wavelength made it possible to fabricate the required nanowires using conventional nanofabrication techniques. However, this led to a small energy difference between the quantum modes, which means that a pronounced conductance quantization in these devices occurs only at liquid helium temperature. For a typical metal (e.g., Au), the electron wavelength is only a few Ångstroms, and so the low temperature is not required in order for this phenomenon to be observed, although the wires must be atomically thin. A number of techniques have been developed to fabricate atomically thin metal wires that exhibit conductance quantization. These methods can be divided into two categories:

- A mechanical approach, in which a nanowire is formed by separating two electrodes from contact [134–136]. During the separation, a metal neck is formed between the two electrodes and this is then stretched into an atomically thin wire before it is completely broken.

- A technique developed by Xu *et al.* [137], in which the nanowires are fabricated electrochemically [138,139]. Recent TEM data have confirmed clearly that a metal wire with quantized conductance consists of a string of a few metal atoms [140,141].

In a semiconductor nanowire or quantum point contact, the width and electron density of the nanowire can be controlled flexibly by using gate electrodes, although this flexibility does not exist for metallic nanowires in vacuum or in air. In electrolytes, however, the potential of a metallic nanowire can be controlled in a similar fashion to the gate voltage. By controlling the electrical potential, it has been possible to study a variety of processes which take place on the nanowires, such as double layer charging, potential-induced stress, ionic and molecular adsorption.

2.3.1.1 Template Procedures to Prepare Nanowires

Many fabrication methods, including thermal decomposition [142], surfactant-assisted hydrothermal process [143], and vapor–liquid–solid (VLS) techniques, have been developed for the preparation of nanowires.

The *template method* is considered to be the most convenient for nanowire growth, where porous polycarbonate (PC) and PAA membranes are typically used as templates. Electroless plating (i.e., the electroplating process conducted without any

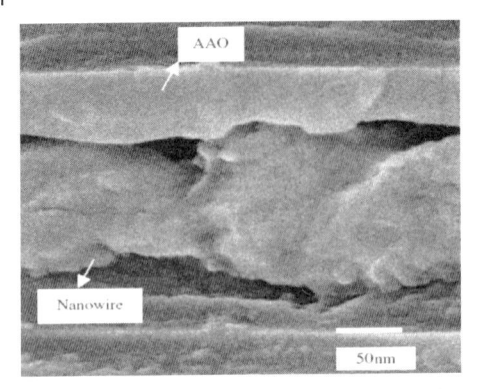

Figure 2.30 Field emission SEM cross-section of 4 V electroplated nanowire in a PAA template. AAO = anodic porous alumina. (Reprinted from Ref. [150].)

applied electrical field) leads to a chemical reduction of the cations in the pores of the template to form the nanowire [144,145]. Nanowire crystal growth processes using templates with different pore sizes have also been investigated [146,147], while many studies have also addressed the modification of nanowires after their growth [148,149].

Silver nanowires with variable surface roughness can be prepared using electroplating with a PAA template [150], and different types of nanostructure, including nanowires with nanoparticles or nanorods on their surfaces can be grown in this way. A pulsed DC voltage can be applied to grow branched structures on the surfaces of Ag nanowires. A cross-section of a nanowire in the initial stage of growth under high electroplating voltage, and with a coarse surface morphology, is shown in Figure 2.30. Of note, some bubbles which were trapped due to the high electroplating potential were considered to be the main cause of the rough surface.

Electrochemical silver infiltration into self-ordered porous alumina templates with pores in the sub-100 nm range has been studied recently by Sauer *et al.* [151]. The achievement of a high degree of electrochemical pore filling with metals in a PAA with 400 nm-diameter pores is difficult due to the thick barrier layer that is generally formed in these templates. This situation leads to instability during pore growth, such that the filling becomes highly inhomogeneous. In 2006, Sauer *et al.* developed a unique imprint stamp consisting of Si_3N_4 pyramids which were wafer bonded to a silicon wafer [152]. After nanoindentation of the aluminum, monodomain porous alumina templates were fabricated via anodization. The results achieved for silver infiltration produced by DC deposition are shown in Figure 2.31, where the average length of the silver rods is approximately 22 μm [153]. Over large areas, the filling of silver in the monodomain porous alumina template with straight long channels was almost 100%, which permitted their use as metallodielectric photonic crystals.

Chu *et al.* [154] studied the formation process of PAA, Ni nanowires arrays, and TiO_2-RuO_2/Al_2O_3 composite nanostructures on glass substrates. A highly pure aluminum layer deposited by radiofrequency sputtering [155,156] onto a glass

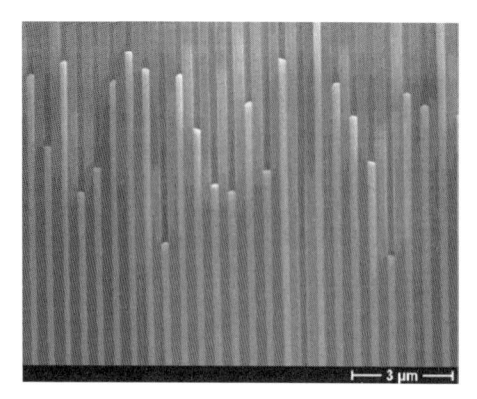

Figure 2.31 SEM image of a cross-sectional view of silver wires infiltrated in the monodomain porous alumina template with a pore diameter of 180 nm. The bright strips are silver wires. (Reprinted from Ref. [153].)

substrate and coated with an ITO film, was used as the starting sample. The PAA is then formed by anodization of the aluminum layer. Owing to its thin dimensions, the barrier layer of anodic alumina films formed on the ITO/glass substrates can be removed by using an appropriate chemical dissolution, thereby preserving the porous layers on the substrates. This feature is important because the conductive ITO film can then be exposed to the electrolyte. Finally, the pores are completely filled with metals (or other materials) by cathodic deposition until a continuous layer is formed. The morphologies of nickel nanowire arrays embedded in porous alumina films on ITO/glass substrates, after being removed the alumina films, are shown in Figure 2.32. It can be seen that the pores of the alumina are uniformly filled with nickel nanowires, even though the heights of the nanowires are uneven (arrows in Figure 2.32a, c, and e).

The electrosynthesis of titanium and ruthenium oxide nanostructures on PAA were also described by Chu *et al.* [154]. These authors first deposited, electrochemically, a peroxicompound of the metals, and this was then converted to the oxide using a calcination step at 600°C. The field emission scanning electron microscopy (FES-EM) images of the fracture sections, and the surface and the bottom morphologies of the as-electrodeposited specimens are shown in Figure 2.33. The deposits are composed of tiny grains ranging from 10 to 50 nm size, which completely fill the pores and then form a continuous layer on the anodic alumina films.

2.3.1.2 Magnetic Nanowires

Currently, the complete characterization of magnetic nanowires is the subject of intense research investigations, the main motivations being led by analyses of magnetization processes based on micromagnetic theory, and their potential use as media for high-density magnetic storage. The most cost-effective technique for producing magnetic nanowires is the electrodeposition of transition metals inside the nanometric channels of porous membranes.

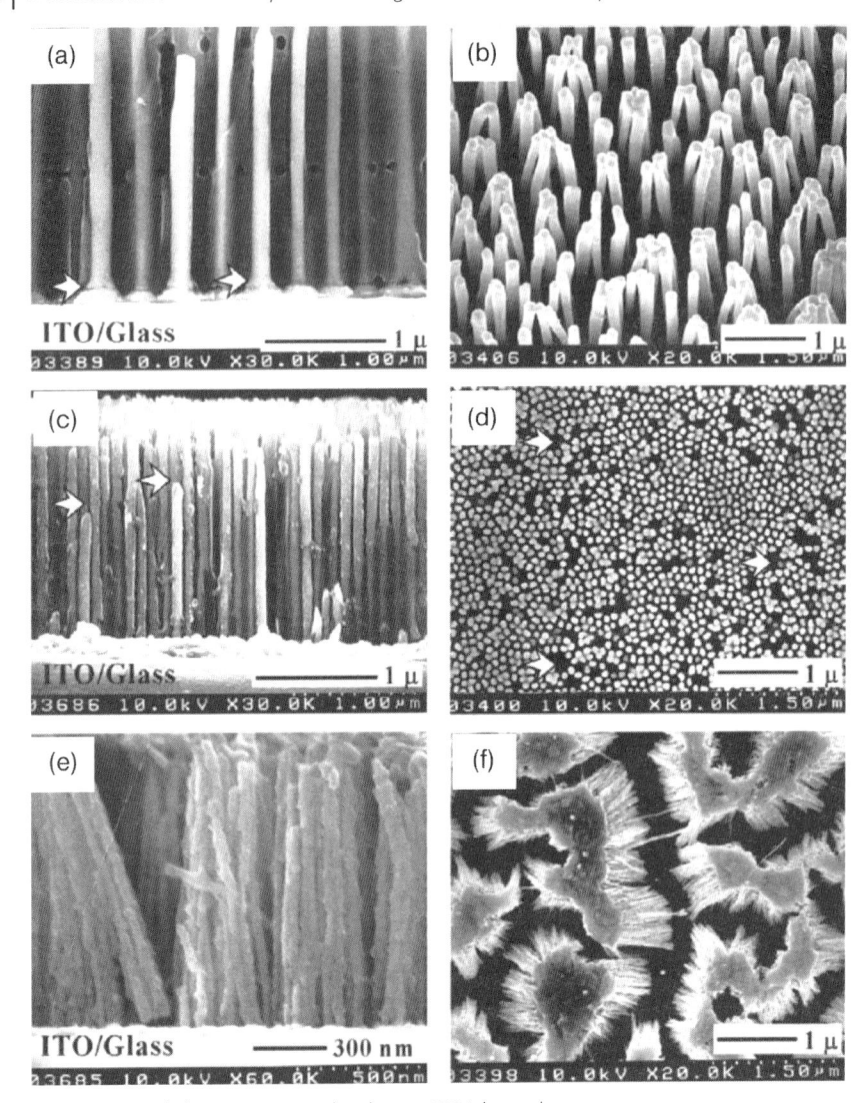

Figure 2.32 Nickel nanowire arrays bending on ITO/glass substrates fabricated in anodic alumina film obtained in H_3PO_4 (a, b), $H_2C_2O_4$ (c, d) and H_2SO_4 solutions (e, f). Before imaging, the specimens were selectively etched in a solution of 10% NaOH at room temperature for 10 s to 5 min. (Reprinted from Ref. [154].)

Garcia *et al.* [157] investigated both the growth and magnetic properties of an array of cobalt nanowires which had been electrodeposited inside the pores of track-etched polymer membranes from Nucleopore™. In order to follow the nanowire electrodeposition under potentiostatic conditions, it is necessary to monitor the current–time curve. As shown in Figure 2.34, three zones can be distinguished: Zone I

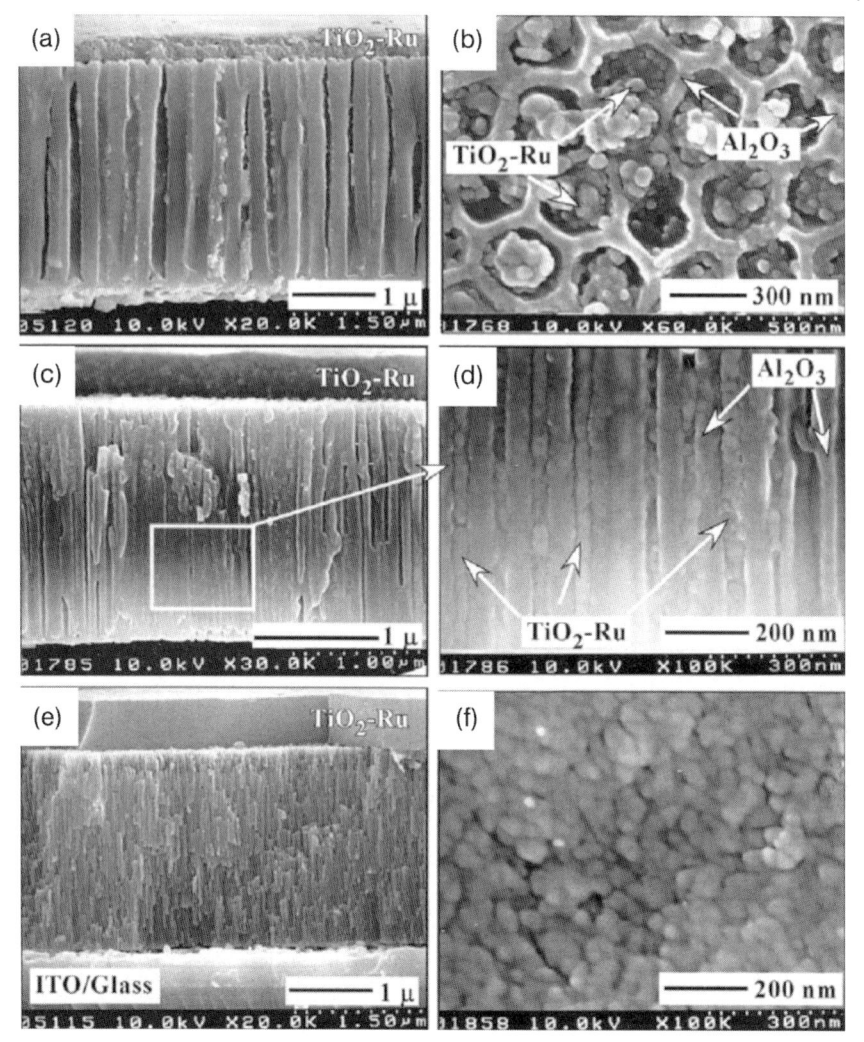

Figure 2.33 Field emission SEM images of vertical fracture section and morphology of TiO_2-RuO_2/Al_2O_3 composite nanostructures on ITO/glass. The porous alumina films are obtained in H_3PO_4 (a, b), $H_2C_2O_4$ (c, d), and H_2SO_4 (e, f) solution. (Reprinted from Ref. [154].)

corresponds to the deposition of Co inside the pores, while zone II arises from the complete filling of some pores which gives rise to the formation of hemispherical Co caps over the end of the longest wires. It has been realised that as the effective cathodic surface is larger, the current increases. Finally, when all of the pores are completely filled, zone III is reached, the caps coalesce into a continuous metallic top layer, and the current saturates with a value higher than that obtained for deposition into the pore membranes.

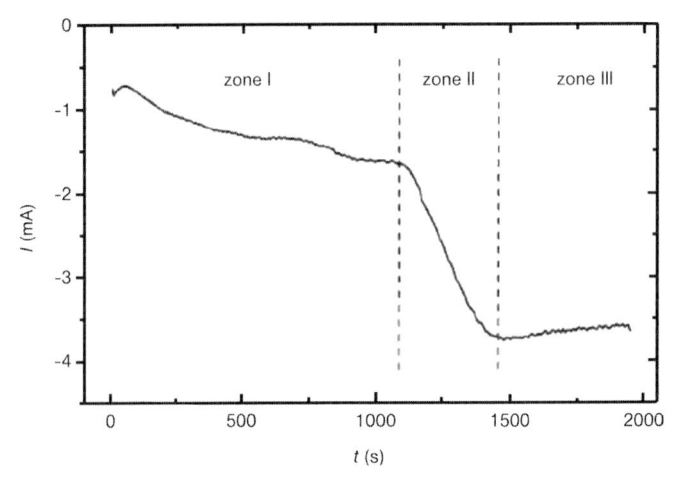

Figure 2.34 Current (*I*) versus time (*t*) plot during the electrodeposition process. (Reprinted from Ref. [157].)

The structural characterization of such arrays has been achieved using X-ray diffraction, thereby revealing that Co grew with a hexagonal structure. Subsequent AFM images were recorded of isolated nanowires after the membrane had been dissolved with appropriate chemicals. A typical AFM image of a 3.1 μm-long nanowire (diameter 90 nm) is shown in Figure 2.35.

It has been proposed that these electrodeposited magnetic nanowires are used as ultra-high-density magnetic recording media [158–161]. This type of use is based on the perpendicular anisotropy of nanowires, which derives from the shape anisotropy of these systems. However, in some cases (e.g., as in Co nanowires), due to the competition of magnetocrystalline and shape anisotropies, there may be no perpendicular anisotropy [161]. Ge *et al.* [162] produced nanowires with different microstructures and, consequently, with different magnetic properties. These authors [162] perfected the conditions for obtaining adequate perpendicular anisotropy by changing the depositing voltage or applying a magnetic field during the process. These different magnetic behaviors can be related to differences in their microstructures, as both X-ray diffraction and electron diffraction studies have confirmed that the former sample is amorphous, whereas the latter is polycrystalline. In the

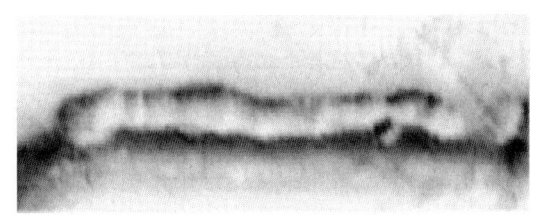

Figure 2.35 AFM image of a 3.1 μm-long, 90 nm-diameter Co nanowire. (Reprinted from Ref. [157].)

polycrystalline sample, due to the competition of shape and magnetocrystal anisotropies, the sample does not display any perpendicular anisotropy. However, magnetocrystal anisotropy is minimal in the amorphous sample, and hence shape anisotropy plays a dominant role which leads to strong perpendicular anisotropy. In alternative experiments, the application of a magnetic field during deposition caused Co grains to grow preferentially along the c-axis, and this also led to a strong perpendicular anisotropy.

Direct current electrodeposition into PAA templates is most often performed by removing the remaining aluminum substrate, opening the pore bottoms by etching in phosphoric acid, and finally depositing a conducting layer (e.g., gold) onto one face of the template [163–166]. Choi et al. [167] have reported a method where the barrier layer was reduced to a thickness which was proportional to a 1 V final anodization potential, at which point the barrier layer was sufficiently thin to enable DC electrodeposition, without prior removal of the aluminum substrate. Unfortunately, however, the barrier layer thinning process is time-consuming and must be accurately controlled to ensure that the resultant barrier layer is thinned uniformly. By utilizing this procedure, the electrodeposition of 30 μm-long silver nanowires was achieved [167]. Recently, Sander et al. [168] also reported DC electrodeposition into PAA formed by depositing silver onto one side of a thin (<100 μm-thick) Al foil, preparing an electrode, and anodizing the Al completely. However, this procedure requires the adhesion of a very thin foil to a substrate (e.g., glass), and it is difficult to uniformly anodize the entire foil. For industrial-scale processes, or applications where a rapid method of template production is required, it is still preferable to develop an AC deposition process as this produces wires of comparable length, without any need for a barrier layer thinning process.

In general, AC electrodeposition through the barrier layer involves many variables that must be controlled, including the electrolyte concentration, composition, temperature, deposition potential, frequency, waveform (sine, square, and triangle), and pulse polarity [169–171]. Moreover, the optimal deposition conditions appear to depend on the metal or compound to be deposited [170].

The feasibility of the electrodeposition of copper into PAA templates under AC conditions from an aqueous metal salt solution was reported by Davydov et al. [172], although the experimental details were not clearly described and the resultant wires were only half-filled in the case of the 5 μm-long pores. Details of the electrodeposition of different metals (including Fe [173], Ni [174], Co [175], Cd [176], Bi [177] and Au [178]) into PAA templates under AC conditions have also been published in the literature. Almost all of the materials prepared by AC deposition used <25 nm-diameter pore anodic alumina grown in sulfuric acid. It is known that the structure and chemical properties of the barrier layer grown in oxalic acid differs significantly from those using PAA grown in sulfuric acid solutions [179], and this difference impacts on AC electrodeposition into the template, as might be expected.

The frequency of the deposition signal had the most significant effect on pore-filling. In one report investigating the copper nanowires formation [180], the authors noted that the best results were obtained using 200 Hz, and that pulsed (non continuous) or continuous deposition also modified the nanostructure characteristics, while continuous wave depositions resulted in improved pore-filling. The results also indicated that final anodization voltage during alumina formation (and consequently the barrier layer thickness) should be set at the low value in order to maximize pore filling. The best-identified conditions led to a complete pore filling with monodisperse wires. Some SEM images of nanostructures prepared under different conditions are presented in Figure 2.36a and b.

Different research groups [181,182] have suggested that silver particles deposited using AC electrolysis reproduce the shape of the alumina pores. For example, Figure 2.37 shows the TEM images of the reverse side of alumina grown in different acid solutions, and the contours of silver particles deposited within these pores [183]. The calculated values of the alumina average pore diameters were 13.5, 40 and 80 nm for sulfuric, oxalic and phosphoric acid baths, respectively. Therefore, alumina templates with pore densities of about 0.81×10^{11}, 1.25×10^{10} and 3.08×10^{9} pores cm^{-2} were used in this investigation. It can also clearly be seen from the experimental images that alumina templates grown in oxalic acid bath exhibit the most uniform array of cells and pores. As can be observed, the Ag nanowire diameters were proportional to the template pore size.

The procedure for synthesizing platinum nanowires (PtNWs) using an alumina template is simple and easy to control [184]. After the deposition, the alumina template is immersed into, for example, an 8% phosphoric solution to dissolve the template membrane [185,186]. Figure 2.38 shows: (a) the template pristine surface; (b) the PtNW bundles obtained after template dissolution; and (c) an individual PtNW. The high-density and well-aligned PtNWs present a comparatively even distribution of diameters of about 250 nm and average length of 5 μm.

2.3.1.3 Nanotubes

Carbon nanotubes (CNTs) have been extensively studied because of their unique electronic, chemical and mechanical properties, which leads in turn to a very large number of potential applications [187–190]. These materials can promote electron-transfer reactions with enzymes, and this has resulted in a series of investigations of their use in biosensor interface fabrication [191–193]. Indeed, the integration of CNTs and some other materials, such as conducting polymers [194], redox mediators and metal nanoparticles with a synergistic effect, has shown particular promise as chemical sensors [195–198]. The group of Professor Yang has reported a synergistic effect for biosensor fabrication, such as using CNTs and cobalt [199] or platinum [200,201] nanoparticles. These authors also showed that a combination of platinum nanoparticles with CNTs produces interesting electrode materials for catalytic applications [201].

Platinum is important for many applications due to its chemical, catalytic and electrocatalytic properties [202]. Platinum particles are used in many different applications [203–205], and many research groups are currently investigating PtNWs and composites of these materials with carbon nanotubes (Pt-CNT). These materials have been prepared using different templates such as alumina [206] and silica [207], or directly over CNT networks [208].

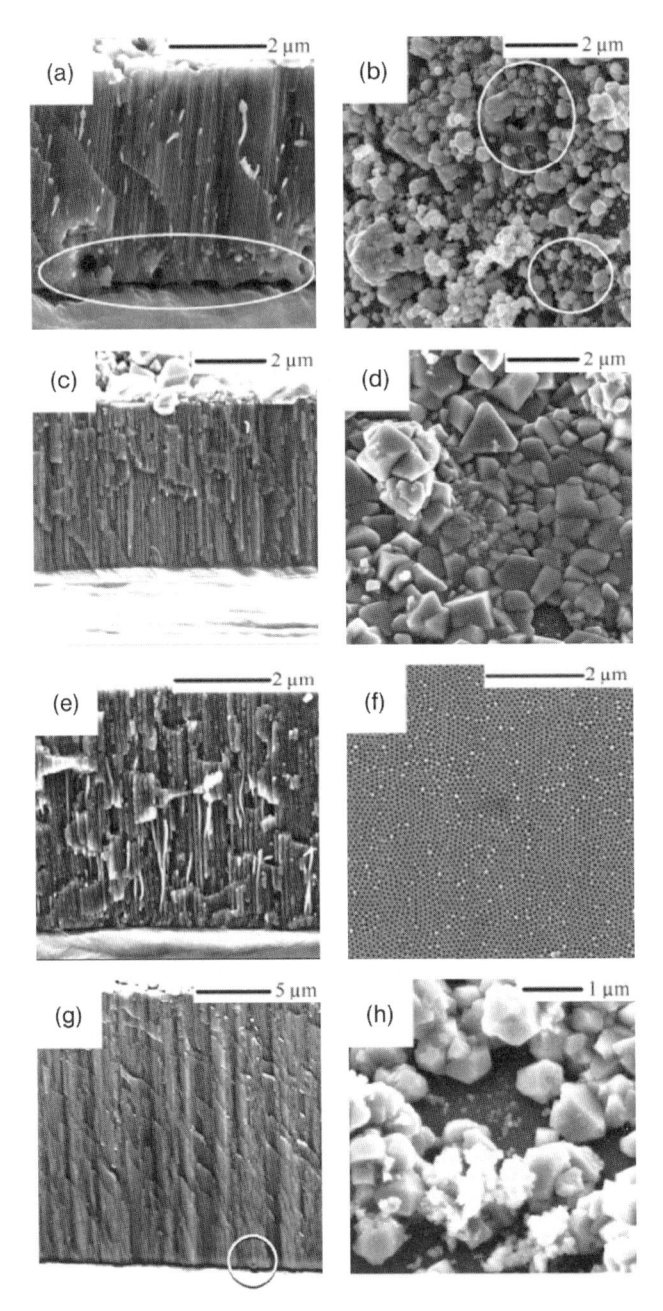

Figure 2.36 SEM images of oxalic acid-anodized templates filled using continuous size wave deposition with (a, b) bulk growth, (c, d) pulsed sine wave deposition with bulk growth, and (e, f) continuous sine wave deposition without bulk growth. (g, h) SEM images of sulfuric acid-anodized templates filled using continuous sine wave deposition. In (a–f) the pores are 40 μm deep; in (g) and (h) the pores are 25 μm deep. Circles in (a), (b) and (g) identify the area of template damage. (Reprinted from Ref. [180].)

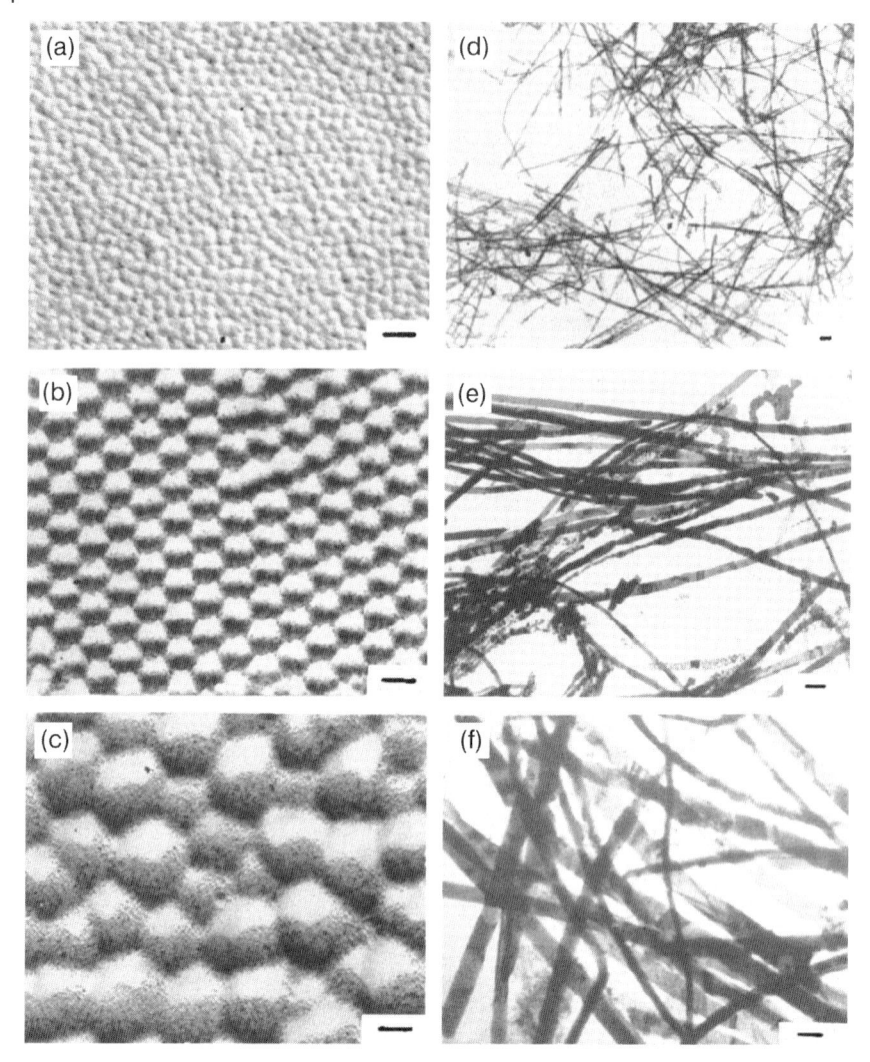

Figure 2.37 TEM images of the alumina/metal interface (a–c) and silver nanowires (d–f) deposited within the alumina template pores. Alumina was grown in sulfuric (a, d), oxalic (b, e) and phosphoric (c, f) acid baths at 15, 40, and 80 V for 1, 2, and 3 h, respectively, with a subsequent decrease in E to 10 V. The nanowires were grown in a solution of 10 mol dm^{-3} AgNO$_3$ + 50 mol dm^{-3} MgSO$_4$, under an AC voltage 15 V control for 10 min. Scale bars = 100 nm. (Reprinted from Ref. [183].)

A few reports have also been made regarding the applications of PtNWs on biosensor fabrication. Qu et al. [184] studied the template strategy combined with potentiostatic electrodeposition techniques to produce PtNW-CNT composites for fabricating biosensors. In order to improve even more the composite electrochemical properties, these authors [184] prepared solutions of the composites with chitosan (CHIT) to create a PtNW–CNT–CHIT system. The resulting PtNW–CNT–CHIT material introduces

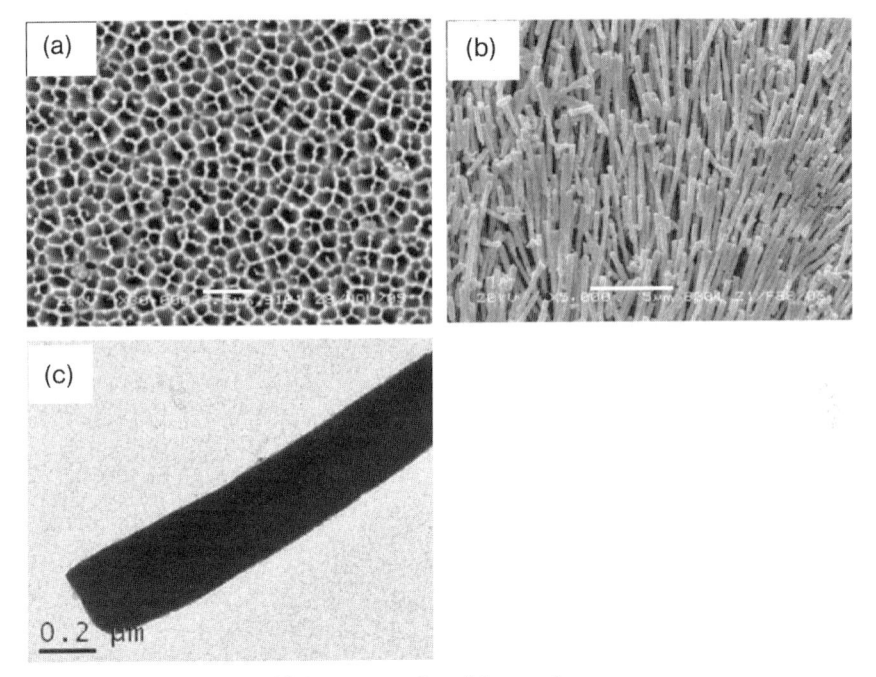

Figure 2.38 (a) SEM image of the primitive surface of the sample. (b) SEM and (c) TEM images of platinum nanowires synthesized by the method of templating. (Reprinted from Ref. [184].)

new capabilities for electrochemical devices due to the interaction of PtNWs and CNTs, which facilitates the electron-transfer process. The behavior of PtNW–CHIT is similar to that expected for polycrystalline platinum [209]. The PtNW–CNT–CHIT film presents a sharp increase in the current peak, but without any increase in peak potential. The fact that there was virtually no change in the cyclic voltammetry (CV) shape and peak potentials suggests that the introduction of CNTs did not affect the electron-transfer process between the PtNWs and the electrode support [210].

The deposition of semi-conducting material over CNT was proposed by Rajeshwar et al. [211], who deposited ZnSe on fibers via cathodic electrosynthesis. The contrast between the "before" and "after" morphology of the nickel-coated carbon fiber after electrodeposition of ZnSe is shown in Figure 2.39.

The cathodic electrosynthesis of metal oxides has a more recent history relative to their chalcogenide counterparts [212–214]. Two types of cathodic deposition processes can be envisioned from a mechanistic perspective. The first process involves a change in the oxidation state, as exemplified by the cathodic electrosynthesis of Cu_2O from Cu(II) precursor species [215]. The second process is based on an electrochemical reaction involving oxygen or other additives (e.g., nitrate) [216–218]. The net effect is an electrochemical generation of OH^- ions and a subsequent precipitation of the oxide/hydroxide phase. A post-deposition thermal annealing then converts the hydroxide phase to the desired oxide material. Table 2.2 contains

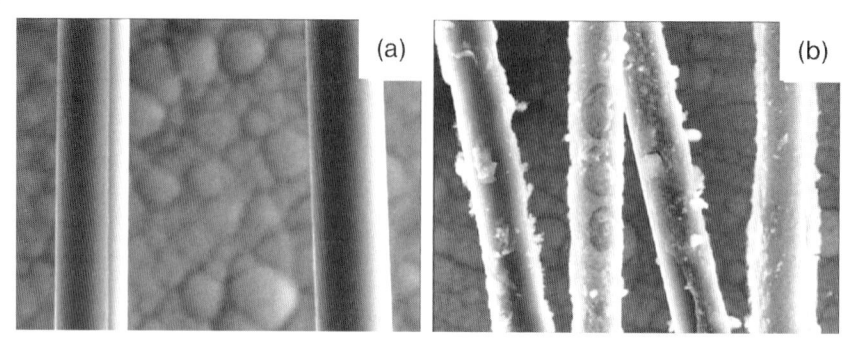

Figure 2.39 SEM image of nickel-coated carbon microfibers (INCOFIBER® 12k20) before (a) and after (b) the cathodic electrosynthesis of ZnSe on their surfaces. (Reprinted from Ref. [211].)

Some examples of oxide semiconductors that have been prepared via the cathodic deposition route are listed in Table 2.2.

In general, recent innovations in oxide electrosynthesis have included the fabrication of superlattices [232], epitaxial growth on selected substrates [233,234], and the preparation of large-scale, uniform nanowire arrays [235].

ZnO is an important oxide semiconductor material because of its good optical, electrical, and piezoelectrical properties. The preparation of well-ordered nanowire arrays is an important step towards fabricated micro-optoelectronic devices [236]. Ordered ZnO nanowire arrays embedded in PAA templates have already been fabricated by electrodepositing Zn into PAA nanopores to form metallic nanowire arrays and then oxidizing them thermally to ZnO [237]. The long time necessary for thermal oxidization ($\sim$35 h) at 300 °C limits its application. This material has also been synthesized by one-step electrodeposition conducted in aqueous zinc nitrate solutions [238], again using PAA as the template. However, the aqueous solution induces the deposition of $Zn(OH)_2$ competing with the formation of ZnO [239]. The problem here is that the $Zn(OH)_2$ compound can quench the near band emission of ZnO [240]. Gal *et al.* [241] reported a new approach to prepare the ZnO film involving non-aqueous dimethyl sulfoxide (DMSO), without $Zn(OH)_2$ formation. Furthermore, a higher deposition temperature can be used in non-aqueous baths, which usually leads to better crystallinity and larger crystal size [242], and avoids blocking of

Table 2.2 Examples of oxide semiconductors prepared by cathodic electrodeposition [211].

Semiconductor	Reference(s)
CuO_2	[219]
ZnO	[220–222]
TiO_2	[223–225]
MoO_3	[226–228]
WO_3	[229–231]

the nanopores of the PAA template during the electrodeposition process. By using a variation of the method developed by Gal et al. [241], Wang et al. [243] prepared ZnO nanowire arrays from a DMSO solution containing $ZnCl_2$, and a molecular oxygen precursor. A typical surface view of a PAA template is shown in Figure 2.40a. Highly

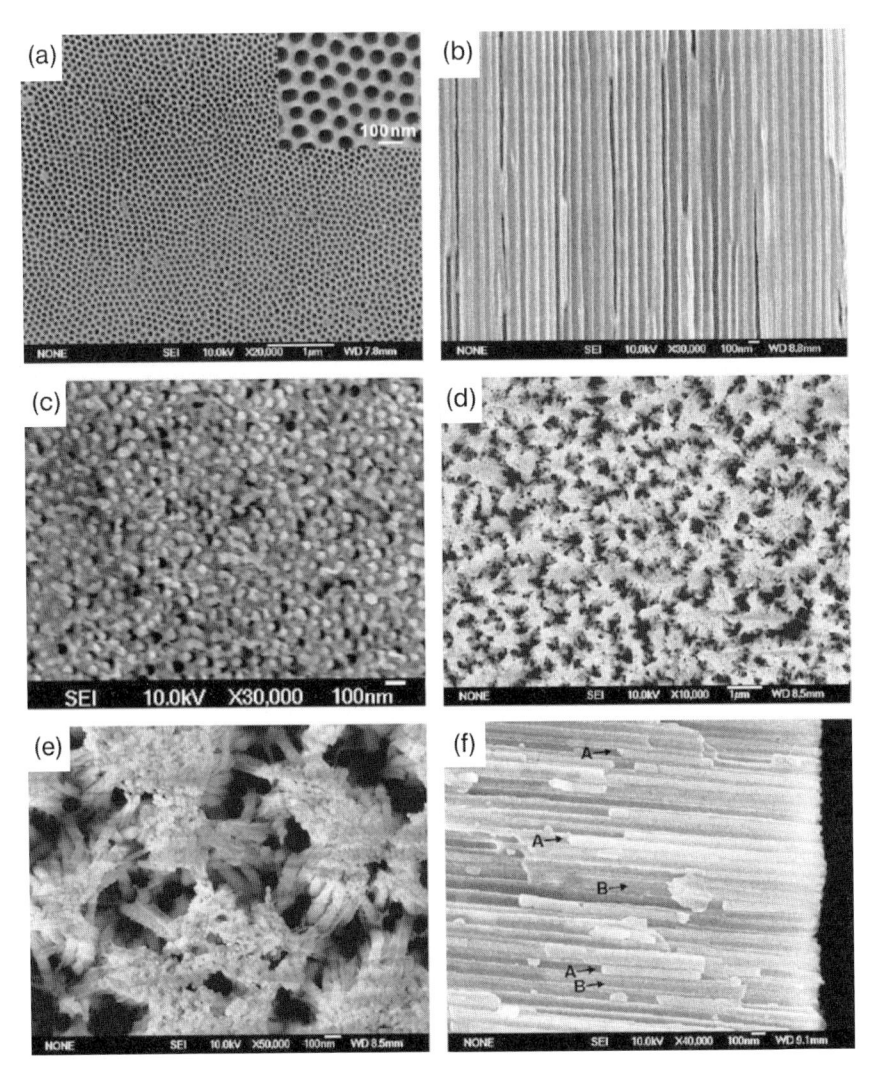

Figure 2.40 SEM images of a PAA template and ZnO nanowire arrays. (a) Surface view of the PAA template. The inset at upper right shows a close-up view of the hexagonal arrangement of the nanopores. (b) Cross-sectional view of the PAA template. (c) Bottom view of ZnO nanowires embedded in a PAA template after the Au-evaporated electrode was mechanically polished off. (d, e) Large-scale uniform ZnO nanowires with the PAA template partly dissolved in $1\,mol\,dm^{-3}$ NaOH solution. (f) Cross-sectional view of ZnO nanowire arrays embedded in an PAA template, for which deposition was carried out for 2 h. A (arrowed) indicates the ZnO nanowires; B (arrowed) indicates the pores of the PAA template. (Reprinted from Ref. [243].)

ordered pores with diameters about 60 nm and an interpore spacing of ∼100 nm can be observed. Figure 2.40c shows the SEM image of the bottom of the ZnO nanowires and PAA assembly with the Au-evaporated electrode mechanically polished off. The bright regions show nanowire-filled pores, while the dark regions show unfilled pores. The ZnO nanowires with 60 nm in diameter are uniformly embedded in the PAA nanopores, and more than 90% of the PAA nanopores are filled. Figure 2.40d and e show the surface images of the ZnO nanowires.

It can be seen that the ZnO nanowires are uniform, smooth, and the diameters are about 60 nm, which is basically equal to that of PAA template nanopores. The cross-sectional image of ZnO nanowire arrays deposited for 2 h is shown in Figure 2.40f. Each individual ZnO nanowire is dense, and completely fills the pore.

2.3.2
Multilayers

Superlattices and multilayers are especially well suited to devices applications, as the confinement dimension (i.e., the individual layer thickness) can be maintained in nanometer range. They are composed of alternating layers of materials, with a bilayer thickness known as the "modulation wavelength". Although multilayers and superlattices are both modulated materials, superlattices have the additional constraint that they are crystallographically coherent [244]. Because of this constraint, superlattices are usually produced with alternate layers of materials with very low lattice mismatch, whereas multilayers can be produced using even amorphous materials.

One reason for the interest in superlattices and multilayers is their enhanced mechanical properties, and today multilayered materials are being considered for protective coatings, as their hardness greatly exceeds the values for comparable bulk alloys [245–249]. With two alternating phases, the resistance to plastic deformation and hardness increase as the modulation wavelength decreases [245]. Although many of these studies have focused on metallic multilayers, single-crystal TiN/VN strained-layered superlattices with extremely high mechanical hardness have been produced by reactive magnetron sputtering [250]. With a modulation wavelength of 5.2 nm, the multilayer had a hardness that was more than 2.5-fold that of TiN, VN, or a $Ti_{0.5}V_{0.5}N$ alloy. Tench and White have used electrodeposition to prepare Ni–Cu multilayers with an ultimate tensile strength (1300 MPa) that was a factor of three greater than that measured for pure Ni metal, and more than twice as large as the value for the corresponding alloy of Ni and Cu [251].

Most studies performed on the fabrication of nanometer-scale superlattices and multilayers have focused on the use of chemical and physical vapor phase deposition processes in an ultra-high vacuum. Electrodeposition can be applied to the synthesis of such materials, and has several advantages [252,253], including: (i) a low processing temperature (often room temperature) which minimizes atom interdiffusion between the layers; (ii) the film thickness can be controlled by monitoring the deposition charge; (iii) the composition and defect chemistry can be controlled through the applied potential; (iv) complex shaped film depositions can be made; and

(v) non-equilibrium phases can be deposited. Possible disadvantages of this procedure are the requirement that the substrate and film have electrical conductivity, and that there is the possibility of contamination of the film from the solution foreign species.

The electrodeposition of compositionally modulated nanostructures such as superlattices and multilayers has been reviewed elsewhere [254,255]. Useful reviews on the general electrochemical synthesis and modification of materials, including metal oxide ceramics, are also available [256,257]. There are two general techniques for the electrochemical deposition of superlattices and multilayers, namely *dual bath* and *single bath* deposition:

- Dual bath deposition is the simpler of the two methods to design, but it is more difficult to implement when large number of layers are necessary, or even with small modulation wavelengths. The electrode is alternated between two deposition solutions containing the chemical precursors of the different layers.

- In single bath deposition, the precursors for both layers are in the same deposition solution. Single bath deposition requires a much more accurate control of the system chemistry, but it is ideal for producing nanostructures with an almost unlimited number of layers. In addition, the electrode is not exposed to the atmosphere between deposition cycles, and the electrical potential control can be maintained throughout the process.

Multilayered metallic structures were first produced by Blum in 1921 by depositing alternating Cu and Ni layers using the dual bath method [258]. The layers were fairly thick (≥ 24 µm), and the films were examined for their higher tensile strengths than the corresponding pure metals or alloys. Brenner first used single bath deposition in 1963 to produce multilayers of Cu and Bi [259]. Several years later, in 1987, Cohen *et al.* produced multilayers with layers as thin as 50 nm of Ag and Pd by pulsing either the current or the potential in a single plating bath [260]. During the same time period, several groups used single bath deposition to prepare multilayers of Cu and Ni [261,262,251]. The Cu/Ni system has been extensively investigated due to its low lattice mismatch (2.6%), and the electrochemical deposition of the metals are almost irreversible – that is, the Ni layer becomes passive and does not re-dissolve during the lower-potential deposition of Cu [263,264].

With regards to the Cu–Co multilayers nanostructures, Kelly *et al.* [265,266] have clearly shown that sodium dodecyl sulfate promotes the displacement reaction of Co by Cu^{2+} during pulse off-times. Gundel *et al.* [267] assumed that the displacement reaction could be negligible under galvanostatic conditions. Fricoteaux *et al.* [268] studied the electrodeposition of Cu–Co multilayers without brightener and leveler, the aim being to quantify the displacement reaction between cobalt and copper.

Tench and White [269] have used the electroless displacement reaction to establish a new process of Cu–Ag multilayered alloy production. These authors suggested that the electroless process might go on beyond few nanometers because of the Cu^{2+} transport via the pores of the growing Ag deposit.

The superimposition of an external magnetic field offers new possibilities to influence the deposition process, mainly by magnetohydrodynamic (MHD) effects,

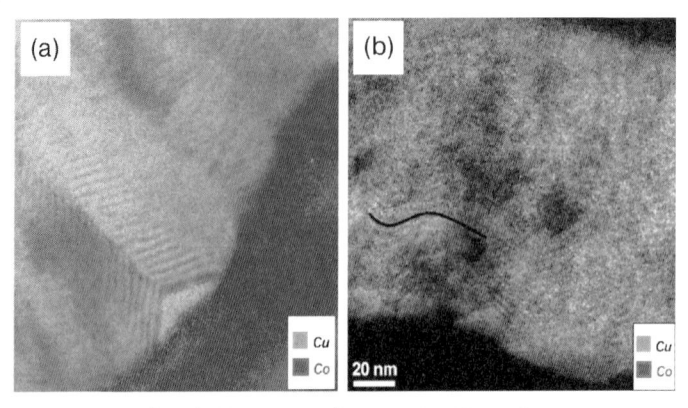

Figure 2.41 Analytical TEM images of [1.9-nm Cu/2.5-nm Co-Cu]*30 multilayers deposited on Si with a Py/Cu seedlayer without a magnetic field (a) and in a non-uniform field generated by an SmCo permanent magnet (b). (Reprinted from Ref. [274].)

and thus to change the microstructure of the grown layers in correlation to the electrical and magnetic properties [270–273]. Furthermore, magnetic fields applied during electrodeposition induce an orientation of grains in the direction of easiest magnetization axis. An uniaxial anisotropy was found for Permalloy layers [272] investigating the deposition of Cu/Co–Cu multilayers. Uhlemann *et al.* [274] studied the superimposition of magnetic fields during deposition and correlated this to microstructure and giant magnetoresistance (GMR) properties. A TEM image showing columnar growth (layer-by-layer) with angularly aligned grains is shown in Figure 2.41a, where each single layer consists of only 10–14 atomic monolayers. The kinks at the grain boundary may lead to a short circuit and reduce the coupling effect, and therefore the GMR effect, as discussed by Shima *et al.* [275]. The detailed TEM investigation (Fig. 2.41b) indicates a more wave-like and near-parallel arrangement of the single layers with a low degree of roughness. As a result, the measured GMR effect increases from 2.8 to 3.9% (at 150 mT). The alignment of the grains in the 100% (1 1 1) direction seems to be caused by the preferred orientation of Co during the deposition in an external magnetic field. The applied magnetic field oriented parallel to the electrode surface generates the classical, aforementioned MHD effect, caused by Lorentz force. The convection in the diffusion layer is enhanced, which causes a reduction in the diffusion layer thickness that in turn results in an increase of the limiting current density under the conditions in which the process is transport-controlled.

The discovery of perpendicular magnetic anisotropy, the magneto-optical Kerr effect [276–278], and GMR [279–281] in metallic multilayers has greatly stimulated interest in magnetic multilayered nanostructures. Jyoko's group presented the first evidence for composition modulation across successive layers in a Co/Pt nanometer-multilayered structure grown by electrodeposition [282]. Jyoko demonstrated the presence of perpendicular magnetic anisotropy in an electrodeposited Co/Pt or

CoNi/Pt multilayered nanostructure, as well as GMR in Co/Cu nanostructures. Multilayered Co/Pt, CoNi/Pt, and Co/Cu thin films were grown on a Cu (1 1 1) substrate, respectively, from two separate electrolytes for the deposition of the constituents of the multilayer by transferring the substrate repeatedly from one to the other under potential control, and from a single electrolyte by repeatedly controlling the electrode (substrate) potentials for the alternate deposition of both constituents. For a heterogeneous Co–Cu alloy, which consists of ultrathin FCC Co-rich clusters in a non-magnetic Cu matrix, a large saturation magnetoresistance of more than 20% at room temperature was obtained, together with a much higher saturation field and a smaller remanent magnetization, which suggested the presence of large antiferromagnetic coupling.

In granular magnetic alloys the largest GMR has been reported in Co_xAg_{1-x} ($x \sim 0.2$) [283,284]. These authors suggested that such effect is related to the cobalt and silver immiscibility [285], which favors a good granular structure with abrupt interfaces. Moreover, the Co–Ag system is known to have a percolation threshold concentration [286] which is larger than in other systems. This favors a larger concentration of magnetic scattering centers contributing to magnetoresistance. Usually, granular magnetic films are deposited by sputtering techniques [287,288]. Granular Co–Ag films have already been produced by electrochemical deposition and show 5% GMR at room temperature [289,290].

The structure of metal atoms deposited on a metal surface is also matter of investigation [291], due to the enhanced properties of the deposited metal layers, alloys and multilayers [292–294]. Consequently, they have received great attention as a new class of magnetic and catalysts [295,296]. Multi-component catalyst systems consisting of two or more active metal components or both metal and oxide constituents [297,298] are often used to carry out electrocatalytic reactions.

Vukovic [299] studied the electrochemical behavior of rhodium in $HClO_4$, and showed that the surface roughness depended on the current density and electrodeposition time. From a theoretical point of view, Norskov et al. [300–304] investigated the effect on the associated electrocatalytic properties of the monolayer deposition over a metal to afford a new material. Oliveira et al. was the first to study the use of nanometric multilayers as electrocatalysts investigating the Rh/Pt [305,306] and Ru/Pt [307] multilayers deposited on a Pt substrate and their electrocatalytic activity for small organic molecule oxidation. The voltammetric profiles for methanol, ethanol, formaldehyde and formic acid oxidation on Pt and the Pt/Rh/Pt bilayers are shown in Figure 2.42. Methanol oxidation at the Pt substrate (dotted line) and the Pt/Rh/Pt bilayer (solid line) is shown in Figure 2.42a. Over the Pt/Rh/Pt bilayers, an increase of 295% in the current density peak density for methanol oxidation was observed. An important increase in the current density was observed during the oxidation of all molecules investigated. The increase in the current observed cannot be explained by an increase of the surface when the roughness factor is the same as shown in Figure 2.43 [307] for Pt/Ru multilayers where the same type of results were obtained. The mechanisms of these effects are currently under investigation in the present authors' group.

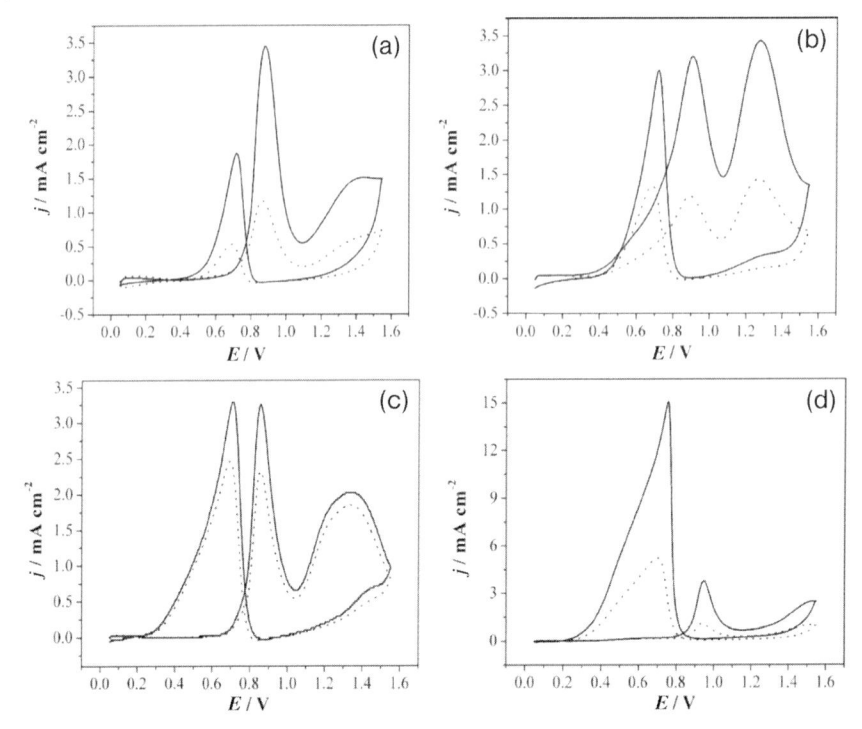

Figure 2.42 Cyclic voltammogram for an oxidation of small organic molecules in 0.1 M $HClO_4$ at a polycrystalline Pt electrode (dotted line) and Pt–Rh(2.3 monolayers)/Pt(1.7 monolayers) (solid line). (a) 0.5 mol dm^{-3} H_2COH; (b) 0.5 mol dm^{-3} H_2CCH_2OH; (c) 0.1 mol dm^{-3} HCOH; (d) 0.1 mol dm^{-3} HCOOH. Sweep rates $= 0.1 V s^{-1}$. (Reprinted from Ref. [305].)

2.3.3
Other Materials

Electrochemical metal deposition is one of the oldest subjects within the framework of electrochemistry. Metal electrodeposition takes place at electrode/electrolyte interfaces, under an electric field, and includes the phase nucleation and growth phenomena.

Deposition from a solution of M^{z+} on the substrate of the corresponding metal, M (e.g., Ag^+ on Ag), begins at the thermodynamically reversible potential. However, if an attempt is made to deposit M^{z+} on S, a substrate differing from M (e.g., Ag^+ on Au), some kind of layer formation on the S surface begins to occur when it is held at a potential many hundreds of millivolts positive to the reversible potential for M^{z+} onto M. When deposition occurs at a potential more *positive* than the thermodynamically reversible potential, the process is called *underpotential deposition* (UPD) [308]. A characteristic of UPD process is the deposition of only one monolayer or

(a)

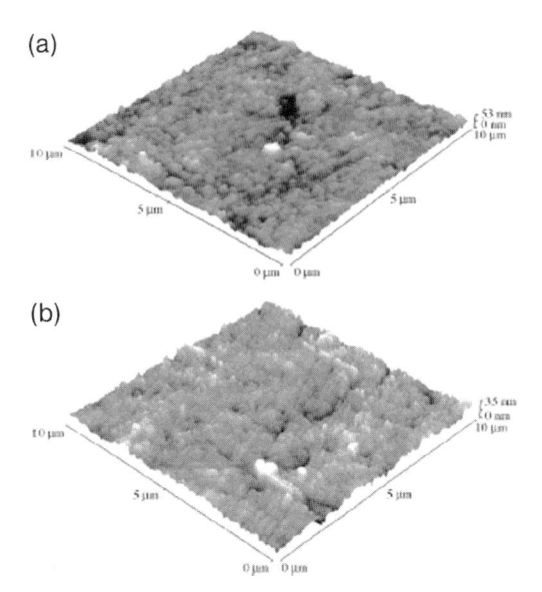

(b)

Figure 2.43 Atomic force microscopy images for: (a) Pt substrate and (b) Ru/Pt bilayer electrodeposited over Pt. (Reprinted from Ref. [307].)

even a submonolayer. Figure 2.44 shows the UPD of 2D nucleation process of Ni on Au (1 1 1) studied by Freyland et al. [309].

In the case of Ni deposition, a complete monolayer is formed after several minutes of the potential. In that case, a regular hexagonal superstructure (Moiré pattern) with a lattice constant of 23 ± 1 Å and with modulation amplitude of 0.6 Å is observed in Figure 2.44. This interpretation is supported by simultaneously measured current transients, which yield an integrated value of the charge of $530 \pm 50\,\mu C\,cm^{-2}$, corresponding to a 0.9 monolayer of Ni [310,311].

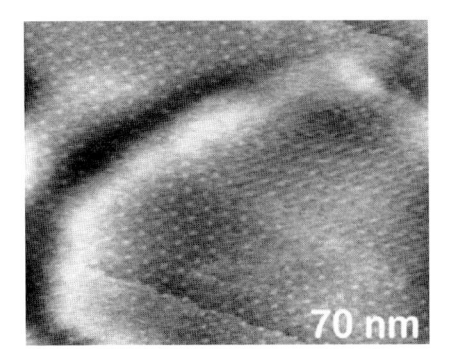

Figure 2.44 Scanning tunneling microscopy images of 2D nucleation of Ni on Au(1 1 1). (a) Moiré pattern of a Ni monolayer electrodeposited at $E = 0.11$ V versus Ni/Ni(II), $E_{tip} = 0.2$ V, $I_{tun} = 3$ nA. (Reprinted from Ref. [309].)

During the 1960s, Schmidt et al. [312–314] were first to recognize the importance of the UPD process in the overall electrocrystallization, and started a systematic study on this subject by introducing a number of thin-layer techniques. In 1973, Lorenz et al. [315–317] performed UPD experiments on single-crystal substrates and proposed an explanation for the experimental results by considering the formation of well-ordered 2D metal overlayer with a different "superlattice structure". Subsequently, extensive studies were performed on the thermodynamics and kinetics of UPD of metals in various systems by Yeager et al. [318,319], Bewick et al. [320], Schultze et al. [321,322], Lorenz, Schmidt and coworkers [323–326], and Kolb et al. [327,328]. Staikov, Lorenz and Budevski [329–333] introduced the use of "quasi-perfect" silver single crystal faces as substrates in the UPD investigation, and established the first important correlation between the processes of UPD and OPD (normal bulk overpotential deposition) of metals under a strong theoretical base. During the past two decades, new important information on the atomic structure and morphology of substrates and metal electrodeposits was obtained by the in-situ application of different modern surface analytical techniques such as extended X-ray absorption fine structure (EXAFS), grazing incident X-ray scattering (GIXS), and scanning probe microscopy (SPM) [334,335].

Solid/liquid interfaces play an important role in nanotechnology, as they provide advantages for the preparation of well-defined nanostructures, without irreversible modifications due to the preparation process [336]. Both, the current flow and supersaturation of species at this interface can be accurately adjusted during electrochemical nucleation and growth, although the supersaturation is usually a fluctuating parameter in ultra-high vacuum deposition processes. This feature results in well-defined nucleation processes at solid/liquid interfaces [337] which are important for the growth control of nanostructures. In addition, electrochemical nucleation and growth can be performed near thermodynamic equilibrium, whereas laser ablation or sputtering processes involve high energies of the involved particles. Nucleation processes at electrochemical interfaces can be deposited directly onto nanoelectrodes using scanning tunneling microscopy (STM) as a building technique instead of an imaging one. Thus, electrochemistry allows the bottom-up growth of nanostructures, thereby avoiding irreversible modifications during the preparation process. This is of particular importance in nanotechnology as the properties of the material are determined by their surface and interface atoms, and the surface modification by defects or passivation may result in a material with completely different physical or chemical properties.

There are many reports where nanostructures have been deposited at electrochemical interfaces, using STM tips as preparation tools [338–340]. The advantage of STM is its accurate control, which allows the species to be placed at the exact point in a nanoscale range below 10 nm. Use of the tip of a scanning tunneling microscope as a "nanoelectrode" provides a higher resolution than with the scanning electrochemical microscope (SECM), which uses a capillary to provide a locally high Me^{z+} ion concentration [341]. This consists of a two-step procedure which generates, in the first step, a nanoelectrode by the deposition of Me^{z+} onto the STM tip, and in a second step the required a local supersaturation condition is reached by an abrupt

2.3 Cathodic Synthesis | 165

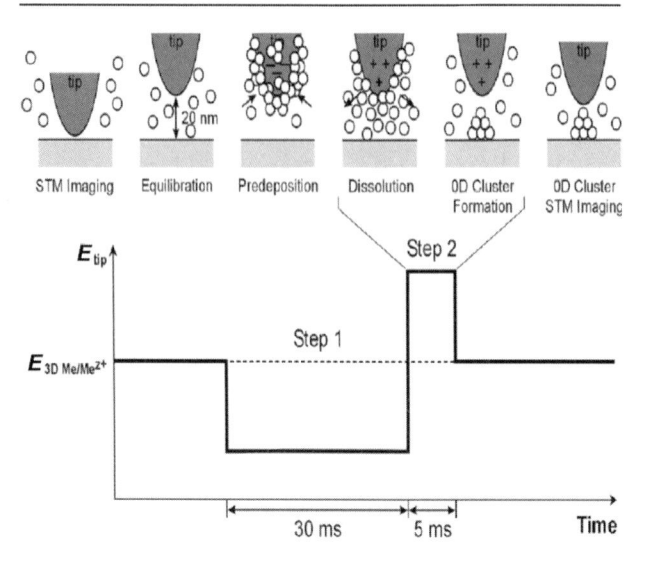

Figure 2.45 Schematic of the two-step process of localized electrodeposition utilizing a STM tip as a nanoelectrode. The distance between the tip apex and substrate surface during the deposition process is of the order of 20 nm (i.e., much larger than a tunneling gap). Step 1: Electrodeposition of Me^{z+} from the electrolyte onto the STM tip and generation of the nanoelectrode. Step 2: Dissolution of Me^{z+} from the STM tip, generation of local supersaturation conditions, and nucleation of Me^{z+} on the substrate surface underneath. (Reprinted from Ref. [342].)

dissolution of Me^{z+} from the STM tip (Figure 2.45), as described by Hulgemann et al. [342–345].

2.3.3.1 Semiconductors

Several reviews and books have been produced on the electrodeposition of semiconductors (e.g., Refs. [346–348]). One specific book, written by Pandey et al. in 1996 [346], provided the first complete view of the field, including references dating back to the late 19th century. The electrodeposited materials range from elemental semiconductors (Si, Ge) to binaries (Groups III–V, II–VI) and ternary compounds. Hodes focused on the preparation of Group II–VI semiconductors, with a specific survey of the period between 1987 and 1992 [347]. In his article, Hodes cited about 100 references, mainly devoted to CdTe, CdSe, CdS, ZnS, ZnSe, ZnTe and their ternary alloys. The case of chalcopyrite semiconductors, based on $CuInSe_2$, for photovoltaics was reviewed by Vedel in 1998 [349], and by Daniel Lincot in 2004 and 2005 [350,351]. Information relating to oxides is available in two reviews [352,353]. Figure 2.46 shows, schematically, the onset of an impetuous development until 2002 of the oxide cathodic electrodeposition, mainly ZnO, which was introduced in 1996 [354,355].

Despite many methods having been developed for the synthesis of ZnO nano/microtubes, no reports were made on the synthesis of ZnO nanotube arrays directly onto the substrate using an electrodeposition method. The electrodeposition

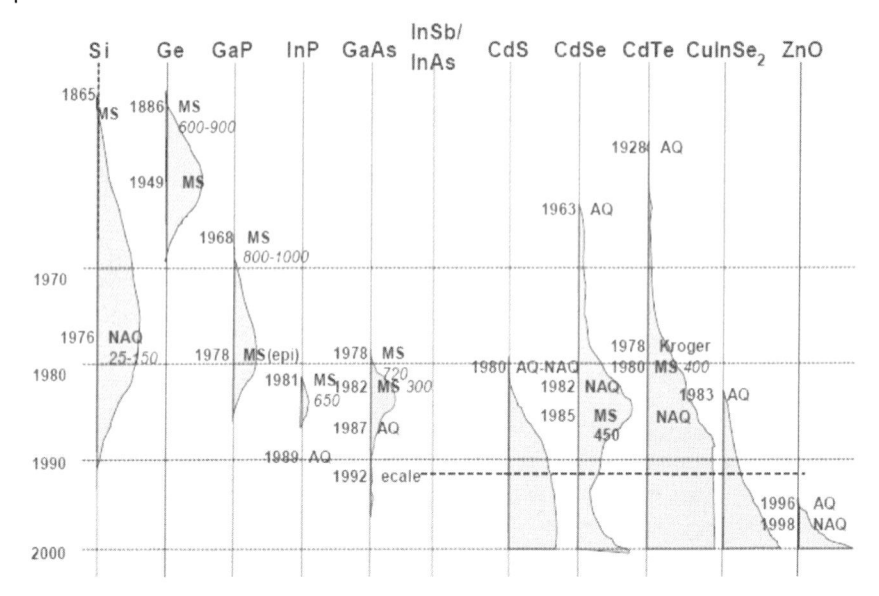

Figure 2.46 Road map of electrodeposition of main inorganic semiconductors established from the analysis of Refs. [347–352] and *Current Contents* database (between 2000 and April 2002). The gray areas correspond to research intensity. MS: Molten salts; AQ: aqueous solvent; NAQ: non-aqueous solvent; ECALE: electrochemical atomic layers electrodeposition. Numbers in bold letters are years; numbers in italic letters are temperatures in °C. (Reprinted from Ref. [351].)

method takes advantages of the possibility of preparing large-area thin films, and the accurate control of film thickness. The electrodeposition of ZnO films has been reported by several groups [356,358], and used in the fabrication of oriented arrays, such as nanowires and nanorods [359–364]. Single-crystalline ZnO nanotube arrays have been fabricated directly onto F-doped SnO_2 (TCO) glass substrates via electrochemical deposition from an aqueous solution by Y. Tang *et al.* [365], although these authors did not use any seed during the preparation. Some images of the morphological evolution of ZnO nanotube arrays obtained at different deposition times are shown in Figure 2.47.

As seen in Figure 2.47, the surface was rather rough because the F–SnO_2 film had a coarse grain structure. After 1 min of electrochemical reaction, nano-scale grains of 10–15 nm could be observed over the whole substrate (see Figure 2.47b). The formation of totally isolated ZnO grains resulted from a total incompatibility between ZnO and the substrate, in contrast to Au [366]. As the reaction time increased, it can be seen from Figure 2.47f that the length of nanotubes increased, and the "broken" parts of nanotubular structure became filled, resulting in the formation of high-quality nanotubes. The thickness of the ZnO nanotube array film was about 300 nm. The key step in this process is the formation of oriented nanowires and their self-assembly to hexagonal circle shapes, as reported by Zhang and colleagues [367]. The nanowires grew subsequently in nanotubular structure, although not all of the

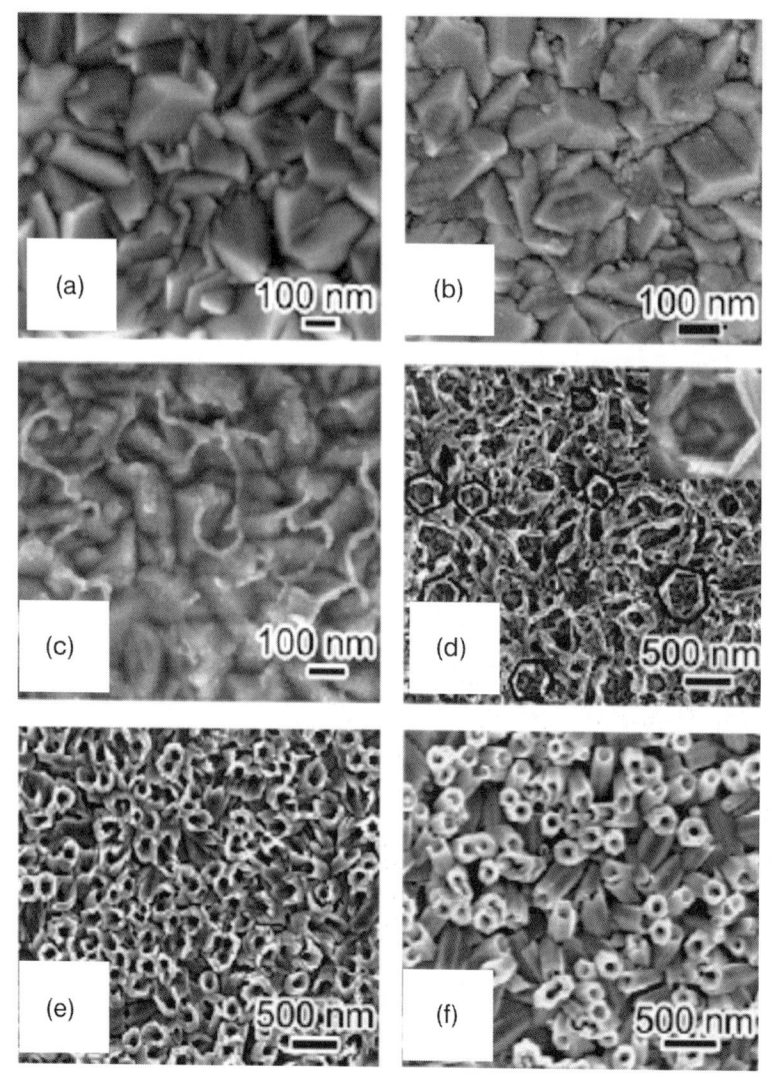

Figure 2.47 SEM images of products prepared on the TCO substrate after different deposition times: (a) 0; (b) 1; (c) 10; (d) 15; (e) 30; and (f) 60 min. (Reprinted from Ref. [365].)

nanowire circle planes were parallel to the horizontal plane, due to the roughness of F–SnO$_2$ surface. ZnO film with remarkable structural quality can be prepared using this method, with different morphologies ranging from arrays of single crystalline microcolumns to continuous films depending on the substrate activation, solution composition and deposition time (at different stages before or after coalescence) [368]. As an illustration, Figure 2.48a and b show the microcolumns of ZnO electrodeposited epitaxially on GaN, both from the surface, with a tilted angle of 45°, and from the cross-section [351].

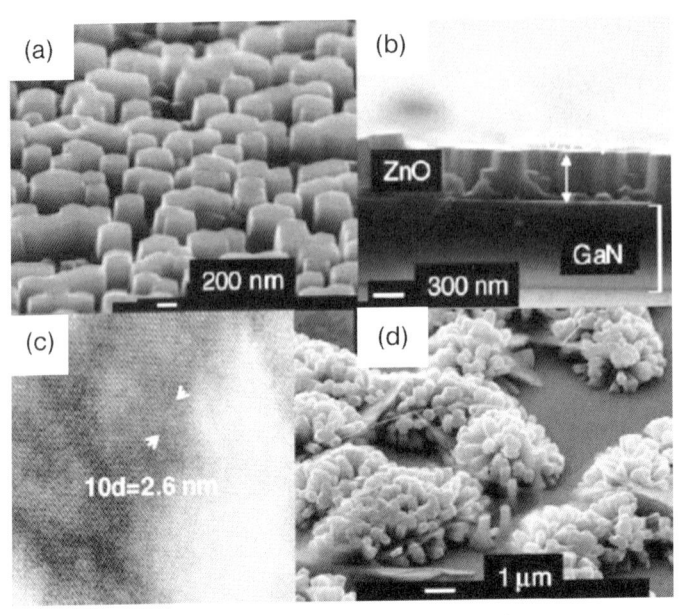

Figure 2.48 Views of ZnO electrodeposited on GAN single crystalline substrates. (a) Epitaxial columns (45° angle). (b) Corresponding cross-section. (c) Internal structure showing high-quality material without extended defects and stacking faults. (d) ZnO "flowers" obtained in non-control conditions. (Reprinted from Ref. [351].)

Structural studies using high-resolution (HR) TEM show that the grains are free of extended defects as stacking faults, as illustrated in Figure 2.48c. This behavior is different from that for CdTe and ZnSe, and indicates that very favorable conditions are present during the growth process. Controlling the shape of electrodeposited zinc oxide is also possible. In addition, by changing the deposition conditions we can modify the aspect ratio of the individual columns. An explanation for this fact comes from the different interactions between the individual surfaces (crystallographic orientation and polarity) and species from the solution. Nucleation effects can also play a role in determining the subsequent growth mode, as illustrated in Figure 2.48d, which shows the formation of ZnO flowers on a GaN substrate.

2.3.3.2 Oxides

Recently, electrochemical capacitors (ECs) have received much attention due to their higher power density and longer cycle life compared to secondary batteries. Also, they have higher energy densities than conventional electrical double-layer capacitors [369–371]. In particular, electrochemical capacitors based on hydrous ruthenium oxides exhibit higher specific capacitance than conventional carbon materials, and better electrochemical stability than electronically conducting polymer materials. Capacitance values up to 760 $F g^{-1}$ (from a single electrode) have been reported [372,373]. However, the high cost of this noble metal limits its commercialization.

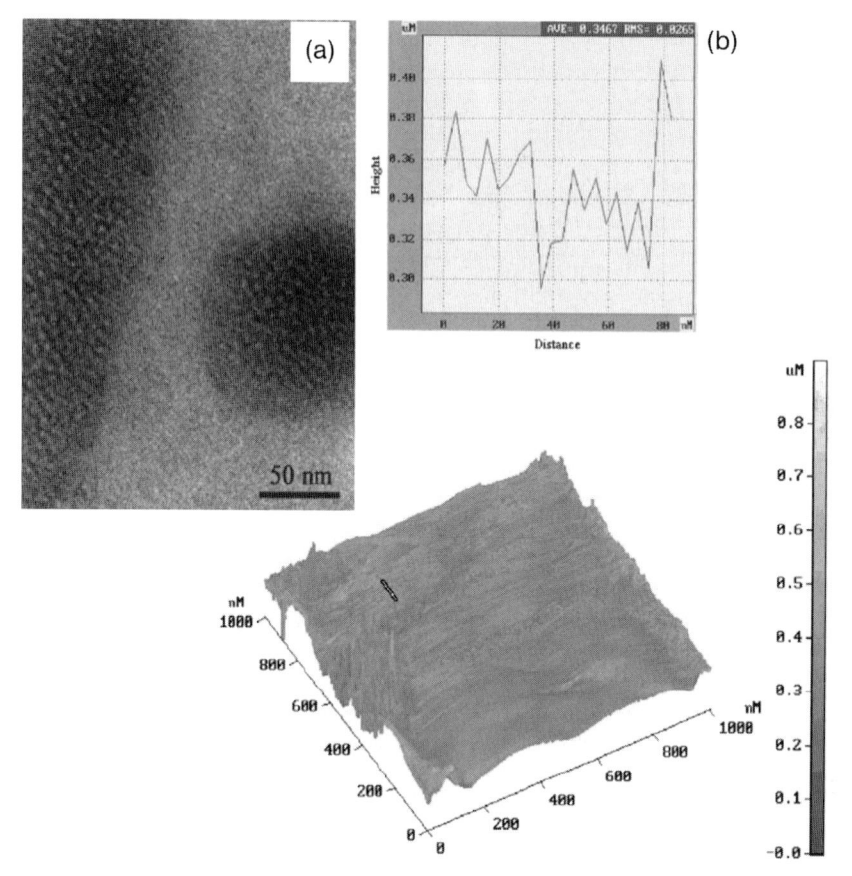

Figure 2.49 TEM and AFM images for the H_I-e Ni(OH)$_2$ films electrodeposited from the hexagonal liquid crystal template consisting of 50 wt.% Brij 56 and 50 wt.% aqueous solutions of 1.8 mol dm^{-3} Ni(NO$_3$)$_2$ and 0.075 mol dm^{-3} NaNO$_3$. (a) TEM image of an H_I-e Ni(OH)$_2$ film, from end-on view of pores. (b) 3D AFM image of an H_I-e Ni(OH)$_2$ film electrodeposited on evaporated gold on mica (inset, cross-section profile along the line); scan range, $1 \times 1\,\mu m^2$. (Reprinted from Ref. [381].)

Hence, much effort has been aimed at searching for alternative inexpensive electrode materials with good capacitive characteristics, such as NiO [374–376], Ni(OH)$_2$ [377], CoO$_x$ [378], Co(OH)$_2$ [379], and MnO$_2$ [380].

Zhao et al. [381], in 2007, proposed a simple method to directly electrodeposit a hexagonal nanoporous Ni(OH)$_2$ film from nickel nitrate dissolved in the aqueous domains of a Brij 56 liquid crystal template (designated H_I-e Ni(OH)$_2$). A specific capacitance as high as 578 F g^{-1} was obtained using these materials. TEM and AFM images of the H_I-e Ni(OH)$_2$ films are shown in Figure 2.49. The TEM image confirms the presence of the expected HI nanostructure (Figure 2.49a), where the bright regions correspond to the cylindrical pores left after removal of the surfactant, and the dark regions correspond to the electrodeposited nickel hydroxide film.

Inspection of these pores shows that they are continuously hexagonally arranged and approximately straight over their whole length. From the TEM image it is estimated that the pore center-to-pore center distance is about 7.0 nm, the pore diameter is about 2.5 nm, and the nickel hydroxide wall thickness about 4.5 nm. These values are comparable with those observed for other mesoporous materials electrodeposited from the HI phase of the same surfactant Brij 56 [382–385].

The 3D AFM image of the surface topography of H_1-e Ni(OH)$_2$ films is shown in Figure 2.49b. The film anisotropic nanostructure is clearly visible, and the long-ranged ordered channels from the side view of the pore array are estimated to have an uniform distance of about 7.0 nm. This is consistent with results from the low-angle X-ray diffraction and TEM studies.

The γ-MnO$_2$ is used predominantly in the aqueous Zn/MnO$_2$ cells which dominate the primary battery market [386,387]. On the other hand, MnO$_2$ appears to be a promising material to prepare pseudocapacitors due to its superior electrochemical performance, environmental friendliness, and low cost [388–392]. Electrode active materials with small crystalline particle size usually show high electrochemical activities and adequate discharge performance due to their high specific surface areas. Furthermore, as the development of chemical and physical technologies, extensive interests have been focused on developing MnO$_2$ nanostructures [393,394]. The synthesis of γ-MnO$_2$ films with carambola-like nanoflakes by applying the combination of pulse galvanostatic or potentiostatic (PS) and cyclic voltammetry (CV) was obtained by S. Chou et al. [395]. The as-obtained nanostructured films were used directly in primary alkaline Zn/MnO$_2$ batteries and electrochemical supercapacitors. The results showed that the γ-MnO$_2$ nanoflake films exhibited high potential plateau in primary Zn/MnO$_2$ batteries at the discharge current density of 500 mAg^{-1}, and a high specific capacitance of 240 F g^{-1} at the current density of 1 mA·cm^{-2}. In Figure 2.50a, it can be seen that MnO$_2$ nanoflakes of 20 nm thickness and at least 200 nm width were electrodeposited by the combination of CV and PS techniques. The as-prepared MnO$_2$ nanoflakes have similar structures to that of *carambolas*, an ornamental evergreen tree that is native to Southeast Asia and has a star-shaped, ridged character (Figure 2.50b).

In-situ TEM images (Figure 2.50c and d) were performed by directly electrodepositing MnO$_2$ onto a carbon-coated copper grid using a combination of CV and PS techniques. From Figure 2.50c, it can be seen that agglomerated nanoflakes of 10 nm thickness are formed. Figure 2.50e shows the SEM image of MnO$_2$ thin film electrodeposited by the PS technique, indicating the formation of crossed needle-like nanostructures with diameters of 5–10 nm and lengths of 50–100 nm. Figure 2.50f shows the top view of the oriented MnO$_2$ nanorod arrays with diameters of 50–70 nm. The present nanorods are analogous to that reported by Wu et al. [396].

2.3.3.3 Metals

Conductive surface templates have also been employed to control the local electrochemical deposition of metals on nanoscale dimensions, for example via preferred nucleation at defects and step edges [397]. Electrochemical surface templates can also be fabricated by patterning of self-assembled monolayers (SAMs). By

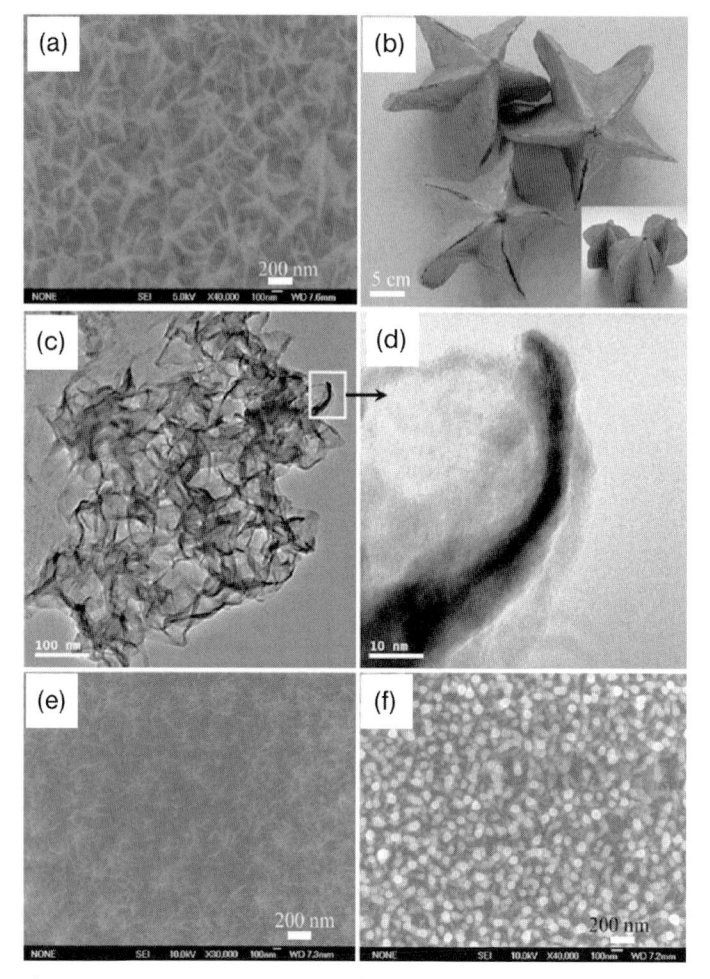

Figure 2.50 Representative SEM (a) and TEM (c, d) images of γ-MnO$_2$ prepared by combined potentiostatic and cyclic voltammetric electrodeposition techniques. (b) The carambola fruit, which is native to South-east Asia. (e, f) SEM images of γ-MnO$_2$ prepared by potentiostatic (e) and cyclic voltammetric (f) techniques. (Reprinted from Ref. [395].)

modification of the molecular structure, SAMs provide a flexible route to control the electron transfer [398,399], and thus the electrochemical properties of the coated surface [400,401]. Patterning of SAMs can be achieved by using conventional lithographic tools (light, electrons, and ions), as well as by soft lithography [402,403]. Nanoscale SAM structures with lateral dimensions of ∼10 nm have been fabricated by scanning probe [404–406], as well as by electron beam lithography [407,408]. For the patterning of large arrays with nanoscale elements, proximity printing with low-energy electrons has been demonstrated as a rapid parallel method [409,410]. SAMs of aliphatic and aromatic thiols are most commonly used to modify gold surfaces. When irradiated with electrons, 1-octadecanethiol (ODT) acts as a positive resist and 1-10-biphenyl-4-thiol (BPT) films as a negative resist [411,412]. The non-irradiated ODT films and the irradiated BPT films, respectively, resist an aqueous cyanide etching solution and thereby generate contrast on the locally irradiated surfaces.

Recently, the electrodeposition of copper was investigated in patterned thiol SAMs on gold electrodes. Aliphatic and aromatic SAMs showed distinct differences. For example, patterned alkanethiols acted as a "positive" template (i.e., Cu was deposited only in the irradiated areas [413,414]), whereas SAMs of ω-(40-methylbiphenyl- 4-yl)-dodecyl thiol and of 1-10-biphenyl- 4-thiol acted as "negative" templates where Cu was deposited only on the non-irradiated areas [415]. The electrodeposition of copper has attracted enormous interest for microelectronics as a reliable tool to deposit high-conductive material for on-chip interconnections [416]. Völker *et al.* [417] studied the fabrication of copper nanostructures, and demonstrated a material preparation procedure combining e-beam lithography on self-assembled monolayers with selective electrochemical deposition. In order to investigate the defect density, Völker's group used proximity printing with a stencil mask to expose a large area (diameter 3 mm) of a BPT-SAM with reasonable exposure time. Figure 2.51 shows an SEM-micrograph of a copper structure that was produced using such a mask. The image shows no visible defects within the shown area of $36\,\mu m^2$.

The roundness and edge accuracy of the circles are limited only by the mask. Since the masks have an active diameter of 3 mm, it is possible to check the homogeneity of the Cu pattern over the whole surface.

For a detailed investigation of the resolution that can be achieved with BPT-SAM templates, Völker's group designed a test pattern for e-beam lithography. This

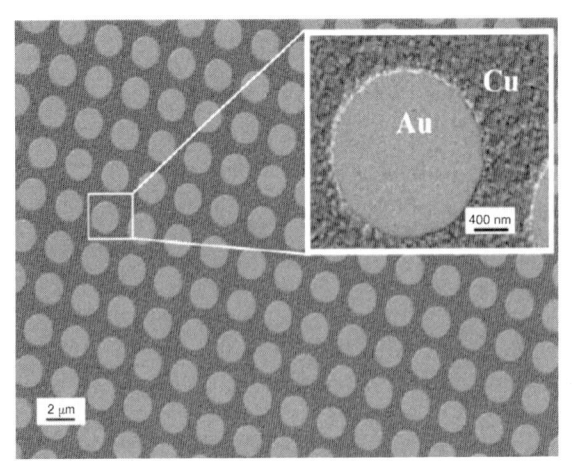

Figure 2.51 SEM image of Cu structure deposited on a BPT SAM template patterned by proximity printing (300 eV, 64 000 μC cm^{-2}). Deposition parameters: 10 mmol dm^{-3} CuSO$_4$, 0.05 V (versus NHE), 120 s. (Reprinted from Ref. [417].)

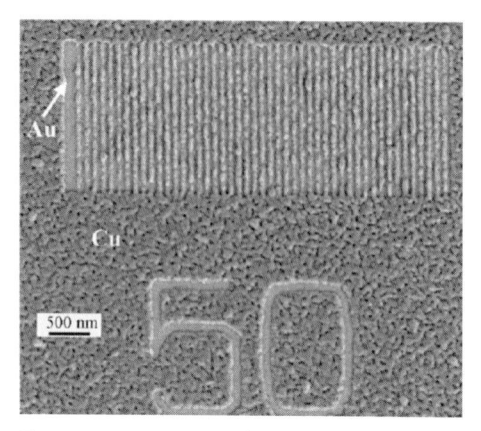

Figure 2.52 SEM image of a Cu line grating deposited on a BPT SAM template patterned by direct electron-beam writing. Deposition parameters: 10 mmol dm^{-3} CuSO$_4$, 0.05 V (versus NHE), 120 s. (Reprinted from Ref. [417].)

pattern contains line gratings, single lines, and isolated dots. Figure 2.52 shows an SEM-micrograph of a copper-electrodeposited grating with 50 nm width line and 100 nm periodicity. Within the non-irradiated areas, the Cu film is continuous and has a thickness of approximately 120 nm.

2.4
Final Remarks

As outlined above, electrochemical tools are today being increasingly used in the preparation of diverse nanostructured materials, such as metals, alloys, semiconductors and oxides. This derives from the fact that both, anodic and cathodic polarization, can be applied to the synthesis of these structures. In addition, they provide the ability to control the shape of deposited materials, as has been presented for the case of metallic nanowires, dot and films with thicknesses as low as submonolayers. The formation of accurately shaped nanostructures may be performed with or without the use of a template. In the former case, the template (e.g., of porous anodic alumina) can also be prepared electrochemically, whereas in the latter case the preparation of ZnO nanotubes has been extensively investigated, in recent years, by using scanning probe microscopy as a fabrication procedure, thereby rendering the field of nanostructured, formed electrochemical materials even more versatile.

Of note, oxidation procedures are used mainly to prepare oxides or salts, and TiO$_2$ presents important optical and photocatalytic properties, as well as biocompatibility. Whilst the main use of porous anodic alumina is as a template for several electrodeposition or non-electrochemical deposition techniques, this system has also been applied to photonic materials, to drug delivery devices, and to microreactors. By contrast, a reduction reaction approach may be used for the electrodeposition of metals, alloys, multilayers, polymers, semiconductors and oxides and, again, the shape of the nanostructured materials can be controlled accurately.

Finally, it is important to stress that although several groups have used electrochemical procedures to prepare different types of material, the most important point is to develop a clear understanding of the mechanism of their formation. In this way, these procedures will surely become the synthetic tools of the 21st century.

References

1 Feynman, R. (1992) There's Plenty of Room at the Bottom, Transcript of the talk given by Richard Feynman on December 26, 1959 at the Annual Meeting American Physical Society. *J. Microelectromech. Systems*, **1**, 60–66.
2 Hodes, G. (1995) in *Physical Electrochemistry: Principles, Methods and Applications*, (ed. I. Rubinstein), Marcel Dekker, p. 515.
3 Gregory, B.W. and Stickney, J.L. (1991) *J. Electroanal. Chem.*, **300**, 543.
4 Gregory, B.W., Norton, M.L. and Stickney, J.L. (1990) *J. Electroanal. Chem.*, **293**, 85.
5 Cerdeira, F., Torriani, I., Motisuke, P., Lemos, V. and Deker, F. (1988) *Appl. Phys.*, **46**, 107.
6 Hodes, G., Fonash, S.J., Heller, A. and Miller, B. (1985) in *Advances in Electrochemistry and Electrochemical Engineering*, (ed. H. Gerisher), Vol.13, J. Wiley & Sons, New York, p. 113.
7 Kressin, A.M., Doan, V.V., Klein, J.D. and Sailor, M.J. (1991) *Chem. Mater.*, **3**, 1015.
8 Suggs, D.W., Villegas, I., Gregory, B.W. and Stickney, J.L. (1992) *J. Vaccum Sci. Technol. A*, **10**, 886.
9 Breyfogle, B.E., Hung, C.J., Shumsky, M.G. and Switzer, J.A. (1996) *J. Electrochem. Soc.*, **143**, 2741.
10 Golden, T.D., Shumsky, M.G., Zhou, Y.C., Vanderwerf, R.A. and Switzer, J.L. (1996) *Chem. Mater.*, **8**, 2499.
11 Van Leeuwen, R.A., Hung, C.J., Kammler, D.R. and Switzer, J.A. (1995) *J. Phys. Chem.*, **99**, 15247.

12 Wade, T., Park, J.M., Garza, E.G., Ross, C.B. and Crooks, R.M. (1992) *J. Am. Chem. Soc.*, **14**, 9457.
13 Wade, T., Ross, C.B. and Crooks, R.M. (1997) *Chem. Mater.*, **9**, 248.
14 Lauerhaas, J.M., Credo, G.M., Heinrich, J.L. and Sailor, M.J. (1992) *J. Am. Chem. Soc.*, **114**, 1911.
15 Doan, V. and Sailor, M.J. (1992) *Appl. Phys. Lett.*, **60**, 619.
16 Switzer, J.A., Shane, M.J. and Philips, R.J. (1990) *Science*, **247**, 444.
17 Switzer, J.A., Hung, C.J., Breyfogle, B.E. and Shumsky, M.G. (1994) *Science*, **264**, 1573.
18 Golden, T.D., Raffaelle, R.P. and Switzer, J.A. (1993) *Appl. Phys. Lett.*, **63**, 1501.
19 Switzer, J.A., Shumsky, M. and Bohannan, E. (1999) *Science*, **284**, 293.
20 Datta, M. and Landolt, D. (2000) *Electrochimica. Acta*, **45**, 2535.
21 Bandyopadhyay, S., Muller, A.E., Chang, H.C., Banerjee, G., Yazhakov, Y., Yue, D.-F., Ricker, R.E., Jones, S., Eastman, J.A., Baugher, E. and Chandrasekhar, M. (1996) *Nanotechnology*, **7**, 360.
22 Bandyopadhyay, S. and Roychowdhury, V.P. (1995) *Physics Low Dimensional Structures*, **8/9**, 29.
23 Chiu, R.-L., Chang, P.-H. and Tung, C.-H. (1995) *Thin Solid Films*, **260**, 47.
24 Landolt, D., Chauvy, P.-F. and Zinger, O. (2003) *Electrochim. Acta*, **48**, 3185.
25 Li, F., Zhang, L. and Metzger, R.M. (1998) *Chem. Mater.*, **10**, 2470.
26 Patermarakis, G. and Karayannis, H. (1995) *Electrochim. Acta*, **40**, 2647.

27 Raimundo, D.S., Stelet, A.B., Fermnandez, F.J.R. and Salcedo, W.J. (2005) *Microelectronics. J.*, **36**, 207–211.
28 Bensadon, E.O., Nascente, P.A.P., Olivi, P., Bulhões, L.O.S. and Pereira, E.C. (1999) *Chem. Mater.*, **11**, 227.
29 Asahi, R., Morikawa, T., Ohwaki, T., Aoki, K. and Taga, Y. (2001) *Science*, **293**, 269.
30 Albrektsson, T., Branemark, P.I., Hansson, H.A. and Lindstrom, J. (1981) *Acta Orthop. Scand.*, **52**, 155.
31 Available in www.azom.com, accessed in January 15,((2007) .
32 Huang, H.H., Pan, S.J. and Lu, S.H. (2005) *Scripta Materialia*, **53**, 1037.
33 Oh, S.-H., Finõnes, R.R., Daraio, C., Chen, L.-H. and Jin, S. (2005) *Biomaterials*, **26**, 4938.
34 Bayoumi, F.M. and Ateya, B.G. (2006) *Electrochem. Commun.*, **8**, 38.
35 Raja, K.S., Misra, M. and Paramguru, K. (2005) *Electrochim. Acta*, **51**, 154.
36 Choi, J., Wehrspohn, R.B., Lee, J. and Gösele, U. (2004) *Electrochim. Acta*, **49**, 2645.
37 Mozalev, A., Mozaleva, I., Sakairi, M. and Takahashi, H. (2005) *Electrochim. Acta*, **50**, 5065.
38 Mozalev, A., Surganov, A. and Magaino, S. (1999) *Electrochim. Acta*, **44**, 3891.
39 Mozalev, A., Sakairi, M., Saeki, I. and Takahashi, H. (2003) *Electrochim. Acta*, **48**, 3155.
40 Martelli, F.H., Trivinho-Strixino, F. and Pereira, E.C. (2006) 5th Brazilian Material Research Science Meeting E540,103.
41 Cheng, H.C., Lee, S.Y., Chen, C.C., Chyng, Y.C. and Ou, K.L. (2006) *Appl. Phys. Lett.*, **89**, 173902/1-3.
42 Sul, Y.T., Johansson, C.B., Jeong, Y. and Albrektsson, T. (2001) *Med. Eng. Phy.*, **23**, 329.
43 Fujishima, A., Rao, T.N. and Tryk, D.A. (2000) *Electrochim. Acta*, **45**, 4683.
44 Gratzel, M. (2001) *Nature*, **414**, 338.
45 Tan, O.K., Cao, W., Hu, Y. and Zhu, W. (2004) *Ceramics Int.*, **30**, 1127.
46 Arabatzis, I.M., Antonaraki, S., Stergiopoulos, T., Hiskia, A., Papaconstantinou, E., Bernard, M.C. and Falaras, P. (2002) *J. Photochem. Photobiol. A-Chemistry*, **149**, 237.
47 Reddy, G.R., Lavanya, A. and Anjaneyulu, C. (2004) *Bull. Electrochem.*, **20**, 337.
48 Hore, S., Palomares, E., Smit, H., Bakker, N.J., Comte, P., Liska, P., Thampi, K.R., Kroon, J.M., Hinsch, A. and Durrant, J.R. (2005) *J. Mater. Chem.*, **15**, 412.
49 Xie, Y.B. and Li, X.Z. (2006) *J. Appl. Electrochem.*, **36**, 663.
50 Rigo, S. and Siejka, J. (1974) *Solid-State Commun.*, **15**, 259.
51 Perriere, J., Rigo, S. and Siejka, J. (1978) *J. Electrochem. Soc.*, **125**, 1549.
52 Perriere, J., Rigo, S. and Siejka, S. (1980) *Corrosion Sci.*, **20**, 91.
53 Perriere, J. and Siejka, J. (1983) *J. Electrochem. Soc.*, **130**, 1267.
54 Shimizu, K., Kobayashi, K., Skeldon, P., Thompson, G.E. and Wood, G.C. (1997) *Thin Solid Films*, **295**, 156.
55 Shimizu, K., Habazaki, H., Skeldon, P., Thompson, G.E. and Wood, G.C. (1999) *J. Surface Finishing Soc. Jap.*, **50**, 2.
56 Thompson, G.E., Wood, G.C. and Shimizu, K. (1981) *Electrochim. Acta*, **26**, 951.
57 Skeldon, P., Shimizu, K., Thompson, G. E. and Wood, G.C. (1990) *Philos. Magn. B*, **61**, 927.
58 Schwartz, G.C. and Platter, V. (1976) *J. Electrochem. Soc.*, **123**, 34.
59 Lazaruk, S., Baranov, I., Maiello, G., Proverbio, E., De Sesare, G. and Ferrari, A. (1994) *J. Electrochem. Soc.*, **141**, 2556.
60 Surganov, V. and Mozalev, A. (1997) *Microelectronics Eng.*, **37/38**, 329.
61 Surganov, V. (1994) *IEEE Trans. Components Packing Manuf., Technol. B*, **17**, 197.
62 Tatarenko, N. and Mozalev, A. (2001) *Solid-State Electronics*, **45/46**, 1009.
63 Maissel, L. and Glang, R. (1970) *Handbook of Thin Film Technology*, McGraw-Hill, New York.
64 Sharp, D.J. and Norwood, D.P. (1987) *Thin Solid Films*, **153**, 387.

65 Liang, J., Chik, H. and Xu, J. (2002) *IEEE J. Selected Topics in Quantum Electronics*, **8**, 9998.

66 Surganov, V. (1994) *IEEE Trans. Components, Packaging and, Manufacturing Technol., Part B*, **17**, 197.

67 Lazarouk, S., Katsouba, S., Demianovich, A., Stanovski, V., Voitech, S., Vysotski, V. and Ponomar, V. (2000) *Solid-State Electronics*, **44**, 815.

68 Diggle, J.W., Downie, T.C. and Goulding, C.W. (1969) *Chem. Rev.*, **69**, 365.

69 Keller, F., Hunter, M.S. and Robinson, D.L. (1953) *J. Electrochem. Soc.*, **100**, 411.

70 Masuda, H. and Fukuda, K. (1995) *Science*, **268**, 1466.

71 Voronoi, G. and Reine, Z. (1908) *Angew. Math.*, **134**, 198.

72 Masuda, H., Asoh, H., Watanabe, M., Nishio, K., Nakao, M. and Tamamura, T. (2001) *Adv. Mater.*, **13**, 189.

73 Asoh, H., Ono, S., Hirose, T., Nakao, M. and Masuda, H. (2003) *Electrochim. Acta*, **48**, 3171.

74 Choi, J., Nielsch, K., Reiche, M., Wehrspohn, R.B. and Gösele, U. (2003) *J. Vacuum Sci. Technol. B*, **21**, 763.

75 Choi, J., Wehrspohn, R.B. and Gösele, U. (2005) *Electrochim. Acta*, **50**, 2591.

76 Masuda, H., Yamada, H., Satoh, M., Asoh, H., Nakao, M. and Tamamura, T. (1997) *Appl. Phys. Lett.*, **71**, 2770.

77 Masuda, H., Yada, K. and Osaka, A. (1998) *Jpn. J. Appl. Phys.*, **37**, L1340.

78 Zou, J., Pu, L., Bao, X. and Feng, D. (2002) *Appl. Phys. Lett.*, **80**, 1079.

79 Pu, L., Bao, X., Zou, J. and Feng, D. (2001) *Angew. Chem. Int. Ed.*, **40**, 1490.

80 Touratier, M. and Béakou, A. (1992) *Composite Sci. Technol.*, **40**, 369.

81 Dragone, T. and Nix, W. (1992) *Acta Metall. Mater.*, **40**, 2781.

82 Cooke, T.F. (1991) *J. Am. Ceram. Soc.*, **74**, 2959.

83 Valcárcel, V., Souto, A. and Guitián, F. (1998) *Adv. Mater.*, **10**, 138.

84 Yuan, Z., Huang, H. and Fan, S. (2002) *Adv. Mater.*, **14**, 303.

85 Pang, Yan-Tao, Meng, Guo-Wen, Zhang, Li-De, Shan, Wen-Jun, Zhang, Chong, Gao, Xue-Yun, Zhao, Ai-Wu and Mao, Yong.-Qiang (2003) *J. Solid State Electrochem.*, **7**, 344.

86 Tian, Y.T., Meng, G.W., Gao, T., Sun, S.H., Xie, T. and Peng, X.S. (2004) *Nanotechnology*, **15**, 189.

87 Parkhutik, V.P. and Shershulsky, V.I. (1992) *J. Physics D: Appl. Phys.*, **25**, 1258.

88 Al Mawlawi, D., Combs, N. and Moskovits, M. (1991) *J. Appl. Phys.*, **40**, 4421.

89 Li, F., Zhang, L. and Metzger, R.M. (1998) *Chem. Mater.*, **10**, 2470.

90 Oliveira, C.P., Cardoso, M.L., Oliveira, A.J.A. and Pereira, E.C. (2006) 5th Brazilian Materials Research Science Meeting C503,167.

91 Cardoso, M.L. (2006) PhD Thesis, Universidade Federal de São Carlos - UFSCar, Programa de Pós-Graduação em Física, São Carlos-SP, Brasil, 124p.

92 Lee, W., Ji, R., Gösele, U. and Nielsch, K. (2006) *Nature Mater.*, **5**, 741.

93 Hillebrenner, H., Buyukserin, F., Kang, M., Mota, M.O., Stewart, J.D. and Martin, C.R. (2006) *J. Am. Chem. Soc.*, **128**, 4236.

94 Luo, D. (2005) *Mater. Res. Soc. Bullg.*, **30**, 654.

95 Raman, C., Berkland, C., Kim, K. and Pack, D.W. (2005) *J. Controlled Rel.*, **103**, 149.

96 Ding, G.Q., Zheng, M.J., Xu, W.L. and Shen, W.Z. (2006) *Thin Solid Films*, **508**, 182.

97 Masuda, H., Yanagishita, T., Yasui, K., Nishio, K., Yagi, I., Rao, T.N. and Fugishima, A. (2001) *Adv. Mater.*, **13**, 247.

98 Che, G., Lakshmi, B.B., Martin, C.R., Fisher, E.R. and Ruoff, R.S. (1998) *Chem. Mater.*, **10**, 260.

99 Qiu, T., Wu, X.L., Huang, G.S., Siu, G.G., Mei, Y.F., Kong, F. and Jiang, M. (2005) *Thin Solid Films*, **478**, 56.

100 Yanagishita, T., Sasaki, M., Nishio, K. and Masuda, H. (2004) *Adv. Mater.*, **16**, 429.

101 Raimundo, D.S., Stelet, A.B., Fernandez, F.J.R. and Salcedo, W.J. (2005) *Microelectronics J.*, **36**, 207.

102 Li, A..-P., Müller, F., Birner, A., Nielsch, K. and Gösele, U. (1999) *Adv. Mater.*, **11**, 483.
103 Yablonovitch, E. (1987) *Phys. Rev. Lett.*, **58**, 2059.
104 Kim, H.K. Available in http://www.nano.pitt.edu/papers/kim-web-2-03-archive.prn.pdf Accessed November, 10, 2005.
105 Li, A.P., Müller, F. and Gösele, U. (2005) (2000) *Electrochem. Solid State Lett.*, **3**, 131.
106 Mikulskas, I., Juodkazis, S., Tomasiunas, R. and Dumas, J.G. (2001) *Adv. Mater.*, **13**, 1574.
107 Masuda, H., Ohya, M., Nishio, K., Asoh, H., Nakao, M., Nohtomi, M., Yokoo, A. and Tamamura, M. (2000) *Jpn. J. Appl. Phys.*, **39**, L1039.
108 Choi, J., Luo, Y., Wehrspohn, R.B., Hillebrand, R., Schilling, J. and Gösele, U. (2003) *J. Appl. Phys.*, **54**, 4757.
109 Jung, M., Kim, H.-G., Lee, J.-K., Joo, O.-S. and Mho, S.-I. (2004) *Electrochim. Acta*, **50**, 857.
110 Li, J., Moskovits, M. and Haslett, T.L. (1998) *Chem. Mater.*, **10**, 1963.
111 Jeong, S.-H., Hwang, H.-Y., Lee, K.-H. and Jeong, Y. (2001) *Appl. Phys. Lett.*, **78**, 2052.
112 Mozalev, A., Magaiano, S. and Imai, H. (2001) *Electrochim. Acta*, **46**, 2825.
113 Li, F., Tang, H., Anderegg, J. and Shinar, J. (1997) *Appl. Phys. Lett.*, **70**, 1233.
114 Kukhta, A.V., Gorokh, G.G., Kolesnik, E.E., Mitkovets, A.I., Taoubi, M.I., Koshin, Yu.A. and Mozalev, A.M. (2002) *Surface Sci.*, **507–510**, 593.
115 Ganley, J.C., Riechmann, K.L., Seebauer, E.J. and Masel, R.I. (2004) *J. Catal.*, **227**, 26.
116 Popat, K.C., Swan, E.E.L., Mukhatyar, V., Chatvanichkul, K.-I., Mor, G.K., Grimes, C.A. and Desai, C.A. (2005) *Biomaterials*, **26**, 4516.
117 Burg, K.J.L., Porter, S. and Kellam, J.F. (2000) *Biomaterials*, **21** (23), 2347.
118 Karlson, M. (2004) PhD Thesis,Uppsala - Suécia, Acta Universitatis Uppsaliensis. Comprehensive Summaries of Uppsala Dissertations from the Faculty of Science and Technology,77p.
119 Gong, D., Yadavalli, V., Paulose, M., Pishko, M. and Grimes, CA. (2003) *Biomedical Microdevices: Therapeutic Micro and Nanotechnology*, **5:1**, 75.
120 Itoh, N., Tomura, N., Tsuji, T. and Hongo, M. (1998) *Microporous Mesoporous Mater.*, **20**, 33.
121 Matsumoto, F., Nishio, K. and Masuda, H. (2004) *Jpn. J. Appl. Phys. Part 2*, **43**, L640.
122 Matsumoto, F., Harada, M., Nishio, K. and Masuda, H. (2005) *Adv. Mater.*, **17**, 1609.
123 Matthias, S., Schilling, J., Nielsch, K., Müller, F., Wehrspohn, R.B. and Gösele, U. (2002) *Adv. Mater.*, **14**, 1618.
124 Liang, J., Chik, H., Yin, A. and Xu, J. (2002) *J. Appl. Phys.*, **91**, 2544.
125 Alivisatos, A.P. (1996) *Science*, **271**, 933.
126 Brus, L. (1994) *J. Phys. Chem.*, **98**, 3575.
127 Krans, J.M., van Rutenbeek, J.M., Fisun, V.V., Yanson, I.K. and Jongh, L.J. (1995) *Nature*, **375**, 767.
128 Díaz, I., Hernández-Vélez, M., Martin Palma, R.J., Villavicencio García, H., Pérez Pariente, J. and Martínez-Duart, J.M. (2004) *Appl. Phys. A*, **79**, 565.
129 Yoffe, A.D. (1993) *Adv. Phys.*, **42** (2), 173.
130 Dobrzynski, L. (2004) *Phys. Rev. B*, **70**, 193307.
131 Datta, S. (1995) *Electronic Transport in Mesoscopic Systems*, Cambridge University Press, Cambridge.
132 Wharam, D.A., Thornton, T.J., Newbury, R., Pepper, M., Ahmed, H., Frost, J.E.F., Hasko, D.G., Peacock, D.C., Ritchie, D.A. and Jones, G.A.C. (1988) *J. Phys. C*, **21**, 20.
133 Wees, B.J.V., Houten, H.V., Beenakker, C.W.J., Williams, J.G., Kouwenhowen, L.P., Marel, D.V.d. and Foxon, C.T. (1988) *Phys. Rev. Lett.*, **60**, 848.
134 Muller, C.J., Van Ruitenbeek, J.M. and de Jongh, L.J. (1992) *Phys. Rev. Lett.*, **69**, 140.
135 Agrait, N., Rodrigo, J.G. and Vieira, S. (1993) *Phys. Rev. B*, **52**, 12345.
136 Pascual, J.I., Mendez, J., Gomez-Herrero, J., Baro, A.M., Garcia, N. and Binh, V.T. (1852) *Phys. Rev. Lett.*, **1993**, **71**.

137 Xu, B., He, H., Boussaad, S. and Tao, N.J. (2003) *Electrochim. Acta*, **48**, 3085.
138 Li, C.Z. and Tao, N.J. (1998) *Appl. Phys. Lett.*, **72**, 894.
139 Li, C.Z., Bogozi, A., Huang, W. and Tao, N.J. (1999) *Nanotechnology*, **10**, 221.
140 Ohnishi, H., Kondo, Y. and Takayanagi, K. (1998) *Nature*, **395** (1), 780.
141 Rodrigues, V., Fuhrer, T. and Ugarte, D. (2000) *Phys. Rev. Lett.*, **85**, 4124.
142 Xu, C. and Xu, G. (2002) *Solid State Commun.*, **122**, 175.
143 Lopes, W.A. and Jaeger, H.M. (2002) *Nature*, **414**, 735.
144 Martin, C.R. (1995) *Acc. Chem. Res.*, **28**, 61.
145 Martin, C.R. (1994) *Science*, **266**, 1961.
146 Zheng, M. and Zhang, L. (2001) *Chem. Phys. Lett.*, **334**, 298.
147 Peng, X.S. and Zhang, J. (2001) *Chem. Phys. Lett.*, **343**, 470.
148 Kroll, M., Benfield, R.E. and Grandjean, D. (2001) *Mater. Res. Soc. Symp. Proc.*, **678**, 1620.
149 Maurice, J.L. and Imhoff, D. (1994) *J. Magn. Mater.*, **184**, 1.
150 Cheng, Y.H. and Cheng, S.Y. (2004) *Nanotechnology*, **15**, 171.
151 Sauer, G., Brehm, G., Schneider, S., Nielsch, K., Wehrspohn, R.B., Choi, J., Hofmeister, H. and Gosele, U. (2002) *J. Appl. Phys.*, **91**, 3243.
152 Choi, J., Schilling, J., Nielsch, K., Hillebrand, R., Reiche, M., Wehrspohn, R.B. and Gösele, U. (2002) *Mater. Res. Soc. Symp. Proc.*, **722**, L5.2.
153 Choi, J., Sauer, G., Nielsch, K., Wehrspohn, R.B. and Gösele, U. (2003) *Chem. Mater.*, **15**, 776.
154 Chu, S.Z., Wada, K., Inoue, S. and Todoroki, S. (2003) *Electrochim. Acta*, **48**, 3147.
155 Chu, S.Z., Wada, K., Inoue, S. and Todoroki, S. (2002) *Chem. Mater.*, **14**, 266.
156 Chu, S.Z., Wada, K., Inoue, S. and Todoroki, S. (2002) *J. Electrochem. Soc.*, **149**, B321.
157 Garcia, J.M., Asenjo, A., Vazquez, M., Aranda, P. and Ruiz-Hitzky, E. (2000) *IEEE Trans. Magnetics*, **36** (5), ((2981) .
158 Fert, A. and Piraux, L. (1999) *J. Magnet. Magnet. Mater.*, **200**, 338.
159 Sun, L. and Searson, P.C. (1999) *Appl. Phys. Lett.*, **74** (19), ((2803) .
160 Whitney, T.M., Jiang, J.S., Searson, P.C. and Chien, C.L. (1993) *Science*, **261** (3), 1316.
161 Piraux, L., Dubois, S. and Ferain, E. (1997) *J. Magnet. Magnet. Mater.*, **165**, 352.
162 Ge, S., Ma, X., Li, C. and Li, W. (2001) *J. Magnet. Magnet. Mater.*, **226–230**, 1867.
163 Martin, C.R. (1994) *Science*, **266**, 1961.
164 Hulteen, J.C. and Martin, C.R. (1997) *J. Mater. Chem.*, **7**, 1075.
165 Shingubara, S. (2003) *J. Nanoparticle Res.*, **5**, 17.
166 Zhang, X.Y., Zhang, L.D., Lei, Y., Zhao, L.X. and Mao, Y.Q. (2001) *J. Mater. Chem.*, **11**, 1732.
167 Choi, J., Sauer, G., Nielsch, K., Wherspohn, R.B. and Gösele, U. (2003) *Mater. Chem.*, **15**, 776.
168 Sander, M.S., Prieto, A.L., Gronsky, R., Sands, T. and Stacy, A.M. (2002) *Adv. Mater.*, **14**, 665.
169 Yin, A.J., Li, J., Jian, W., Bennett, A.J. and Xu, J.M. (2001) *Appl. Phys. Lett.*, **79**, 1039.
170 Sun, M., Zangari, G. and Metzger, R.M. (2000) *IEEE Trans. Magnetism*, **36**, 3005.
171 Routkevitch, D., Bigioni, T., Moskovits, M. and Xu, J.M. (1996) *J. Phys. Chem.*, **100**, 14037.
172 Davydov, D.N., Sattari, P.A., Al Mawlawi, D., Osika, A., Haslett, T.L. and Moskovits, M. (1999) *J. Appl. Phys.*, **86**, 3983.
173 Al Mawlawi, D., Coombs, N. and Moskovits, M. (1991) *J. Appl. Phys.*, **70**, 4421.
174 Nielsch, K., Muller, F., Li, A. and Gösele, U. (2000) *Adv. Mater.*, **12**, 582.
175 Sun, M., Zangari, G., Shamsuzzoha, M. and Metzger, R. (2001) *Appl. Phys. Lett.*, **78**, 2964.
176 Preston, C. and Moskovits, M. (1993) *J. Phys. Chem.*, **97**, 8495.
177 Sauer, G., Brehm, G., Schneider, S., Nielsch, K., Wherspohn, R.B., Choi, J., Hofmeister, H. and Gösele, U. (2002) *J. Appl. Phys.*, **91**, 3243.

178 Al Mawlawi, D., Liu, C. and Moskovits, M. (1994) *J. Mater. Res.*, **9**, 1014.
179 Thompson, G.E. and Wood, G.C. (1983) Anodic Films on Aluminum, in *Corrosion: Aqueous Processes and Passive Films. Treatise on Materials Science and Technology*, (ed. J.C. Scully), Vol. **23**, Academic Press, New York. pp. 205–329.
180 Gerein, N.J. and Haber, J.A. (2005) *J. Phys. Chem. B*, **109**, 17372.
181 Goad, D.G.W. and Moskovits, M. (1978) *J. Appl. Phys.*, **49**, 2929.
182 Clebny, J., Doudin, B. and Ansermet, J.-Ph. (1993) *Nanocryst. Mater.*, **2**, 637.
183 Jagminiene, A., Valincius, G., Riaukaite, A. and Jagminas, A. (2005) *J. Crystal Growth*, **274**, 622.
184 Qu, F., Yang, M., Shen, G. and Yu, R. (2007) *Biosensors and Bioelectronics*, **22**, 1749.
185 Piao, Y.Z., Lim, H.C., Chang, J.Y., Lee, W.Y. and Kim, H.S. (2005) *Electrochim. Acta*, **50**, 2997.
186 Li, H., Xu, C., Zhao, G. and Li, H. (2005) *J. Phys. Chem. B*, **109**, 3759.
187 Dyke, C.A. and Tour, J.M. (2003) *J. Am. Chem. Soc.*, **125**, 1156.
188 Wang, J., Liu, G.D. and Jan, M.R. (2004) *J. Am. Chem. Soc.*, **126**, 3010.
189 Zhao, W., Song, C. and Pehrsson, P.E. (2002) *J. Am. Chem. Soc.*, **124**, 12418.
190 Day, T.M., Wilson, N.R. and Macpherson, J.V. (2004) *J. Am. Chem. Soc.*, **126**, 16724.
191 Gong, K.P., Zhang, M.N., Yan, Y.M., Su, L., Mao, L.Q., Xiong, S.X. and Chen, Y. (2004) *Anal. Chem.*, **76**, 6500.
192 Wang, J., Musameh, M. and Lin, Y. (2003) *J. Am. Chem. Soc.*, **125**, 2408.
193 Lin, Y., Lu, F., Tu, Y. and Ren, Z. (2004) *Nano. Lett.*, **4**, 191.
194 Simoes, F.R., Mattoso, L.H.C. and Pereira, E.C. (2006) 17th Brazilian Congress on Engineering and Materials Science.
195 Wang, J., Dai, J. and Yarlagadda, T. (2005) *Langmuir*, **21**, 9.
196 Zhang, M. and Gorski, W. (2005) *J. Am. Chem. Soc.*, **127**, 2058.
197 Quinn, B.M., Dekker, C. and Lemay, S.G. (2005) *J. Am. Chem. Soc.*, **127**, 6146.
198 Medeiros, S., Martinez, R.A., Fonseca, F.J., Leite, E.R., Manohar, S.K., Gregório, R. Jr. and Mattoso, L.H.C. (2006) Proceedings of the World Polymer Congress, Macro 2006, July 16–21, Rio de Janeiro, Brazil.
199 Yang, M.H., Jiang, J.H., Yang, Y.H., Chen, X.H., Shen, G.L. and Yu, R.Q. (2006) *Biosens. Bioelectron.*, **21**, 1791.
200 Yang, M.H., Yang, Y., Yang, H.F., Shen, G.L. and Yu, R.Q. (2006) *Biomaterials*, **27**, 246.
201 Hrapovic, S., Liu, Y., Male, K.B. and Luong, J. (2004) *Anal. Chem.*, **76**, 1083.
202 Chen, J.Y., Herricks, T., Geissler, M. and Xia, Y.N. (2004) *J. Am. Chem. Soc.*, **126**, 10854.
203 Basnayake, R., Li, Z.R. and Katar, S. (2006) *Langmuir*, **22** (25), 10446.
204 Um, Y.Y., Liang, H.P. and Hu, J.S. (2005) *J. Phys. Chem. B*, **109** (47), 22212.
205 Narayanan, R. and El-Sayed, M.A. (2005) *J. Phys. Chem. B*, **109** (26), 12663.
206 Piao, Y.Z., Lim, H.C., Chang, J.Y., Lee, W.Y. and Kim, H.S. (2005) *Electrochim. Acta*, **50**, 2997.
207 Sakamoto, Y., Fukuoka, A., Higuchi, T., Shimomura, N., Inagaki, S. and Ichikawa, M. (2004) *J. Phys. Chem. B*, **108**, 853.
208 Day, T.M., Unwin, P.R., Wilson, N.R. and Macpherson, J.V. (2005) *J. Am. Chem. Soc.*, **127**, 10639.
209 Kicela, A. and Daniele, S. (2006) *Talanta*, **68**, 1632.
210 Joshi, P.P., Merchant, S.A., Wang, Y.D. and Schmidtke, D.W. (2005) *Anal. Chem.*, **77**, 3183.
211 Rajeshwar, K., de Tacconi, N.R. and Chenthamarakshan, C.R. (2004) *Curr. Opin. Solid State Mater. Sci.*, **8**, 173.
212 Switzer, J.A. (1987) *Am. Ceram. Bull.*, **66**, 1521.
213 Matsumoto, Y. (2000) *MRS Bull.*, 47.
214 Therese, G.H.A. and Kamath, P.V. (2000) *Chem. Mater.*, **12**, 1195.
215 (a) Golden, T.D., Shumsky, M.G., Zhou, Y., Van der Werf, R.A., Van Leeuwen, R.A. and Switzer, J.A. (1996) *Chem. Mater.*, **8**, 2499. (b) Switzer, J.A., Hung, C.-J.,

Bohannan, E.W., Shumsky, M.G., Golden, T.D. and Van Aken, D.C. (1997) *Adv. Mater.*, **9**, 334. (c) Switzer, J.A., Hung, C.-J., Huang, L.Y., Miller, F.S., Zhou Y. and Raub E.R., et al. (1998) *J. Mater. Res.*, **13**, 909. (d) Switzer, J.A., Hung, C.-J., Huang, L.-Y., Switzer, E.R., Kammler, D.R. and Golden, T.D., et al. (1998) *J. Am. Chem. Soc.*, **120**, 3530. (e) Switzer, J.A., Maune, B.M., Raub, E.R. and Bohannan, E.W. (1999) *J. Phys. Chem. B*, **103**, 395. (f) Bohannan, E.W., Huang, L.-Y., Miller, F.S., Schumsky, M.G. and Switzer, J.A. (1999) *Langmuir*, **15**, 813.

216 (a) Peulon, S. and Lincot, D. (1996) *Adv. Mater.*, **8**, 166. (b) Peulon, S. and Lincot, D. (1998) *J. Electrochem. Soc.*, **145**, 864, (c) Pauporté, T. and Lincot, D. (2001) *J. Electroanal. Chem.*, **517**, 54, (and references therein). (d) Pauporté, T., Yoshida, T., Goux, A. and Lincot, D. (2002) *J. Electroanal. Chem.*, **534**, 55.

217 (a) Izaki, M. and Omi, T. (1996) *Appl. Phys. Lett.*, **68**, 2439. (b) Izaki, M. and Omi, T. (1996) *J. Electrochem. Soc.*, **143**, L53. (c) Izaki, M. and Omi, T. (1997) *J. Electrochem. Soc.*, **144**, 1949.

218 (a) Gu, Z.H., Fahidy, T.Z., Hornsey, R. and Nathan, A. (1997) *Can. J. Chem.*, **75**, 1437. (b) Gu, Z.H. and Fahidy, T.Z. (1999) *J. Electrochem. Soc.*, **146**, 156.

219 (a) Golden, T.D., Shumsky, M.G., Zhou, Y., Van der Werf, R.A., Van Leeuwen, R.A. and Switzer, J.A. (1996) *Chem. Mater.*, **8**, 2499. (b) Switzer, J.A., Hung, C.-J., Bohannan, E.W., Shumsky, M.G., Golden, T.D. and Van Aken, D.C. (1997) *Adv. Mater.*, **9**, 334. (c) Switzer, J.A., Hung, C.-J., Huang, L.Y., Miller, F.S., Zhou, Y. and Raub, E.R., et al. (1998) *J. Mater. Res.*, **13**, 909. (d) Switzer, J.A., Hung, C -J., Huang, L.-Y., Switzer, E.R., Kammler, D.R. and Golden, T.D., et al. (1998) *J. Am. Chem. Soc.*, **120**, 3530. (e) Switzer, J.A., Maune, B.M., Raub, E.R. and Bohannan, E.W. (1999) *J. Phys. Chem. B*, **103**, 395. (f) Bohannan, E.W., Huang, L.-Y., Miller, F.S., Schumsky, M.G. and Switzer, J.A. (1999) *Langmuir*, **15**, 813.

220 (a) Peulon, S. and Lincot, D. (1996) *Adv. Mater.*, **8**, 166. (b) Peulon, S. and Lincot, D. (1998) *J. Electrochem. Soc.*, **145**, 864. (c) Pauporté, T. and Lincot, D. (2001) *J. Electroanal. Chem.*, **517**, 54, (and references therein). (d) Pauporté, T., Yoshida, T., Goux, A. and Lincot, D. (2002) *J. Electroanal. Chem.*, **534**, 55.

221 (a) Izaki, M. and Omi, T. (1996) *Appl. Phys. Lett.*, **68**, 2439. (b) Izaki, M. and Omi, T. (1996) *J. Electrochem. Soc.*, **143**, L53. (c) Izaki, M. and Omi, T. (1997) *J. Electrochem. Soc.*, **144**, 1949.

222 (a) Izaki, M. and Omi, T. (1996) *Appl. Phys. Lett.*, **68**, 2439. (b) Izaki, M. and Omi, T. (1996) *J. Electrochem. Soc.*, **143**, L53. (c) Izaki, M. and Omi, T. (1997) *J. Electrochem. Soc.*, **44**, 1949.

223 Natarajan, C. and Nogami, G. (1996) *J. Electrochem. Soc.*, **143**, 1547.

224 (a) Shitomirsky, I. (1997) *Nanostruct. Mater.*, **8**, 521. (b) Shitomirsky, I. (1999) *J. Eur. Ceram. Soc.*, **19**, 2581.

225 Karuppuchamy, S., Amalnerkar, D.P., Yamaguchi, K., Yoshida, T., Sugiura, T. and Minoura, H. (2001) *Chem. Lett.*, 78.

226 Gourgaud, S. and Elliott, D. (1997) *J. Electrochem. Soc.*, **124**, 102.

227 (a) Guerfi, A. and Dao, L.H. (1989) *J. Electrochem. Soc.*, **136**, 2435. (b) Guerfi, A., Paynter, R.W. and Dao, L.H. (1995) *J. Electrochem. Soc.*, **142**, 3457.

228 Stevenson, K.J., Hurtt, G.J. and Hupp, J.T. (1999) *Electrochem. Solid-State Lett.*, **2**, 175.

229 Yamanaka, K. (1987) *Jpn. J. Appl. Phys.*, **26**, 1884.

230 (a) Shen, P.K., Syed-Bokhari, J. and Tseung, A.C.C. (1991) *J. Electrochem. Soc.*, **138**, 2778. (b) Shen, P.K. and Tseung, A.C.C. (1992) *J. Mater. Chem.*, **2**, 1141.

231 Meulenkamp, E.A. (1997) *J. Electrochem. Soc.*, **144**, 1664.

232 (a) Switzer, J.A., Shane, M.J. and Phillips, R.J. (1990) *Science*, **247**, 444. (b) Switzer, J.A., Raffaelle, R.P., Phillips, R.J., Hung, C.-J. and Golden, T.D. (1992) *Science*, **258**, 1918. (c) Switzer, J.A., Hung, C.-J., Breyfogle, B.E. and Golden, T.D. (1994)

Science, **264**, 505. (d) Phillips, R.J., Golden, T.D., Shumsky, M.G., Bohannan, E.W. and Switzer, J.A. (1997) *Chem. Mater.*, **9**, 1670.

233 Liu, R., Vertegel, A.A., Bohannan, E.W., Sorenson, T.A. and Switzer, J.A. (2001) *Chem. Mater.*, **13**, 508.

234 Pauporté, T. and Lincot, D. (1999) *Appl. Phys. Lett.*, **75**, 3817.

235 Zheng, M.J., Zhang, L.D., Li, G.H. and Shen, W.Z. (2002) *Chem. Phys. Lett.*, **363**, 123.

236 Liu, C.H., Zapien, J.A., Yao, Y., Meng, X. M., Lee, C.S., Fan, S.S., Lifshitz, Y. and Lee, S.T. (2003) *Adv. Mater.*, **15**, 838.

237 Li, Y., Meng, G.W., Zhang, L.D. and Phillip, F. (2000) *Appl. Phys. Lett.*, **76**, 2011.

238 Zheng, M.J., Zhang, L.D., Li, G.H. and Shen, W.Z. (2002) *Chem. Phys. Lett.*, **363**, 123.

239 Peulon, S. and Lincot, D. (1998) *J. Electrochem. Soc.*, **145**, 864.

240 Zhou, H., Alves, H., Hofmann, D.M., Kriegseis, W. and Meyer, B.K. (2002) *Appl. Phys. Lett.*, **80**, 210.

241 Gal, D., Hodes, G., Lincot, D. and Schock, H.W. (2000) *Thin Solid Films*, **79**, 361.

242 Peulon, S. and Lincot, D. (1996) *Adv. Mater.*, **8**, 166.

243 Wang, Q., Wang, G., Xua, B., Jiea, J., Han, X., Li, G., Li, Q. and Hou, J.G. (2005) *Mater. Lett.*, **59**, 1378.

244 Switzer, J.A. (2001) Electrodeposition of Superlatices and Multilayers, in *Electrochemistry of Nanomaterials*, (ed. G. Hodes), Wiley-VCH, Weinheim. Chapter 3.

245 Weins, T.P., Barbee, T.W. Jr. and Wall, M. A. (1997) *Acta Mater.*, **45**, 2307.

246 Was, G.S. and Foedke, T. (1996) *Thin Solid films*, **1**, 286.

247 Shih, K.K. and Bove, D.B. (1992) *Appl. Phys. Lett.*, **61**, 654.

248 Mirkarimi, P.B., Hultman, L. and Barneit, S.A. (1990) *Appl. Phys. Lett.*, **57**, 2654.

249 Cammarata, R.C., Schlesinger, T.E., Kim, C., Qadri, S.B. and Edelstein, A.S. (1990) *Appl. Phys. Lett.*, **56**, 1862.

250 Helmersson, U., Todorova, S., Barnett, S. A., Sundgren, J.E., Market, L.C. and Greene, J.E. (1987) *J. Appl. Phys.*, **62**, 481.

251 Tench, D. and White, J. (1984) *Metallurg. Trans. A*, **15**, 2039.

252 Switzer, J.A. (1998) Electrodeposition of Superlattices and Multilayers, in *Nanoparticles and Nanostructured Films*, (ed. J.H. Fendler),Wiley-VCH, Weinheim, p. 53. Chapter 3.

253 Switzer, J.A. (1997) Electrodeposition of Nanoscale Architectures, in *Handbook of Nanophase Materials*, (ed. A.N. Goldstein), Marcel Dekker, New York, Chapter 4.

254 Ross, C.A. (1994) *Annu. Rev. Mater. Sci.*, **24**, 159.

255 Haseeb, A., Celis, J.P. and Roos, J.R. (1994) *J. Electrochem. Soc.*, **141**, 230.

256 Searson, P.C. and Moffat, T.P. (1994) *Crit. Rev. Surface Chem.*, **3**, 272.

257 Therese, G.H.A. and Kamath, P.V. (2000) *Chem. Mater.*, **12**, 1195.

258 Blum, W. (1921) *Trans. Am. Electrochem. Soc.*, **40**, 307.

259 Brenner, A. (1963) *Electrodeposition of Alloy*, Vol.2, Academic Press, New York, p. 589.

260 Cohen, U., Koch, F.B. and Sard, R. (1983) *J. Electrochem. Soc.*, **138**, 1987.

261 Ogden, C. (1986) *Plating Surface Finishing*, **5**, 133.

262 (a) Yahalon, J. and Zadok, O. (1987) *J. Mater. Sci.*, **22**, 499.(b) Yahalon, J. and Zadok, O. (1987) U.S. Patent, 4, 652, 348.

263 Menezes, S. and Anderson, D.P. (1989) *J. Electrochem. Soc.*, **136**, 1651.

264 Lashmore, D.S. and Dariel, M.P. (1988) *J. Electrochem. Soc.*, **135**, 1218.

265 Kelly, J.J., Bradley, P.E. and Landolt, D. (2000) *J. Electrochem. Soc.*, **147**, 2975.

266 Kelly, J.J., Kern, P. and Landolt, D. (2000) *J. Electrochem. Soc.*, **147**, 3725.

267 Guntel, A., Chassaing, E. and Schmidt, J. E. (2001) *J. Appl. Phys.*, **90**, 5257.

268 Fricoteaux, P. and Douglade, J. (2004) *Surface and Coatings Technology*, **184**, 63.

269 Tench, D.M. and White, J.T. (1992) *J. Electrochem. Soc.*, **139**, 443.

270 Fahidy, T.Z. (1999) in *Modern Aspects of Electrochemistry*, (eds B.E. Conway, *et al.*), Vol.**32**, Kluwer Academic Publishers, New York, p. 333.
271 Tacken, R.A. and Janssen, L.J. (1995) *J. Appl. Electrochem.*, **25**, 1.
272 Coey, J.M.D. and Hinds, G. (2001) *J. Alloys Compd.*, **326**, 238.
273 Chopart, J.P., Douglade, J., Fricoteaux, P. and Olivier, A. (1991) *Electrochim. Acta*, **36**, 459.
274 Uhlemann, M., Gebert, A., Herrich, M., Krause, A., Cziraki, A. and Schultz, L. (2003) *Electrochim. Acta*, **48**, 3005.
275 Shima, M., Salamanca-Riba, L.G., McMichael, R.D. and Moffat, T.P. (2001) *J. Electrochem. Soc.*, **148**, C518.
276 Carcia, P.F. (1988) *J. Appl. Phys.*, **63**, 5066.
277 Lee, C.H., Farrow, R.F.C., Hermsmeier, B.D., Marks, R.F., Bennett, W.R., Lin, C. J., Marinero, E.E., Kirchner, P.D. and Chien, C.J. (1991) *J. Magn. Magn. Mater.*, **93**, 592.
278 Zeper, W.B., Greidanus, F.J.A.M., Carcia, P.F. and Fincher, C.R. (1989) *J. Appl. Phys.*, **65**, 4971.
279 Baibich, M.N., Broto, J.M., Fert, A., Nguyen Van Dau, F., Petro, F., Etienne, P., Creuzet, G., Friederich, A. and Chazelas, J. (1988) *Phys. Rev. Lett.*, **61**, 2472.
280 Mosca, D.H., Petro, F., Fert, A., Schroeder, P.A., Pratt, W.P. Jr. and Laloee, R. (1991) *J. Magn. Magn. Mater.*, **94**, L1.
281 Parkin, S.S.P., Marks, R.F., Farrow, R.F. C., Harp, G.R., Lam, Q.H. and Savoy, R.J. (1992) *Phys. Rev. B*, **46**, 9262.
282 Jyoko, Y., Kashiwabara, S. and Hayashi, Y. (1996) *J. Magn. Magn. Mater.*, **156**, 35.
283 Wang, J.Q. and Xiao, G. (1994) *Phys. Rev. B*, **49**, 3982.
284 Parkin, S.S.P., Farrow, R.F.C., Rabedeau, T.A., Marks, R.F., Harp, G.R., Lam, Q.H., Toney, M., Savoy, R. and Geiss, R. (1993) *Euro. Phys. Lett.*, **22**, 455.
285 Kubaschewski (ed.), *Iron-Binary Phase Diagrams*, Springer, Berlin, (1982) , p. 5.
286 Gavrin, A., Kelley, M.H., Xiao, J.Q. and Chien, C.L. (1996) *J. Appl. Phys.*, **79**, 5306.
287 Berkowitz, A.E., Mitchell, J.R., Carey, M. J., Young, A.P., Zhang, S., Spada, F.E., Parker, F.T., Hutten, A. and Thomas, G. (1992) *Phys. Rev. Lett.*, **68**, 3745.
288 Xiao, J.Q., Jiang, J.S. and Chien, C.L. (1992) *Phys. Rev. B*, **46**, 9266.
289 Zaman, H., Ikeda, S. and Ueda, Y. (1997) *IEEE Trans. Magnet*, **33**, 3517.
290 Zaman, H., Yamada, A., Fukuda, H. and Ueda, Y. (1998) *J. Electrochem. Soc.*, **145**, 565.
291 Biberian, J.P. and Somorjai, G.A. (1979) *J. Vacuum Sci. Technol.*, **16**, 2973.
292 Lash, K., Jörinssen, L. and Garche, J. (1999) *J. Power Sources*, **84**, 225.
293 Ueda, Y., Nikuchi, N., Ikeda, S. and Houga, T. (1999) *J. Magnet. Magnet. Mater.*, **198**, 740.
294 Christoffersen, E., Liu, P., Ruban, A., Skriver, H.L. and Norskov, J.K. (2001) *J. Catal.*, **199**, 123.
295 Garcia, P.F., Meinhaldt, A.D. and Suna, A. (1985) *Appl. Phys. Lett.*, **47**, 178.
296 Grunberg, P., Cherieber, R., Pang, Y., Brodsky, M.B. and Sower, H. (1986) *Phys. Rev. Lett.*, **57**, 2442.
297 Burke, L.D. (1981) in *Electrodes of Conductive Metallic Oxides, Part A*, (ed. S. Trasatti),Elsevier, Amsterdam, p. 141.
298 Tilak, B.V., Lu, P.V.T., Colman, J.E. and Srinivasan, S. (1981) in *Comprehensive Treatise of Electrochemistry*, (eds J.O'.M., Bockris, B.E., Conway and R.E., White), Vol. **2** Plenum Press, New York, p. 1.
299 Vukovic, M. (1988) *J. Electroanal. Chem.*, **242**, 97.
300 Hammer, B. and Norskov, J.K. (1995) *Surf. Sci.*, **343**, 211.
301 Jacobsen, K.W., Stolze, P. and Norskov, J. K. (1996) *Surf. Sci.*, **366**, 394.
302 Edersen, P.M.O., Helveg, S., Ruban, A., Stensgaard, I., Laegsgaard, E., Norskov, J. K. and Besenbacher, F. (1999) *Surf. Sci.*, **426**, 395.
303 Christoffersen, E., Liu, P., Ruban, A., Skriver, H.L. and Norskov, J.K. (2001) *J. Catal.*, **199**, 123.
304 Hammer, B., Morikawa, Y. and Norskov, J.K. (1996) *Phys. Rev. Lett.*, **76**, 2141.

305 Oliveira, R.T.S., Santos, M.C., Marcussi, B.G., Nascente, P.A.P., Bulhões, L.O.S. and Pereira, E.C. (2005) *J. Electroanal. Chem.*, **575**, 177.

306 Oliveira, R.T.S., Santos, M.C., Marcussi, B.G., Tanimoto, S.T., Bulhões, L.O.S. and Pereira, E.C. (2006) *J. Power Sources*, **157**, 212.

307 Lemos, S.G., Oliveira, R.T.S., Santos, M.C., Nascente, P.A.P., Bulhões, L.O.S. and Pereira, E.C. (2007) *J. Power Sources*, **163**, 695.

308 Bockris, J.O'M., Reddy, A.K.N. and Gamboa-Aldeco, M. (2002) *Electrochemistry 2A Fundamentals of Electrodics*, 2nd edn., Kluwer Academic Publishers, New York.

309 Freyland, W., Zell, C.A., Zein El Abedin, S. and Endres, F. (2003) *Electrochim Acta*, **48**, 3053.

310 Zell, C.A., Freyland, W. and Endres, F. (2000) in *Molten Salt Chemistry*, (eds R.W. Berg and H.A. Hjuler), Vol.1, Elsevier, Paris. p. 597.

311 Kleinert, M., Waibel, H.-F., Engelmann, G.E., Martin, H. and Kolb, D.M. (2001) *Electrochim. Acta*, **46**, 3129.

312 Schmidt, E. and Gygax, H.R. (1966) *J. Electroanal. Chem.*, **12**, 300; Schmidt, E. and Gygax, H.R. (1967) *J. Electroanal. Chem.*, **14**, 126.

313 Schmidt, E. (1969) *Helv. Chim. Acta*, **52**, 1763.

314 Schmidt, E. and Siegenthaler, H. (1969) *Helv. Chim. Acta*, **52**, 2245; Schmidt, E. and Siegenthaler, H. (1970) *Helv. Chim. Acta*, **53**, 321.

315 Lorenz, W.J. (1973) *Chem. Ing. Technol.*, **45**, 175.

316 Hilbert, F., Mayer, C. and Lorenz, W.J. (1973) *J. Electroanal. Chem.*, **47**, 167.

317 Lorenz, W.J., Hermann, H.D., Wüthrich, N. and Hilbert, F. (1974) *J. Electrochem. Soc.*, **121**, 1167.

318 Adzic, R., Yeager, E. and Cahan, B.D. (1974) *J. Electrochem. Soc.*, **121**, 474.

319 Horkens, J., Cahan, B.D. and Yeager, E. (1975) *J Electrochem Soc*, **122**, 1585.

320 Bewick, A. and Thomas, B. (1975) *J. Electroanal. Chem.*, **65**, 911; Bewick, A. and Thomas, B. (1976) *J. Electroanal. Chem.*, **70**, 239; Bewick, A. and Thomas, B. (1977) *J. Electroanal. Chem.*, **84**, 127; Bewick, A. and Thomas, B. (1977) *J. Electroanal. Chem.*, **85**, 329.

321 Schultze, J.W. and Dickertmann, D. (1976) *Surf. Sci.*, **54**, 489.

322 Schultze, J.W. and Dickertmann, D. (1977) *Faraday Symposia Chem. Soc.*, **12**, 36.

323 Lorenz W.J., Schmidt E., Staikov G. and Bort H. (1977) *Faraday Symposia Chem. Soc.*, **12**, 14.

324 Staikov, G., Jüttner, K., Lorenz, W.J. and Schmidt, E. (1978) *Electrochim. Acta*, **23**, 305.

325 Siegenthaler, H., Jüttner, K., Lorenz, W.J. and Schmidt, E. (1978) *Electrochim Acta*, **23**, 1009.

326 Jüttner, K. and Lorenz, W.J. (1980) *Z. Phys. Chem.*, **NF 122**, 163.

327 Kolb, D.M. (1978) in *Advances in Electrochemistry and Electrochemical Engineering*, (eds. H. Gerischer and C.W. Tobias), Vol.11, Wiley, New York, p. 125.

328 Beckmann, H.O., Gerischer, H., Kolb, D.M. and Lehmpfuhl, G. (1977) *Faraday Symposia Chem. Soc.*, **12**, 51.

329 Staikov, G., Lorenz, W.J. and Budevski, E. (1978) *Comm. Dep. Chem. Bulg. Acad. Sci.*, **11**, 474.

330 Staikov, G., Juttner, K., Lorenz, W.J. and Budevski, E. (1978) *Electrochim. Acta*, **23**, 319.

331 Jüttner, K., Lorenz, W.J., Staikov, G. and Budevski, E. (1978) *Electrochim. Acta*, **23**, 741.

332 Budevski, E., Bostanov, V. and Staikov, G. (1980) *Annu. Rev. Mater. Sci.*, **10**, 85.

333 Budevski, E. (1983) in *Comprehensive Treatise of Electrochemistry*, (eds B.E. Conway, J.O'.M. Bockris, E. Yeager, S.U.M. Khan and R.E. White), Vol. 7, Plenum Press, New York, p. 399.

334 Abruna H. (ed.)((1991) *Electrochemical Interfaces: Modern Techniques for In situ Interface Characterization*, VCH, Weinheim.

335 A.A. Gewirth and H. Siegenthaler (eds) (1995) *Nanoscale Probes of the Solid: Liquid Interface*, NATO ASI Series E: Appled Science, Kluwer, Dordrecht, Vol. 288.

336 Roher, H. (1995) Nanoscale probes of the solid/liquid interface, in Proceedings NATO ASI Sophia Antipolis, France, July 10–20, 1993, in: A.A. Gewirth, H. Siegenthaler (Eds.), NATO Science Series E, Vol. 288. Kluwer, New York.

337 Budevski, E., Staikov, G. and Lorenz, W.J. (1996) *Electrochemical Phase Formation and Growth*, VCH, Weinheim.

338 Staikov, G. and Lorenz, W.J. (1998) in *Electrochemical Nanotechnology - In situ Local Probe Techniques at Electrochemical Interfaces*, (eds W. Plieth and W.J. Lorenz), Wiley-VCH, Weinheim, p. 13.

339 Staikov, G., Lorenz, W.J. and Budevski, E. (1999) in *Imaging of Surfaces and Interfaces - Frontiers in Electrochemistry*, (eds P. Ross and J. Lipkowski), Vol.5, Wiley-VCH, Weinheim, p. 1.

340 Li, W., Virtanen, J.A. and Penner, R.M. (1992) *Appl. Phys. Lett.*, **60**, 1181.

341 Husser, O.E., Craston, D.H. and Bard, A.J. (1989) *J. Electrochem. Soc.*, **136**, 3222.

342 Hugelmann, M., Hugelmann, P., Lorenz, W.J. and Schindler, W. (2005) *Surface Sci.*, **597**, 156.

343 Hofmann, D., Schindler, W. and Kirschner, J. (1998) *Appl. Phys. Lett.*, **73**, 3279.

344 Schindler, W., Hofmann, D. and Kirschner, J. (2001) *J. Electrochem. Soc.*, **148**, C124.

345 Hofmann, D. (1999) PhD Thesis, Universität Halle-Wittenberg.

346 Pandey, R.K. Sahu, S.N. and Chandra, S. (1996) *Handbook of Semiconductor Electrodeposition*, Marcel Dekker, New York.

347 Hodes, G. (1995) in *Physical Electrochemistry*, (ed. I. Rubinstein), Marcel Dekker, New York, p. 515.

348 Schlesinger, M. (2000) in *Modern Electroplating*, (eds M. Schlesinger and M. Paunovic), John Wiley & Sons New York p. 585.

349 Vedel, J. (1998) *Inst. Phys. Conf. Ser.*, **152**, 261.

350 Lincot, D., Guillemoles, J.F., Taunier, S., Guimard, D., Sicx-Kurdi, J., Chomont, A., Roussel, O., Ramdani, O., Hubert, C., Fauvarque, J.P., Bodereau, N., Parissi, L., Panheleux, P., Fanouill, P., Ore Naghavi, N., Grand P.P., Benfarah M., Mogensen P. and Kerrec O. (2004) *Sol. Energy*, **77**, 725.

351 Lincot, D. (2005) *Thin Solid Films*, **487**, 40.

352 Helen Annal Therese, G. and Vishnu Kamath, P. (2000) *Chem. Mater.*, **12**, 1195.

353 Matsumoto, Y. (2000) *MRS Bull.*, 47.

354 Peulon, S. and Lincot, D. (1996) *Adv. Mater.*, **8**, 166.

355 Izaki, M. and Omi, T. (1996) *Appl. Phys. Lett.*, **68**, 2439.

356 Gao, Y.F., Nagai, M., Masuda, Y., Sato, F. and Kuomoto, K.J. (2006) *Crystal Growth*, **286**, 445.

357 Gu, Z.H. and Fahidy, T.Z. (1999) *J. Electrochem. Depos.*, **146**, 156.

358 Chen, Z.G., Tang, Y.W., Zhang, L.S. and Luo, L.J. (2006) *Electrochim. Acta*, **51**, 5870.

359 Mari, B., Cembrero, J., Manjon, F.J., Mollar, M. and Gomez, R. (2005) *Phys. Status Solidi A*, **202**, 1602.

360 Konenkamp, R., Boedecher, K., Lux-Steiner, M.C., Poschenrieder, M. and Wagner, S. (2000) *Appl. Phys. Lett.*, **77**, 2575.

361 Levy-Clement, C., Katty, A., Bastide, S., Zenia, F., Mora, I. and Munoz- Sanjose, V. (2002) *Physica*, **14**, 229.

362 Cui, J.B. and Gibson, U.J. (2005) *Appl. Phys. Lett.*, **87**, 133108.

363 Xu, L.F., Guo, Y., Liao, Q., Zhang, J.P. and Xu, D.S. (2005) *J. Phys. Chem. B*, **109**, 13519.

364 Cao, B.Q., Li, Y., Duan, G.T. and Cai, W.P. (2006) *Cryst. Growth Design*, **6**, 1091.

365 Tang, Y., Luo, L., Chen, Z., Jiang, Y., Li, B., Jia, Z. and Xu, I. (2007) *Electrochem. Commun.*, **9**, 289.

366 Hou, T.N., Li, J., Smith, M.K., Nguyen, P., Cassell, A., Han, J. and Meyyappan, M. (2003) *Science*, **300**, 1249.

367 Zhang, P., Binh, N.T., Wakatsuki, K., Segawa, Y., Yamade, Y., Usami, N.,

Kawasaki, M. and Koinuma, H.J. (2004) *Phys. Chem. B*, **108**, 10899.

368 Peulon, S. and Lincot, D. (1996) *Adv. Mater.*, **8**, 166.

369 Conway, B.E. (1999) *Electrochemical Supercapacitors*, Kluwer Academic/Plenum Publishers, New York.

370 Trasatti, S. and Kurzweil, P. (1994) *Platinum Met. Rev.*, **38**, 46.

371 Sarangapani, S., Tilak, B.V. and Chen, C.P. (1996) *J. Electrochem. Soc.*, **143**, 3791.

372 Sugimoto, W., Iwata, H., Yasunaga, Y., Murakami, Y. and Takasu, Y. (2003) *Angew. Chem. Int. Ed.*, **42**, 4092.

373 Zheng, J.P., Cygan, P.J. and Jow, T.R. (1995) *J. Electrochem. Soc.*, **142**, 2699.

374 Srinivasan, V. and Weidner, J.W. (1997) *J. Electrochem. Soc.*, **144**, L210.

375 Wang, Y.G. and Xia, Y.Y. (2006) *Electrochim. Acta*, **51**, 3223.

376 Xing, W., Li, F., Yan, Z.F. and Lu, G.Q. (2004) *J. Power Sources*, **134**, 324.

377 Cao, L., Kong, L.B., Liang, Y.Y. and Li, H.L. (2004) *Chem. Commun.*, **14**, 1646.

378 Lin, C., Ritter, J.A. and Popov, B.N. (1998) *J. Electrochem. Soc.* **145**, 4097

379 Cao, L., Xu, F., Liang, Y.Y. and Li, H.L. (2004) *Adv. Mater.*, **16**, 1853.

380 Pang, S.C., Anderson, M.A. and Chapman, T.W. (2000) *J. Electrochem. Soc.*, **147**, 444.

381 Hao, D.D.Z., Bao, S.J., Zhou, W.J. and Li, H.L. (2007) *Electrochem. Commun.*, **9**, 869.

382 Nelson, P.A., Elliott, J.M., Attard, G.S. and Owen, J.R. (2002) *Chem. Mater.*, **14**, 524.

383 Nelson, P.A. and Owen, J.R. (2003) *J. Electrochem. Soc.*, **150**, A1313.

384 Bartlett, P.N., Gollas, B., Guerin, S. and Marwan, J. (2002) *Phys. Chem. Chem. Phys.*, **4**, 3835.

385 Luo, H.M., Zhang, J.F. and Yan, Y.S. (2003) *Chem. Mater.*, **15**, 3769.

386 Chabre, Y. and Pannetier, J. (1995) *Prog. Solid State Chem.*, **23**, 1.

387 Winter, M. and Brodd, R.J. (2004) *Chem. Rev.*, **104**, 4245.

388 (a) Pang S.C., Anderson M.A. and Chapman T.W. (2000) *Electrochem. J. Soc.*, **147**, 444. (b) Pang S.C. and Anderson M.A. (2000) *J. Mater. Res.*, **15**, 2096.

389 (a) Lee H.Y., Kim S.W. and Lee H.Y. (2001) *Electrochem. Solid-State Lett.*, **4**, A19. (b) Hong M.S., Lee S.H. and Kim S.W. (2002) *Electrochem. Solid-State Lett.*, **5**, A227.

390 Chin, S.F., Pang, S.C. and Anderson, M.A. (2002) *J. Electrochem. Soc.*, **149**, A379.

391 (a) Prasad, K.R. and Miura, N. (2004) *Electrochem. Solid-State Lett.*, **7**, A425. (b) Prasad, K.R. and Miura, N. (2004) *Electrochem. Commun.*, **6**, 1004,

392 Kuzuoka, Y., Wen, C.J., Otomo, J., Ogura, M., Kobayashi, T., Yamada, K. and Takahashi, H. (2004) *Solid State Ionics*, **175**, 507.

393 Subramanian, V., Zhu, H.W., Vajtai, R., Ajayan, P.M. and Wei, B.Q. (2005) *J. Phys. Chem. B*, **109**, 20207.

394 (a) Cheng, F.Y., Chen, J., Gou, X.L. and Shen, P.W. (2005) *Adv. Mater.*, **17**, 2753. (b) Cheng, F.Y., Zhao, J.Z., Song, W.E., Li, C.S., Ma, H., Chen, J. and Shen, P.W. (2006) *Inorg. Chem.*, **45**, 2038.

395 Chou, S., Cheng, F. and Chen, J. (2006) *J. Power Sources*, **162**, 727.

396 (a) Wu M.S. (2005) *Appl. Phys. Lett.*, **87**, 153102-1. (b) Wu M.S., Lee J.T., Wang Y.Y. and Wan C.C. (2004) *J. Phys. Chem. B*, **108**, 6331.

397 Zach, M.P., Ng, K.H. and Penner, R.G. (2000) *Science*, **290**, 2120.

398 Rampi, M.A., Schueller, O.J.A. and Whitesides, G.M. (1998) *Appl. Phys. Lett.*, **72**, 1781.

399 Holmlin, R.E., Haag, R., Chabinyc, M.L., Ismagilov, R.F., Cohen, A.E., Terfort, A., Rampi, M.A. and Whitesides, G.M. (2001) *J. Am. Chem. Soc.*, **123**, 5075.

400 Ulman, A. (1996) *Chem. Rev.*, **96**, 1533.

401 Finklea, H.O. (1996) *Electroanal. Chem.*, **19**, 109.

402 Xia, Y.N., Rogers, J.A., Paul, K.E. and Whitesides, G.M. (1999) *Chem. Rev.*, **99**, 1823,

403 Xia, Y.N. and Whitesides, G.M. (1998) *Angew. Chem. Int. Ed.*, **37**, 551.

404 Lercel, M.J., Redinbo, G.F., Craighead, H.G., Sheen, C.W. and Allara, D.L. (1994) *Appl. Phys. Lett.*, **65**, 974.

405 Liu, G.Y., Xu, S. and Cruchon-Dupeyrat, S. (1998) in *Self-Assembled Monolayers of Thiols*, (ed. A. Ulman), San Diego, p. 81.

406 Piner, R.D., Zhu, J., Xu, F., Hong, S.H. and Mirkin, C.A. (1999) *Science*, **283**, 661.

407 Gölzhauser, A., Eck, W., Geyer, W., Stadler, V., Weimann, T., Hinze, P. and Grunze, M. (2001) *Adv. Mater.*, **13**, 806.

408 Geyer, W., Stadler, V., Eck, W., Gölzhäuser, A., Grunze, M., Sauer, M., Weimann, T. and Hinze, P. (2001) *J. Vacuum Sci. Technol. B*, **19**, 2732.

409 David, C., Müller, H.U., Völkel, B. and Grunze, M. (1996) *Microelectron Eng.*, **30**, 57.

410 Felgenhauer, T., Yan, C., Geyer, W., Rong, H.T., Gölzhäuser, A. and Buck, M. (2001) *Appl. Phys. Lett.*, **79**, 3323.

411 Geyer, W., Stadler, V., Eck, W., Zharnikov, M., Gölzhäuser, A. and Grunze, M. (1999) *Appl. Phys. Lett.*, **75**, 2401.

412 Gölzhäuser, A., Geyer, W., Stadler, V., Eck, W., Grunze, M., Edinger, K., Weimann, T. and Hinze, P. (2000) *J. Vacuum Sci. Technol. B*, **18**, 3414.

413 Sondag-Huethorst, J.A.M. and Fokkink, L.G.J. (1992) *Langmuir*, **8**, 2560.

414 Sondag-Huethorst, J.A.M., Van Helleputte, H.R.J. and Fokkink, L.G.J. (1994) *Appl. Phys. Lett.*, **64**, 285.

415 Kaltenpoth, G., Völkel, B., Nottbohm, C.T., Gölzhäuser, A. and Buck, M. (2002) *J. Vacuum Sci. Technol. B*, **20**, 2734.

416 Whitman, C., Moslehi, M.M., Paranjpe, A., Velo, L. and Omstead, T. (1999) *J. Vacuum Sci. Technol. A*, **17**, 1893.

417 Völkel, B., Kaltenpoth, G., Handrea, M., Sahre, M., Nottbohm, C.T., Küller, A., Paul, A., Kautek, W., Eck, W. and Gölzhauser, A. (2005) *Surface Sci.*, **597**, 32.

3
Top-Down Approaches to the Fabrication of Nanopatterned Electrodes
Yvonne H. Lanyon and Damien W.M. Arrigan

3.1
Introduction

The transition from macro- to micro-sized electrodes for use in electroanalytical chemistry first began around three decades ago [1,2], although their application in neurophysiological science had already been established by then [3]. A major driving force behind this transition was based on the need for portable and *in-vivo* sensing devices with increased sensitivity, capable of measuring trace concentrations of analytes in ultra-small sample volumes. Electrodes based on nanometer dimensions further facilitate the miniaturization of sensors for such applications, allowing real-time measurement of a number of physiological analytes such as dopamine release from single living vesicles [4]. A variety of microtechnologies have emerged that are based upon the adaptation of electronic device manufacturing methods, and have facilitated the development of miniaturized devices and sensors [5]. The fabrication of microelectrodes has largely been based on methods developed by the semiconductor industry using thin-film microfabrication technologies [6], which have been widely used for the production of microelectronic chips and Micro Electro Mechanical Systems (MEMS).

A vast number of publications on the fabrication of microelectrodes and their applications can be found in the literature. Most report a number of advantages arising from the use of these reduced-sized microelectrodes, giving rise to a number of benefits for electrochemical sensing applications. The advantages of microelectrodes over macroelectrodes are largely attributed to the rapid formation of radial (three-dimensional) diffusion fields and reduced ohmic resistance effects, resulting in increased mass transport rates, increased current densities and enhanced signal-to-noise ratios [7]. These effects are proposed to be even greater with the transition from micro- to nano-sized electrodes [8].

In recent years, with the research and development of highly specialized instrumentation, there has been an increased transition from the micro- to nano-dimension electrode scale. Aside from the benefits aforementioned for microelectrodes,

---

Nanostructured Materials in Electrochemistry. Edited by Ali Eftekhari
Copyright © 2008 WILEY-VCH Verlag GmbH & Co. KGaA, Weinheim
ISBN: 978-3-527-31876-6

electrodes with nanometer dimensions offer great opportunities for fundamental studies of diffusion characteristics, electron-transfer kinetics, and double-layer structure at electrode surfaces [9]. The combination of these factors can greatly enhance the overall sensitivity of a sensing device, offering the possibility of single molecule detection, as well as opening up the wide arena of portable miniaturized electrochemical sensing using ultra-low sample volumes. Nanoelectrodes have found applications in a number of areas such as biomolecular sensors for medical diagnostics [10], single molecule detection [11–13], and real-time monitoring of cell exocytosis [4].

The wide arena of nanofabrication makes use of a range of techniques to deposit, etch, grow or manipulate materials at the nanoscale. *Nanofabrication* refers to the processes and methods employed in the preparation of nanoscale objects and nanopatterned substrates. The two main methods for nanoelectrode fabrication can be divided into top-down and bottom-up approaches.

The *bottom-up approach* is generally based on the spontaneous formation of objects induced by a rational design of the chemical and physical interactions of various materials, and is thus in the hands of the chemist. It makes use of the interactions between molecules or colloidal particles in order to assemble discrete nanoscale structures in two and three dimensions [14]. It includes monolayer self-assembly, supramolecular assembly, and nanoparticle formation, and has been reported largely for the fabrication of random nanoelectrode ensembles such as those produced by Wang and coworkers [15].

Top-down approaches are extensions of conventional lithography used in microelectronics fabrication, and refer to the size reduction of a material down to nanoscale by use of various patterning, etching, and deposition steps. This can largely be defined as extensions of micro-fabrication techniques, including lithography, soft-lithography, and evaporation techniques [16]. In essence, top-down fabrication involves the progressive thinning or alteration of a bulk material to obtain even smaller features, where bulk refers to a starting material that is continuous in all directions at a scale much larger than the structures of interest, typically in a thin film [17].

Top-down and bottom-up approaches each have advantages and disadvantages. For example, top-down approaches are often easier to integrate into existing fabrication processes, whereas bottom-up methods have fewer restrictions regarding the need for expensive facilities. Nevertheless, a combination of the two approaches into the realization of functional devices is an active research endeavor (e.g., Ref. [18]).

Other common approaches to nanoelectrode fabrication have included:

- the creation of band nanoelectrodes by the deposition of insulator–metal–insulator layers, followed by etching or polishing to expose one of the edges [19]
- the deposition of metal into nanoporous filtration membranes [20]
- the insulation of a conductive wires or fibers with electrophoretic paint, followed by exposure of the very tip by heat-shrinking or etching [21,22].

The first method led to the production of the first types of nanoelectrode, but offers little scope for mass production or *in-vivo* measurements. The second method creates nanoelectrode ensembles where there is no control over the electrode dimensions or spacing, while the nanoelectrodes produced by the third method have more appropriate applications as tips for scanning electrochemical microscopy (SECM).

The aim of this chapter is to review recent developments in nanoelectrode fabrication using top-down approaches, including electron-beam (e-beam) lithography, focused ion beam (FIB) milling, and other methods such as nano-imprint lithography. These have been the most widely studied methods for the preparation of nanoelectrode arrays. Each of the techniques will be assessed on the types of nanoelectrode that can be produced, the electrochemical performance of the electrodes, and their suitability for use as analytical sensors. The fabrication of nanoelectrodes by bottom-up and the other aforementioned approaches is beyond the scope of this review, which will focus largely on the fabrication of nanoelectrode array sensors for analytical applications. Although research on nanoelectrodes is growing annually, top-down approaches remain the most widely investigated, and have been the subject of a number of applications studies. However, it must be emphasized that to date, focus has been primarily on the fabrication rather than on the demonstration of applications.

3.2
Considerations for Choosing a Nanoelectrode Fabrication Strategy

Nanoelectrodes can be defined as electrodes with at least one dimension in the nanometer scale, which is referred to as the *critical dimension*, and that which controls the type of electrochemical response [8]. The classification of nanoelectrodes has been the subject of much debate, but most commonly their definition has been based on the critical dimension within the sub-100 nm range [8]. Despite many of the proposed advantages of nanoelectrodes, one of the shortcomings of a single nanoelectrode is the very small current that is generated from it. Single nanoelectrodes have more appropriate applications as scanning tips for SECM microscopy, although for sensing applications, arrays of nanoelectrodes can be used to enhance the sensor signal. In this case, all of the electrodes in the array act in parallel, with each electrode having its own independent diffusion profile (provided that there is sufficient spacing between the electrodes), thus greatly enhancing the current density. A relatively small area of a sensor chip can be patterned with vast numbers of nanoelectrodes, thereby facilitating the development of ultra-sensitive miniaturized sensors.

In an array format, the ratio between the critical dimension and the inter-electrode spacing is of great importance. Where the spacing is insufficient, this creates overlapping diffusion zones between the electrodes, generating a response typical of a macro-sized electrode. Characterization by cyclic voltammetry can be a means of identifying the type of diffusion that occurs in a given array layout, where peak-shaped voltammograms are indicative of overlapping diffusion zones, and hence insufficient spacing between neighboring electrodes. As the spacing between the electrodes increases sufficiently to allow each electrode to have its own radial diffusion profile, sigmoid-shaped voltammograms are prevalent. This is a result of the overall faster mass transport of the electroactive species to the electrode surface [7].

The choice of a particular (top-down) fabrication technique largely depends on the end application of the nanoelectrodes. Where high sensitivity for sensing applications is required, the design of the nanoelectrodes can be customized in order to maximize their benefits. In this case, electrode fabrication with controlled dimensions and separation is of great importance, as is the reproducible fabrication of large arrays to greatly enhance the sensor signal. In such cases, top-down methods such as e-beam and ion beam lithography have been most commonly employed. Where the end application of the nanoelectrodes is trace analysis, then interdigitated array (IDA) nanoelectrodes are a popular choice where redox cycling between adjacent electrodes allows amplification of the low signals arising from trace concentrations of analytes (see Sections 3.3.1 and 3.3.3). Such electrodes can be produced by e-beam lithographic methods, but also more cost-effectively by nanoimprint lithography. Single molecule detection requires the fabrication of nanoelectrodes with dimensions of comparable size to the molecule of interest. Nanopores have become a popular approach to single molecule detection, where nanopores have been prepared using high-resolution e-beam and ion beam lithography. However, when the overall dimensions of the electrode approach the size of the molecule of interest, the fabrication process becomes more difficult. Hence, there is now increasing interest in a combination of top-down methods with bottom-up approaches and alternative electrode formats such as nanoelectrodes with nanogaps as a means of analyzing single molecules (see Section 3.3.4).

The following sections review some of the recent top-down approaches to nanoelectrode fabrication of various types, including their characterization and end applications.

3.3
Nanoelectrode Fabrication Using Top-Down Approaches

Over the past decade, although an increasing number of reports has been published on the fabrication of nanoelectrodes, only a limited number of these have related to fabrication using top-down methods. This largely reflects the difficulty in their fabrication, particularly within the sub-100 nm scale. For sensor applications – and especially where the mass production of arrays of nanoelectrodes is required – the reproducibility of the fabrication technique becomes important. Depending on the end application of the sensor, the capacity of a fabrication technique for the desired geometry, dimensions of the electrodes and design of the array layout must be assessed. Techniques that are used to make nanoelectrodes with controlled dimensions, spacings and designs are typically created using "top-down" methods, which are the focus of this chapter. While photolithographic techniques have been widely used in microelectrode fabrication, the smallest feature that can be made must be larger than half the wavelength of the light that is used [23]. The miniaturization of electrodes from the micro- to nano-scale (particularly in the tens of nanometers range) requires the use of advanced instrumentation with high resolution, using smaller wavelengths for mask exposure [17]. The use of advanced optical and e-beam lithographic methods are common approaches which, in turn, have increased enormously the cost of nanoelectrode fabrication.

Scanning beam lithography is one of the most common approaches to nanoelectrode array fabrication. It is a serial technique which can generate high-resolution features with a variety of patterns. There are three main classes of scanning beam lithography [14]:

- scanned laser beams with ~250 nm resolutions
- focused electron beams with sub-50 nm resolution (depending on tool settings, choice of resist and aspect ratios)
- focused ion beams, some with sub-50 nm resolution (depending on similar parameters to the e-beam).

The latter technique is most commonly used for research purposes, whereas the electron beam methods can be used for the larger-scale production of nanoelectrodes.

3.3.1
E-Beam Lithography

The first electron-beam lithography (EBL) systems, based on the scanning electron microscope, were developed during the late 1960s. Because of its very high resolution and flexibility, EBL it is now widely used in many universities and research institutes for nanoscale patterning [24]. The use of this technology has become popular in recent years for the fabrication of nanoelectrodes. The technique is based on the high-resolution etching of substrate materials using a finely focused beam of electrons. Because of its short electron wavelength, e-beam writing is not diffraction-limited, allowing resolution down to the nanometer range [24]. EBL can be applied to nanoelectrode preparation by using either direct or indirect pattern techniques, and can also be used in combination with other techniques such as nanoimprint lithography. Most of the prepatterning of substrates, such as contact pad formation, can be achieved using standard microfabrication techniques such as photolithography, followed by a final EBL step to create nanoscale electrodes.

In the drive towards a better understanding of the functions of transport mechanisms across the biological cell membrane, the fabrication of nanopores approaching the dimensions of single biomolecules has gained increasing interest in recent years. Many reports can be found on the creation of nanopores on membranes through which single molecules such as DNA can be passed and analyzed [11–13,25], although fewer exist on the creation of nanopores incorporating an electrode of similar dimensions. Some techniques have been applied to circumvent this problem by the creation of nanogap electrodes (see Section 3.3.4).

The fabrication of nanopore electrodes using EBL has been achieved in recent years, typically where the nanoelectrode is recessed at the bottom of a pore within an insulating material. Lemay and coworkers [26] detailed the use of silicon-on-insulator wafers on which a number of standard microfabrication processing steps were conducted, before e-beam patterning of a 40 nm-thick silicon membrane within a 400 nm single square. This pattern was transferred to a silicon oxide layer of the wafer by CHF_3 plasma etching, following by further etching and deposition steps to create a recess with angled walls containing a pore at the bottom, with sizes ranging from completely closed to 200 nm diameter. Finally, the electrodes were formed by deposition and evaporation of gold into the pores, forming a recessed indented nanoelectrode. The advantage of this technique lies in the metal electrode layer being the last fabrication step, so that it is never exposed to contaminants during the fabrication process. Electrochemical characterization allowed validation of the electrodes and determination of the geometry using the limiting current obtained from cyclic voltammetry with mathematical models developed for that electrode geometry type. More recently, the group have modified the technique to create nanoelectrodes with geometry opposite to those made previously, in which the resultant electrode is pyramidal in shape with radii ranging from 0.4 nm to 100 nm (Figure 3.1) [27]. The nanoelectrodes gave characteristic steady-state cyclic voltammograms (Figure 3.2), where the effects of non-faradaic current were more evident with decreasing electrode size, particularly

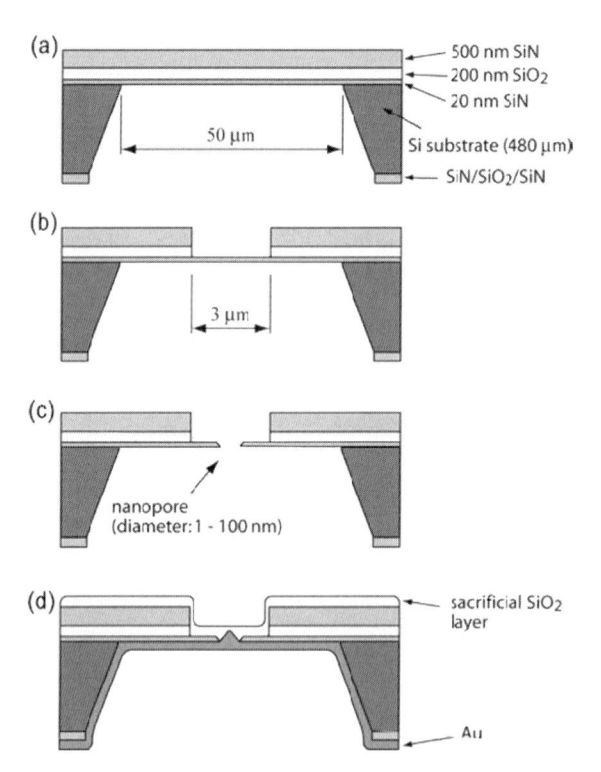

Figure 3.1 Preparation of nanoelectrode by electron-beam lithography. (a) Chemical vapor deposition three-layer deposition on silicon, backside lithography and KOH etch. (b) Front-side lithography and etch of the silicon nitride and silicon dioxide layers forming a 20 nm silicon nitride free-standing membrane. (c) Nanopore drilling in the silicon nitride membrane with a focused electron beam. (d) Sacrificial silicon dioxide layer sputtering and gold evaporation. The sacrificial layer is then removed in buffered hydrofluoric acid. (Reprinted from Ref. [27].)

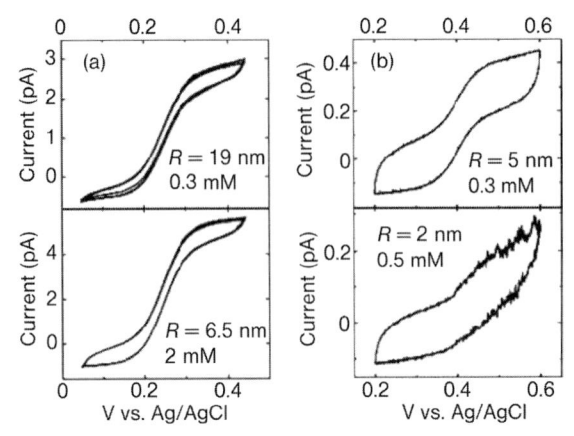

Figure 3.2 Cyclic voltammograms recorded at nanopore electrodes, using (a) ferrocenedimethanol and (b) ferrocenylmethyltrimethylammonium at scan rates of (a) $10\,mV\,s^{-1}$ and (b) $0.5\,mV\,s^{-1}$ (top) and $0.1\,mV\,s^{-1}$ (bottom). (Reprinted from Ref. [27].)

where the current was in the sub-pA range. This was reported to be due to the dielectric relaxation of the membrane material, rather than to the metal–electrolyte double-layer capacitance [27].

A less complex technique employed by Sandison and Cooper [28] involved the direct EBL patterning of nanopores in a layer of insulating photoresist covering a silicon wafer containing exposed $1\,mm^2$ metal electrode areas. In this case, the patterned resist layer was not removed from the wafer, but acted as the insulation material separating individual nanopores. The resultant electrodes were recessed gold nanodiscs with pore radii in the range 50 to 500 nm, where greater peak current densities were obtained with decreasing electrode radii, in accordance with theoretical calculations. However, preliminary electrochemical characterization of the nanoelectrodes revealed initially low electrochemical signals which the authors reported to be due to resist contamination of the electrodes. This was resolved by treatment with oxygen plasma, although this etching process also removed some of the insulating material around the pores, hence increasing the pore radii. Despite the increase in peak current density with decreasing electrode size, the close proximity of the pores resulted in diffusion overlap and peak-shaped voltammograms [28].

The development of nanopore electrodes for sensing applications was undertaken by Jung and coworkers [29], in which nanopore electrodes fabricated by EBL were functionalized by antibodies for the creation of an immunosensor. The nanopores were created by e-beam patterning of a standard polymer resist (ZEP520) layer on top of an underlying layer of gold. A total resist area of $100 \times 100\,\mu m^2$ was exposed by the e-beam onto which biotinylated functional lipid vesicles (FLVs) were immobilized on the insulating material around the pores (Figure 3.3). The FLVs were used to bind specific antibody receptors onto the surface. The immunosensor was

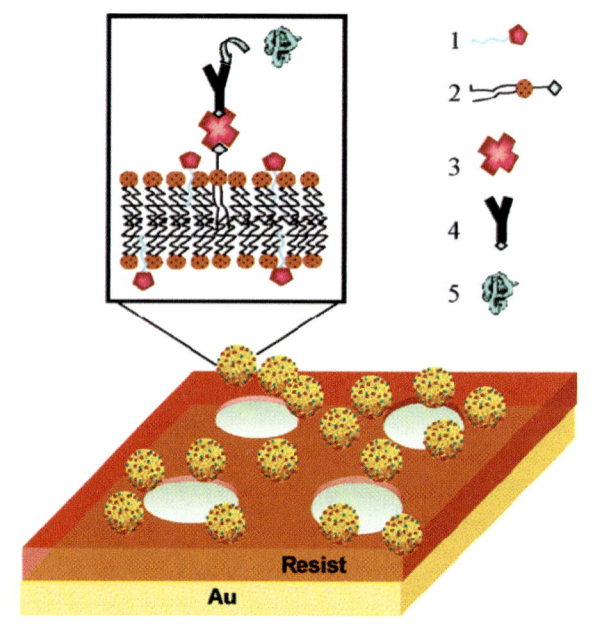

Figure 3.3 Nanowell electrode approach to functional lipid vesicles (FLV) immunosensing. (1) N-(10,12-pentacoasdiynoic) acetylferrocene (Fc-PDA); (2) 1,2-dioleoylphosphatidylethanolamine-N-caproly-amine (Cap-PE); (3) streptavidin; (4) biotinylated capture antibody; (5) target protein. (Reprinted from Ref. [29].)

characterized by model protein analytes including human serum albumin (HSA) and carbonic anhydrase of bovine source (CAB). A decrease in current was found upon binding of the capture antibody with the analyte [29].

An increasingly popular format of nanoelectrodes created by EBL are interdigitated arrays (IDAs) of nanoelectrodes. These are typically bands of micrometer length and nanometer width, separated by nanometer gaps. The electrochemical detection mechanism with these electrodes differs from the conventional format using the aforementioned nanopore electrodes, where the working electrode is an array of nanoelectrodes working in parallel, and is used to detect an electroactive species by application of potential by amperometry or voltammetry. The most common application of nanoelectrode IDAs is *redox cycling*, and which is commonly used for trace analyses to greatly enhance the sensor signal. With this technique, one band of the IDA electrodes is set to a particular potential to drive the reduction of a redoxactive species, while the other band is set to another potential to drive the oxidation. Hence, the species diffuses back and forth between the two bands (known as redox cycling) [30] (Figure 3.4). This process is made possible by decreasing the distance between adjacent electrodes towards the diffusion layer thickness. Redox

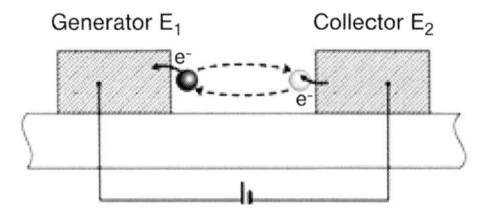

Figure 3.4 Redox cycling between closely-spaced electrodes. (Reproduced from *Analyst*, 2007, *132*, 365–370, by permission of the Royal Society of Chemistry.)

cycling greatly enhances the limiting current, and moreover, can be improved by reducing the band width and spacing [30].

Ueno and coworkers [31] reported the fabrication of IDAs on glass substrates using a combination of photolithography and EBL. Band electrodes with widths of 500 nm and 100 nm, and gaps of 500 nm or 30 nm, were fabricated by EBL, while the large defining patterns and bond pads were fabricated by ultraviolet (UV) lithography and Ti/Au lift-off. Gap dimensions of 10 nm were also achieved, though these were not as sharply defined as the wider gap electrodes. The electrochemical characterization of the IDAs using a ferrocene derivative solution, revealed steady-state sigmoidal-shaped cyclic voltammograms (CVs) indicating rapid diffusion of reduced and oxidized forms of the redox couple across the nanoscale gap between the electrodes. This trend was evident even at the smallest gap spacings between the bands. Moreover, a significant increase in the current density was observed with the 100 nm-wide bands with respect to the 1 mm bands (with similar spacing ratios). This increase was confirmed by numerical simulations and was related to the significant decrease in the size of the nanoband lateral dimension, with respect to the scale of the diffusion layer. The authors also noted an increase in sensitivity using electrodes with a Au layer thickness of 140 nm with respect to a thinner 70 nm layer [31].

A similar approach was adopted by Zhu and Ahn [30], where fabrication of their IDAs was also achieved by a combination of e-beam nanolithography and photolithography. In this case, a layer of polymethyl methacrylate (PMMA) with a thickness of 300 nm was spin-coated onto the oxidized silicon wafer surface, onto which the e-beam patterns were written. This was followed by the deposition of a Ti–Au layer using an e-beam evaporator, after which the sample was dipped in acetone for lift-off [30]. On-chip reference and counter electrodes were also formed by standard photolithographic techniques. The electrochemical characterization of the IDA nanoelectrode sensors with *p*-aminophenol revealed a detection capability in the range of 1 to 10 nM.

This approach was also adopted by Finot *et al.* [32] for the detection of DNA with IDA nanoelectrodes. First, a bulk silicon wafer was covered with a 200 nm layer of SiO_2, deposited by cathodic pulverization to reduce the surface roughness. Next, a gold layer (5 nm thickness, required for the EBL) was thermally deposited before a layer of PMMA was spun on top of the plate and patterned by the e-beam. The electrode material was then deposited (4 nm Cr/60 nm Au) by thermal evaporation, before the PMMA film was removed. This process was used to create 64 interdigitated electrodes of 100 nm width and spacing of 200 nm, covering an area of 100 µm × 50 µm. The authors reported the application of these IDAs for the detection of DNA based on the use of hexaammineruthenium(III) chloride as a redox mediator. The reaction between the ruthenium complex and DNA is an intercalation process in which both the oxidized and reduced forms of ruthenium are bound to the single strand of the DNA. Hybridization events between complementary DNA strands were detected by a 10–40% increase in the electrochemical signal. Moreover, the nanoelectrodes were found to be more sensitive than macroelectrodes, and with faster response times [32].

Another reported application of IDA nanoelectrodes involves the manipulation and immobilization of biomolecules between or around the nanoelectrode edges [33]. In this case, a PMMA resist was spun onto a dehydrated silicon wafer with a thin layer of silicon dioxide (ca. 500 nm). The nanoelectrode patterns were exposed using a scanning electron microscope equipped with a pattern generator. After development of the PMMA resist, the wafer was coated with a 3 nm thin layer of chromium by e-beam evaporation, and gold by thermal evaporation. The devices were completed by lift-off processes in acetone, resulting in IDAs of nanoelectrodes of 60 nm bands with inter-band spacings of 380 nm. Dielectrophoresis was used to trap biomolecules such as bovine serum albumin (BSA) and antibodies on the edges of the nanoelectrodes, which may find useful applications in biosensor studies [33].

Despite the high resolution of the EBL technique and application for the mass production of nanoelectrodes of several types, the main drawbacks of the technique are associated with its high cost and number of processing steps. Methods of reducing the number of processing steps by direct patterning of a photoresist layer acting as an electrode insulator, proved unsuccessful due to electrode contamination by degradation of the resist [28]. The more aggressive etching by focused ion beams can also be used for the direct patterning of more robust electrode insulating layers, and this is discussed in the following section.

3.3.2
Focused Ion Beam Lithography

A somewhat lesser used technique for the fabrication of nanoelectrodes is that of focused ion beam (FIB) lithography. This technique is based on the maskless etching of substrate materials by a finely focused beam of ions (most commonly Ga^+). It can mill substrates by selectively removing material through ion bombardment or create patterns in an additive process by ion deposition or a localized chemical vapor deposition [14]. During etching applications, the focused beam touches a sample surface and dislodges the material, which is redeposited locally in or around the patterned structure. This is a technique in which nanometer patterns can be directly written onto the substrate, thus involving fewer processing steps than EBL. Moreover, the fabrication step can be monitored *in situ* by secondary electron images

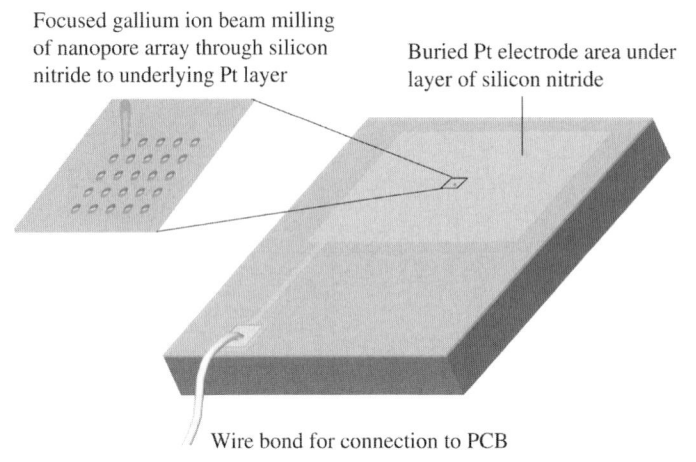

Figure 3.5 General schematic of nanoelectrode fabrication strategy by focused ion beam milling. A silicon nitride layer on top of a buried Pt electrode is milled by a focused ion beam to open up nanopores through to the underlying Pt. (Reprinted from Ref. [36].)

induced by the FIB [33]. One of the principal applications of FIB is the physical sputtering of material by ion impact for micro- and nano-patterning [35].

The FIB milling technique has been used to create recessed nanopore and nanoband electrodes on oxidized silicon wafer substrates [35,36]. Platinum electrodes (100 nm thick, and typically microbands of 2 mm × 0.6 mm) were initially fabricated using UV lithography, metal deposition and lift-off processes. A silicon nitride film (500 nm thick) was then deposited to form an insulating layer over the buried Pt microelectrodes. The FIB (30 keV Ga ions, 10 nm nominal spot diameter, 10 pA beam current) was then employed for direct-write nanoscale milling of the silicon nitride passivation layer (Figure 3.5). The FIB technique was used to produce pores and bands with controlled dimensions, spacing, and array layouts (Figure 3.6). Moreover, the technique was also found to produce a cone-shaped recess in which the bottom of the pore or band was smaller than the mouth. Figure 3.7 shows the steady-state CVs produced by arrays of nanopores (with radii of 225 nm) with increasing electrode number. These authors concluded from the characterization of the nanoelectrodes, that the electrochemical response was controlled by diffusion to the recess mouth rather than to the electrode at the bottom. With the cone-shaped nature of the recess, and since the aspect ratio of the nano-recess did not allow direct imaging by scanning electron microscopy (SEM), the steady-state data from the CVs was used with mathematical models to enable calculation of the electrode dimensions at the bottom of the pore [36,37].

The FIB lithography approach has also been used to create interdigitated titanium nanoelectrodes by Santschi and coworkers [35]. In this case, photolithography and plasma etching of a silicon-based wafer to create bond pads for the nanoelectrodes were initially conducted. This was followed by titanium deposition by sputtering at

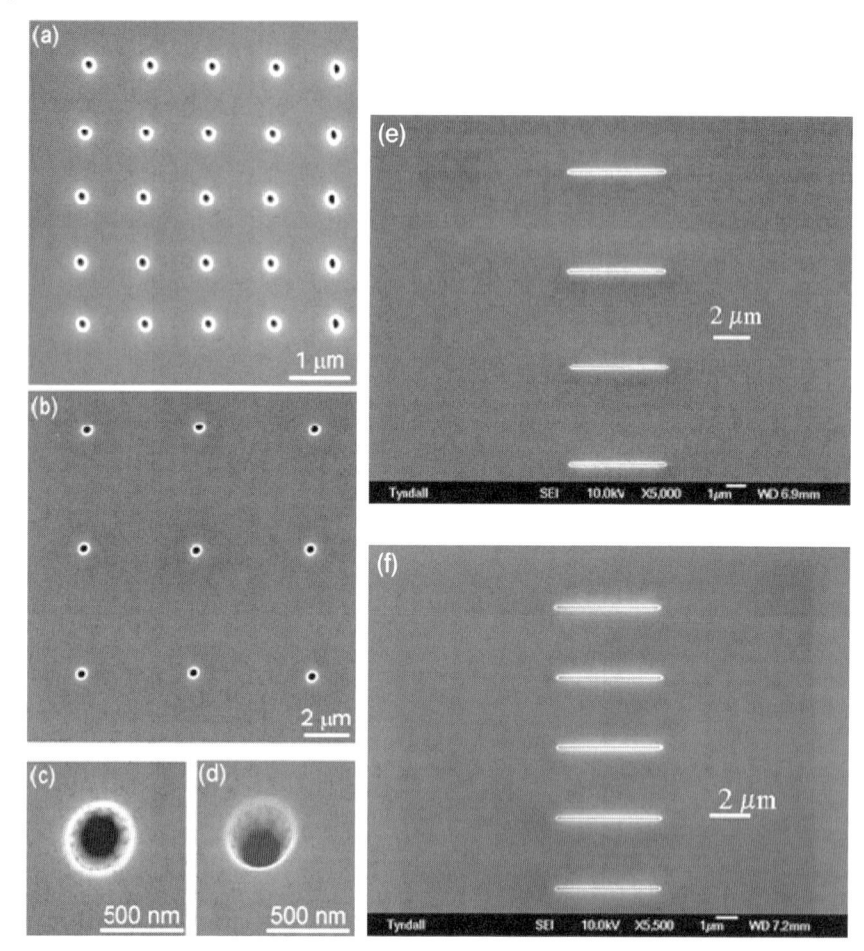

Figure 3.6 Nanopores and nanobands prepared by focused ion beam (FIB) milling through silicon nitride. SEM images of: (a) 5×5 array; (b) 3×3; (c) and (d) are single nanopore electrodes. Pores are truncated cone-shaped. (e) and (f) are images of nanoband arrays fabricated by FIB milling, of width 117 nm and 101 nm, respectively. [(a)–(d) reprinted from Ref. [36]; © American Chemical Society; (e) and (f) reprinted from Ref. [37].)

200 °C, providing a 40 nm coating thickness. The focused ion beam (Ga^+) was then used to etch the interdigitated nanoelectrodes in combination with gas enhancement using XeF_2 in order to reduce material redeposition during the etching process, which can affect adjacent structures milled previously. The nanoelectrodes were 1 μm bands in length with 50 nm band widths and 50 nm band gaps. Although complete electrochemical characterization of the interdigitated nanoelectrodes was not conducted, the resistance between adjacent electrodes was higher than 1 GΩ, and thus was reported to be suitable for a wide range of applications [35].

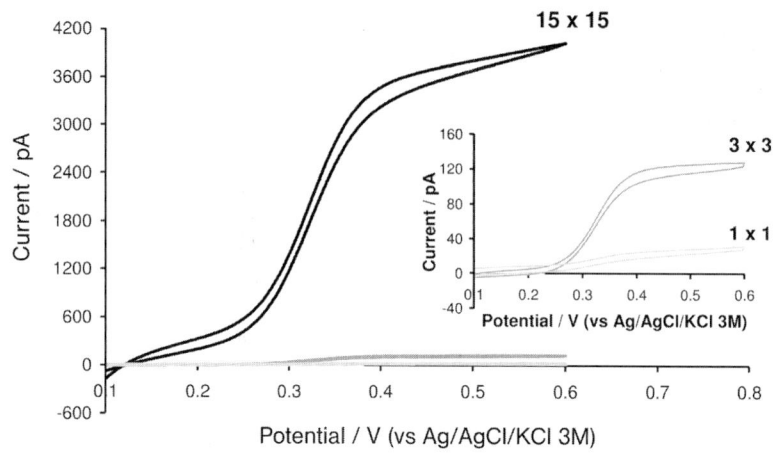

Figure 3.7 Voltammetric characterization of selected nanopore electrode arrays using 1 mM FcCOOH in phosphate-buffered saline. The influence of increasing nanoelectrode numbers on the cyclic voltammogram response at 5 mV s^{-1} based on arrays with pore radii of 225 nm. (Reprinted from Ref. [36].)

Nagase and coworkers [38] have also used FIB lithography to create nanoelectrode structures with nanogaps, which are becoming an increasingly popular means of creating electrode structures of comparable size to single molecules (see Section 3.3.4). The technique was applied to a metal film of Au (10–30 nm thick) on Ti (1–2 nm thick) deposited on 200 nm-thick, thermally grown oxides on silicon wafers. The nanogap electrodes were formed by two etching steps; the first step to produce nanowire structures in the metal film, followed by a second milling pattern by single line scanning to create nanogaps between the electrodes. The minimum sizes of the structures that could be formed using this technique were nanowire electrodes of 30 nm width and nanogaps of 3 nm. The resistance of most of the gaps was reported to be higher than a few GΩ, with the highest around 80 GΩ. Some leak current was found at the gap structures, most likely related to the re-deposition of Au during the FIB etching process. These authors concluded that some optimization of the wafer layer thicknesses and etching procedures would be required to correct this problem [38].

Despite the high resolution of e-beam and ion-beam lithography, one of the main drawbacks of these techniques is the high cost. Research into lower-cost, top-down methods of nanoelectrode fabrication is of great interest, and is outlined in the following section.

3.3.3
Nano-Imprint Lithography

The high operating costs of the e-beam and ion-beam etching techniques have driven the search for lower-cost, high-resolution alternatives. Among those techniques

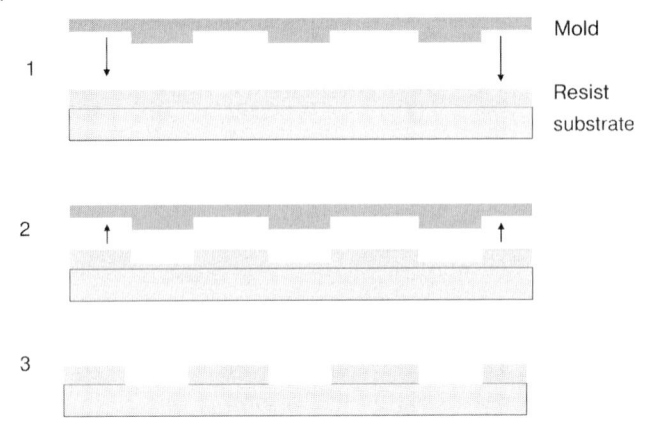

Figure 3.8 Summary of the nanoimprint lithography (NIL) process. A mold with desired nanoscale features is pressed into a polymer (resist) layer (step 1), the mold is removed (step 2), and the pattern imprinted is then transferred to the substrate using a suitable etching process (step 3), such as reactive ion etching (RIE). (Adapted from Ref. [39].)

that can be classified as top-down approaches are nano-imprint or soft lithography. Techniques that prepare a soft mold or stamp by casting a liquid polymer precursor against a topographically patterned master are commonly referred to as soft lithography [14]. Nano-imprint lithography (NIL) is a high-resolution, high-throughput, and low-cost parallel lithographic technique [39,40]. It refers to the pressure-induced transfer of a topographic pattern from a rigid mold (typically silicon) into a thermoplastic polymer film heated above its glass-transition temperature [14]. As illustrated in Figure 3.8, NIL consists primarily of two steps: an imprinting step, followed by pattern transfer to the substrate using an etching process. Most NIL processes require a high-resolution fabrication of the mold or stamp employed, and this is usually reliant on e-beam lithography in order to achieve a stamp with the desired high-resolution features. The most common electrode type fabricated by this method is interdigitated array (IDA) nanoelectrodes, although other nano-array structures have also been reported using this technique.

Montelius and coworkers [41] reported the use of a single imprint step on 2-inch (7 cm) silicon wafers to create nanometer-structured IDA electrodes. Stamp production was undertaken on another 7-cm silicon wafer using a double-layer resist system which was transferred to a projection mask aligner before deposition and lift-off of chromium (30 nm) and gold (20 nm). The double metal layer was necessary to achieve a good contrast between the metal and insulating parts of the pattern for successful alignment during the e-beam lithography and associated processing of the stamp. For imprinting of the nanopatterned stamp, another 7-cm silicon wafer was used with a double layer resist system. After imprinting, 50 Å of titanium and 150 Å of gold were deposited by lift-off, followed by further processing steps for mounting and bonding. The nanoelectrode widths and spacings achievable using this method ranged from 100 nm to 800 nm.

Electrochemical characterization of the IDA electrodes was undertaken by redox cycling of ferrocene dicarboxylic acid using chronoamperometry. The results showed that steady-state current conditions were achieved after 10 s, based on the largest nanoelectrodes and spacings (both 800 nm). Characterization of the smaller widths and spacings was not reported. The authors reported a slight asymmetry between the generator and collector current which resulted in a collection efficiency of 80% [41].

An approach adopted by Jiao and coworkers [40] reported the use of negative nanoimprint lithography (N-NIL) to form IDA nanoelectrodes. In this case, a metallic film was deposited onto a silicon substrate, followed by spin-coating of a PMMA resist layer. This layer was imprinted by a stamp followed by further reactive ion etching and wet etching of the metal film. The technique resulted in metallic bands that were negative replications of the stamp protrusion structures, creating band widths and spacings of 170 nm and 370 nm, respectively. The stamps were created on SiO_2/Si substrates by e-beam lithography, CHF_3 reactive ion etching (RIE) and lift-off (see Figure 3.9). The authors reported advantages of this technique over conventional NIL, including: (i) variation of the band dimensions and spacing with chemical etching time; (ii) the creation of bimetallic structures; and (iii) the fabrication of complex nanopatterns by a single imprint. Electrochemical characterization of the IDA nanoelectrodes was not reported [40].

Carcenac and coworkers [42] reported the use of e-beam lithography to create patterns of nanoelectrodes (200–400 nm band widths), which are situated in a circular layout where the tapered ends of the bands point inwards forming a central circular gap of 30 to 100 nm. Silicon substrates with 100 nm of thermally grown oxide were used, followed by spin-coating of a 50 nm-thick layer of PMMA. This PMMA layer was patterned by the e-beam and the pattern then transferred to the silicon substrate by etching. Once fabricated, there followed the creation of negative molds for future replication of the nanoelectrodes by NIL. In this case, polydimethylsiloxane (PDMS) was used as a thermocurable resist, but some problems associated with unresolved inter-electrode gaps were reported. The electrochemical characterization of the nanoelectrodes was not reported [42].

Tallal and coworkers [43] reported the reproducible replication of sub-40 nm electrodes on an 8-in. (20 cm) silicon wafer using a combination of top-down methods. First, a hybrid mold was created by a combination of optical and e-beam lithography, featuring different nanoelectrode geometries and gap sizes. Photolithography was used to define the connection pads. A layer of resist was then developed and the patterns transferred to a depth of 150 nm in silicon by reactive ion etching. Each pattern was contained in a $2\times 2\,cm^2$ area, consisting of 150 band electrodes with gap variations from 100 nm to less than 30 nm. Second, the silicon mold replication was achieved by spin-coating the substrate with a 240 nm layer of polymer. The combination of pressure applied between the mold and substrate, as well as using a printing temperature above the polymer glass transition temperature, caused the polymer to flow into the mold cavities and to reproduce the mold patterns. The entire

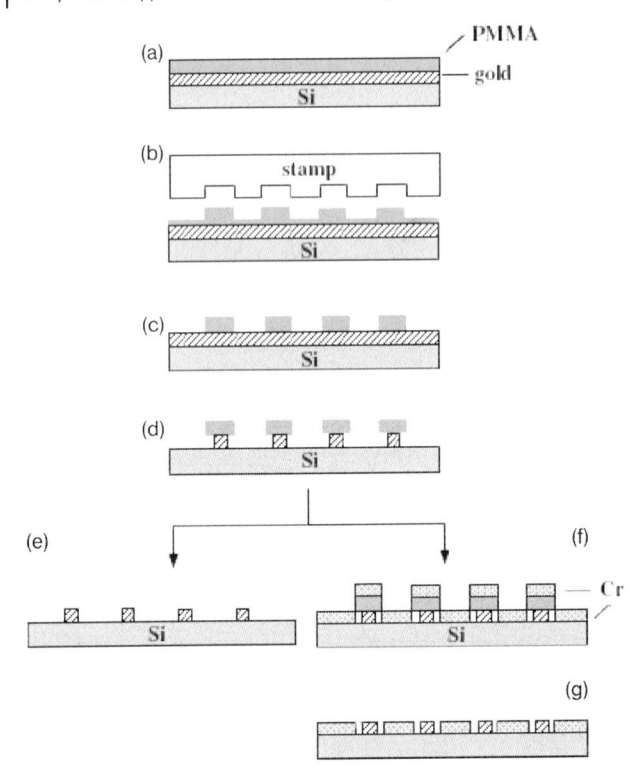

Figure 3.9 Use of negative nanoimprint lithography (N-NIL) for preparation of nanoelectrodes. (a) Silicon substrate precoated with a metal film and a thin poly methylmethacrylate (PMMA) film is spin-coated on the metal. (b) After imprinting, the stamp pattern is replicated in the PMMA film. (c) Residual PMMA in the trenches is removed by reactive ion etching. (d) The exposed metal film is dissolved with wet etchant. (e) Metallic nanostructures after lift-off of residual PMMA. (f) A second metal is deposited using a PMMA pattern as a mask. (g) The bimetallic nanostructures which result, after removing PMMA. (Reproduced from Ref. [40].)

system was then cooled before demolding. Finally, metal layers (5 nm Ti and 35 nm Au) were evaporated by plasma-enhanced chemical vapor deposition (PECVD) and lift-off processes. These authors reported the reproducible production of metallic nanoelectrodes and nanogaps based on an optimal choice of NEB22A2 polymer. No electrochemical studies were reported, however [43].

Subsequent to their studies on nanopore electrodes using EBL, Sandison and Cooper [28] reported the fabrication of similar nanopore electrodes using NIL. These studies were based on the EBL approach described Section 3.3.1, using silicon wafers with a top metal electrode layer covered by a layer of resist. Rather than using EBL to open up nanopores in the resist, in this case a silicon-based nanostructured stamping tool consisting of an array of appropriately sized pillars (250 nm radius, 500 nm height) was pressed into a polymer film covering the underlying metal electrode layer. The stamp (prepared by a combination of EBL and reactive ion etching) was pressed into a polymer film that was heated above its glass transition temperature, where the pressure was maintained until the polymer cooled. After removal of the stamp, the sample was treated with oxygen plasma to remove any residual polymer at the base of the patterned features. The authors reported some deformation of the nanopore structures by the imprinting process, resulting in an increase in the electrode radii, from 250 nm to 340 nm. However, comparable electrochemical results were obtained to similar nanopore electrodes formed by EBL, with the added advantage of lower production costs. The close proximity of the pores both in the EBL and NIL production (with an electrode spacing/radius ratio of 10) resulted in peak-shaped voltammograms, indicative of overlapping diffusion fields [28].

Today, NIL has become an attractive alternative for the low-cost mass production of nanoelectrode arrays, and finds most application in the production of IDA electrodes. The fabrication of alternative nanopatterns can be limited by the deformations caused by the printing process, resulting in irreproducible or larger electrode structures.

3.3.4
Nanogap Electrodes

An increasing number of reports on the fabrication of nanogap electrodes are now becoming apparent. In this case, the electrodes are not interdigitated (as discussed above), but rather are simply two closely spaced electrodes, where the gap is of nanoscale dimensions. These electrodes are of great importance for the development of sensors for molecular electrochemical analysis. The use of nanoelectrodes to study events at the molecular level requires electrodes of comparable size to the molecules under investigation [41]. The direct fabrication of the recessed nanopore electrodes described in Section 3.3.1 is somewhat limited by the aspect ratios of the pores, and thus dimensions approaching the single molecule scale are difficult to produce using this approach. Moreover, for single molecule detection, the distance between the electrodes should be as small as possible to enable trapping or manipulation of the molecules. The role of the nanoelectrodes in this case is to convert biological information contained in the affinity between molecular species into a measurable electrical signal. This type of nanosystem does not rely on target amplification systems requiring complex procedures, and hence the main challenge that it faces is the sensitivity required to generate information from a few biomolecules [44]. Nevertheless, an essential requirement for the use of very narrow gaps for molecular and nanoelectronics applications is that they do not allow any leakage currents to flow between them [45]. Therefore, the first characterization step for most fabricated nanogap electrodes is the measurement of resistance. This subsection focuses on the preparation of nanogap electrodes, and some of the applications associated with them.

Nanogap electrodes were briefly described in each of the preceding subsections, where methods such as e-beam [42] and focused ion-beam lithography [38] as well as NIL [43] have been used for their fabrication. Reports of nanoelectrodes of this type are highlighted in the following paragraphs.

Fischbein and Drndić [46] reported the fabrication of nanogaps by EBL. Silicon wafers were processed with photolithographic techniques to define larger patterning windows (in this case a free-standing Si_3N_4 membrane) for subsequent e-beam patterning. A layer of PMMA resist was spun onto the membrane, which was patterned by the e-beam and then used as a template to create nanogap structures. The metallization of the structures to create the electrodes was created by the deposition of gold on nickel or chromium adhesion layers, followed by lift-off of the undesired metal areas by dissolving the resist beneath it. The resulting device was composed of metal features defined by the EBL patterns, with electrode gaps ranging from 0.7 nm to 6 nm. The gaps were characterized using PbSe nanocrystals capped with oleic acid localized within the electrode gap using standard voltage–current curves; capacitance measurements were also undertaken [46].

Waser and coworkers [47] reported the use of two EBL steps to create nanoelectrodes with nanogaps on silicon wafers. On each chip, 30 nanoelectrodes representing 15 pairs of nanogaps were fabricated. Using silicon-based wafers, the surface was covered with a two-layer e-beam resist system consisting of PMMA and PMMA/MAA, the latter as a positive resist. The first e-beam step was needed to define the overlap between the optical and e-beam part, and to create pairs of nanoelectrodes with nanogaps. The electrodes were fabricated by the deposition of Ti/Pt/Au layers, followed by lift-off. A protective resist layer was then spin-coated onto the wafer and patterned by EBL, followed by reactive ion-beam etching. The nanoelectrodes were characterized by capacitance measurements in sulfuric acid, which was used to study the capacitance. Additionally, the electrodeposition of copper ions on the nanogap electrodes was demonstrated as a promising strategy to narrow the gap down to a few Angstroms for contacting small clusters and tailored molecules in metal–molecule–metal hybrid assemblies [47].

Sun and coworkers [45] demonstrated a further application of their nanogap electrodes for the study of functionalized cobalt particles. Starting with a silicon-based substrate, these authors fabricated electrode structures defined by EBL, followed by gold deposition and lift off. The gap sizes were controlled by adjusting the e-beam dose and pattern development time, producing gaps which ranged from 7 nm to 25 nm, where the fabrication yield was almost 100% for 8- to 9-nm gaps, but much less (~15%) for 3- to 4-nm gaps. The authors reported the formation of a cobalt particle junction structure between the nanogap electrodes, where its transport properties could be measured by using voltammetry.

Despite the successful creation of the nanogap electrodes thus far reviewed, one notable limitation of the high-resolution techniques is the reproducibility of the nanogaps produced, the size of which is comparable to a single molecule (0.5–2 nm) [48]. A number of methods have been reported for the fabrication of nanogaps, many using EBL as a primary step for the definition of nanoscale dimensions, followed by a further technique for the reduction of the gap dimensions down to the molecular scale. Ah et al. (2006) used an initial EBL step to create interdigitated nanoelectrodes on a silicon-based wafer with gaps of 50 nm. The gaps were then narrowed to several nanometers by the deposition of Au onto the surface of the initial Au band electrodes via the surface-catalyzed chemical reduction of Au ions with hydroxylamines. The tunneling current between two gap electrodes was measured by voltammetry and used with numerical calculations to estimate the gap distance where this was difficult to resolve by field emission scanning electron microscopy (FESEM). The authors reported the fabrication of arrays comprising 20 000 nanogaps within a $1 \times 1\,mm^2$ area, where over 90% of the gap distances were less than 5 nm [48].

A more specific application of nanogap electrodes for DNA detection was achieved by Chang and coworkers [49]. The gold nanogap electrodes in these studies were fabricated by EBL, where a gap distance of 300 nm and electrode height of 65 nm was created. Functionalization of the electrodes was achieved by the self-assembly of gold nanoparticles and bio-bar-code-based amplification (BCA) DNA onto the gold electrodes. More specifically, the surface between the electrodes and multilayer of gold nanoparticles was established by the hybridization of complementary DNA. Measurable current between the nanogap electrodes was achieved through the multilayer of gold nanoparticles, followed by current amplification using magnetic nanoparticles and bio-bar-code DNA. The authors reported the detection of DNA down to 1 fM by voltammetric detection using this method [49].

3.3.5
Non-High-Resolution Techniques

The fabrication of nanoelectrodes without the use of high-resolution instrumentation is very rarely reported. However, an approach adopted by Kleps *et al.* [50] took advantage of the shape of microstructures formed by standard microfabrication techniques to fabricate nanoelectrodes with radii of 50 to 250 nm, without using any high-resolution instrumentation. In this case, pyramidal structures were formed on a silicon wafer by Si etching through a silicon dioxide mask. An additional oxidation process was applied to sharpen the structure before deposition of either a gold or platinum film by metal organic chemical vapor deposition (MOCVD). A final layer of insulating SiO_2 or SiO_2/Si_3N_4 sandwich was deposited, before a short period of maskless UV exposure to remove the very tops of the insulating material covering the pyramid, thus creating nanocone electrodes [50,51]. The electrochemical characterization of these types of nanoelectrodes was recently reported [52].

The fabrication of nanogap electrodes has also been achieved without using high-resolution instrumentation by Chen and coworkers [53]. Initially, gold electrode pairs were fabricated on thermally oxidized silicon wafers using conventional photolithography with a spacing of 5 µm. Copper was then electrodeposited onto the gold electrodes by the application of a 10 µA DC current. This technique was used to reduce the gap separation between the electrodes to 9 nm, and the authors reported that the technique could also be applied to other metals. The gap separation was controlled by changing parameters such as the frequency and the series resistances of the control circuit on a home-made electrochemical set-up during the plating process. The gap separation was reported to be impedance-dependent, originating from the changing ratio between the capacitive and resistive parts of the materials. Chen *et al.* suggested that this method was an effective way of controlling gap separations even down to the atomic scale, was low-cost, and had a high throughput. Possible applications include the investigation of the transport properties of various nanomaterials and organic molecules [53].

3.4
Applications

The types of nanoelectrode fabricated by top-down methods, and reviewed in this chapter, have generally been fabricated with various applications in mind, and which determine the electrode dimensions, spacing, and array layout. For those parameters, three main types of top-down fabrication approaches have been applied: EBL, FIB milling, and NIL. Such applications have not yet generally been investigated exhaustively, however, and remain the subject of further investigations still to be reported. Thus, the applications summarized here are illustrations of future possible uses of the devices described.

Currently, there is much interest in the detection of single molecules, such as DNA, using electrochemical detection methods, due to its suitability to device miniaturization and *in-vivo* measurements. However, one of the main challenges facing this application is the reproducible fabrication of electrodes, the dimensions of which are comparable to the size of the molecule of interest. Much research has been conducted on the fabrication of nanogap electrodes for this purpose, though few have reported the study of applications such as the detection of DNA [49] and cobalt particles [45]. The former case reports detection levels down to 1 fM of DNA. Some reports have also been made on the electrodeposition of metals in order to narrow the gaps between the electrodes even down to atomic scale dimensions, where possible applications include the contacting of small clusters and tailored molecules in metal–molecule–metal hybrid assemblies [47], and investigation of the electrical transport properties of various nanomaterials and organic molecules [53].

Some of the nanoelectrodes reviewed in Sections 3.3.1 and 3.3.2 (EBL and FIB milling) are examples of nanoelectrodes where all of the dimensions of the electrode's active surface area are in the nanometer scale (rather than only one lateral dimension of a nanoband electrode). In such cases, the fabrication of these electrodes has been more challenging, and hence fewer reports exist of sensor applications using these electrodes. The most promising of this type has been developed by Jung and coworkers [29] by the functionalization of the insulating material around their recessed disc nanopore electrodes with antibodies for the creation of an immunosensor (e.g., Figure 3.3). One particular advantage of this type of surface modification is the utilization of a larger surface area for the capture elements (i.e., antibody–vesicle complex) and an uninhibited electrode surface; both will result in increased sensitivity, as well as the fundamental benefits of nanoelectrodes (as described in Sections 3.1 and 3.2). As shown by Jung *et al.* [29], electrochemical detection of the antigen to the antibody [in this example, human serum albumin (HSA) as the antigen and anti-HSA as the antibody] was achieved using square-wave voltammetry. However, further experimentation was planned as part of a multiple protein sensing strategy. Similar devices have also been used by this group for DNA sensing [54]. In this case, single-stranded DNA was immobilized on the metal electrodes at the base of the nanowells formed by EBL in the polymeric resist layer. Hybridization detection was achieved by inhibition of the electrochemistry of a diffusing redox probe species in solution [54].

A further illustration of a possible immunosensor application of nanoelectrodes is a report by Zhu and Ahn [30] that the detection of 4-aminophenol in the 10^{-8} to 10^{-9} M range was possible with IDA nanoelectrodes. (4-Aminophenol is the product of an enzyme reaction often used in enzyme-labeled electrochemical immunoassays [55].)

One type nanoelectrode which been reported predominantly is the interdigitated array (IDA) nanoelectrode. These electrodes function on the basis of the cycling of redox species between adjacent electrodes, and are used to create a significant amplification of the currents obtained from low analyte concentrations. The main application for IDA nanoelectrodes is in the trace analysis of very low analyte concentrations. An interesting application of IDA nanoelectrodes was reported by Finot et al. [32], in which DNA was detected via a hexammineruthenium(III) chloride redox mediator. In this case, single-stranded DNA immobilized on e-beam patterned gold IDA nanoelectrodes was detected by the interaction of the DNA with $Ru(NH_3)_6^{3+}$. Hybridization of sample DNA to single-stranded DNA on the electrode surface was also detected by reaction with $Ru(NH_3)_6^{3+}$, where the Ru $(NH_3)_6^{3+}$ intercalates into the DNA double strand. These authors reported a greater sensitivity than was achieved with macro-sized electrodes, and faster response times. A further sensor application was reported by Luo et al. [33], where dielectrophoresis was used with IDA nanoelectrodes to trap biomolecules such as BSA and antibodies on the edges of the nanobands. This has promising applications for several sensor types for trace analyte detection. Currently, IDA microelectrodes have found wide use for biosensor applications, and are particularly suitable for use with microfluidic systems [54]. Future requirements for nanoscale IDAs lie in the optimization of a reproducible fabrication technique for the large-scale production of these electrodes, particularly for the lower-cost alternative approaches such as NIL.

3.5 Conclusions

This chapter has reviewed the recent advances in nanoelectrode fabrication using top-down approaches. A range of applications of these electrodes exist, from single molecule detection to signal amplification for trace analysis. A particular fabrication technique is selected depending on the dimensions, spacing, and number of electrodes required. Many of the techniques based on the three lithography types

Table 3.1 A comparison of the nanoelectrode fabrication methods.

Technique	Advantages	Disadvantages
Electron-beam lithography	Design flexibility High resolution	Low throughput Cost
Focused ion beam lithography	Flexibility Direct writing of pattern	Low throughput Cost
Nano-imprint lithography	High throughput Reproducible Low cost	Inflexible Patterning method

reviewed (EBL, FIB, and NIL) have been adapted and modified to achieve the desired nanoelectrode geometries and dimensions. The reproducible fabrication of nanoelectrodes below the 10 nm scale is becoming increasingly difficult, and hence the fabrication of nanogap electrodes has become a promising alternative in approaching the dimensions of the single molecules they are designed to detect. Although, currently e-beam and ion beam high-resolution fabrication techniques are widely used, the cost implications of these methods has driven forward the search for lower-cost alternatives, such as NIL. Indeed, NIL is today more widely applied to IDA nanoelectrodes, where deformations to adjacent band structures caused by the imprinting process are less prominent using this type of electrode arrangement. Some of the advantages and disadvantages of these three top-down fabrication methods, as applied to nanoelectrodes and nanoelectrode arrays, are summarized in Table 3.1. Today, EBL, and especially FIB, are the research tools of choice for the fabrication of novel structures such as experimental prototypes, whereas NIL represents a low-cost approach for the fabrication of arrays of proven device designs. Thus, the different fabrication methods surveyed here offer benefits in alternative regions of the biosensor/bioanalytical system development pathway.

Among the nanoelectrodes fabricated by top-down methods and reviewed in this chapter, one of the main shortcomings is their reproducibility, particularly in the tens of nanometers or sub-10 nm ranges, and where mass production is required. Moreover, many of the high-resolution techniques are costly and require complex procedures, thus further complicating the fabrication process. Future research into the fabrication of nanoelectrodes should focus on methods to improve the reproducibility of the electrodes, particularly where high-resolution techniques are used. For the range of applications studied with nanoelectrodes thus far, greater sensitivity and faster response times have been reported, thereby indicating their potential for future sensors based on miniaturized, ultra-sensitive devices. However, systematic investigations of possible applications of nanoscale electrodes remain substantially to be demonstrated. In order to achieve this, these nanoelectrodes must be sufficiently robust to be placed in the hands of experimental chemists and biochemists, who can apply the necessary protocols to test the electrodes' capabilities.

References

1 Zoski, C.G. (2002) *Electroanalysis*, **14**, 1041–1051.
2 Dayton, M.A., Brown, J.C., Stutts, K.J. and Wightman, R.M. (1980) *Anal. Chem.*, **52**, 946–950.
3 Loeb, G.E., Peck, R.A. and Martynuik, J. (1995) *J. Neurosci. Meth.*, **63**, 175–183.
4 Wu, W.-Z., Huang, W.-H., Wang, W., Wang, Z.-L., Cheng, J.-K., Xu, T., Zhang, R.-Y., Chen, Y. and Liu, J. (2005) *J. Am. Chem. Soc.*, **127**, 8914–8915.
5 Leggett, G.J. (2005) *Analyst*, **130**, 259–264.
6 Sandison, M.E., Anicet, N., Glidle, A. and Cooper, J.M. (2002) *Anal. Chem.*, **74**, 5717–5725.
7 Berduque, A., Lanyon, Y.H., Beni, V., Herzog, G., Watson, Y.E., Rodgers, K., Stam, F., Alderman, J. and Arrigan, D.W.M. (2007) *Talanta*, **71**, 1022–1030.
8 Arrigan, D.W.M. (2004) *Analyst*, **129**, 1157–1165.
9 Nagale, M.P. and Fritsch, I. (1998) *Anal. Chem.*, **70**, 2908–2913.
10 Delvaux, M., Demoustier-Champagne, S. and Walcarius, A. (2004) *Electroanalysis*, **16**, 190–198.
11 Choi, Y., Baker, L.A., Hillebrenner, H. and Martin, C.R. (2006) *Phys. Chem. Chem. Phys.*, **8**, 4976–4988.
12 Vlassiouk, I., Takmakov, P. and Smirnov, S. (2005) *Langmuir*, **21**, 4776–4778.
13 Fologea, D., Gershow, M., Ledden, B., McNabb, D.S., Golovchenko, J.A. and Li, J.L. (2005) *Nano. Lett.*, **5**, 1905–1909.
14 Gates, B.D., Xu, Q., Stewart, M., Ryan, D., Willson, C.G. and Whitesides, G.M. (2005) *Chem. Rev.*, **105**, 1171–1196.
15 Cheng, W., Dong, S. and Wang, E. (2002) *Anal. Chem.*, **74**, 3599–3604.
16 Spatz, J.P., Chan, V.Z.H., Mossmer, S., Kamm, F.M., Plettl, A., Ziemann, P. and Moller, M. (2002) *Adv. Mater.*, **14**, 1827–1832.
17 Cojocaru, C.V., Ratto, F., Harnagea, C., Pignolet, A. and Rosei, F. (2005) *Microelectron. Eng.*, **80**, 448–456.
18 Hah, J.H., Mayya, S., Hata, M., Jang, Y.-K., Kim, H.-W., Ryoo, M., Woo, S.-G., Cho, H.-K. and Moon, J.-T. (2006) *J. Vac. Sci. Technol. B*, **24**, 2209–2213.
19 Caston, S.L. and McCarley, R.L. (2002) *J. Electroanal. Chem.*, **529**, 124–134.
20 Menon, V.P. and Martin, C.R. (1995) *Anal. Chem.*, **67**, 1920–1928.
21 Slevin, C.J., Gray, N.J., Macpherson, J.V., Webb, M.A. and Unwin, P.R. (1999) *Electrochem. Commun.*, **1**, 282–288.
22 Gray, N.J. and Unwin, P.R. (2000) *Analyst*, **125**, 889–893.
23 Li, G., Xi, N., Chen, H. and Saeed, A. (2004) *IEEE Conference on Nanotechnology Proceedings*, pp. 352–354.
24 Pain, L., Tedesco, S. and Constancias, C. (2006) *C. R. Physique*, **7**, 910–923.
25 Deamer, D.W. and Akeson, M. (2000) *Trends Biotech.*, **18**, 147–151.
26 Lemay, S.G., van den Broek, D.M., Storm, A.J., Krapf, D., Smeets, R.M.M., Heering, H.A. and Dekker, C. (2005) *Anal. Chem.*, **77**, 1911–1915.
27 Krapf, D., Wu, M.-Y., Smeets, R.M.M., Zandbergen, H.W., Dekker, C. and Lemay, S.G. (2006) *Nano Lett.*, **6**, 105–109.
28 Sandison, M.E. and Cooper, J.M. (2006) *Lab Chip*, **6**, 1020–1025.
29 Jung, H.S., Kim, J.M., Park, J.W., Lee, H.Y. and Kawai, T. (2005) *Langmuir*, **21**, 6025–6029.
30 Zhu, X. and Ahn, C.H. (2006) *IEEE Sensors J.*, **6**, 1280–1286.
31 Ueno, K., Hayashida, M., Ye, J.-Y. and Misawa, H. (2005) *Electrochem. Commun.*, **7**, 161–165.
32 Finot, E., Bourillot, E., Meunier-Prest, R., Lacroute, Y., Legay, G., Cherkaoui-Malki, M., Latruffe, N., Siri, O., Braunstein, P. and Dereux, A. (2003) *Ultramicroscopy*, **97**, 441–449.
33 Luo, C.P., Heeren, A., Henschel, W. and Kern, D.P. (2006) *Microelectron. Eng.*, **83**, 1634–1637.

34 Nagase, T., Gamo, K., Kubota, T. and Mashiko, A. (2005) *Microelectron. Eng.*, **78–79**, 253–259.

35 Santschi, C., Jenke, M., Hoffmann, P. and Brugger, J. (2006) *Nanotechnology*, **17**, 2722–2729.

36 Lanyon, Y.H., De Marzi, G., Watson, Y.E., Quinn, A.J., Gleeson, J.P., Redmond, G. and Arrigan, D.W.M. (2007) *Anal. Chem.*, **79**, 3048–3055.

37 Lanyon, Y.H. and Arrigan, D.W.M. (2007) *Sens. Actuat. B-Chem.*, **121**, 341–347.

38 Nagase, T., Gamo, K., Ueda, R., Kubota, T. and Mashiko, S. (2006) *J. Microlith. Microfab. Microsyst.*, **5**, 011009-1-6.

39 Chou, S.Y., Krauss, P.R. and Renstrom, P.J. (1996) *J. Vac. Sci. Technol. B*, **14**, 4129–4133.

40 Jiao, L., Gao, H., Zhang, G., Xie, G., Zhou, X., Zhang, Y., Zhang, Y., Gao, B., Luo, G., Wu, Z., Zhu, T., Zhang, J., Liu, Z., Mu, S., Yang, H. and Gu, C. (2005) *Nanotechology*, **16**, 2779–2784.

41 Beck, M., Persson, F., Carlberg, P., Graczyk, M., Maximov, I., Ling, T.G.I. and Montelius, L. (2004) *Microelectron. Eng.*, **73–74**, 837–842.

42 Carcenac, F., Malaquin, L. and Vieu, C. (2002) *Microelectron. Eng.*, **61–62**, 657–663.

43 Tallal, J., Peyrade, D., Lazzarine, F., Berton, K., Perret, C., Gordon, M., Gourgon, C. and Schiavone, P. (2005) *Microelectron. Eng.*, **78–79**, 676–681.

44 Malaquin, L., Vieu, C., Geneviève, M., Tauran, Y., Carcenac, F., Pourciel, M.L., Leberre, V. and Trévisiol, E. (2004) *Microelectron. Eng.*, **73–74**, 887–892.

45 Liu, K., Avouris, P., Bucchignano, J., Martel, R., Sun, S. and Michl, J. (2002) *Appl. Phys. Lett.*, **80**, 865–867.

46 Fischbein, M.D. and Drndić, M. (2006) *Appl. Phys. Lett.*, **88**, 063116-1-3.

47 Kronholz, S., Karthäuser, S., Mészáros, G., Wandlowski, T., van der Hart, A. and Waser, R. (2006) *Microelectron. Eng.*, **83**, 1702–1705.

48 Ah, C.S., Yun, Y.J., Lee, J.S. and Park, H.J. (2006) *Appl. Phys. Lett.*, **88**, 133116.

49 Chang, T.-L., Tsai, C.-Y., Sun, C.-C., Uppala, R., Chen, C.-C., Lin, C.-H. and Chen, P.-H. (2006) *Microelectron Eng.*, **83**, 1630–1633.

50 Kleps, I., Angelescu, A., Avram, M., Miu, M. and Simion, M. (2002) *Microelectron. Eng.*, **61–62**, 675–680.

51 Kleps, I., Angelescu, A., Vasilco, R. and Dascalu, D. (2001) *Biomed. Microdev.*, **3**, 29–33.

52 Daniele, S., De Faveri, E., Kleps, I. and Angelescu, A. (2006) *Electroanalysis*, **18**, 1749–1756.

53 Chen, F., Qing, Q., Ren, L., Wu, Z. and Liu, Z. (2005) *Appl. Phys. Lett.*, **86**, 123105.

54 Lee, H.Y., Park, J.W., Kim, J.M., Jung, H.S. and Kawai, T. (2006) *Appl. Phys. Lett.*, **89**, 113901.

55 Bange, A., Halsall, B. and Heineman, W.R. (2005) *Biosens. Bioelectron.*, **20**, 2488–2503.

4
Template Synthesis of Magnetic Nanowire Arrays

Sima Valizadeh, Mattias Strömberg, and Maria Strømme

4.1
Introduction

During the late 20th century, the growing interest in creating advanced materials using nanoscaled building blocks was employed to develop synthetic techniques for re-engineering existing materials to novel functional materials such as quantum dots, nanodots, nanorods, nanotubes, nanofibrils and quantum wires. These nanowires can act as one-dimensional conductors with interesting electronic and optical properties. Moreover, nanowires may find applications in ultrasensitive chemical and biological sensors.

To date, a physical vapor deposition (PVD) technique followed by lithography is usually considered to represent the ultimate method for producing nanoscaled materials. Although a number of successful efforts have been reported on the fabrication of freestanding nanopillars or nanowells using lithographic techniques, there is a whole range of potential obstacles for these approaches. These arise mainly from the fact that, when the technologies are pushed to smaller sizes, the cost rapidly escalate and the tolerances are more difficult to maintain. The lithographic processes used to pattern nanoelectronic structures also currently limit the dimensions of the electronic components such as nanowires. Optical lithography is limited by the wavelength of the light used to produce the wires. Consequently, the wavelength of light used is moving into the deeper regions of the ultraviolet (UV) spectrum, and the cost of fabricating optical components to achieve this is expected to rise at an astronomic rate. These challenges have forced many to seek new nanofabrication techniques, especially those which grow nanostructures directly in the openings of a template. One of the most successful approaches to these nanofabrication methods is known as the *membrane-based synthesis* of nanomaterials [1–4], where the membrane acts as a template for the electrodeposition of multilayered nanowires, as well as for single metallic nanowires [5,6]. The advantage of using template synthesis of nanowires is that it can be considered as an alternative solution to overcome the difficulty in fabricating fibrils with very small diameters by lithographic methods [7].

Nanostructured Materials in Electrochemistry. Edited by Ali Eftekhari
Copyright © 2008 WILEY-VCH Verlag GmbH & Co. KGaA, Weinheim
ISBN: 978-3-527-31876-6

This improvement may offer a new and drastic change for future computer memories and magnetic sensors [8]. Moreover, by using template synthesis, mass production of nanostructures is possible. Examples of such templates for the electrodeposition of multilayered nanowires, as well as single metallic nanowires, are track-etched polycarbonate membranes [9,10], nanoporous alumina [11–13], and nanoporous mica [14].

Historically, Possin et al. [15] were the first to develop a potentiostatic synthesis method for producing different single metal nanowires in a 15 μm-thick mica sample with a pore density in the order of 104 pores per cm^2. During the late 1980s, Baibich et al. and Binasch et al. [16,17] were the first to observe the giant magnetoresistance (GMR) phenomenon in Fe/Cr magnetic structures. Since that time, artificially fabricated magnetic materials associated with transition metals, produced either by molecular beam epitaxy (MBE) or sputtering deposition techniques [18–23], have become an important research field. These GMR structures have magnetic layers separated by a non-magnetic spacer metal. A magnetized multilayer system with either antiferromagnetic or ferromagnetic moment alignment exhibits GMR when the electrical resistivity is reduced by an applied magnetic field that induces parallel magnetization alignment. In the past, most experiments on the GMR effect in magnetic multilayer structures have been focused on the measurement of the current in the plane of the magnetic multilayer system, the so-called "current in the plane" (CIP) geometry. The alternative is the "current perpendicular to the plane" (CPP) geometry. Valet and Fert [20] found that the magnetoresistance of magnetic multilayers in CPP geometry is larger than that in a corresponding CIP geometry. However, with CPP measurement, equipment using the Superconducting QUantum Interference Device (SQUID) is needed. The reason for using such an extremely sensitive technique is that, when the current flows perpendicular to the plane of the magnetic multilayer, the length of the sample is just the film thickness, which is usually in the nanometer scale. As a consequence, these low CPP resistance changes can only be measured using SQUID, which acts as a picovoltmeter and detects a resistance of pico-ohms. Although this method was initially developed by Pratt Jr. et al. [24], the CPP-type resistance measurement has subsequently opened up an entirely new avenue in the field of GMR research during recent years.

Nevertheless, electrodeposition has become the potential candidate in many manufacturing technologies. One of the key applications for the electrodeposition of magnetic multilayers is based on the cylindrical narrow pores in track-etched polycarbonate membranes or porous alumina templates. This has become an important route for GMR measurements in the CPP geometry. More recently, several groups have improved this method to investigate the magnetic properties of template-synthesized structures, which may offer a significant change for future computer memories and magnetic sensors [1,25–29].

In this chapter, attention is focused mainly on electrochemical synthesis techniques for preparing magnetic single and multilayered nanowire arrays using ion-track-etched polycarbonate films as templates. Finally, recent advances in the preparation of electrical contacts of electrochemically deposited Au nanowires in a four-point probe configuration using the dual beam focused ion beam (FIB/SEM) technique, are reported.

4.2
Electrochemical Synthesis of Nanowires

This section includes three parts. In the first part, attention is focused on the fabrication of nanoelectrodes. The growth and nucleation of Co nanowires is discussed in the second part. Finally, electrodeposition of magnetic multilayered nanowire arrays is described in the third part.

4.2.1
Fabrication of Nanoelectrodes

Nanowires and arrays of nanowires are usually obtained by filling a porous template that contains a large number of parallel-sided cylindrical pores that are aligned with a narrow size distribution. Today, polycarbonate membranes or electrochemically etched aluminum metal with nanochannels are used as templates for either the chemical or electrochemical deposition of metals, conductive polymers and other materials for the generation of nanofibers or nanotubes. As the nanochannels in these membranes are very uniform in size, the diameter and aspect ratio of the nanofibers (or nanotubes) synthesized by the membrane–template technique can be precisely controlled. Choosing nanoporous materials as templates provides the opportunity of creating very thin cylinders that have uniform dimensions, with a controlled diameter as small as a few tenths of a nanometer. Arrays of metallic nanowires exhibit interesting properties such as one-dimensional (1D) quantum phenomena, and can therefore be used to construct new nanowire-based electronic devices and circuits [5,30–37].

Track-etched membranes are made when heavy charged particles from a nuclear radiation source pass through a 5- to 20-µm-thick polymer film. The tracks can be selectively etched in a suitable media, so as to introduce pores with a diameter depending on the etching time [28,5,31]. It has been reported that the electrodeposited nanowires using the commercially available porous polycarbonate membranes usually exhibit "toothpick" or "cigar-like" shapes, with a non-homogeneous surface rather than the expected cylindrical shape [30,32]. The explanation for this defect may be related to the acceleration of high-energy particles onto the membranes, which gives rise to regions with point defects in polymer crystals and breakage of polymeric bonds in the membranes [5,31].

Recently, a reproducible technology has been improved to supply accurate membranes with cylindrical arrangements of pores having well-defined shapes. Various groups [33,34] have investigated the electrochemical template synthesis conditions where the membranes were usually sputter-deposited with a 500 nm-thick Ag, Au, or Cu film onto one side, to serve as the working electrode. In order to find analytical expressions for the diffusion and concentration variation of electroactive species in the vicinity and within the pores of a porous membrane as a function of time, the theory of microelectrodes is turned to nanodes, and the authors preferably use the term "nanode" instead of nanoelectrode [35]. One important property of microelectrodes functioning as miniaturized working electrodes is the high signal-to-noise ratio due to the non-linear diffusion of electroactive species towards the electrode surface. The hemispherical diffusion fluxes opposed to linear diffusion for a planar substrate are expressed for a microelectrode in Refs. [36–41]. In this hemispherical diffusion mode, more electroactive species reach the electrode surface per unit time and area than in the linear diffusion mode, due to the small diffusion layer thickness. A steady-state can be reached within a few milliseconds [42].

4.2.2
Reactions, Diffusion and Nucleation in the Electrochemical Deposition of Co Nanowires

Generally, metal deposition from aqueous electrolytes is not only a reaction of great technological importance, but is also a classic example of a nucleation and growth process for which nucleation sites play a decisive role. On the other hand, the fundamental aspects of electrocrystallization of metals are directly related to the problems of nucleation and crystal growth. However, a full control over – and a satisfactory understanding of – the physical phenomena requires a detailed knowledge of the synthesis process involving pore filling during electrolytic growth.

By decreasing the dimensions of fabricated components and devices, the influence of individual interfaces and defects will evidently dominate the physical behavior of the nanowires. Therefore, a both experimental and theoretical study of electrochemical synthesis of nanodes as recessed electrodes is crucial.

Nanocrystals made of metals such as Ni, Fe, and Co display a wealth of size–dependent structural, magnetic, electronic, and catalytic properties [43]. In particular, the exponential dependence of the magnetization relaxation time on volume has spurred intensive studies of Co nanocrystal synthesis for magnetic storage purposes. Therefore, an experimental set-up for template synthesis of Co nanowires was chosen to study the electrode kinetics of the nanode functioning as a working electrode [35].

4.2.2.1 Theoretical Considerations of Spherical Diffusion at a Nanode Array
The expression for the diffusion-controlled limiting current, which can be described as the sum of the Cottrell equation and a correction term for the steady-state (non-linear diffusion), is given in Ref. [44]

$$i(t) = i_d nFDAc^b \sqrt{\frac{D}{\pi t}} + \frac{nFDAc^b}{r} \tag{1}$$

where n is number of electrons involved in the reaction, F is Faraday's constant, r is the pore radius, D is the diffusion coefficient, and c^b is the bulk concentration of electroactive species.

It is important to note that the diffusion-controlled limiting current at the nanode array is essentially time-independent. By assuming the nanodes as recessed microelectrodes, we can consequently state that, as time elapses, the recessed electrode obeys the Cottrell equation for a short time – that is, the current is proportional to $1/(r+L)$ where L is the membrane thickness or length of the nanodes. Under these conditions, the expression for the diffusion-controlled current at the recessed nanodes reads

$$i(t)_{pores} = i_{d\ pores} = \frac{4\pi nFDc^b r^2}{4L + \pi r} \quad (2)$$

The important consideration is on the transition of diffusion variations, starting from each individual nanode, with time. Eventually, a situation is approached where the diffusion zones overlapping from a series of nanodes become insulated from each other in the array. At this state, the diffusion zones from all nanodes are effectively covering the total electrode surface area.

The concentration variation with time during the pore-filling process is illustrated in Figure 4.1(a–d). By applying the potential when the initial nucleation process begins, two types of diffusion layer are formed, with one zone in the pores and another zone outside the pores. The diffusion zones are indicated as D_2 and D_1 in Figure 4.1a. The diffusion within the pores is linear, whereas outside the diffusion is linear for a short time of only millisecond duration (i.e., Cottrellian), whilst for longer time the diffusion is spherical and time-independent. During pore filling, the inner diffusion zone (D_2) becomes thinner (see Figure 4.1b). At the moment of complete pore filling, only the outer diffusion zone will remain (Figure 4.1c). During further deposition, the whole surface outside the pores is covered and the diffusion then becomes linear (Figure 4.1d).

Increasing the radii of the spherical diffusion zones increases the volume of the depleted region. This is revealed from an extended region into solution for a high population of electroactive species to diffuse "down" to the electrode surface by linear diffusion [44].

The rate of the electrochemical processes, apart from the effect of mass transport by the diffusion-controlled limiting current, is also dependent on the effective electrode surface area. Thus, the analytical expression for diffusion to the effective surface area of a random distribution of recessed nanodes, which are established during ion track etching processing of the membrane, is complicated. However, solutions are available by assuming the fact that the same amount of material is entering through the boundary of the diffusion zones by linear diffusion as that from the radial spherical diffusion to the nanodes [44–46].

In order to evaluate the fractional surface area of the nanode array, we can formulate the changes in concentration with time by using a special case of Fick's second law in one dimension

$$\frac{\partial c(x,t)}{\partial t} = D\frac{\partial^2 c(x,t)}{\partial x^2} \quad (3)$$

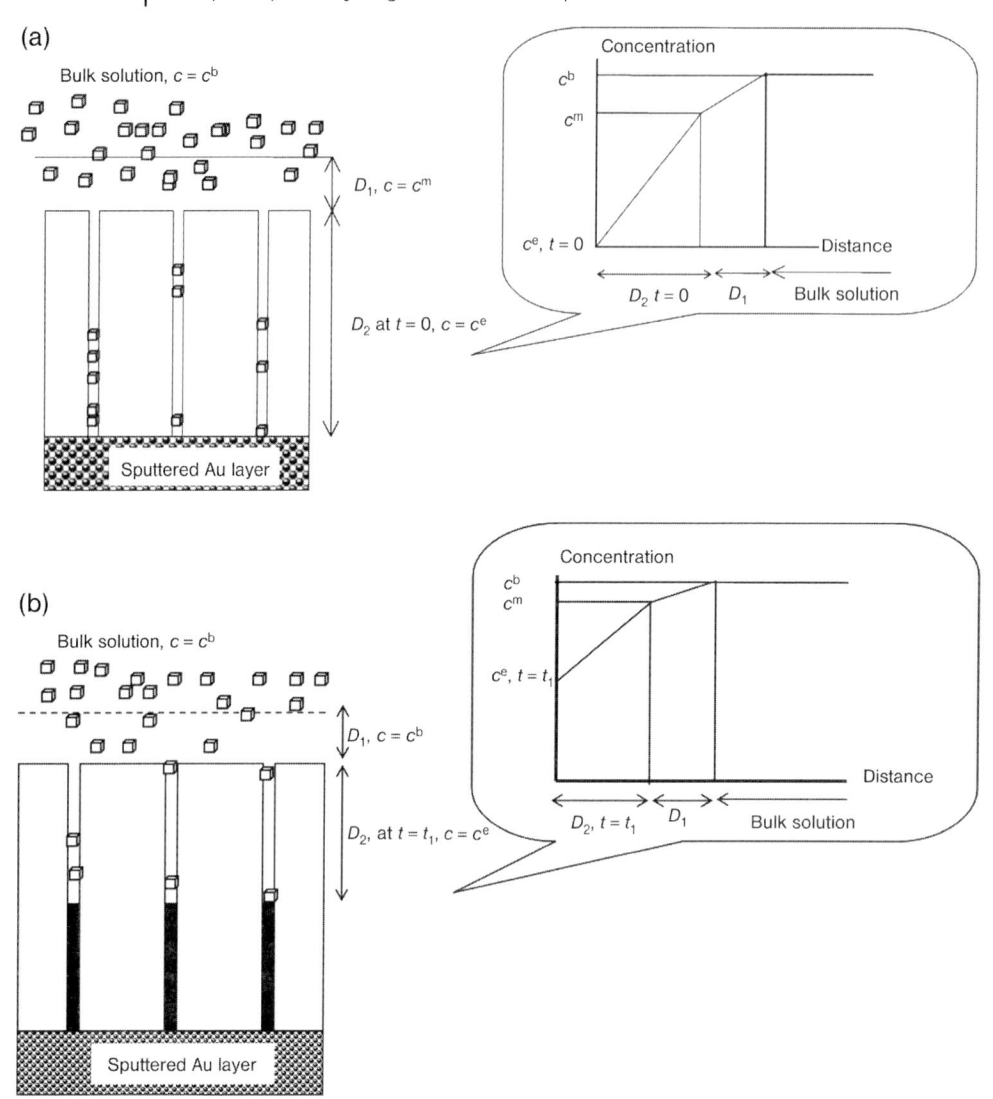

Figure 4.1 (a–d) Concentration variation with time during the pore-filling process. (a) Initial pore filling; (b) half-filled pores; (c) complete pore filling; (d) pores completely filled and deposited material is emerging on the upper membrane surface.

under the following boundary conditions: Before the experiment starts at $t = 0$, we have for all x that $c(x, 0) = c^b$. As $x \to \infty$, for all t we have that $\lim c(x, t) = c^b$. Finally, $c(0, t) = c^m$ holds. In these boundary conditions, c^b is the bulk concentration and c^m is the concentration at the pore mouth which is provided to be constant for a short period of time. If c^m is assumed to be unchanged for a short period, the diffusion

(c)

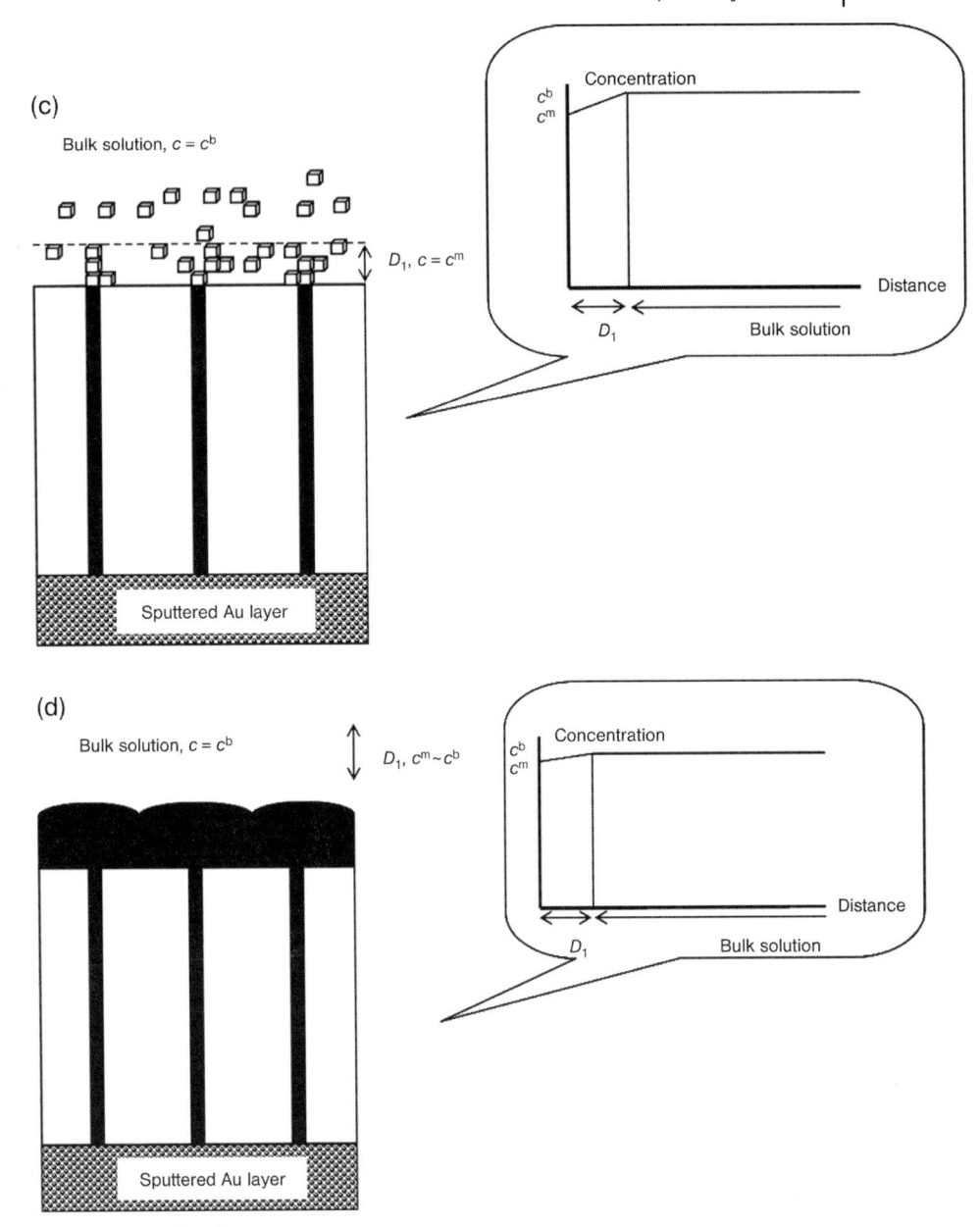

Bulk solution, $c = c^b$

D_1, $c = c^m$

Sputtered Au layer

Concentration
c^b
c^m
D_1 — Bulk solution
Distance

(d)

Bulk solution, $c = c^b$

D_1, $c^m \sim c^b$

Sputtered Au layer

Concentration
c^b
c^m
D_1 — Bulk solution
Distance

Figure 4.1 (Continued)

current, I_{eq}, for a short time period of milliseconds, can be obtained from the modified Cottrell equation

$$i_{eq}(t) = \frac{nF\pi r_d^2(c^b - c^m)}{\pi^{1/2} t^{1/2}} \tag{4}$$

However, from Faraday's law, the amount of deposited material is directly proportional to the current or current density [13]. Thus, we can write $I_{eq} = I_{pore}$ or

$$\pi r_d^2 j_{eq} = \pi r^2 j_{pore}. \tag{5}$$

The equivalent surface area of the nanode, $S_{eq} = \pi r_d^2$, where r_d is the radius of the equivalent surface, can then be written using Eqs. (3)–(5) as:

$$S_{eq} = \frac{4r^2 \pi (Dt\pi)^{1/2}}{(4L + \pi r)} \frac{c^b}{(c^b - c^m)} \tag{6}$$

As the pores are randomly distributed, due to the nuclear etching process of the membrane, the total equivalent area with respect to the pore density, N (pores cm^{-2}), is NS_{eq} [8].

By using the Avrami theorem [48], the fractional area is

$$S = 1 - \exp(-NS_{eq}) \tag{7}$$

so that

$$S = 1 - \exp\left[\frac{4N\pi \pi^2 (Dt\pi)^{1/2}}{(4L + \pi r)} \frac{c^b}{(c^b - c^m)}\right] \tag{8}$$

The numerical value of the growth current for the total electrode surface area can be evaluated as the fractional surface area multiplied by the current density based on Eqs. (4) and (8); that is:

$$I = nF\left(\frac{D}{\pi r}\right)^{1/2} (c^b - c^m) \left\{1 - \exp\left[\frac{4N\pi \pi^2 (Dt\pi)^{1/2}}{(4L + \pi r)} \frac{c^b}{(c^b - c^m)}\right]\right\} \tag{9}$$

As stated above, if the diffusion equation for hemispherical growth is solved for this particular situation, and if this result is used to determine the variation of the plating current with time, one can predict and extract parameters to determine the concentration of ions at the vicinity and at the mouth of the pores.

Figure 4.2 illustrates both experimental and theoretical current transient curves. There is a clear agreement between the theoretical and experimental curves [35]. The slight differences in shape between the experimental and theoretical transients most likely originate from experimental conditions such as progressive growing diffusion zones and diffusive flux of Co species through the randomly distributed pores. Figure 4.2 also provides an illustration of the growth mechanism for Co nanowires on four different stages:

1. For the initial deposition, the current decreases due to mass transport limitation.
2. The metal is growing in the pores and an almost stagnated regime is reached.

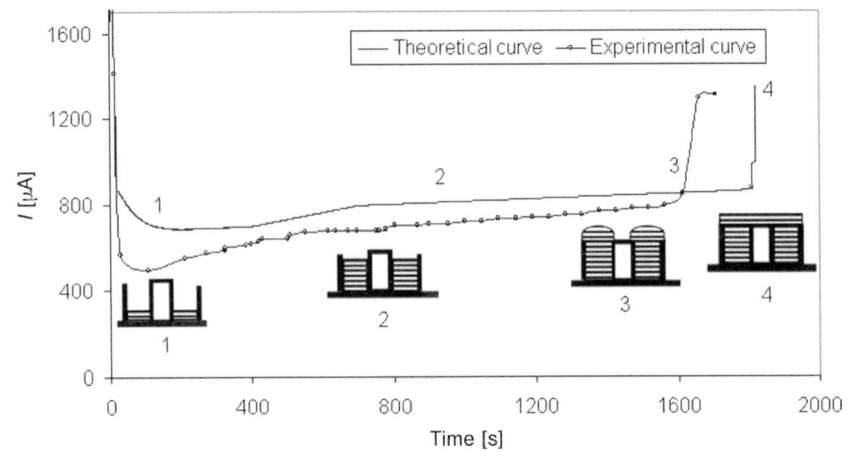

Figure 4.2 Experimental and theoretical [using Eq. (9)] curves showing the current transient response. The schematic insets illustrate the growth mechanism for Co nanowires on four different stages, as explained in the text.

3. The pores are filled up to the top of the membrane surface, which gives rise to the cap formation associated with three-dimensional (3D) deposition.
4. The deposition current increases rapidly due to the simultaneous growth of some wires that emerge from the membrane surface and change the effective electrode area.

After dissolving the membrane in dichloromethane (DCM) and collecting the wires on holey carbon grids, the morphology and evidence of well-defined Co nanowires from completely filled pores was established by a scanning electron microscopy (SEM) investigation, as illustrated in Figure 4.3. Figure 4.4 shows a

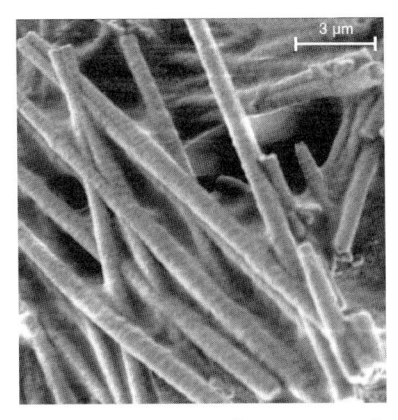

Figure 4.3 SEM image of Co nanowires from completely filled pores after dissolving the membrane in dichloromethane and collecting the wires on holey carbon grids.

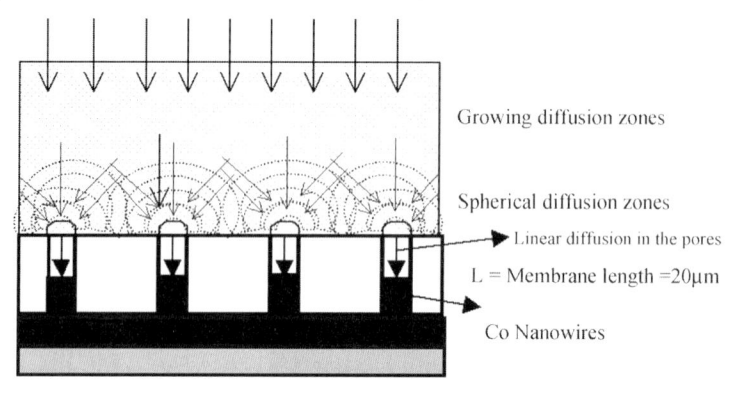

Figure 4.4 Schematic view of the evolving growing diffusion zones as time elapses. The extended vertical lines show the direction of fluxes towards the diffusion zones.

schematic view of the evolving diffusion zones [35]. At the initial stage, the steady-state rate of mass transport is reached within a few milliseconds. In addition, after a longer time, a spherical diffusion zone develops from each individual nanode and increases with time. Consequently, an extended 3D diffusion volume, arising from the overlapping of neighboring diffusion volumes from each individual nanode, is stabilized in the vicinity of the solution. Thus, more electroactive species diffuse in the direction normal to the region where the electrode surface, including both the insulating and active surface area, covers symmetrically with a spherical diffusion volume.

It is important to note that the diffusion-controlled limiting current at the nanode array essentially facilitates the non-linear steady state – that is, time-independence at longer times. As reduced electroactive species instantaneously establish a continuous depleted region at the electrode surface and in the vicinity of the electrode, a concentration gradient is formed in the electrolyte between the nanode mouths and the bulk concentration region. Thus, a continuous linear diffusion normal to the spherical diffusion zones at the electrode surface prevails. As a result, the concentration of the electroactive species in the mouth of the nanodes will differ from that in the vicinity of the electrode. Consequently, for any constructed nanode – whether it is planar, spherical, disc, or an array of each geometry – the diffusion-controlled limiting current is concentration-dependent. Following this further, by assuming a linear concentration profile inside the pore and knowing c^m from Eq. (4) for an unstirred solution, the concentration in the vicinity of the electrode, c^e, can be solved using the following relationships:

$$i(t) = 4nFD(c^b - c^m)r \qquad (10)$$

$$\frac{i_{pore}}{n\pi r^2 F} = D\frac{\partial c(x,t)}{\partial x} = D\frac{c^m}{L} \qquad (11)$$

By solving Eqs. (13) and (14) for the same current diffusion, we obtain

$$c^m = \frac{4Lc^b}{4L + r\pi} \qquad (12)$$

4.2 Electrochemical Synthesis of Nanowires

Table 4.1 Calculated c^m and c^e values based on four different thicknesses.

Current [μA]	Membrane thickness after Co deposition [μm]	Time [s]	Co layer thickness [μm]	c^m [M]	c^e [M]
160	4	1300	16	1.1826	0.9644
270	6	800	14	1.1656	0.3403
395	14	480	6	1.1351	0.1646
630	16	350	4	1.0681	0.0781

The expressions for the concentration at the vicinity of the electrode can be calculated from the following equations:

$$\frac{i_{pore}}{n\pi r^2 F} = D\frac{\partial c(x,t)}{\partial x} = D\frac{c^m - c^e}{L} \tag{13}$$

$$c^e = c^m - \frac{i_{pore} L}{D\pi r^2 nF} \tag{14}$$

The microscopic variations of c^m and c^e for the conducted experiments evaluated from Eqs. (13) and (14), with the diffusion coefficient, $D = 2.5 \times 10^{-5}$ cm^2 s^{-1} are summarized in Table 4.1.

Figure 4.5 illustrates the relationship between the concentration of electroactive Co species at the mouth and in the vicinity of the electrode when the deposited Co layer

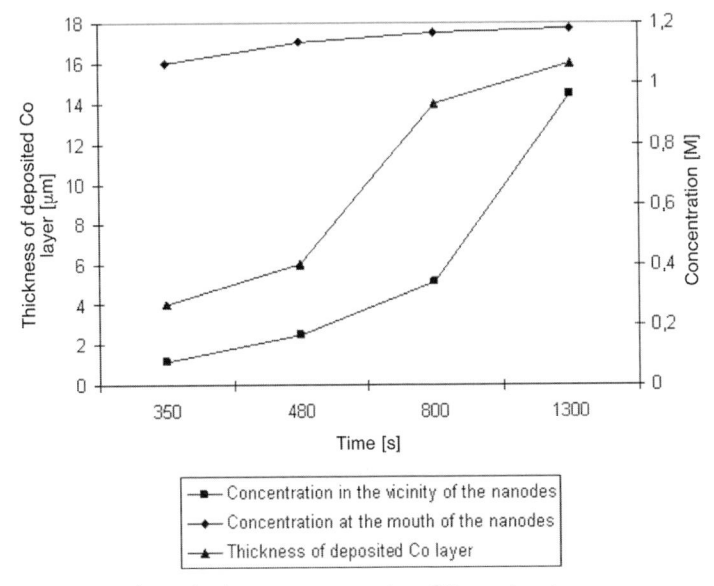

Figure 4.5 Relationship between concentration of Co species at the mouth (c^m), and in the vicinity of the electrode surface (c^e), when the deposited layer increased through the nanodes.

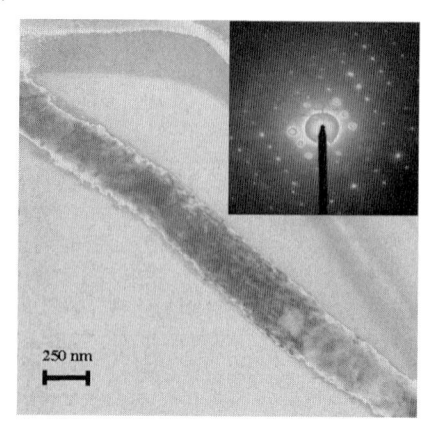

Figure 4.6 Bright-field TEM image and selected area electron diffraction pattern of a Co nanowire with a diameter of 250 nm.

thickness is varied in the nanode array. Hence, an increase in the Co deposited layer thickness through the nanodes leads to a decreasing thickness of the porous membrane as a recessed electrode [35]. Thus, an increasing diffusion-limited current due to entrance of a larger ion flux towards the electrode surface can be observed. Additionally, c^m and c^e show increasing trends when the deposited layer thickness is increased. In other words, by growing the deposit into the nanodes, the amount of electroactive species entering to the diffusion zones increases. Consequently, an increasing current which passes through the maximum and then approaches the diffusion-limited current, can be observed during the growth process of the Co nanowires.

After deposition, transmission electron microscopy (TEM) samples were prepared by dissolving the polycarbonate membrane in DCM, and collecting the wires on holey carbon grids. Figure 4.6 shows a bright-field TEM image and the corresponding selected area electron diffraction pattern of a Co nanowire with a diameter of 250 nm after removal of the polycarbonate membrane. The wire has even edges with grains elongated in the growth direction of the wire [35].

As a general conclusion, the electrodeposition of dense and continuous nanowires requires a full control of the time-dependent transient current and concentration gradient variation during the pore-filling process.

4.2.3
Electrodeposition of Magnetic Multilayered Nanowire Arrays

Electrochemical synthesis is not only suitable for the fabrication of single-layered nanowire arrays, but it is also an efficient technique for depositing multilayered nanowires in nanometer-wide cylindrical pores of a porous material, where conventional deposition processes would have severe limitations in the pore-filling efficiency.

One essential application for the electrodeposition of magnetic multilayered nanowires is based on a new technique, that of nanomaterial membrane-based synthesis [25,26,29,39]. Template-synthesized magnetic nanowire arrays exhibit extremely anisotropic magnetic properties, characterized by a square hysteresis loop when the applied field is perpendicular to the deposited film; that is, parallel to the wire axis (the easy axis). The advantage of such an arrangement lies in its ability to detect tiny magnetic fields. This effect can be exploited in a range of applications in magnetic data storage and magnetic sensors [20,24,47]. In this connection, a variety of multilayered nanowires including Co/Cu, Fe/Cu, NiFe/Cu have been fabricated by electrodeposition techniques. In these methods, magnetic multilayered nanowire arrays are fabricated by filling the pores of ion track-etched membranes with layers of a magnetic material in a single domain state separated by a non-ferromagnetic material as spacer layer. This ability to deposit on complex surfaces also allows electrochemical deposition to be used in conjunction with state-of-the-art nanofabrication methods.

Pulse plating can be used to grow multilayer structures of metals and alloys from a single electrolyte containing two or more elements. In this way, different metals are deposited in an alternating sequence by varying the applied potential to match the deposition potentials of the different constituents. The deposition of a nobler metal is arrested during the deposition of a more active constituent layer by lowering the concentration of the nobler metal ions in the bath. By repetition of this process, composition-modulated alloys are formed. The more noble metal is kept in diluted concentration compared to that of the less noble metal, so that the rate of reduction of the nobler metal is slow and limited by diffusion. However, the disadvantage of the pulsed electrodeposition technique is that, when depositing a less noble metal from a bath containing also a more noble metal (e.g., Co in the presence of Au or Ag), the noble metal is co-deposited. This effect may be reduced by having large concentration differences between the two metals, although a further reduction is desirable at least in some systems. One comment should be made at this point: After the deposition of a less noble metal, the slow growth rate of the more noble metal usually cause the simultaneous dissolution of the less noble metal layer. This in turn clearly increases the interface roughness and the non-homogenous layer thickness along the whole wire.

In addition, during deposition, many other unexpected reactions such as the co-deposition of impurities and hydrogen evolution occur at the electrode, and this may lead to a decreased amount of deposited metal compared to the calculated value. Indeed, the quantity of deposited metal, bilayer thickness uniformity all along the filament and current efficiency, which is a fraction of the deposition current density, can be improved by using a coulometer to integrate the charge that passes during each layer deposition.

For ideal electrochemical reactions with 100% efficiency, proportionality exists between the amounts of deposited or dissolved metal and the electrical charge passed through the cell. For this purpose, an adequate calculation based on Faraday's law can be performed for such a periodic electrodeposition of multicomponent metal A and B using a single electrolyte, as follows:

$$m_B = \frac{M_B Q_B}{n_B F} \chi_B \qquad m_A = \frac{M_A Q_A}{n_A F} \chi_A \qquad (15)$$

Here, m_B and m_A are the weights of the deposited metals A and B, respectively, M_B and M_A are the molecular weights of A and B, n is the charge number of both metals, F is Faraday's constant, Q_B and Q_A are the transition charges corresponding to the desired layer thickness and χ_B and χ_A are the current efficiencies during deposition. Additionally, the weights of deposited films (m_B and m_A) can be expressed in terms of current densities; that is:

$$\rho_B * S * d_B = m_B \qquad \rho_A * S * d_A = m_A \tag{16}$$

where S is the cathode surface area, which depends on the pore density and the diameter of the membrane and d is the thickness of the deposited layer. By insertion of Eq. (15) into Eq. (16), the layer thickness of each deposited layer can be determined according to

$$d_B = \underbrace{\frac{M_B}{2*F*\rho_B*S}}_{\alpha} * Q_B * \chi_B \tag{17}$$

$$d_A = \underbrace{\frac{M_A}{F*\rho_A*S}}_{\beta} * Q_A * \chi_A \tag{18}$$

As the periodic multilayers of both corresponding layers are formed by sequential growth of these two layers, the modulation wavelength (d_{bilayers}) can be defined as the sum of the layers thickness, d_A and d_B; that is:

$$d_{\text{bilayers}} = \alpha(Q_B * \chi_B) + \beta(Q_A * \chi_A) \tag{19}$$

where α and β are the constants indicated in Eqs. (17) and (18).

In conclusion, a linear relationship between d_{bilayer} and Q_A can be expressed as

$$d_{\text{bilayers}} = d_B + \gamma(Q_A) \tag{20}$$

where γ is a constant.

4.2.3.1 Electrodeposition of 8 nm Ag/15 nm Co Multilayered Nanowire Arrays (Wire Diameter 120 nm)

Until now, few investigations have been conducted on sputter-deposited multilayered Ag/Co that exhibits the GMR effect [49–51]. For the electrodeposition technique, there have been two prior attempts to fabricate Ag/Co. For example, Zaman et al. [52] reported evidence of magnetoresistance in an electrodeposited $(Co_{70}Ag_{30})/(Co_8Ag_{92})$ multilayered film on a planar substrate, while Fedosyuk et al. [53] reported on granular deposited Ag/Co, using porous anodic alumina as template, with magnetoresistance at room temperature.

For the latter situation, the challenge was to synthesize, electrochemically, Ag/Co multilayered nanowire arrays using a track-etched polycarbonate membrane as template by utilizing the above-mentioned theoretical calculations to enhance the sharp interface between the sublayers. The electrodeposition began with applying a pulsed potential, using the new single bath, which produced Ag/Co$_{97}$Ag$_3$ multilayers on a planar substrate [54]. Ag nanowires were deposited at -600 mV versus Ag/AgCl, while Co-rich (97 wt.% Co) metallic nanowires were deposited at -1100 mV. Due to the low concentration of Ag$^+$ ions, the reduction rate was limited by mass transport. By measuring the charge passed during the Co and Ag layer deposition, respectively, and also the number of repeated layers, the average layer thickness for both Co and Ag could be derived. We therefore expected a linear relationship between $d_{Co,Ag}$ and the charge passed during the time interval for each metal layers, and bilayer thickness. It was observed that 31% of the nominal Co layer thickness was dissolved when switching the potential as discussed above. As a conclusion, the Co dissolution presents a severe problem during the deposition process. In order to avoid the interface roughness and pinholes in the deposited multilayers, the Co layer thickness should always be dimensioned a few nanometers more than the desired value. However, it should be noted that, probably, the condition $\chi_{Co} < 100\%$ may not only be related to Co dissolution, but also that hydrogen evolution during the Co deposition may contribute to the low Co current efficiency. Further details of these studies can be found elsewhere [55].

X-ray powder diffraction (XRD) diffraction patterns (Figure 4.7) of 20 μm-long Ag, Co, and Ag/Co multilayered nanowires showed that the Ag deposits exhibited a fcc <1 1 1> texturing, while the Co deposits exhibited basal plane diffraction from hcp (0 0 2) and fcc (1 1 1) planes. The Ag deposits exhibited polycrystalline diffraction from fcc (1 1 1), (2 0 0), and (2 2 0) planes, with preferential <1 1 1> crystallite orientation. As can be seen in the diffraction pattern (Figure 4.7), the pure Co deposits exhibited a rather weak peak from (0 0 2) lattice planes of hcp Co and/or (1 1 1) lattice planes of fcc Co stacked parallel to the film. The strong peak broadening observed is interpreted as small grain sizes in combination with different tilts around the surface normal for the nanowire array. As a reference, the interatomic distance in Ag(2 0 0) planes is 2.044 Å, which is very close to the spacing of Co fcc (1 1 1) planes and Co hcp (0 0 2) planes (i.e., 2.023 Å).

Evidently, well-defined 8 nm Ag/15 nm Co multilayered nanowires have been obtained using Faraday's law to convert the deposition charge into individual layer thicknesses. The pulse sequence, defined by $U_{Co} = -1100$ mV, $t_{Co} = 450$ ms, and $U_{Ag} = -600$ mV, $t_{Ag} = 100$ s, were repeated for 870 cycles to deposit Co and Ag, respectively. A TEM image of these nanowires is shown in Figure 4.8 [55].

4.2.3.2 Template Synthesis of 2 nm Au/4 nm Co Multilayered Nanowire Arrays (Wire Diameter 110 nm)

Currently, tremendous effort is being expended to develop new methods for preparing magnetic multilayered nanowires with unique material properties that originate at the nanoscale. During the past decade, it was not only sputter-deposited Au/Co multilayered films with large perpendicular magnetic anisotropy properties that attracted great attention [56–60], but also Au–Co alloys, the so-called "hard gold" material with low cobalt content, which are recognized as important alloys because

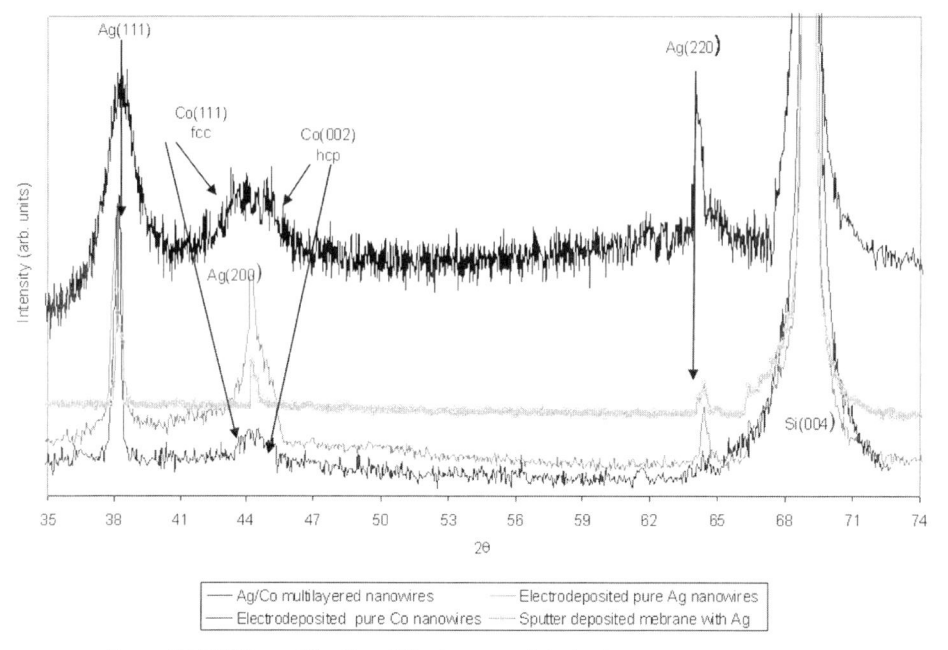

Figure 4.7 θ-2θ X-ray diffraction diffractograms of electrochemically deposited Ag/Co multilayered nanowires, as well as Ag and Co nanowires within a porous polycarbonate membrane. The peak positions for Ag(1 1 1), (2 0 0) and the mixture of fcc Co (1 1 1) and hcp Co(0 0 2) are indicated in the figure.

of their high resistance. These structures have a wide applicability in plating electrical contacts in printed circuit board technology [61–63].

To date, the electrodeposition of Au/Co magnetic multilayered nanowire arrays has not been reported, although electrodeposited Au/AuCo multilayers on planar

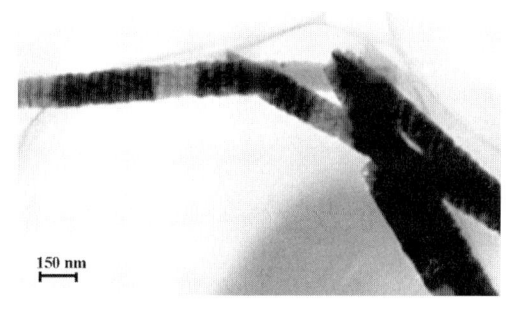

Figure 4.8 TEM image of Ag/Co multilayered nanowires with a diameter of 120 nm deposited with a pulse sequence defined by $U_{Co} = -1100$ mV, $t_{Co} = 450$ ms, and $U_{Ag} = -600$ mV, $t_{Ag} = 100$ s. The pulse sequence was repeated for 870 cycles to deposit Co and Ag, respectively. The bilayer thickness was 23 ± 1 nm.

substrates have been produced from a commercial hard gold plating bath with high Au concentration [64].

For the latter reasons, attempts were made to electrodeposit Au/Co multilayered nanowires from a newly developed single bath with low gold concentration – which is a crucial factor from an economical point of view. The electrochemical deposition of Au/Co multilayers was studied in single bath containing 0.285 M Co ($CoSO_4 \cdot 7H_2O$), 0.8 M $C_6H_8O_6$ and 0.3×10^{-3} M Au ($KAu(CN)_2$) at pH 3.5–4 [65]. Cyclic voltammetry, chronoamperometry and pulse-potential experiments were used to determine the deposition conditions for pure Au and 98 wt.% Co layers [66]. The Co-rich metallic nanowires were deposited at $U_{Co} = -1100$ mV, and the Au nanowires at $U_{Au} = -490$ mV versus Ag/AgCl [66].

Similar strategies as described above were used to deposit Au/Co multilayered nanowires by using a pulse-potential sequence defined by $U_{Co} = -1100$ mV, $t_{Co} = 400$ ms and $U_{Au} = -490$ mV, $t_{Au} = 15$ s. The pulse sequence was repeated for 1250 cycles, giving an approximate total membrane film thickness of 20 μm [66].

Figure 4.9 shows a dark-field TEM image of 2 nm Au/4 nm Co multilayered nanowires, where the apparent layer definition is similar to that observed in Au/Co multilayers on a planar substrate [66]. As can be seen in the image, the interface between the Au spacer layer and the first Co layer appears to be flat, but as the number of repeated layers increases, the interfaces between the layers become ambiguous and irregular. This can be attributed to the fact that Co and Au are not entirely deposited layer by layer.

In order to improve the surface roughness of the deposited nanowires and the thickness uniformity of the layer interfaces, it is necessary to refine the operating parameters in the pulse-potential deposition. In this connection, a new pulse sequence was designed based on the so-called "periodic displacement reaction", to deposit Au/Co multilayered nanowires with sharp layer interfaces between the Co and Au layers. Under the so-called "open-circuit condition", the deposition of the less noble metal (Co) is periodically interrupted to permit the displacement reaction of part of the deposition metal by the more noble metal (Au). The open-circuit

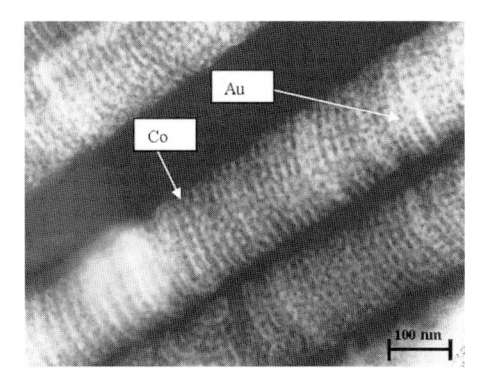

Figure 4.9 Dark-field TEM image of 2 nm Au/4 nm Co multilayered nanowires. The Co layers appear with a dark contrast.

potential is the reset potential where no external current is applied. As Co is a less noble metal than Au, a displacement reaction takes place (i.e., $Co + 2Au^+ \rightarrow Co^{2+} + 2Au$). As soon as the deposited Co layer is completely covered with Au, the deposition process stops. Therefore, the process under open-circuit conditions is defined as a diffusion-controlled reaction that yields no faradic reactions [8,67]. In order to estimate the Au-layer thickness under the open circuit condition, first, a series of Co nanowire samples with a length of 4 µm was deposited during chronoamperometry experiments. Second, the sample was kept in the electrolyte under the open-circuit condition for time intervals of between 1 and 20 s. Finally, the thickness of the deposited nanowires was investigated using atomic force microscopy (AFM). The thickness of the Au layer deposited under open-circuit conditions was calculated from the average chemical composition (wt.% Au) obtained from energy-dispersive X-ray (EDX) analysis in combination with TEM studies. Figure 4.10 shows the wt.% Au values and the calculated thicknesses based on these data [66]. It should be noted that the 2 wt.% Au content in the Co deposition is subtracted from the total wt.% Au deposits during the displacement reactions. Consequently, during the 20-s delay for open-circuit, the Co layer is dissolved and Co^{2+} ions pass into solution, while Au deposits on the electrode surface. Hence, the Co dissolution and Au deposition occur simultaneously, with the average thickness of the deposited Au layer at an open circuit time of 20 s being estimated to 1.5 nm. The open-circuit condition was applied in combination with a steep ramp between the Co and Au potential pulses in order to avoid the otherwise severe Co dissolution when the potential was switched to the less-negative potential for Au deposition.

Briefly, this deposition scheme is based on pulse-potential and periodic displacement deposition in combination with ramp pulsating potentials. Apparently, other

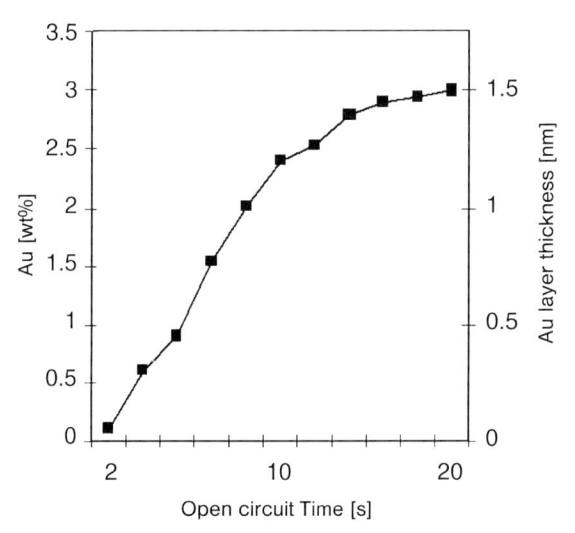

Figure 4.10 Thickness of deposited Au layer, during displacement deposition of Au on Co, as a function of open circuit time.

groups have not reported the details of this deposition process. The deposition sequence was as follows: deposit Co at −1100 mV for 400 ms; sweep the potential to −800 mV (at 80 mV s^{-1}); wait for 3 s; apply open-circuit conditions and wait for 20 s; deposit Au at −490 mV for 15 s; ramp down the potential to −1100 mV (at 80 mV s^{-1}); repeat. Ramping up the potential from −1100 mV to −800 mV was used to tone down the effect of hydrogen evolution which occurs simultaneously with Co deposition. As hydrogen evolution is accompanied by a local increase in pH at the cathode surface, a precipitation of Co hydroxide may occur, which then incorporates into the deposits; this phenomenon is known as "burning". Another important point is that at −800 mV, an Au–Co alloy is deposited, containing 98 ± 1 wt.% Co and 2 wt. % Au. In this case, with a 20 s delay, the deposition of an Au–Co alloy precedes the displacement reaction, such that the sharpness of the transition zone between the bilayers is improved [66]. In conclusion, during this step, a 1.5 nm-thick Au layer can be obtained (Figure 4.10). However, a displacement reaction – that is, a simultaneous dissolution of Co and deposition of Au – may present severe problems, such as pinhole formation during the Au deposition. Therefore, electrodeposition of Au layers at −490 mV is used to enhance the quality of the Au deposits and also to obtain the desired Au layer thickness. This process limits the Au-ion concentration in the vicinity of the electrode. Finally, ramping the potential to the Co potential avoids replenishment of the depleted diffusion layer of Au ions. Figure 4.11 shows the current reprocesses corresponding to four pulse cycles, recorded during the deposition process, using the new pulse sequence [66]. As can be seen, at different stages of the pore-filling process, increasing the deposited layer thickness increases the current level.

Conclusively, two series of Au/Co multilayered nanowires with bilayer thicknesses of 16 nm and 42 nm were successfully deposited into a 20 μm-thick track-etched polycarbonate membrane. The layer interfaces are well defined, as shown in Figure 4.12, which shows TEM images of: (a) 4 nm Au/12 nm Co; and (b) 10 nm

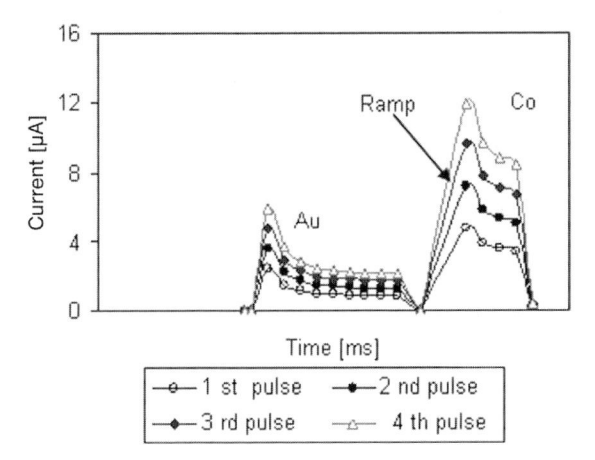

Figure 4.11 Current response curves recorded during a four-pulse sequence.

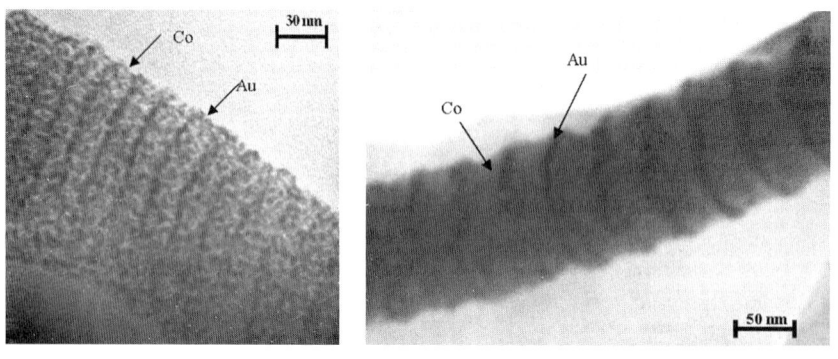

Figure 4.12 (a, b) TEM images of (a) 4 nm Au/12 nm Co and (b) 10 nm Au/32 nm Co multilayered nanowires. The bright contrast represents the Co layers (lower atomic number than Au).

Au/32 nm Co. Figure 4.13 shows a characteristic feature of the initial growth process for Au/Co multilayered nanowires. After the nucleation of Au during the initial electrochemical deposition of Co [66], subsequent layers of each period form semi-concentric shells. For steady-state layer growth, however, the layers appear more or less perpendicular to the pore walls; clearly, the nucleation sites need not be at the bottom of the pores. Wang et al. [68] speculated that, during the evaporation of a seed layer of Au on the backside of the membrane, Au may enter into the pores and stick preferentially to the edges of the pore walls, prior to complete covering of the pores. When starting the electrodeposition process with Co, this metal grows on Au particles at the edge of the pore wall, which gives rise to a concave growth formation.

Figure 4.13 Characteristic features of the initial growth process for Au/Co multilayered nanowires. After the nucleation of Co onto surface irregularities of the inner pore wall during the initial electrochemical deposition, the Co and subsequent layers of each period form semi-concentric shells.

4.3
Physical Properties of Electrodeposited Nanowires

4.3.1
Magnetic Properties of Nanowire Arrays

Template-synthesized magnetic nanowire arrays exhibit extremely anisotropic magnetic properties, characterized by square hysteresis loops when the applied field is perpendicular to the deposited film – that is, parallel to the wire axis (the easy axis). The advantage of such an arrangement lies in its ability to detect tiny magnetic fields – an effect which can be exploited in a range of applications involving magnetic data storage and magnetic sensors [13–15].

The magnetic nanowire arrays can be fabricated either as a single ferromagnetic material or as a binary magnetic materials system (multilayers). The binary system is usually based on ferromagnetic and antiferromagnetic materials such as Ni–R and Co–R alloys (where R represents Au, Ag, Cu or other metals from the Periodic Table).

The importance of fabricating 1D magnetic nanowires lies in their preferred magnetization direction perpendicular to the film and parallel to the wire axis. One-dimensional magnetic materials with perpendicular anisotropy having large coercivity and high remanence magnetization, may allow for a smaller bit size and thus an increase in the recording density. Therefore, membranes filled with a ferromagnetic material that have a strong perpendicular magnetic anisotropy, are suitable for perpendicular recording media.

For electrodeposited Co nanowires, the structure is influenced by different factors such as composition and pH. In general, the c-axis of hcp Co is the direction of easy magnetization at room temperature. Co nanowires exhibits an hcp crystal structure with <1 0 0> orientation. The magnetic saturation field for both perpendicular and parallel applied fields was in good agreement with the demagnetization field for bulk Co (i.e., $2\pi M_s = 8796$ Oe) (Figure 4.14).

The magnetization of Ag/Co multilayered nanowires was measured with the applied field parallel and perpendicular to the wires. Electrodeposition was stopped before complete membrane filling and magnetization measurements were performed at room temperature, with the wires inside the membrane, using a commercial Princeton Measurements Corporation AGM. By varying the Co layer thickness from 50 m to 15 m, and having an Ag layer thickness of 8 nm, hysteresis loops showed the change of the easy direction from parallel to perpendicular to the wire. This change was due to the competition between shape anisotropy and dipolar interaction where, in the case of small cylinders ($d_{Co} = 15$ nm), it can be expected in the ideal case that the magnetization of each cylinder is aligned in the direction perpendicular to the axis of the wire. Furthermore, the wire magnetizations form an anti-parallel arrangement from one cylinder to another one due to the dipolar interaction. This can explain the low remanence observed in both directions [55]. In the case of longer cylinders ($d_{Co} = 50$ nm), the axis along the wire starts to become the easy axis (lower saturation field) as expected in the limit where $d_{Ag} = 0$ (single Co nanowires), due to the shape anisotropy. Again, in both directions, a low remanence

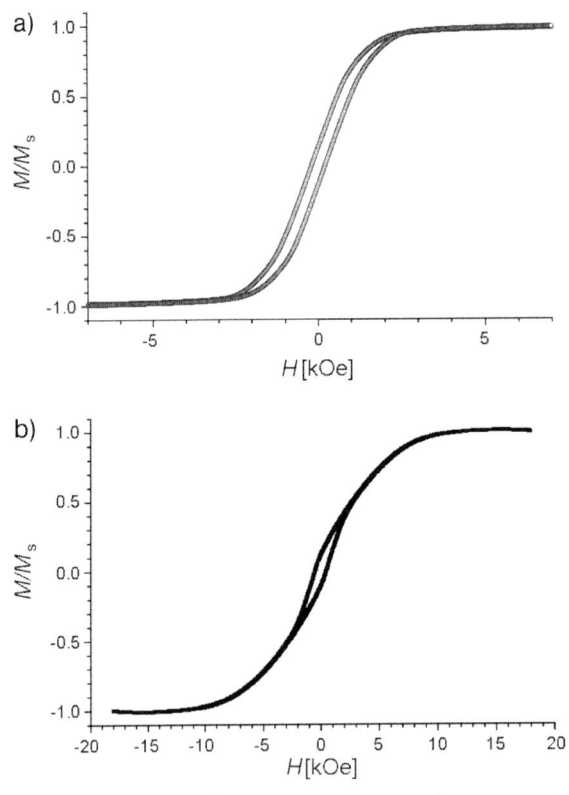

Figure 4.14 Hysteresis loop at room temperature for an array of Co nanowires (diameter 250 nm) with the magnetic field applied (a) along the axis of the wires and (b) perpendicular to the axis of the wires.

was measured leading to an intricate magnetic domain configuration at zero fields [67–71] which requires more characterizations for its elucidation (Figure 4.15).

The magnetization of Au/Co multilayered nanowires, made from a single electrolyte, was also measured with the applied field parallel and perpendicular to the wires [66]. The difference in the saturation field between the two directions (parallel and perpendicular) is smaller than the expected value of $2\pi M_s$ (8796 Oe for Co) due to shape anisotropy (Figure 4.16a). This result can certainly be related to the observation made on the Co system made from a Co/Cu solution, where the competition between shape anisotropy and magnetocrystalline anisotropy leads to a multidomain configuration at zero fields for this range of diameters [66,72,73]. The magnetocrystalline anisotropy term perpendicular to the axis of the wire, in combination with the difference in the saturation field between the parallel and perpendicular directions (ca. 5000 Oe), allows a rough estimation of the first magnetocrystalline anisotropy constant ($K_1 = 2.7 \times 10^6 \, \text{erg cm}^{-3}$), which can be compared to the theoretical bulk value for Co ($5 \times 10^6 \, \text{erg cm}^{-3}$). Magnetic measurements on 12 nm

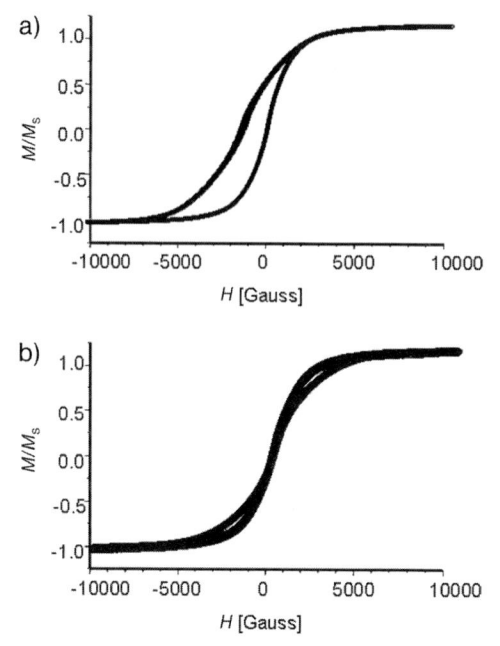

Figure 4.15 Hysteresis loops of Co/Ag nanowires. (a) Co 15 nm/ Ag 8 nm with the magnetic field applied perpendicular and parallel to the axis of the wires. (b) Co 50 nm/Ag 8 nm with the magnetic field applied perpendicular and parallel to the axis of the wires.

Co/4 nm Au multilayered nanowires provide evidence of a low value of the magnetic moment. Measurements were carried out at zero fields, both parallel and perpendicular to the direction of easy magnetization along the axis (Figure 4.16b).

Magnetoresistance measurements were also made on Co 12 nm/Au 4 nm multilayered nanowires in a two-point contact mode using a standard Linear Research AC

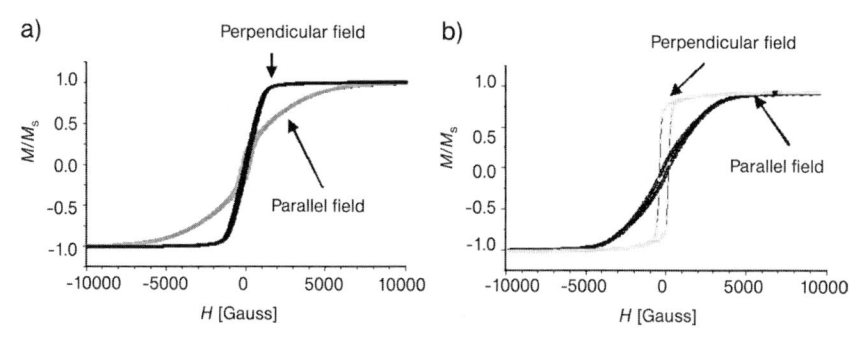

Figure 4.16 (a, b) Hysteresis loops of Co and Co/Au multilayered nanowires at room temperature. The magnetic field is applied perpendicular to the wires or parallel to the wires for: (a) Co nanowires and (b) Co 12 nm/Au 4 nm multilayered nanowires.

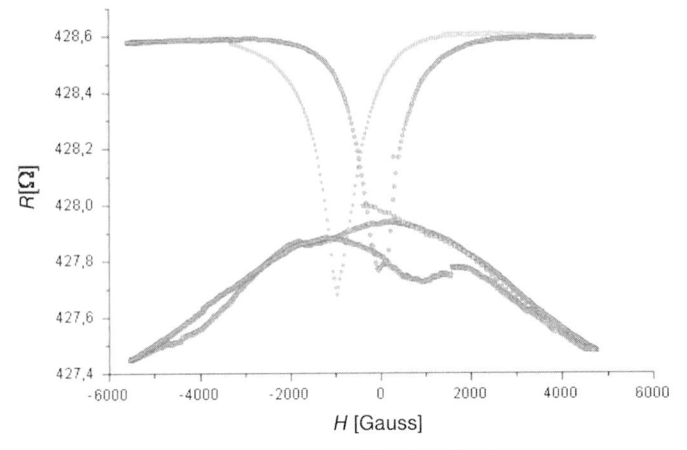

Figure 4.17 Magnetoresistance curves of Co 8 nm/Au 4 nm multilayered nanowires with the applied field perpendicular and parallel to the axis of the wires.

bridge. To prepare the samples for this measurement, the electrodeposition was stopped after the emergence of the wires on the membrane surface. One contact was then made on the bottom electrode, and a second contact on top of the polycarbonate film. No GMR effect was observed on this system. Nevertheless, anisotropic magnetoresistance (AMR) was observed (Figure 4.17). The origin of AMR comes from the spin orbit coupling, while the resistivity depends on the relative orientation of the magnetization and the current [72]. The classical behavior of a magnetic system with a random magnetic state at zero fields was observed. The high value estimation of the resistivity (which has a minimum value of 50 $\mu\Omega$cm if it is supposed that just one wire was connected for the high resistance value measured) indicates a consequent number of intrinsic defects which can perhaps average the two spin directions and explain the absence of GMR. Electrodeposited 1D Au/Co multilayered nanowires may have prospective applications in microelectronics and, particularly, in magnetic recording media.

4.3.2
Electrical Transport Measurements on Single Nanowires Using Focused Ion Beam Deposition

The electrical properties of 1D nanostructures, such as individual nanowires or nanotubes, have recently received considerable attention due to their electronic and optical properties, which are of interest for both fundamental research and technological applications. Electron transport in typical metallic nanowires (e.g., Au nanowires) obeys a classical behavior at room temperature; this is because the wires are much longer than the electron mean free path. In addition, electron transport through these nanowires is strongly affected by collisions with phonons, defects, and impurities. Therefore, a thorough understanding of structure–property relationships and electron transport along nanowires is technologically relevant, notably because exploitation of this fundamental knowledge is essential when constructing new nanowire-based electronic devices.

Recently, different techniques based on scanning probe microscopy (SPM) and AFM have been developed for both the fabrication and electrical transport measurements of Au nanowires [74–77]. Measurement of the electrical characteristics of a single nanowire having a nanoscopic diameter and macroscopic length, necessitates not only specialized tools but also a wealth of technical knowledge.

To date, several groups have used focused ion beam (FIB) deposition for direct Pt deposition to form electrical contacts on Ag and Au nanowires [78,79] and carbon nanotubes [80]. This is often achieved by dispersing the nanowire solution on a surface and locating the positions of the nanowires for the lithographic [81,82] deposition of electrodes.

FIB techniques have been used for many applications, including the creation of electrical contacts to metallic nanowires with localized Pt deposition [83,84]. FIB technology provides reliable means of fabricating nanostructures, of patterning, and of the deposition of materials for rapid prototyping, with resolution in the range of tenths of a nanometer. Ion beam-assisted deposition (IBAD) is performed as follows. When the surface of a sample is exposed to organometallic gas molecules with a FIB, the gas molecules adsorbed onto the substrate decompose, the organic constituents or fragments with a high vapor pressure vaporize, and the metal atoms remain on the surface such that a metal film is deposited. Today, computer-aided beam scanning control has made 3D fabrications possible [85–88]. A major advantage of the FIB system is associated with its unique ability to perform, *in situ*, direct-write electron- and ion-beam nanopatterning, without breaking the vacuum condition.

The possibility of fabricating metallic stripes using IBAD and electron-beam assisted deposition (EBAD) techniques can be used to make electrical contacts between nanowires and conventional microelectrodes, thus enabling measurement of their electrical characteristics.

In these studies, four-point microprobes – fabricated using conventional Si microfabrication – were used to characterize the conducting properties of electrochemically synthesized Au nanowires. Figure 4.18 shows a bright-field TEM image of electrodeposited Au nanowires with a diameter of 100 nm. An array of planar Cr/Au microelectrodes with a total layer thickness of 100 nm were patterned on a Si/SiO$_2$ (thermally oxidized) chip substrate using photolithography, metal evaporation, and lift-off techniques. The four contact pads of $50 \times 50\,\mu m^2$, with electrodes having 5 μm widths and 15 μm lengths, were separated by a gap of $45 \times 45\,\mu m^2$, as shown in Figure 4.19a. The Au electrodeposited membrane was dissolved and small drop of the solution then dispersed on the prepatterned substrate. Electrical measurements of the contacted Au nanowires were conducted at room temperature using a four-point probe set-up. All measured resistances were subsequently converted into resistivities by using the simple expression $R = \rho L/A$, where A and L are the area and length of the wire, respectively. The characterization of a metallized Au nanowire was first attempted by connecting only two electrodes to the edges of the nanowire using Pt-IBAD. In two experimental attempts performed, the results

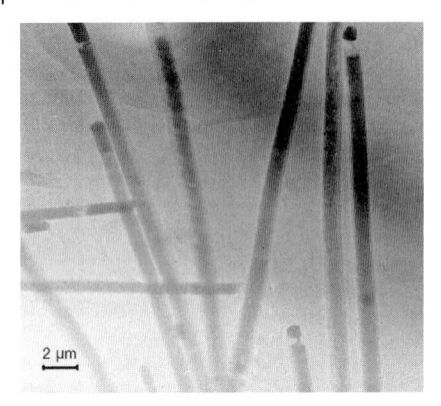

Figure 4.18 Typical bright-field TEM image of electrodeposited Au nanowires with a diameter of 100 nm.

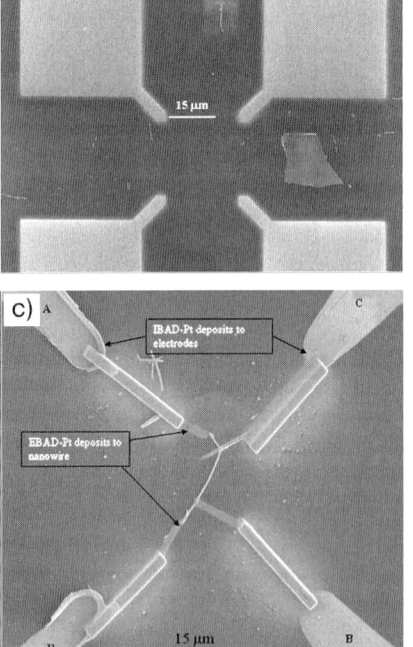

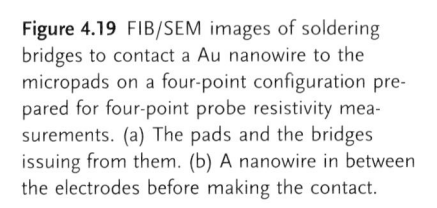

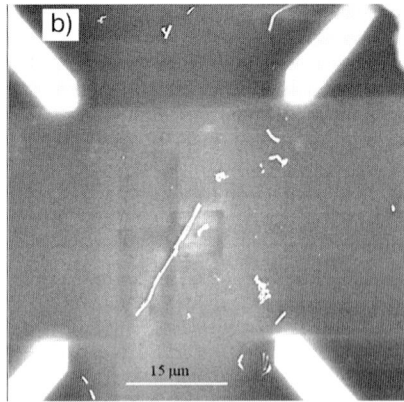

Figure 4.19 FIB/SEM images of soldering bridges to contact a Au nanowire to the micropads on a four-point configuration prepared for four-point probe resistivity measurements. (a) The pads and the bridges issuing from them. (b) A nanowire in between the electrodes before making the contact. (c) Fabrication of the microelectrode by IBAD and EBAD, where the contacts closer to the nanowire have been prepared using the electron beam; the wider contacts between the above-mentioned contacts, and the bridges emerging from the pads have been prepared using the ion beam.

showed resistivity values from 1.5 to 5.2 Ω·cm for stripes with cross-sections larger than 25 μm² to 0.25 μm², respectively. The resistivity values obtained were higher than that of bulk Pt (0.01 Ω·cm), which was in good agreement with results reported elsewhere [81]. This high resistance value is related to an extremely noisy voltage signal, which makes the measurement unfeasible.

The chemical composition analysis of deposited IBAD stripes using auger electron spectroscopy (AES) also showed a high concentration of impurity elements, such as C and Ga (65% C, 27% Pt, 8% Ga). These impurities originated from the organometallic decomposition of the precursor during the Pt deposition. However, the resistivity of the Pt-EBAD deposits present 100- to 1000-fold higher than that of the Pt-IBAD stripes [89]. In the present case, the preference was to use the four-point probe measurement technique to avoid both contact resistance between the Pt-IBAD stripes and the gold microelectrodes, and between the Pt nanocontacts and the nanowire during measurements.

Figure 4.19 shows FIB/SEM images of the process employed for the fabrication of contacts, where Figure 4.19a shows the pads and the bridges issuing from them; Figure 4.19b shows a specific nanowire that is in between the electrodes before making the contacts; and the fabrication of the microelectrodes by electron and IBAD is shown in Figure 4.19c, where the contacts closer to the nanowire have been prepared using the electron beam. The wider contacts between the above-mentioned ones and the bridges emerging from the pads have been prepared using the ion beam. Furthermore, Figure 4.19 also shows a halo contrast around the Pt-deposited contacts, especially around the ion-beam deposited contacts, which is a lateral spread of the beam due to secondary electrons emerging from the sample. For the four-point probe

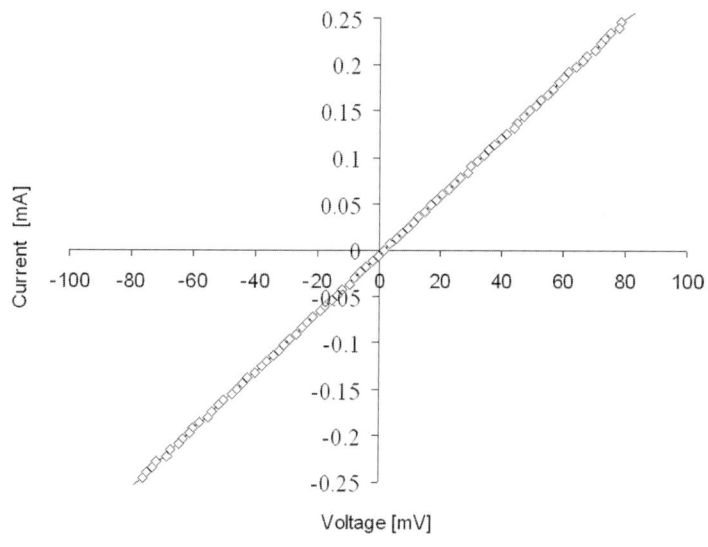

Figure 4.20 Current–voltage (I–V) curve for determining the resistivity of Au nanowires (diameter 100 nm), recorded using a four-point set-up at room temperature.

measurements, contacts A and B in Figure 4.19c have been used to drive the current into the nanowire, while the voltage drop was measured between contacts C and D. As can be seen in Figure 4.20, the measured resistivity of the Au nanowire based on the four-point probe method is $2.8 \times 10^{-4}\,\Omega\cdot\text{cm}$. This value was found to be higher than that of bulk Au ($2.8 \times 10^{-6}\,\Omega\cdot\text{cm}$) at room temperature. The relatively high value can be related to the nanowire wall roughness (see Figure 4.13), which in turn results in an increased electron scattering at the grain boundaries of the nanowire [90–93].

4.4 Summary

Template-synthesized Co nanowires, including well-defined multilayered Co/Ag and Au/Co nanowires, have been fabricated. The method consists of a pulse-plating method in which two metals are deposited from a single solution bath using either potentiostatic or galvanostatic control and a pulsed deposition technique. The electrodeposition process is controlled by a computer which continuously integrates the charge during each layer deposition. The potential is switched when the deposition charges for the nonmagnetic and the magnetic layers reach the set value.

In conclusion, these research investigations have clearly demonstrated that electrodeposition from aqueous solutions in a nanoporous media is an efficient and relatively simple technique to fabricate arrays of magnetic nanowires and magnetic multilayered nanowires with extremely large aspect ratios. In spite of the relative simplicity of growth by electrolysis, these multilayers are indeed superlattices with single-crystal grains including several tens of layers. It has been shown that the electrodeposition process in a confined media is capable of producing magnetic multilayers with predesignable, variable and controllable composition and structure on an atomic scale.

Finally, *in-situ* FIB ion and IBAD techniques to contact Au nanowires on a microscaled four-point probe set-up were successfully demonstrated. In addition, the current–voltage characteristics of a single Au nanowire were investigated by using a four-point microprobe set-up, and showed that the resistivity of a single Au nanowire with a diameter of 100 nm was an order of magnitude higher than the bulk value. An electrical characterization of the four-point probe set-up demonstrated the feasibility of the approach.

References

1 Martin, C.R. (1994) *Science*, **266**, 1961–1965.
2 Cai, Z. and Martin, C.R. (1991) *J. Electroanal. Chem.*, **300**, 35–50.
3 Al Mawlawi, D., Coombs, N. and Moskovits, M. (1991) *J. Appl. Phys.*, **70**, 4421–4425.
4 Lubin, J.A., Huber, T.E. and Huber, C.A. (1993) *Bull. Am. Phys. Soc.*, **38**, 2178.
5 Ferain, E. and Legras, R. (1993) *Nucl. Instrum. Methods Phys. Res. B*, **82**, 539–548.
6 Trautman, A.C. (1994) International Symposium on Ionizing Radiation Pol, Guadeloupe.

7 Hasegawa, N. and Saito, M. (1990) *J. Magn. Soc. Jpn.*, **14**, 313–318.
8 Whitney, T.M., Searson, P.C., Jiang, J.S. and Chien, C.L. (1993) *Science*, **261**, 1316–1319.
9 Cohen, U., Koch, F.B. and Sard, R. (1983) *J. Electrochem. Soc.*, **130**, 1987–1991.
10 Cohen, U. and Tau, M. (1987) Electrochemistry Society 172nd Meeting, 18–23 October.
11 Cohen, U. (1987) US Patent 4-678-722.
12 Kashiwabara, S., Jyoko, Y. and Hayashi, Y. (1996) *Mater. Trans. JIM*, **37**, 289.
13 Tsuya, N., Saito, Y., Nakamura, H., Hayano, S., Furugohri, A., Ohta, K., Wakui, Y. and Tokushima, T. (1986) *J. Magn. Magn. Mater.*, **54–57**, 1681–1682.
14 Cavallotti, P.L., Bozzini, B., Nobili, L. and Zangari, G. (1994) *Electrochim. Acta*, **39**, 1123–1131.
15 Possin, G.E. (1970) *Rev. Sci. Instrum.*, **41**, 772–774.
16 Baibich, M.N., Broto, J.M., Fert, A., Nguyen Van Dau, F., Petroff, F., Eitenne, P., Creuzet, G., Friederich, A. and Chazelas, J. (1988) *Phys. Rev. Lett.*, **61**, 2472–2475.
17 Binasch, G., Grünberg, P., Saurenbach, F. and Zinn, W. (1989) *Phys. Rev. B*, **39**, 4828–4830.
18 Krebs, J.J., Lubitz, P., Chaiken, A. and Prinz, G.A. (1989) *Phys. Rev. Lett.*, **63**, 1645–1648.
19 Krishnan, R., Cagan, V., Porte, M. and Tessier, M. (1990) *J. Magn. Magn. Mater.*, **83**, 65–66.
20 Valet, T. and Fert, A. (1993) *Phys. Rev. B*, **48**, 7099–7113.
21 Mosca, D.H., Barthelemy, A., Petroff, F., Fert, A., Schroeder, P.A., Pratt, W.P., Laloee, R. and Cabanel, R. (1991) *J. Magn. Magn. Mater.*, **93**, 480–484.
22 Xiao, J.Q., Jiang, J.S. and Chien, C.L. (1992) *Phys. Rev. Lett.*, **68**, 3749–3752.
23 Gijs, M.A.M., Lenczowski, S.K.J. and Giesbers, J.B. (1993) *Phys. Rev. Lett.*, **70**, 3343–3346.
24 Pratt, W.P., Jr Lee, S.-F., Slaughter, J.M., Loloee, R., Schroeder, P.A. and Bass, J. (1991) *Phys. Rev. Lett.*, **66**, 3060–3063.
25 Blondel, A., Meier, J.P., Doudin, B. and Ansermet, J.-Ph. (1994) *Appl. Phys. Lett.*, **65**, 3019–3021.
26 Piraux, L., Dubois, S. and Fert, A. (1996) *J. Magn. Magn. Mater.*, **159**, L287–L292.
27 Yahalom, J. and Zadok, O. (1987) *J. Mater. Sci.*, **22**, 499–503.
28 Piraux, L., George, J.M., Despres, J.F., Leory, C., Ferain, E., Legras, R. Ounadjela, K. and Fert, A. (1994) *Appl. Phys. Lett.*, **65**, 2484–2486.
29 Valizadeh, S. (2001) PhD Thesis (Dissertation No. 685), Linköping University, Sweden.
30 Chlebny, I., Doudin, B. and Ansermet, J.-Ph. (1993) *Nanostruct. Mater.*, **2**, 637–642.
31 Ferain, E. and Legras, R. (2001) *Nucl. Instrum. Methods Phys. Res. B*, **174**, 116–122.
32 Schonenberger, C., van der Zande, B.M.I., Fokkink, L.G.J., Henny, M., Schmid, C., Kruger, M., Bachtold, A., Huber, R., Birk, H. and Staufer, U. (1997) *J. Phys. Chem. B*, **101**, 5497–5505.
33 Wang, L.W., Liu, Y. and Zhang, Z. (2002) *Handbook of Nanophase and Nanostructured Materials*, Kluwer Academic Publishers, Dordrecht.
34 Sulitanu, N.D. (2002) *Mater. Sci. Eng. B*, **95**, 230–235.
35 Valizadeh, S., George, J.M., Leisner, P. and Hultman, L. (2001) *Electrochim. Acta*, **47**, 865–874.
36 Widrig, C.A., Porter, M.D., Ryan, M.D., Strein, T.G. and Ewing, A.G. (1990) *Anal. Chem.*, **62**, 1R–20R.
37 Kissinger, P.T. (1996) *J. Pharm. Biom. Anal.*, **14**, 871–880.
38 Johnson, D.C., Weber, S.G., Bond, A.M., Wightman, R.M., Shoup, R.E. and Krull, I.S. (1986) *Anal. Chim. Acta*, **180**, 187–250.
39 Aoki, K. (1993) *Electroanalysis*, **5**, 627–639.
40 Edmonds, T.E. (1985) *Anal. Chim. Acta*, **175**, 1–22.
41 Kim, Y.T., Scarnulis, D.M. and Ewing, A.G. (1986) *Anal. Chem.*, **58**, 1782–1786.
42 Wightman, R.M. and Wipt, D.O. (1989) *Electroanal. Chem.*, **15**, 267–353.

43 Chiriac, H., Moga, A.E., Urse, M. and Óvári, T.A. (2003) *Sensors and Actuators A*, **106**, 348–351.

44 Gunawardena, G.A., Hills, G.J. and Montenegro, I. (1978) *Electrochim. Acta*, **23**, 693–697.

45 Bond, A.M., Fleischmann, M. and Robinson, J. (1984) *J. Electroanal. Chem.*, **168**, 299–312.

46 Scharifker, B. and Hills, G. (1981) *J. Electroanal. Chem.*, **130**, 81–97.

47 Sun, L., Chien, C.L. and Searson, P.C. (2000) *J. Mater. Sci.*, **35**, 1097–1103.

48 Avrami, M. (1939) *J. Chem. Phys.*, **7**, 1103–1112.

49 Slaughter, J., Bass, J., Pratt, W.P., Jr, Schroeder, P.A. and Sato, H. (1987) *Jpn. J. Appl. Phys.*, **26**, (Suppl 26-3),1451–1452.

50 Szymczak, H., Zuberek, R., Krishnan, R. and Tessier, M. (1990) *IEEE Trans. Magn.*, **26**, 2745–2746.

51 Barnard, J.A., Waknis, A., Tan, M., Haftek, E., Parker, M.R. and Watson, M.L. (1992) *J. Magn. Magn. Mater.*, **114**, L230–L234.

52 Zaman, H., Ikeda, S. and Ueda, Y. (1997) *IEEE Trans. Magn.*, **33**, 3517–3519.

53 Fedosyuk, V.M., Kasyutich, O.I. and Schwarzacher, W. (1999) *J. Magn. Magn. Mater.*, **198–199**, 246–247.

54 Valizadeh, S., Holmbom, G. and Leisner, P. (1998) *Surf. Coat. Tech.*, **105**, 213–217.

55 Valizadeh, S., George, J.M., Leisner, P. and Hultman, L. (2002) *Thin Solid Films*, **402**, 262–271.

56 Ferré, J., Grolier, V., Meyer, P., Lemerle, S., Maziewski, A., Stefanowicz, E., Tarasenko, S.V., Tarasenko, V.V., Kisielewski, M. and Renard, D. (1997) *Phys. Rev. B*, **55**, 15092–15102.

57 Murayama, A., Hyomi, K., Eickmann, J. and Falco, C.M. (1998) *Phys. Rev. B*, **58**, 8596–8604.

58 Forrer, P., Schlottig, F., Siegenthaler, H. and Textor, M. (2000) *J. Appl. Electrochem.*, **30**, 533–541.

59 Cagnon, L., Gundel, A., Devolder, T., Morrone, A., Chappert, C., Schmidt, J.E. and Allongue, P. (2000) *Appl. Surf. Sci.*, **164**, 22–28.

60 Darby, E.C. and Harris, S.J. (1975) *Trans. IMF*, **53**, 115.

61 Clenghorn, W.H., Crossly, J.A., Lodge, K.J. and Gnanasekaran, K.S.A. (1972) *Trans. IMF*, **73**, 50.

62 Holt, L., Ellis, R.J. and Stanyer, J. (1973) *Plating*, **9**, 24.

63 Takahata, T., Araki, S. and Shinjo, T. (1989) *J. Magn. Magn. Mater.*, **82**, 287–293.

64 Celis, J., Cavalloti, P., da Silva, J.M. and Zielonka, A. (1998) *Trans. IMF*, **76**, 163–170.

65 Valizadeh, S., Svedberg, E.B. and Leisner, P. (2002) *J. Appl. Electrochem.*, **32**, 97–104.

66 Valizadeh, S., Hultman, L., George, J.M. and Leisner, P. (2002) *Adv. Funct. Mater.*, **12**, 766–777.

67 Tench, D.M. and White, J.T. (1990) *J. Electrochem. Soc.*, **137**, 3061–3066.

68 Wang, L., Yu-Zhang, K., Metrot, A., Bonhomme, P. and Troyon, M. (1996) *Thin Solid Films*, **288**, 86–89.

69 Dubois, S., Marchal, C., Beuken, J.M., Piraux, L., Duvail, J.L., Fert, A., George, J.M. and Maurice, J.L. (1997) *Appl. Phys. Lett.*, **70**, 396–398.

70 Piraux, L., Dubois, S., Ferain, E., Legras, R., Ounadjela, K., George, J.M., Maurice, J.L. and Fert, A. (1997) *J. Magn. Magn. Mater.*, **165**, 352–355.

71 Belliard, L., Miltat, J., Thiaville, A., Dubois, S., Duvail, J.L. and Piraux, L. (1998) *J. Magn. Magn. Mater.*, **190**, 1–16.

72 Maurice, J.-L., Imhoff, D., Etienne, P., Durand, O., Dubois, S., Piraux, L., George, J.-M., Galtier, P. and Fert, A. (1998) *J. Magn. Magn. Mater.*, **184**, 1–18.

73 Campbell, I.A. and Fert, A. (1982) in *Ferromagnetic Materials*, (ed. E.P. Wohlfarth), North-Holland Publishing Co., Vol. 3, pp. 747–804.

74 Ramsperger, U., Uchihashi, T. and Nejoh, H. (2001) *Appl. Phys. Lett.*, **78**, 85–87.

75 Durkan, C., Schneider, M.A. and Welland, M.E. (1999) *J. Appl. Phys.*, **86**, 1280–1286.

76 Cronin, S.B., Lin, Y.-M., Rabin, O., Black, M.R., Ying, J.Y., Dresselhaus, M.S., Gai, P.L., Minet, J.-P. and Issi, J.-P. (2002) *Nanotechnology*, **13**, 653–658.

77 Molares, M.E.T., Höhberger, E.M., Schaeflein, Ch., Blick, R.H., Neumann, R. and Trautmann, C. (2003) *Appl. Phys. Lett.*, **82**, 2139–2141.

78 Gopal, V., Radmilovic, V.R., Daraio, C., Jin, S., Yang, P. and Stach, E.A. (2004) *Nano Lett.*, **4**, 2059–2063.

79 Smith, P.A., Nordquist, C.D., Jackson, T.N., Mayer, T.S., Martin, B.R., Mbindyo, J. and Mallouk, T.E. (2000) *Appl. Phys. Lett.*, **77**, 1399–1401.

80 Ebbesen, T.W., Lezec, H.J., Hiura, H., Bennett, J.W., Ghaemi, H.F. and Thio, T. (1996) *Nature*, **382**, 54–56.

81 Ziroff, J., Agnello, G., Rullan, J. and Dovidenko, K. (2003) *Mater. Res. Soc. Symp. Proc.*, **772**, M8.8.1.

82 Tao, T., Ro, J., Melngailis, J., Xue, Z. and Kaesz, H.D. (1990) *J. Vac. Sci. Technol. B*, **8**, 1826–1829.

83 van den Heuvel, F.C., Overwijk, M.H.F., Fleuren, E.M., Laisina, H. and Sauer, K.J. (1993) *Microelectron. Eng.*, **21**, 209–212.

84 De Marzi, G., Iacopino, D., Quinn, A.J. and Redmond, G. (2004) *J. Appl. Phys.*, **96**, 3458–3462.

85 Lipp, S., Frey, L., Lehrer, C., Frank, B., Demm, E. and Ryssel, H. (1996) *J. Vac. Sci. Technol. B*, **14**, 3996–3999.

86 Pascual, J.I., Méndez, J., Gómez-Herrero, J., Baró, A.M., García, N. and Binh, V.T. (1993) *Phys. Rev. Lett.*, **71**, 1852–1855.

87 Rubio, G., Agraït, N. and Vieira, S. (1996) *Phys. Rev. Lett.*, **76**, 2302–2305.

88 Ohnishi, H., Kondo, Y. and Takayanagi, K. (1998) *Nature*, **395**, 780–783.

89 Puretz, J. and Swanson, L.W. (1992) *J. Vac. Sci. Technol. B*, **10**, 2695–2698.

90 Hernández-Ramírez, F., Casals, O., Rodríguez, J., Vilà, A., Romano-Rodríguez, A., Morante, J.R., Abid, M. and Valizadeh, S. (2005) *MRS Symp. Proc.*, **872**, J5.2.1.

91 Romano-Rodríguez, A., Hernández-Ramírez, F., Rodríguez, J., Casals, O., Vilà, A., Morante, J.R., Abid, M., Abid, J.-P., Valizadeh, S., Hjort, K., Collin, J.-P. and Jouaiti, A. (2005) EUSPEN'05 France Montpellier 8–11May.

92 Hernández, F., Casals, O., Vilà, A., Morante, J.R., Romano-Rodríguez, A., Abid, M., Valizadeh, S. and Hjort, K. (2005) Microscopy of semiconducting materials conference UK 11–14 April.

93 Valizadeh, S., Abid, M., Hernández-Ramírez, F., Romano Rodríguez, A., Hjort, K. and Schweitz, J.Å. (2006) *Nanotechnology*, **17**, 1134–1139.

5
Electrochemical Sensors Based on Unidimensional Nanostructures

Arnaldo C. Pereira, Alexandre Kisner, Nelson Durán, and Lauro T. Kubota

5.1
Introduction

Today, sensors based on low-dimensional materials are no longer a novelty. In fact, the miniaturization of chemical and biochemical sensors has a long history which dates back to the studies of Wise *et al.* [1] in the late 1960s and of Bergveld [2,3] in the early 1970s, both of whom might be termed "pioneers". Over the past few years, however, a new wave of highly sensitive materials has caused a reactivation in both scientific and industrial interest. These nanostructured materials, as they are now widely recognized, are of special interest on the basis of their promising new applications, in addition to the novel dimension-dependent properties that they exhibit. Consequently, one-dimensional (1D) structures such as nanowires and nanotubes are expected to serve as excellent primary transducers in systems such as biosensors. Moreover, due to the relative ease with which these 1D materials can be integrated with microelectronic techniques, such wires and/or tubes on a nanoscale can serve as active devices to provide a complete set of new nanobiotechnology devices, such as biochips. Thus, although other nanostructured systems have shown much potential for field applications, the focal points of this chapter will be the electrochemical synthesis of nanowires and nanotubes, and their application as sensors to detect biological and chemical species. Initially, some basic concepts and simple electrochemical techniques for preparing metallic and non-metallic wires and tubes will be described, after which some applications and limitations will be discussed.

5.2
Preparation of Nanowires and Nanotubes by Template-Based Synthesis

One general question that arises in the preparation of nanostructured materials is how to assemble atoms or other building blocks into nanometer-scale objects with diameters much smaller than the available lengths (1D structures). Although many

Nanostructured Materials in Electrochemistry. Edited by Ali Eftekhari
Copyright © 2008 WILEY-VCH Verlag GmbH & Co. KGaA, Weinheim
ISBN: 978-3-527-31876-6

approaches are available for the preparation of 1D nanostructures [4–7], in general the "top-down" and "bottom-up" methods represent the strategies that have used intensively. In particular, electrochemical methods based on top-down or bottom-up strategies represent a suitable alternative to obtain well- arranged and well-defined 1D nanostructures.

In this context, *template-based synthesis* represents one of the most successful bottom-up methods for preparing nanowires and nanotubes [6–8]. The method is based on substrate-like materials pursuing a set of poles and boards onto which nanometer-sized structures are generated *in situ* and shaped into the predetermined morphology of the template. Over the past few years, a wide range of templating methods has been exploited to grow and pattern many types of metal and polymeric material [9,10]. One commonly used template approach was inspired for patterned surfaces, where either nano- or microstructural defects are generated positively or negatively on the surface of a solid substrate using lithography and etching techniques.

In the case of *positive-templates*, the wire-like nanostructures, such as DNA and carbon nanotubes (CNTs) [11–13], act as templates to guide the electrodeposition of the intended material onto the template surface In this way, there is no physical restriction due to template size, and the diameter of the nanostructures can be controlled simply by adjusting the quantity of material deposited on the template. An elegant approach based on the positive-template method is also used to create 1D π-conjugated polymer nanostructures [14–17]. Another approach, termed "electrochemical epitaxial polymerization", was based on the double role played by iodine in a solution containing a thiophene monomer [18]. The iodine, which is an oxidizing agent, can lead to the formation of thiophene radical cations, which sequentially may couple to form oligothiophenes. As the reaction takes place at the surface of an Au (111) electrode, an iodine adlayer covers the electrode surface. By applying voltage pulses, however, an electrochemical polymerization occurs which results in the propagation of single poly (thiophene) wires. In this case, the presence of iodine is pivotal because the adlayer on the electrode surface is able to act as a nucleus for nanowire orientation and growth.

In the *negative-template* method, V-like grooved structures are used to deposit, by solution-phase electrochemical plating, metal or semiconductors at the bottom of each V-shaped groove. By using this simple procedure, continuous thin nanowires with lengths up to hundreds of micrometers may be prepared on the surface of solid substrates, which may in turn be released sequentially into free-standing forms or transferred onto another support [19]. In fact, this is one of the best advantages of the template technique, as the template is only used to limit physically the geometric features of 1D nanostructures. Subsequently, it can be selectively removed by etching or calcination on completion of the synthesis, allowing the resultant nanostructures to be collected.

5.2.1
Template-Based Mesoporous Materials

Other successful negative-templates are those based on nanoporous membranes. In that case, the nanometer-scale pores of membranes such as polycarbonate and anodic alumina (or even zeolites) are used to confine and act as a template for the growth of nanostructures [8,20–26]. The pores of these nanoporous membranes can be used as templates for forming the desired material. For example, polycarbonate template membranes were loaded with enzymes, such as glucose oxidase, catalase, subtilisin, trypsin, and alcohol dehydrogenase, to create a new type of enzymatic bioreactor or to allow drug delivery [74]. Specifically, anodic alumina membranes (AAM) are of particular interest because their geometric features (e.g., thickness, pore and cell diameter) and composition can be controlled experimentally. Another interesting characteristic is the growth of arrays of self-organized hexagonal compacted cells with central pores parallel to each other and perpendicularly oriented to the substrate surface [27–29]. By using this approach, aluminum foils may normally be anodized under a constant voltage (Figure 5.1A), using an acid electrolyte such as sulfuric, phosphoric oxalic acid. It is important to note that the pore-size diameter can be controlled by the chosen electrolyte [30]. For example, anodizing performed with sulfuric acid may yield pores with diameters smaller than 25 nm, whereas phosphoric acid may provide larger pores (>200 nm), even before pore widening [69]. By comparison, oxalic acid solutions provide pores with intermediate diameters ranging typically from approximately 30 nm to about 100 nm [29,30,69,70]. After anodizing, the aluminum substrate is removed and the AAM is soaked in an acid solution to remove any barrier layer formed at the pore bottom during anodization. Finally, a conductive layer to serve as a cathode posterior in the electrochemical cell is deposited on one side of AAM. In doing so, wires or tubes with the pore-size diameters and lengths of several micrometers can be grown not only by electrochemical deposition but also from electroless plating solutions.

One of the pioneering groups to experiment with both types of deposition was that of Martin [4,5,31]. Perhaps unsurprisingly, the results of these studies suggested that electroless plating allowed for a more uniform deposition, in this case of Au, than did the electrochemical plating method. The differences in field distribution along the porous membrane were proposed as the main factor for such variation.

The same group also showed that the deposition time plays a pivotal role in the formation of tubes and wires [31]. In this case, it is noteworthy that in the early stage of the electrodeposition/deposition, the material might be trapped preferentially as uniform layers on the walls of those pores to form tubular nanostructures [6]. In addition, nanotubes can be prepared based on the preferential electrodeposition of the desired material along the pore walls. The key requirement in this case is a sensitization or preactivation step of the pore wall before the electrodeposition process. One of the most common sensitizing procedures is to silanize the pore wall with aminopropyltriethoxysilane (APTES), which yields free amino groups on the wall surface that are able to act as anchors for complex metallic nanoparticles [32]. In a subsequent step, the porous modified membrane is immersed in an electrochemical plating solution (Figure 5.1B) and the electrodeposition takes place initially at the pore bottom where a metallic layer was previously deposited. As the pore wall presents immobilized nanoparticles, the electrodeposition continues preferentially along the nanochannel surface, thus, resulting in 1D tubular structures. As the electrodeposition process continues, the diameter of the nanochannel decreases gradually and nanowires may also be obtained. A similar technique has

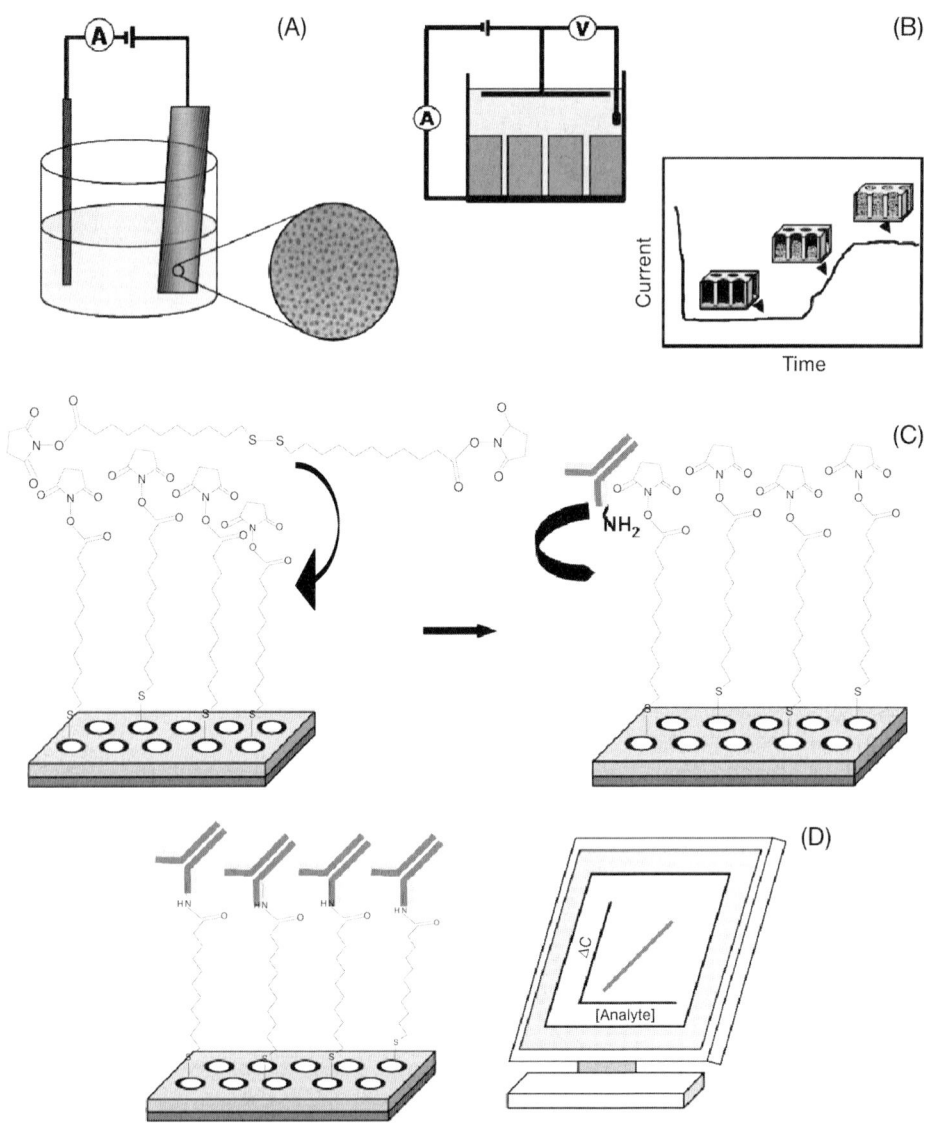

Figure 5.1 (A) Representative view of the anodizing process of an aluminum sheet showing oxide growth. (B) Schematic of metal electrodeposition into a nanoporous membrane and chronoamperometric profile of nanowire growth. (C) Illustration of nanowire surface modification with thiosuccinimidylundecanoate followed by covalent immobilization of anti-E. coli antibody onto the nanowire surface. (D) Detection of E. coli cells and record showing the capacitance change of nanowire nanoelectrodes with different concentrations of cells. (Adapted from Ref. [24].)

also been applied to the formation of tubular polymeric nanostructures, although in this approach wetting methods are most often used [10]. Carbon nanotubes may also be grown by chemical vapor deposition (CVD) into the nanoporous membranes. In this case, the deposition of catalyst clusters on the pore bottom is not necessary as the nanochannel of membrane will limit and direct the growth of CNTs in a 1D manner [33]. An excellent overview of the use of anodic porous alumina in the template-assisted fabrication of nanomaterials is also available (see Chapter 1, this volume).

5.2.1.1 The Memorable Marks of Electrochemical Nanowires

One experimental advantage of the electrodeposition method within nanoporous materials is the growth of nanowires that can be monitored by the reduction current (Figure 5.1B). For example, in the early stages the reduction current is too low as the electroactive species need to diffuse along the pores. When the reduction takes place, however, the current begins to increase and the growth of nanowires or nanotubes proceeds until the pores are totally filled to the top surface of the membrane. Beyond this, the growth can extend in three dimensions, resulting in hemispherical cap shapes on the top of the wires, which grow in size until coalescence of the membrane surface occurs. During this stage, growth proceeds over the entire membrane surface and the reduction current reaches an asymptotic value. The composition of nanowires may also be controlled by adjusting the current intensity and electrolyte composition; thus, multiple segments of different materials can be fabricated simultaneously in a solution containing different metal ions [96]. These multisegment nanowires open the possibility of exploring their application as barcodes in biological assays. For instance, Nicewarner-Pena and coworkers showed that controlled multisegment nanowires could be used like barcodes in DNA and proteins bioassays [97]. The central point in this case is based in the inherent properties of each metal stripe within a single nanowire, which can selectively adsorb different molecules, such as DNA oligomers. In this way, the optical scattering efficiency of the multisegment nanowires can be significantly enhanced by altering the dimensions of the segment, such that different excitations of the surface plasmon occur for each stripe, and this can be used to detect concomitantly different biological molecules.

In conclusion, template-based methods represent a simple, highly-oriented, cost-effective and mass producible means for the formation of nanowires and nanotubes from various materials. An interesting point here is that, if we consider a nanoporous membrane with a constant lattice of 500 nm and a pore diameter of 400 nm, there will be 460 000 000 pores per cm^2 which, in theory, might yield after a deposition procedure the same amount of 1D structures.

5.2.2
Nanowires as Nanoelectrodes

One special advantage of electrochemical nanowires is that they can be charged by applying an external potential. In particular, one of the great advantages of electrodeposited metallic nanowires is that they are highly conductive, as the electrodeposition is based on electron transfer, which is fastest along the highest conductive path

[96]. A most important point here is that their work functions are increased by about 1 eV, which entails a shift of the potential of zero charge by about the same amount. Consequently, at ordinary electrode potentials accessible in aqueous solutions, these wires are always negatively charged. Hence, a very promising application of these 1D nanostructures includes the possibility to use them as nanoelectrodes in highly resistive electrochemical systems. Moreover, because of their small dimensions, these nanowires not only provide opportunities to investigate the kinetics of redox process that occur too rapidly to be studied by conventional macroelectrodes, but also they fall under a new class of electrodes that form a link between nanotechnology and biological systems, allowing the latter to be evaluated in a non-invasive manner. In this context, another remarkable feature of these nanowires is their dimensionality, which is comparable to those of biological macromolecules such as proteins and nucleic acids. Therefore, based on these special characteristics, nanowires represent an attractive new tool for a diverse set of applications in the field of sensor technology.

The first example to mention is the use of template-based nanoporous materials to fabricate metallic nanowires that can act as active nanoelectrodes directly over the nanoporous substrate [4,5,34]. In doing so, a protein film voltammetric (PFV) technique can be performed using these nanowires as nanoelectrodes to study fast electron-transfer in redox proteins. Furthermore, electron catalyst enzymes such as cytochromes, which may have their core center too hindered and thus, are difficult to investigate using conventional macroelectrodes, may now be expected to be sensed as the nanoelectrodes are of comparable size to these biological molecules. Indeed, early studies have shown that the peak current in cyclic voltammetry (CV), which is considered a rather poor electroanalytical technique, when performed with gold nanoelectrodes can reach a range of picoamperes, such that CV would become a powerful tool for measuring the kinetics of electron transfer [5]. One particular problem encountered with nanoelectrodes supported in a matrix, such as polycarbonate, is the large double-layer charging current. This is derived from the solution that creeps between the pore wall and the nanoelectrode. However, in case of polycarbonate, a simple heat-shrink causes the polymer chains to relax to their stretched conformations, in turn shrinking the membrane pore. This shrinking procedure causes the junction between the nanowire and the pore wall to be sealed and, as a consequence, the charging currents become negligible. In the case of AAM, the pores are often filled up by the nanowire and this problem does not occur.

5.2.2.1 Electrochemical Aspects of Nanoelectrodes

A detailed analytical study of the performance of nanowires as nanoelectrodes was made by Menon and Martin [5]. By using $Ru(NH_3)_6Cl_3$ and [(trimethylamine) methyl] ferrocene ($TMAFc^+$) perchlorate as electroactive species in an aqueous media, these authors showed that the diffusion layers at these nanoelectrodes follow a total-overlap mechanism, namely that the diffusion layer is linear to the overall geometric area of the nanoelectrodes. In consequence, peaked-shaped voltammograms were obtained. The effect of the supporting electrolyte concentration was also interesting. Better-defined voltammetric waves for the chosen electroactive species were reached when low concentrations of supporting electrolytes were used.

Another good relationship was the comparison of detection limits between the electrodes containing 10-nm diameter gold discs and a large-diameter (ca. 3.2 mm) gold disc electrode. Results taken from cyclic voltammograms showed that the detectivity for the nanoelectrode ensembles could be as much as three orders of magnitude better than those at macroelectrodes.

The analysis produced results that were often in agreement with quantitative and qualitative theoretical data, and served in most other cases as a good reference point for the prediction of the external effects in nanoelectrode performance.

Taking into account these results, Martin and coworkers exploited the possibility of using these nanoelectrodes as biosensors [35]. Three electron-transfer mediators, two phenothiazines (azure A and B) and methylviologen, which are commonly used for biosensors based on reductase enzymes, were investigated. The findings suggested that the detection limit of the nanoelectrode ensemble depended on the $E_{1/2}$ value of the mediator used. It was also the first time that a nanoelectrode array was used to determine standard heterogeneous rate constants (k^0). Nevertheless, at these electrodes the true value of k^0 could not be determined directly because the gold nanodiscs represented only a small fraction of the overall area of the electrode surface. For this reason, an apparent standard heterogeneous rate constant [26] (k^0_{app}) was determined from

$$\Psi = k^0_{app}[D_0\pi(nFv/RT)]^{-1/2} \qquad (1)$$

where D_0 is the diffusion coefficient for the oxidized species (assuming $D_0 = D_R$), α is the charge transfer coefficient, v is the scan rate, T is the temperature, F is Faraday's constant, and R is the gas constant. The real k^0 value is then obtained by following relationship:

$$k^0_{app} = k^0(1-\theta) = k^0 f \qquad (2)$$

where θ is the fraction of blocked electrode surface and f is the fraction of the electrode surface that are the Au nanoelectrodes. The value of f may be determined from background cyclic voltammograms [5]. Textor et al. [98] also showed that Au nanowire arrays may be applied as electrolytic sensors with large internal surface and with low sensitivity to undesired faradaic processes. Since vertically aligned nanowires exhibit a high surface area in contact with the electrolyte, the faradaic process is restricted to the relatively small surface frontal area of each nanowire, which significantly reduces both capacitive and faradaic currents in comparison with macroscopic flat gold electrodes.

5.2.2.2 Nanoelectrodes Based on Chemically Modified Surface

Because their surface properties are easily modified, nanowires and nanotubes can be modified by potential chemical or biological molecular recognition units, entailing the wires or tubes themselves as selective elements (Figure 5.1C and D). For example, Subhash Basu and coworkers [24] developed, by electrochemical deposition into AAM, an array of Au nanowires that were in a subsequent step chemically modified with an anti *Escherichia coli* antibody. In that way, by using electrochemical impedance spectroscopy (EIS) the authors were able to show that, under the addition of different *E. coli* cell concentrations, a detection limit of 10 cells over a 0.173 cm^2 area can be achieved. The underlying detection in this case ultimate depends upon the antigen–antibody complex formation, which changes the capacitance surface properties of the sensor. In this case, *E. coli* served as an antigen and the detection limit for the capacitance change was between 10^{-9} F and 10^{-12} F. Self-assembled monolayers (SAMs) of mercaptoethylamine or mercaptopropionic acid were also used to immobilize glucose oxidase on the surface of sensor-based Au nanotubular electrode ensembles [71,72]. The sensor showed excellent stability and sensitivity for the amperometric detection of glucose. Indeed, the glucose responses were as large as 400 nA mM^{-1} cm^{-2}, which was considered one of the highest values reported in the literature for comparable systems. A remarkable reproducibility was also achieved after 38 measurements, and no interference from species such as ascorbic and uric acids were observed in potentials up to +0.9 V.

Also of interest was the suggestion that tubular 1D nanostructures might serve to separate molecules on the basis of molecular size and transport properties of the nanotubes. In addition, changing the inner surface of tubular walls by chemisorbing specific analytes, the chemical and/or physical environment within the tubules might be altered. As a result, this changed the transport properties of the nanotubes [6,27,38]. Gold nanotubules deposited in pores of polymeric membranes, with an inside diameter of molecular dimension (<1 nm), were used to separate small molecules on the basis of molecular size. Such a "filter" was applied to separate pyridine (molecular weight 79 g mol^{-1}) and quinine (molecular weight 324 g mol^{-1}) and, as a result of dimensionality, only the smaller pyridine molecule was transported through the nanotubules [73]. These so-called "solid-state nanopores" may also act as sensitive thermometers with single nanometer dimensions because they measure the temperature-dependent conductance of the ionic liquids. A laser heats up at its focal point, and its three-dimensional intensity profile would be mapped out by scanning the nanopore. Thus, the temperature distribution can be measured in aqueous environments with nanometer resolution [75,76].

Today, although we live in an era seeking personalized medicine, the greatest interest in these nanoporous structures is by no means devoted to genomic applications. Currently, DNA size determination and elucidation of the constitution of an individual chromosome are the major applications. Sensors can be built by incorporating a biomolecular recognition complex into the pore [75]. Then, by measuring the transverse electrical current of single-stranded DNA while it translocates through a nanopore, theory-based calculations have estimated that an entire human genome could be monitored with very high accuracy in a matter of hours [77]. On the other hand, first-principles calculations of the transverse conductance across DNA fragments placed between gold nanoelectrodes have revealed that direct current measurements across DNA with gold contacts do not represent a convenient approach to DNA sequencing, as the DNA fragments (the nucleotide bases A, C, G, and T) are limited by geometric factors [78]. By and large, solid-state nanoporous materials may find applications that are very different from biomolecular detection. Indeed, their use as nanoelectrodes has been reported in the literature [99], while high flux analyses of charged species on nanometer-sized porous suggest that the ionic transport would follow a non-linear conventional description [100].

Today, a series of attempts is underway reporting the use of tubular nanostructures as protein biosensors [38–40,99]. In particular, noble metal nanotubes are of special interest due to the well-known spontaneous chemisorption of thiols on their surface. Hence, a broad range of cross-links can be obtained from a thiol-modified surface, and this would allow chemical interactions between drugs and proteins and other systems to be studied on a molecular basis.

5.3
An Electrochemical Step Edge Approach

Following the footsteps of his ex-tutor, Charles Martin, Penner and colleagues at the University of California at Irvine developed a new electrochemical method to synthesize 1D structures [41,42]. This approach, which was called *Electrochemical Step Edge Decoration* (ESED), consists of the electrodeposition of either noble metals or oxides accomplished by three successive voltage pulses on a highly oriented pyrolytic graphite (HOPG) surface (Figure 5.2A). Central to nanowire growth are the sequence of voltage pulses (Figure 5.2B) – that is, a first oxidizing pulse is used to activate the step edges of HOPG. Sequentially, a second negative potential pulse is applied for a few milliseconds for nucleation and metal deposition. A small-amplitude reducing pulse is then applied and particles or wires with different diameters can be grown, depending on the chosen potentials and time of electrolysis. On the other hand, the growth of nanowires by direct metal electrodeposition is not possible. One reasons for this is the low surface free energy of the graphite basal plane, which leads to a low density of metal nuclei along the step edge. In fact, defects on HOPG such as the step edges are able to catalyze electron transfer to metal ions in solution. Thus, nucleation occurs in a preferentially energetic manner at step edges on the graphite surface. Therefore, if the deposition process is prolonged sufficiently for individual particles nucleated along a step edge to coalesce, then nanowires with diameters as small as 13 nm and lengths up to 500 µm may be obtained [41].

5.3.1
The Predeterminant Mechanism

This electrochemical technique shows some advantages over physical methods such as vapor deposition, as the wires can really extend into the third dimension. In contrast, in CVD the dominant process is the adsorption on the terraces, followed by diffusion. This difference may be highlighted mainly by the mechanism of deposition. To date, the growth of structures on a surface is usually preceded by the formation of a stable nucleus [43]. Single adatoms that arrive at the surface and

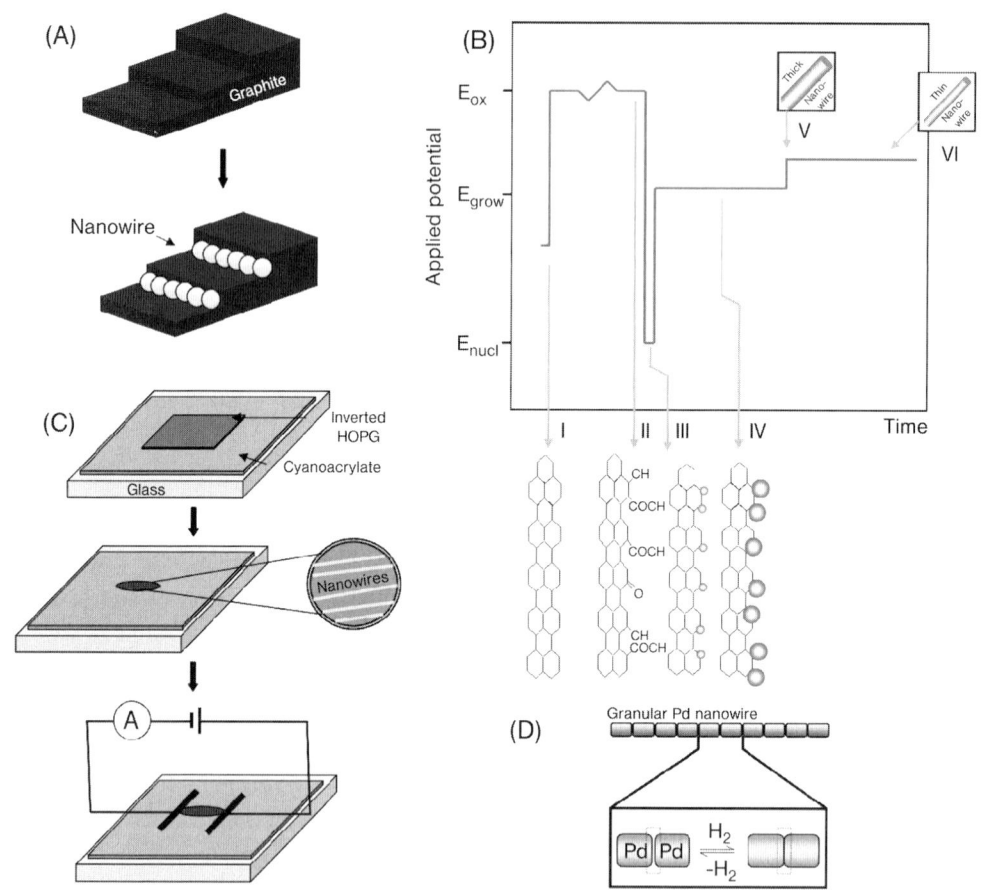

Figure 5.2 (A) Schematic of electrochemical step edge deposition to fabricate 1D nanostructures. (B) Illustrative view of triple pulse voltage used to produce metal nanowires. (I) and (II) are the step edge and the oxidizing stage. (III) Nucleation and growth of nanoparticles followed by (IV) coalescence and nanowire formation. (V) At the final stage, thick nanowires are often produced, but these can be shrunk (VI) by controlled electro-oxidizing. (C) Transfer procedure of nanowires to a non-conductor substrate and lateral electrical contact to make nanowires act as sensors. (D) Illustration of granular Pd nanowires showing the nanogaps and the reversible behavior under H_2 exposure. (Adapted from Refs. [47,52].)

diffuse over the terraces incidentally meet each other and stick together, due to attractive interactions. This concept on a two-dimensional (2D) surface is identical to the nucleation of 3D water droplets from supersaturated water vapor in a cloud [44]. By a careful adjustment of the supersaturation of the 2D gas phase – namely the concentration – the diffusing adatoms on the terraces can be kept very low, and therefore nucleation on the terraces can be effectively suppressed. However, defects such as step edges, where the adatoms are more strongly bound than on the bare terraces, may act as nucleation sites. Hence, at a sufficiently low supersaturation, the growth of islands of the deposited material may start exclusively at the step edges of the surface. In vacuum deposition, the supersaturation is controlled either by the deposition rate or the substrate temperature. Similarly, upon electrochemical deposition the supersaturation is controlled by the electrochemical deposition but, in contrast to vacuum deposition – in which the incoming atoms extend along the surface – electrochemical deposition is balanced by an equilibrium process, as the electrochemical potential steers the balance between the absolute rate of atom deposition and their dissolution [45,46]. One drawback of this procedure (which is also a form of template technique) and of the conventional template-based methods, is that procedures often produce polycrystalline materials with diameters greater than required to observe the effects of quantum phenomena confinement.

The diameter of these nanowires may still be shrunk via a kinetically controlled electrochemical process. By integration of the ESED approach with an electro-oxidation process after the nanowire has been generated, Penner and colleagues [47] showed that, by carrying out the electro-oxidation at low current densities, the oxidation current decreases in proportion to the area of nanowire and also directly proportional to the oxidation time, as would be expected for a kinetically controlled process. The basic requirement for this attempt is the solubility of the oxidation products.

By taking advantage of this strategy, Penner and coworkers [41,42,47,48] were able to fabricate a series of noble metal and other nanowires (e.g., Au, Ag, Cu, Pd, Cd, Mo, Ni, etc). Moreover, semiconductor nanowires such as CdS can also be prepared using this method, but in this case a chemical step after the electrodeposition is necessary to form the semiconductor wires.

5.3.2
Nanowire-Based Gas Sensors

In order to make these nanowires as chemical sensors, the prepared nanostructures can be transferred onto a solid, non-conductive substrate and then contacted laterally using a conductive ink such as silver epoxy (Figure 5.2C). This simple approach has in fact been performed by Penner and colleagues, who fabricated the first hydrogen sensor based on palladium nanowires [42].

In contrast to direct electrodeposition into nanoporous materials, which produces continuous 1D nanostructures, nanowires produced by step edge decoration are narrowly dispersed in diameter and present a granular form. This granular form, however, is principally responsible for resistance changes when nanowires are applied as gas sensors. The investigations pioneered by Penner and coworkers [42] showed that, when an array of palladium nanowires is biased under a constant voltage and then exposed to a 5% H_2 concentration, the response time of the sensor is about 75 ms. Furthermore, these nanowire-based nanosensors were insensitive to gases such as O_2, CO, or CH_4, all of which are considered contaminants in conventional H_2-based palladium sensors. Hence, these authors showed that two of the main problems in palladium-based hydrogen sensors could be overcome by using these nanowires. Interestingly, the response mechanism of the sensor is in contrast to that of conventional hydrogen sensors, namely, that the resistance of palladium nanowires decreases temporarily in the presence of H_2. The reason for this is related to the above-mentioned granular form of the nanowires. Microscopy images reveal that the nanowires contain nanogaps in the absence of H_2; after a first exposure of these nanowires to H_2, palladium metal forms a thermodynamically stable β-phase, that is, $PdH_{0.7}$. Consequently, an expansion occurs of the face-centered-cubic (fcc) Pd lattice by 3.5% (at 25 °C with 1 atm. H_2). This expansion is accommodated by a 3.5% compression of each wire along its axis, which in turn decreases the intergranular resistance. Re-exposing the nanowires to air opens these gaps, which makes the palladium nanowires act as an active switcher (Figure 5.2D). The sensitivity of the sensor is approximately 0.5%, which is eightfold lower than the H_2 lower explosion limit in air, and a good level for the promising energy sources of future [48,49].

An interesting point is the monitoring of the natural sources of H_2 production, that is, hydrogenase enzymes are a class of metalloenzymes produced by archaebacteria that are able to produce, under optimum conditions, up to 10 mmol H_2 min^{-1} (per mg protein) [50,51]; thus, with the miniaturization of this sensor it would be possible to achieve a direct control of hydrogen production in the microorganisms. Other gases than hydrogen were also sensed using noble metal nanowires fabricated by ESED [52]. Finally, due to their granular form, noble metal nanowires electrodeposited in step edges may also have their surface changed by thiol compounds. Biological macromolecules may then be able to bind to the surface and be used as receptors to study their interaction with small drug molecules. As the latter are similar in size to the nanowire, they may sensitize the inelastic scattering of electron conduction by the charge effect at the surface of the sensor. A list of different nanowires and their possible applications as sensors is presented in Table 5.1.

An equally promising application of nanowires as nanoelectrodes was demonstrated by research groups at New York University and MIT, in the development of a nanowire that could be inserted into an artery and threaded up to the brain, and was capable of sending and receiving signals. Although such clinical use is at present premature, the future suggestion is that these nanoelectrodes might be used to deliver high-frequency electrical pulses to specific areas of the brain in patients suffering from neurological disorders, such as Parkinson's disease. The biocompatibility of the nanowire must be proven, however, before actual medical application can be considered. The fact that these nanosensors may transduce and

Table 5.1 Nanowires and their possible use as highly sensitive sensors.

Nanowire	Application
Au, Ag	Biosensors, ion contaminants
Pd	H_2 gas sensor
Bi, Fe, Ni, Co	Magnetic sensors
Al_2O_3, ZnO, SnO_2, In_2O_3	Air quality (detect CO, NH_3, NO, CH_4)
p-Si or n-Si	Biomacromolecules, molecular receptors

enhance biological events supports their potential for interconnecting between controllable nanowires or nanotubes and biological processing systems [101].

5.4
Atomic Metal Wires from Electrochemical Etching/Deposition

If the start of the 20th century was considered the "golden years" for quantum mechanics, then the start of the 21st century has been the golden era for nanoscale science and technology. In this sense, one interesting feature is the novel quantum phenomena exhibited when the materials are sufficiently small. For example, for a metallic wire consisting only of a string of a few atoms, the conductance (G) along the wire become quantized, and is given by the Landauer formula [53]:

$$G = \frac{2e^2}{h} T_n \qquad (3)$$

where e is the electron charge, h is Planck's constant, and T_n is a transmission-probability function. For an ideal wire, $T_n = 1$ and the term $2e^2/h = 77.4\,\mu S$, which is called the *quantum conductance*. In this case, the transport of electrons along the wires is ballistic; that is, the electron wave can travel through the wire without experiencing scattering that may affect its momentum or phase. However, what does the quantized conductance mean physically? Equation (3) describes an ideal model for a 1D wire where the electron motion is in the longitudinal direction of propagation of the wire – that is, the electron transport is confined to a single, much like a person walking on a narrow bridge. Unlike a macroscopic metallic wire, where electron conduction is involved in many elastic-scattering collisions so that the conductance is proportional to the cross-sectional area of the wire and inversely proportional to its length, in an atomically thin metal wire the electron mean free path is lower than the length of the wire, and consequently there is no collision (ballistic transport). According to the de Broglie relationship, for the electron conduction wavelength (λ_F), which is described by $\lambda_F = 2\pi/(2mE_F)^{1/2}$, where E_F is the Fermi energy (the energy of the conduction electrons), for a 1D wire of width W with $W = \lambda_F/2$, there is only one standing wave, namely, $N = 1$. In this case, as the width of the wire is compared to the λ_F, the conduction electrons in the transverse direction form well-defined quantum modes. Additionally, N is an integer that increases with wire width, making the conductance increase – in contrast to classical wires. This phenomenon, which was termed *conductance quantization*, is not only determined by the cross-section of the wire, but is also affected by the intrinsic properties of the material. Experimentally, when the conductance is reduced below $\sim 5 \times 10^{-4}$ S, it begins to decrease in discrete steps of $2e^2/h$; hence, the quantum modes can be strictly controlled [54–58].

However, the question was – how could such fascinating nanowires be produced? One simple method proved to be electrochemical in nature, and in this sense the first to produce electrochemically quantum metal wires was Tao's group [54–56,58,59].

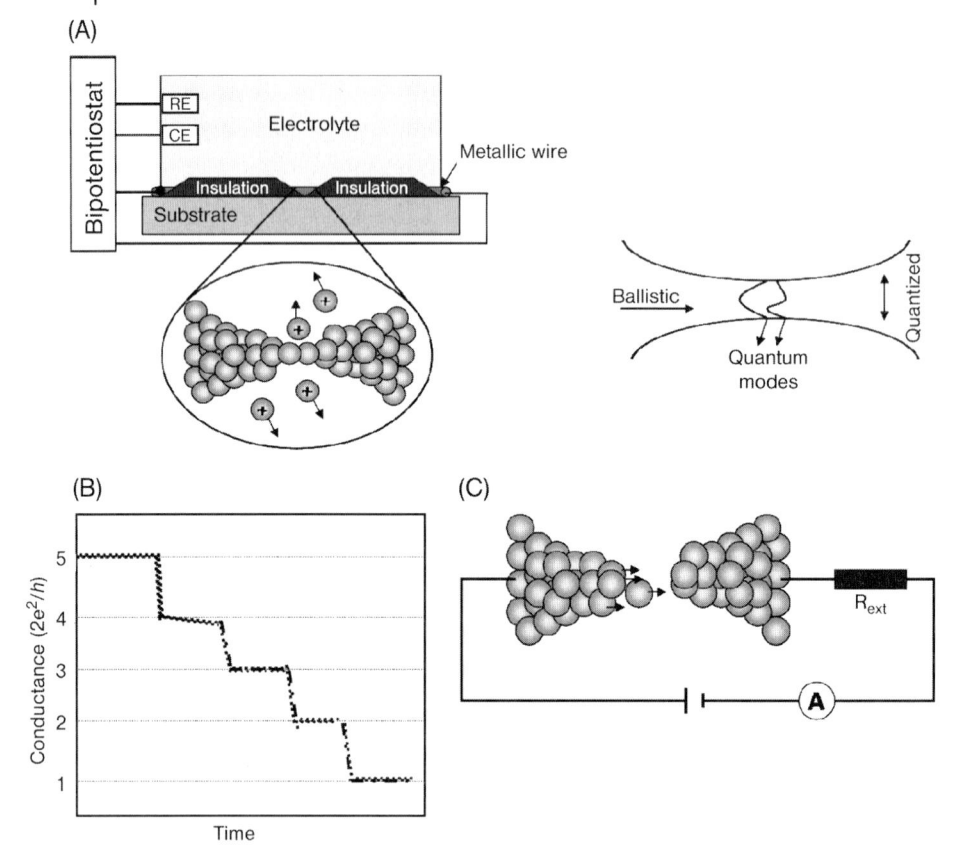

Figure 5.3 (A) Illustrative view of quantum metal wires production and quantization of conductance, where conducting electrons transport ballistically through the nanowire and form well-defined quantum modes in the transverse direction. (B) Quantum conductance versus time recorded for conductance change during the dissolution process of a noble metal. (C) Schematic of quantum metal wire production via the self-repair electrochemical method. (Adapted from Refs. [55,59].)

By using a thin metal wire which was almost totally covered by glue and supported on a solid insulating substrate, the research team exposed the uncovered portion of wire to an electrolyte solution for electrodeposition and etching. This exposed portion, which was less than a few micrometers in length, was reduced to the atomic scale, wherein the conductance became quantized. One advantage of this method was that the process could be reversed by electrodeposition. Moreover, in both cases, the conductance of the wire could be controlled *in situ* by using a bipotentiostat (Figure 5.3A and B).

Boussaad and Tao [59] also demonstrated another method based on a simultaneous etching/deposition mechanism that did not require a bipotentiostat or feedback control to fabricate atomic wires (Figure 5.3C). In this approach, a bias voltage was applied between two electrodes separated by a relatively large gap and immersed in an electrolyte solution. The atoms were etched off from the anode and deposited onto the cathode. Although the etching occurred over all anode surfaces, the deposition took place only on a single point of contact on the cathode. The process was stopped when a point contact was formed between the two electrodes, which were consecutively monitored by the conductance between them (in this case the ionic conductance was negligibly small). The method also involved a self-repair mechanism; that is, once a contact was formed between the two electrodes, after applying a small bias voltage between them for a few minutes, the contact was broken, most likely due to a metastable state of the structure. Consequently, the conductance fell back to zero. Hence, the voltage across the electrodes passed back to a maximum value and the etching and deposition process started over again.

Two essential differences between the conductance sensitivity of these quantum metallic wires and those produced, for example, by template methods are the inherent wavelength and mean free path of electrons. Metals such as Au, Ag and Cu have wavelengths of only a few Angstroms, and mean free paths of a few tens of nanometers, which allow conductance quantization be pronounced even at room temperature, as long as the diameter of the metal wires reach the size of an atom. In contrast, conventional nanowires with diameters of a few tens of nanometers are not expected to present marked conductance sensitivity. In fact, Tao and colleagues observed that the atomically thin metal wires cause the quantized conductance to become sensitive to the adsorption of molecules onto the nanowire [60,61]. Hence, due to this sensitivity, these materials can be applied as highly sensitive chemical sensors, although some limitations such as selectivity still need to be overcome.

5.4.1
Sensing Molecular Adsorption with Quantized Nanojunction

Direct experimental evidence of the quantized conductance sensitivity was reported by Tao and coworkers [54,55], who showed that the binding strengths of the molecules or ions onto the nanowire play a pivotal role in adsorbate-induced conductance changes. For a better understanding of the range of validity of the quantum wires model, and in order to establish how it can be successively applied to the sensors field, some basic principles based on the conductance change upon molecular adsorption are now introduced. A general concept arises from electron transport in which the scattering of the electron conduction determines the conductance change. In this way, the microscopic theory developed by Ishida created a model to study the conductance change in terms of adsorbate–substrate binding length and adsorbate–adsorbate distance [63]. Another valuable approach was made by Persson [64], who developed a simple relationship between the adsorbate-induced conductance change and the density of states of adsorbed molecules, $\rho_a(E_F)$, at the Fermi energy of the conductor. In Persson's model the conductance change per adsorbate molecule is described by

$$\frac{\Delta G}{G} \alpha - \frac{\Gamma \rho_a(E_F)}{d} \tag{4}$$

where d is the thickness of the metal film and Γ is the width of $\rho_a(E)$. Because each adsorbate molecule has a different $\rho_a(E_F)$, the conductance change should be specific for each adsorbate. Indeed, this has been confirmed for classical conductors [36,38]. As can be seen in Eq.(3), in terms of sensor applications, thinner films mean higher sensitivity as the conductance change is inversely proportional to the thickness of the film.

Based on these arguments, Persson's theory opens a window to a new sensors configuration based on dynamic measurements of each molecule. However, the model still describes electron transport in a diffusive way, in contrast to the ballistic regime of quantum wires, for which a theory has not yet been developed. Simulations show that the adsorption of a methyl thiol onto a gold nanowire made of a chain of four atoms, yields a strong bond, making the molecule part of the wire. Interestingly, the interaction distance within these nanowires is larger than in bulk metals, and it tends to be larger upon molecular adsorption. Scanning tunneling microscopy (STM) images of a Au (111) electrode surface has verified these observations [65]. Using molecules with different binding strengths, Tao et al. studied the effects on conductance change of Cu quantum wires under additions of dopamine, 2,2'-bipyridine (22BPy) and mercaptopropionic acid (MPA), the binding strengths of which are ~ 0.6, ~ 8.4 and $\sim 44\,\mathrm{kcal\,mol^{-1}}$, respectively. Their findings indicate that, as the quantum modes of the wire are decreased, the conductance sensitivity increases. In fact, upon addition of molecules quantum wires with a single quantum mode showed a marked decrease in conductance [60].

Dopamine   2,2-bipyridine   3-mercaptopropionic acid

A worthy observation is that the quantized conductance falls to near half. However, it is important to note that the conductance decreases even without adding analytes, but the molecular adsorption often increases their occurrence. As expected, the magnitude of conductance decreases with the increase of binding strength of the molecules. In the case of MPA, the conductance usually falls to zero due to the breakdown of the nanowire in response to the effects mentioned above. The attributions to this conductance change are mainly due to the scattering of electron conduction in quantum wires by the adsorbates and rearrangement in the atomic configuration of metallic wires.

Tao's group also evaluated the conductance change upon addition of charged species. In doing so, anion adsorption was studied using the following electrolytes: NaF, NaCl, NaBr, and NaI in a set of potentials. Unsurprisingly, no adsorption took place when the quantum metallic wires were biased negatively. On the other hand, upon increasing the potential, the conductance increases for all electrolytes. The magnitude of change varies in the order of $F^- < Cl^- < Br^- < I^-$, which is in good agreement with the adsorption strengths of these anions. Again, an electron-scattering mechanism is suggested as being responsible for the conductance change [55], based on this higher sensitivity. Quantum metal wires present a powerful background to sense small species and macrobiological molecules. For example, as proteins are normally charged in aqueous solutions, these nanojunctions have the potential to detect only a single molecule, and even a single or a few small ions may be detected using this 1D nanostructure. The selectivity of the sensor may be further improved by adsorbing specific receptors onto the nanowires, as has been reported earlier [66].

Finally, it is of interest to note that theoretical calculations showed that the sensitivity of these nanowire nanosensors is more heavily dependent on their length rather than on their radius [67]. This reinforces those methods that are capable of producing high aspect-ratio nanowires.

5.5
Future Prospects and Promising Technologies

Although advances in electrochemically synthesized sensors based on 1D nanostructures are still mainly limited to R&D investigations, many companies are working arduously to develop the three major paradigms of nanoscale science, namely the growth, assembly, and integration of these nanomaterials into a functional network array. The high surface area of the nanowires and nanotubes places them as a powerful tool to sense genomics, proteomics, molecular diagnostics and high-throughput screening. Currently, the Palo Alto's company, Nanosys, is close to commercializing its first product to use the large surface area of a dense array of nanowires to increase sensitivity in diagnostics. The product enables a faster analysis of small molecules in complex samples of bio-fluids, without any need for sample preparation, with efficient drug development, or rapid drug detection directly from urine [112]. The Illuminex Corporation is also producing nanowires using electrochemical processes that could be cheaper and more scalable than vapor deposition, and can conveniently leave the vertically aligned wires with one end attached to the substrate, as grown, for its application. Currently, Illuminex is working on a sensor for detecting ovarian cancer markers and other bio-molecules indicative of disease. Illuminex grows electrochemically metallic nanowires directly on the tip of an optical fiber probe, and attaches antibodies and nucleic acids to the wires to bind to target proteins or genes. Binding of the molecules changes the highly light absorbent nanowires's optical properties, so the target material can be identified by optical spectroscopy. This nanobiosensor has applications in the medical diagnostics market which, in 2005, was valued at US$28.5 billion [113]. The multi-plurality of nanowires has been also exploited in the bio-assembled field. For example, Cambrios Technologies Corporation is using biology to connect electronics, whereby a transparent nanowire conductor film is being tested at display makers as a replacement for tin-doped indium oxide (ITO) [114]. Companies currently making progress on the development of a wide range of applications with 1D and other nanomaterials include: Nanosys; Illuminex; Cambrios; Nanosens; Nano Clusters Devices Ltd.; NanoDynamics; Nanomix; Enable IPC; Atomate Corporation; and QuNano.

In the Netherlands, startup Nanosens [115] is gathering together a large variety of nanowires with silicon-on-insulator (SOI) technology. The process etches out 200 to 500 parallel wires with diameters of 5 to 50 nm and lengths ranging from 1 to 1000 µm on its chips. These sensor arrays are considered promising technologies to be used from hydrogen and environmental gas sensors to cancer diagnosis. Furthermore, the scalable semiconductor processes bring production costs down to less than US$1 for a chip with an array of 1000 individually addressed Au nanowires.

In fact, the integration of nanowires into miniaturized systems of SOI technology enables these 1D structures to be configured as electronic devices that can electrically detect pH, DNA sequences, protein markers of cancer and other diseases, and even a single virus molecule [83]. Charles Lieber's group at Harvard University has produced highly sensitive biosensors based on field effect transistors (FETs) by assembling p- and n-silicon semiconductor nanowires as planar gate electrodes. These FETs exhibit electrical performance characteristics comparable to, or better than, those achieved in the microelectronics industry for planar silicon devices. The binding of biomolecules to the surface of the nanowire leads to the depletion or accumulation of carriers in the bulk of the nanometer-diameter structure, versus only the surface region of a planar device. This unique feature of semiconductor nanowires provides sufficient sensitivity to enable the detection of single viruses and of single molecules in solution [79–83]. The underlying conception to detection using these nanowires is based on the classical electrical behavior of FETs, which exhibit a conductivity change in response to variations in the field or potential at their surface. Thus, in the case of a p-silicon, applying a positive gate voltage depletes carriers and reduces the conductance, whereas applying a negative gate voltage leads to an accumulation of carriers and increases the conductance. In this sense, the binding of a charged species to the gate dielectric is analogous to the effect of applying a voltage with a gate electrode. Historically, this is a contemporaneous configuration of sensors inspired in FETs, as the original idea was introduced several decades ago by Bergveld [2] and his ion-sensitive field effect transistor (ISFET) for which sensitivity was limited, thereby precluding any major impact that had been expected in the biomedical field. Lieber's group also showed that these nanowires can act in a non-invasive manner to detect, stimulate, and inhibit neuronal signals with a good spatial and temporal resolution [84]. In the same way, single- and multi-walled CNTs have also been configured as planar FETs to explore, with high sensitivity, the detection of gases [85,86] and biomolecules such as antibodies [87,88]. Valcke *et al.* [89] showed that the label-free detection of DNA hybridization and the discrimination of single-nucleotide polymorphism was possible using this configuration. The transport properties of DNA molecules were also studied by assembling CNTs as FETs [90], and some reports showing the detection of humidity, glucose, cytochrome c and other proteins are now available [91–95]. One practical limitation for the *in-vivo* use of these sensor-based FETs is that the detection sensitivity depends on solution ionic strength. As blood serum samples have a high ionic strength, any diagnostic procedure will require a previous desalting step to be used before the analysis in order to achieve the highest sensitivity.

In a word, the major challenge for the advance of nanowires and nanotubes technology depends on a major integration between leading research groups and companies worldwide. It is also inevitable that people from different sectors of industry, and with widely differing backgrounds, will form the driving force behind these technologies.

5.6
Concluding Remarks

In this chapter a host of examples has been presented to indicate how science and technology, at the nanoscale, have promoted progress during the past few years, and how such advance has stimulated mankind's creativity toward the production and development of new and highly interesting materials. In this new hybrid field of nanosize, attention was focused on an interplay between 1D materials and their potential to act as highly sensitive sensors. The multifunctionality of nanowires and nanotubes has been discussed in a panoramic manner, in which their intrinsic properties are responsible for their applications. Such examples may point towards important, unsolved problems, where the exploitation of a new physics or chemistry is still necessary.

The relatively small number of atoms in an atomically thin metal wire obtained by the dissolution of thick metallic wires is attractive to understanding not only the physical phenomenon on the nanometer scale, but also to construct models capable of predicting quantum-chemical interactions with small and large molecules [116–118]. For example, in the framework of a simplified interaction between a monovalent ion and a string composed of a few atoms, it was corroborated experimentally that the charge dispersion of the ion is pivotal in the conductance change of the wire. In addition, nanowires obtained by step edge approaches may function as active switches when exposed to certain gases, thereby demonstrating their vulnerability to the environment and associated conditions such as temperature and pressure which, from a practical viewpoint, shed light on the applications and interpretations of the phenomena involved.

In summary, dimensionality investigations have shown that characteristic features of the systems studied are described as functions of their size due to the different ways in which electrons can interact in macrosystems. In other words, the properties of 1D structures can be completely changed by their interaction with a few molecules. In this respect, nanowire-based nanosensors prepared by template methods represent an important achievement [119]. These complementary approaches are essential in order to study the detection of only a single molecule, since at this stage both the molecules and the nanowires share a similar size.

References

1 Wise, K.D., Angell, J.B. and Starr, A. (1970) *IEEE Trans. Bio-Med. Eng.*, **17**, 238.
2 Bergveld, P. (1970) *IEEE Trans. Bio-Med. Eng.*, **17**, 70.
3 Bergveld, P. (1972) *IEEE Trans. Bio-Med. Eng.*, **19**, 342.
4 Penner, R.M. and Martin, C.R. (1987) *Anal. Chem.*, **59**, 2625.
5 Menon, V.P. and Martin, C.R. (1995) *Anal. Chem.*, **67**, 1920.
6 Kobayashi, Y. and Martin, C.R. (1999) *Anal. Chem.*, **71**, 3665.
7 Choi, J., Sauer, G., Nielsch, K., Wehrspohn, R.B. and Gösele, U. (2003) *Chem. Mater.*, **15**, 776.
8 Martin, C.R. (1994) *Science*, **266**, 1961.
9 Xia, Y., Yang, P., Sun, Y., Wu, Y., Mayers, B., Gates, B., Yin, Y., Kim, F. and Yan, H. (2003) *Adv. Mater.*, **15**, 353.
10 Steinhart, M., Wehrspohn, R.B., Gösele, U. and Wendorff, J.H. (2004) *Angew. Chem. Int. Ed.*, **43**, 1334.
11 Coffer, J.L., Bigham, S.R., Li, X., Pinizzotto, R.F., Rho, Y.G., Pirtle, R.M. and Pirtle, I.L. (1996) *Appl. Phys. Lett.*, **69**, 3851.
12 Braun, E., Eichen, Y., Sivan, U. and Ben-Yoseph, G. (1998) *Nature*, **391**, 775.
13 Zhang, Y. and Dai, H.J. (2000) *Appl. Phys. Lett.*, **77**, 3015.
14 Liang, W. and Martin, C.R. (1990) *J. Am. Chem. Soc.*, **112**, 9666.
15 De Vito, S. and Martin, C.R. (1998) *Chem. Mater.*, **10**, 1738.
16 Marinakos, S.M., Brousseau, L.C., Jones, A. and Feldheim, D.L. (1998) *Chem. Mater.*, **10**, 1214.
17 Nishizawa, M., Mukai, K., Kuwabata, S., Martin, C.R. and Honeyama, H. (1997) *J. Electrochem. Soc.*, **144**, 1923.
18 Sakaguchi, H., Matsumura, H. and Gong, H. (2004) *Nature*, **3**, 551.
19 Kapon, E., Kash, K., Clausen, E.M., Jr., Hwang, D.M. and Colas, E. (1992) *Phys. Phys. Lett.*, **60**, 477.
20 Brumlik, C.J. and Martin, C.R. (1991) *J. Am. Chem. Soc.*, **113**, 3174.
21 Brumlik, C.J., Menon, V.P. and Martin, C.R. (1994) *J. Mater. Res.*, **9**, 1174.
22 Hulteen, J.C. and Martin, C.R. (1997) *J. Mater. Res.*, **7**, 1075.
23 Tang, B.Z. and Xu, H. (1999) *Macromolecules*, **32**, 2569.
24 Whitney, T.M., Jiang, J.S., Searson, P.C. and Chien, C.L. (1993) *Science*, **261**, 1316.
25 Ajayan, P.M., Stephan, O. and Redlich, P. (1995) *Nature*, **375**, 564.
26 Wu, C.G. and Bein, T. (1994) *Science*, **266**, 1013.
27 Keller, F., Hunter, M.S. and Robinson, D.L. (1953) *J. Electrochem. Soc.*, **100**, 411.
28 Diggle, J.W., Downie, T.C. and Goulding, C.W. (1969) *Chem. Rev.*, **69**, 365.
29 Masuda, H. and Fukuda, K. (1995) *Science*, **268**, 466.
30 Li, A.P., Müller, F., Birner, A., Nielsch, K. and Gösele, U. (1998) *J. Appl. Phys.*, **84**, 6023.
31 Wirtz, M. and Martin, C.R. (2003) *Adv. Mater.*, **15**, 455.
32 Lee, W., Scholz, R., Nielsch, K. and Gösele, U. (2005) *Angew. Chem.*, **117**, 6204.
33 Ahn, H.J., Sohn, J.I., Kim, Y.S., Shim, H.S., Kim, W.B. and Seong, T.Y. (2006) *Electrochem. Commun.*, **8**, 513.
34 Basu, M., Seggerson, S., Henshan, J., Jiang, J., Cordona, R.A., Lefave, C., Boyle, P.J., Miller, A., Pugia, M. and Basu, S. (2004) *Glycoconjugate J.*, **21**, 487.
35 Brunetti, B., Ugo, P., Moretto, L.M. and Martin, C.R. (2000) *J. Electroanal. Chem.*, **491**, 166.
36 Amatore, C., Savéant, J.M. and Tessier, D. (1983) *J. Electroanal. Chem.*, **147**, 39.
37 Martin, C.R., Nishizawa, M., Jirage, K. and Kang, M. (2001) *J. Phys. Chem. B*, **105**, 1925.
38 Siwy, Z., Trofin, L., Kohli, P., Baker, L.A., Trautmann, C. and Martin, C.R. (2005) *J. Am. Chem. Soc.*, **127**, 5000.

39. Kriz, K., Kraft, L., Krook, M. and Kriz, D. (2002) *J. Agric. Food Chem.*, **50**, 3419.
40. Weber, J., Kumar, A., Kumar, A. and Bhansali, S. (2006) *Sens. Actuators B*, **117**, 308.
41. Walter, E.C., Murray, B.J., Favier, F., Kaltenpoth, G., Grunze, M. and Penner, R.M. (2002) *J. Phys. Chem. B*, **106**, 11407.
42. Walter, E.C., Ng, K., Zach, M.P., Penner, R.M. and Favier, F. (2002) *Microelectron. Eng.*, **61**, 555.
43. Brune, H. (1998) *Surf. Sci. Rep.*, **31**, 121.
44. Debenedetti, P.G. (1996) *Metastable Liquids: Concepts and Principles*, Princeton University Press, Princeton, NJ.
45. Bockris, J.O.M. and Reddy, A.K.N. (1970) *Modern Electrochemistry*, Plenum Press, New York.
46. Schmickler, W. (1996) *Interfacial Electrochemistry*, Oxford University Press, New York.
47. Thompson, M.A., Menke, E.J., Martens, C.C. and Penner, R.M. (2006) *J. Phys. Chem. B*, **110**, 36.
48. Walter, E.C., Favier, F. and Penner, R.M. (2002) *Anal. Chem.*, **74**, 1546.
49. Walter, E.C., Penner, R.M., Liu, H., Ng, K.H., Zach, M.P. and Favier, F. (2002) *Surf. Interface Anal.*, **34**, 409.
50. Darensbourg, M.Y., Lyon, E.J. and Smee, J.J. (2000) *Coord. Chem. Rev.*, **206**, 533.
51. Artero, V. and Fontecave, M. (2005) *Coord. Chem. Rev.*, **249**, 1518.
52. Murray, B.J., Walter, E.C. and Penner, R.M. (2004) *Nano Lett.*, **4**, 665.
53. Landauer, R. (1957) *IBM J. Res. Dev.*, **1**, 223.
54. He, H.X., Boussaad, S., Xu, B.Q., Li, C.Z. and Tao, N.J. (2002) *J. Electroanal. Chem.*, **522**, 167.
55. Xu, B., He, H. and Tao, N.J. (2002) *J. Am. Chem. Soc.*, **124**, 13568.
56. Li, C.Z., Bogozi, A., Huang, W. and Tao, N.J. (1999) *Nanotechnology*, **10**, 221.
57. Snow, E.S., Park, D. and Campbell, P.M. (1996) *Appl. Phys. Lett.*, **69**, 269.
58. Li, C.Z. and Tao, N.J. (1998) *Appl. Phys. Lett.*, **72**, 894.
59. Boussaad, S. and Tao, N.J. (2002) *Appl. Phys. Lett.*, **80**, 2398.
60. Bogozi, A., Lam, O., He, H., Li, C., Tao, N.J., Nagahara, L.A., Amlani, I. and Tsui, R. (2001) *J. Am. Chem. Soc.*, **123**, 4585.
61. Li, C.Z., He, H.X., Bogozi, A., Bunch, J.S. and Tao, N.J. (2000) *Appl. Phys. Lett.*, **76**, 1333.
62. Li, J., Kanzaki, T., Murakoshi, K. and Nakato, Y. (2002) *Appl. Phys. Lett.*, **81**, 123.
63. Ishida, H. (1995) *Phys. Rev. B*, **52**, 10819.
64. Persson, B.N.J. (1993) *J. Chem. Phys.*, **98**, 1659.
65. Kim, Y.T., McCarley, R.L. and Bard, A.J. (1992) *J. Phys. Chem.*, **96**, 7416.
66. Forzani, E.S., Zhang, H., Nagahara, L.A., Amlani, I., Tsui, R. and Tao, N.J. (2004) *Nano Lett.*, **4**, 1785.
67. Sheehan, P.E. and Whitman, L.J. (2005) *Nano Lett.*, **5**, 803.
68. Masuda, H. et al. (2006) *Jpn. J. Appl. Phys.*, **45**, L406–L408.
69. Shingubara, S. (2003) *J. Nanoparticle Res.*, **5**, 17.
70. Li, A.-P., Müller, F., Birner, A., Nielsch, K. and Gösele, U. (1999) *Adv. Mater.*, **11**, 483.
71. Delvaux, M., Champagne, S.D. and Walcarius, A. (2004) *Electroanalysis*, **16**, 190.
72. Delvaux, M. and Champagne, S.D. (2003) *Biosens. Bioelectron.*, **18**, 943.
73. Jirage, K.B., Hulteen, J.C. and Martin, C.R. (1997) *Science*, **278**, 655.
74. Martin, C.R. (1996) *Chem. Mater.*, **8**, 1739.
75. Dekker, C. (2007) *Nat. Nanotech.*, **2**, 209.
76. Keyser, U.F., Krapf, D., Koeleman, B.N., Smeets, R.M.M., Dekker, N.H. and Dekker, C. (2005) *Nano Lett.*, **5**, 2253.
77. Lagerqvist, J., Zwolak, M. and Di Ventra, M. (2006) *Nano Lett.*, **6**, 779.
78. Zhang, X.-G., Krstié, P.S., Zikié, R., Wells, J.C. and Cabrera, M.F. (2006) *Biophys. J.*, **91**, L04.
79. Cui, Y., Wei, Q., Park, H. and Lieber, C.M. (2001) *Science*, **293**, 1289.
80. Zheng, G., Patolsky, F., Cui, Y., Wang, W.U. and Lieber, C.M. (2005) *Nat. Biotech.*, **23**, 1294.

81. Wang, W.U., Chen, C., Lin, K.-H., Fang, Y. and Lieber, C.M. (2005) *Proc. Natl. Acad. Sci. USA*, **102**, 3208.
82. Hahm, J. and Lieber, C.M. (2004) *Nano Lett.*, **4**, 51.
83. Patolsky, F., Zheng, G., Hayden, O., Lakadamyali, M., Zhuang, X.W. and Lieber, C.M. (2004) *Proc. Natl. Acad. Sci. USA*, **101**, 14017.
84. Patolsky, F., Timko, B.P., Yu, G., Fang, Y., Greytak, A.B., Zheng, G. and Lieber, C.M. (2006) *Science*, **313**, 1100.
85. Kong, J., Franklin, N.R., Zhou, C., Chapline, M., Peng, S., Cho, K. and Dai, H. (2000) *Science*, **287**, 622.
86. Collins, P.G., Bradley, K., Ishigami, M. and Zettl, A. (2000) *Science*, **287**, 1801.
87. Chen, R.J., Bangsaruntip, S., Drouvalakis, K.A., Wong, N., Kam, S.H., Shim, M., Li, Y., Kim, W., Utz, P.J. and Dai, H. (2003) *Proc. Natl. Acad. Sci. USA*, **100**, 4984.
88. Star, A., Gabriel, J.-C.P., Bradley, K. and Grüner, G. (2003) *Nano Lett.*, **3**, 459.
89. Star, A., Eugene, T., Niemann, J., Gabriel, J.-C.P., Joiner, C.S. and Valcke, C. (2006) *Proc. Natl. Acad. Sci. USA*, **103**, 921.
90. Sasaki, T.K., Ikegami, A., Mochizuki, M., Aoki, N. and Ochiai, Y. (2004) Proceedings 2nd Quantum Transport Nano-Hana International Workshop, Conference Series, **5**, 97.
91. Star, A., Han, T.-R., Joshi, V. and Stetter, J.R. (2004) *Electroanalysis*, **16**, 108.
92. Besteman, K., Lee, J.-O., Wiertz, F.G.M., Heering, H.A. and Dekker, C. (2003) *Nano Lett.*, **3**, 727.
93. Sotiropoulou, S. and Chaniotakis, N.A. (2003) *Anal. Bioanal. Chem.*, **375**, 103.
94. Boussaad, S., Tao, N.J., Zhang, R., Hopson, T. and Nagahara, L.A. (2003) *Chem. Commun.*, **13**, 1502.
95. Chen, R.J., Choi, H.C., Bangsaruntip, S., Yenilmez, E., Tang, X., Wang, Q., Chang, Y.-L. and Dai, H. (2004) *J. Am. Chem. Soc.*, **126**, 1563.
96. He, H. and Tao, N.J. (2003) in *Encyclopedia of Nanoscience and Nanotechnology* (ed. N.S. Nalwa), American Scientific Publishers, p. 1.
97. Nicewarner-Pena, S.R., Freeman, R.G., Reiss, B.D., He, L., Pena, D.J., Walton, I.D., Cromer, R., Keating, C.D. and Natan, M.J. (2001) *Science*, **294**, 137.
98. Forrer, P., Schlottig, F., Siegenthaler, H. and Textor, M. (2000) *J. Appl. Electrochem.*, **30**, 533.
99. Heng, J.B., Ho, C., Kim, T., Timp, R., Aksimentiev, A., Grinkova, Y.V., Sligar, S., Schulten, K. and Timp, G. (2004) *Biophys. J.*, **87**, 2904.
100. Krapf, D., Quinn, B.M., Wu, M.-Y., Zandbergen, H.W., Deeker, C. and Lemay, S.G. (2006) *Nano Lett.*, **6**, 2531.
101. Jones, W.D. (2005) *IEEE Spectrum*, **42**, 20.
102. De Leo, M., Kuhn, A. and Ugo, P. (2007) *Electroanalysis*, **19**, 227.
103. Liu, L., Jia, N.Q., Zhou, Q., Yan, M.M. and Jiang, Z.Y. (2007) *Mater. Sci. Eng. C*, **27**, 57.
104. Cumming, D.R.S., Bates, A.D., Callen, B.P., Cooper, J.M., Cosstick, R., Geary, C., Glidle, A., Jaeger, L., Pearson, J.L., Proupin-Perez, M. and Xu, C. (2006) *J. Vac. Sci. Technol. B: Microelectron. Nanometer Struct.-Process., Meas., Phenom.*, **24**, 3196.
105. Hassel, A.W., Smith, A.J. and Milenkovic, S. (2006) *Electrochim. Acta*, **52**, 1799.
106. Wang, Y.R., Hu, P., Chen, L.X., Liang, Q.L., Luo, G. and Wang, Y.M. (2006) *Chin. J. Anal. Chem.*, **34**, 1348.
107. Wang, C.Y., Shao, X.Q., Liu, Q.X., Mao, Y.D., Yang, G.J., Xue, H.G. and Hu, X.Y. (2006) *Electrochim. Acta*, **52**, 704.
108. Zhu, X.S. and Ahn, C.H. (2006) *IEEE Sens. J.*, **6**, 1280.
109. Pereira, F.C., Bergamo, E.P., Zanoni, M.V.B., Moretto, L.M. and Ugo, P. (2006) *Quim. Nova*, **29**, 1054.
110. Sandison, M.E. and Cooper, J.M. (2006) *Lab Chip*, **6**, 1020.
111. Hermann, M., Singh, R.S. and Singh, V.P. (2006) *Pramana-J. Phys.*, **67**, 93.
112. www.nanosysinc.com.
113. www.illuminex.biz.
114. www.cambrios.com.
115. www.nanosens.nl.

116 He, H.X., Boussaad, S., Xu, B.Q. and Tao, N.J. (2003) *Properties and Devices* (ed. Z.L. Wang), Kluwer Academic Press.
117 He, H.X., Li, C.Z. and Tao, N.J. (2001) *Appl. Phys. Lett.*, **78**, 811.
118 Nguyen, L. and Tao, N.J. (2006) *Appl. Phys. Lett.*, **88**, 043901/1.
119 Lee, W., Ji, R., Gösele, U. and Nielsch, K. (2006) *Nat. Mater.*, **5**, 741.

6
Self-Organized Formation of Layered Nanostructures by Oscillatory Electrodeposition
Shuji Nakanishi

6.1
Introduction

6.1.1
Self-Organized Formation of Ordered Nanostructures

Solid surfaces with ordered nanostructures, such as layers, dots, holes, and grooves (or ridges), have provided unique microelectronic, optical, magnetic, and micromechanical properties, as summarized in a recent review [1]. Focused photon-, electron-, ion-, and molecular-beam lithography (top-down method) have been widely used for creating desired nanostructures at the surface. However, these techniques are now facing serious problems, such as difficulties in mass production and cost increases due to the use of expensive, specialized apparatus. In the hope of overcoming those problems associated with the conventional techniques, a self-organization method (bottom-up method) has recently attracted much attention.

In general, self-organization methods in chemistry can be categorized into two different types. One type is self-organization under thermodynamically equilibrium conditions, in which the ordered structures are formed on the basis of specific properties of intermolecular forces. Self-assembled structures, such as lipid-bilayers, close-packed layers of nanospheres, and monolayers of thiol molecules on gold surfaces, are the representative examples. Many studies have been performed on this type of self-organization, as summarized in reviews [2–10]. At present, the main issues to be tackled for these methods are to improve their regularity and to be able to place nanostructures of desired size at desired locations. In the other type of self-organization, the ordered structures are formed under thermodynamically non-equilibrium conditions. A wide variety of dynamic spatiotemporal orders, such as oscillations and spatiotemporal patterns, appear in a self-organization manner [10–13].

The spatiotemporal patterns in the dynamic self-organization phenomena have some unique and attractive properties for producing materials of ordered structures:

- The patterns appear spontaneously, without any external control.
- The observed patterns have a long-range order.
- Various ordered patterns are obtained simply by changing experimental parameters.

In fact, several studies have been performed on the application of dynamic self-organization for the formation of ordered structures. For example, the self-organized formation of 2-dimensional (2D) ordered patterns on nano- and micro-meter scales have been achieved by use of Risegang-ring [14], etching [15,16], and dewetting processes [17,18]. These ordered 2D patterns on solid surfaces have already been used in practical applications as templates for constructing three-dimensional (3D) structures and substrates for the incubation of biological cells.

6.1.2
Dynamic Self-Organization in Electrochemical Reactions

Electrochemical reactions show a variety of dynamic self-organization phenomena, because the electrochemical reactions themselves proceed under non-equilibrium conditions. An autocatalytic process, which is the key factor for the appearance of a spatiotemporal structure, is easily achieved by a combination of *chemical* and *electric* mechanisms. In fact, various oscillations and spatiotemporal patterns have been reported to date in all types of electrochemical reaction [19–130], as summarized in Table 6.1.

Electrochemical systems have certain major strong advantages for studies of dynamic self-organization phenomena, as compared to other systems:

- The Gibbs energies for reactions can be regulated continuously and reversibly by tuning of the electrode potential.
- The oscillations can be directly observed in electric signals such as current or potential.

Table 6.1 Representative examples of electrochemical oscillations reported to date.

Electrocatalytic reactions		Anodic dissolution		Cathodic deposition	
Reaction	Reference(s)	Anode	Reference(s)	Cathode	Reference(s)
Oxidation of H_2	[19–23]	Cu	[53–62]	Zn	[88–93]
Oxidation of HCHO	[24,25]	Fe	[63–71]	Sn	[94–98]
Oxidation of HCOOH	[26–32]	Co	[72–75]	Au	[99,100]
Oxidation of CO	[33,34]	Ni	[76–78]	Pb	[101]
Oxidation of S^{2-}	[35,36]	Si	[79–81]	Cu	[102–106]
Oxidation of thiourea	[37]	Ag	[82]	AgSb–alloy	[107–114]
Reduction of In^{3+}	[38,39]	Nb	[83,84]	Cu/Cu_2O	[115–125]
Reduction of H_2O_2	[40–47]	Al	[85]	SnCu–alloy	[126,127]
Reduction of $S_2O_8^{2-}$	[48–50]	InP	[86]	NiP–alloy	[129,130]
Reduction of IO_3^-	[51,52]	Ti	[87]	–	–

- The diffusion process can be flexibly controlled by changing the sizes and geometric arrangements of the electrodes in electrochemical cells.
- The modes, period, and amplitude of the spatiotemporal patterns can be tuned easily by changing the geometric arrangements of the electrodes and the applied potential or current.

Figure 6.1 shows a typical example of the spatiotemporal patterns observed during H_2-oxidation reactions [131]. Recent theoretical studies explained rather elegantly the origin of the dynamic patterns and the patterns selection principle in electrochemical systems [131–133]. From the point of view of structure formation, it must be borne in mind that dynamic self-organized patterns can only be sustained while a flow of energy is present under non-equilibrium conditions, as they disappear immediately the flow of energy stops. For example, the beautiful spatiotemporal patterns shown in Figure 6.1 disappear instantly when the power of the electrochemical apparatus is turned off. Thus, a novel strategy is required if the aim is to produce certain ordered structures at solid (electrode) surfaces by use of the spatiotemporal structures in electrochemical systems.

Oscillatory electrodeposition can be used to solve this problem, because histories of ever-changing self-organized spatiotemporal patterns are recorded as architectures of electrodeposits at the electrode (solid) surface. Thus, ordered structures are constructed on solid substrates in bottom-up style. Encouragingly, the present author has focused on the merit of the oscillatory electrodeposition and developed a number of investigations on this subject. With regards to the oscillatory electrodeposition and accompanied formation of layered deposits, various

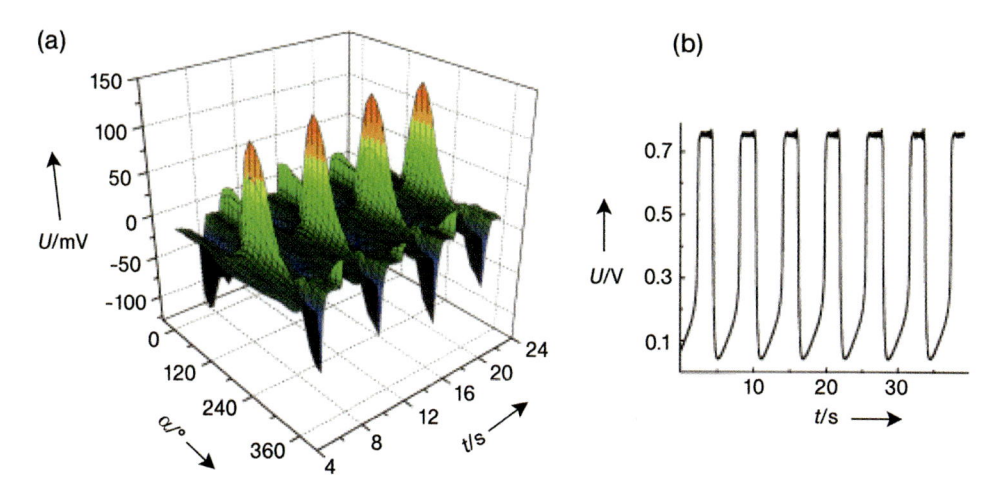

Figure 6.1 (a) Spatiotemporal evolution of the potential at a Pt electrode during the oxidation of H_2 in the presence of Cu^{2+} and Cl^- ions. (b) Temporal evolution of the total current. The first four oscillations correspond to the time interval shown in (a). (Reprinted from Ref. [131].)

studies have reported previously, though only from a phenomenological point of view, and without any detailed understanding of the self-organization mechanisms. Schlitte et al. were the first to report on the formation of a layered architecture via the oscillatory electrodeposition of Cu (Figure 6.2a) [104]. More recently, Krastev et al. reported that the co-deposition of Sb and Ag caused a potential oscillation to yield layered deposits, with dynamic target patterns appearing at the

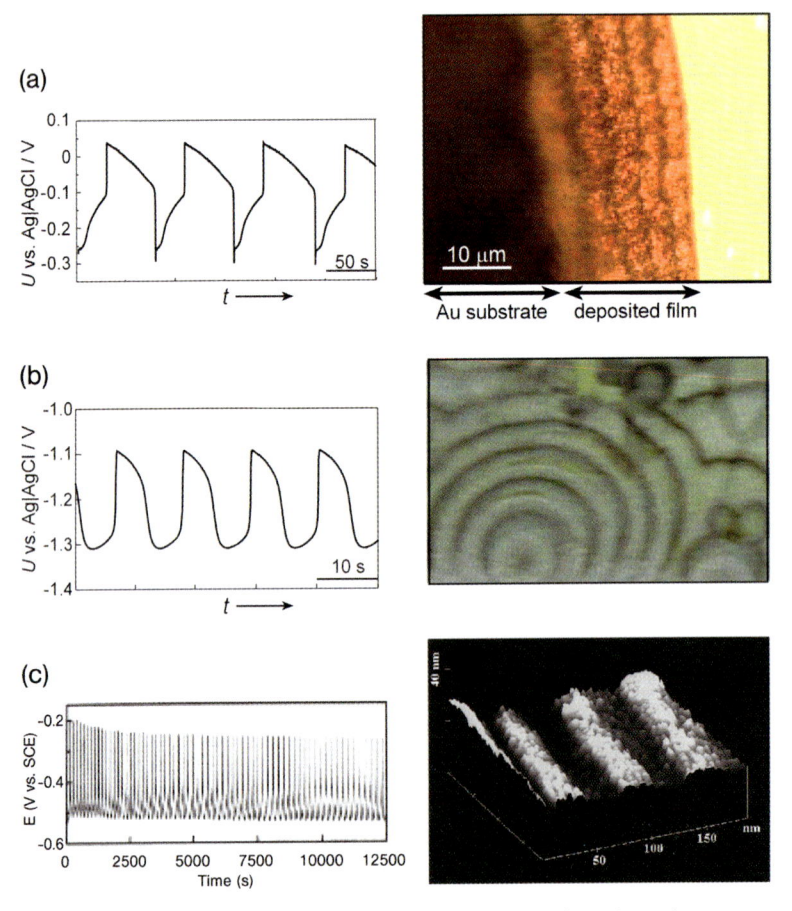

Figure 6.2 (a) Left: Potential oscillation during the Cu electrodeposition from an aqueous acidic solution including o-phenanthroline. Right: A cross-sectional view of a Cu deposit obtained during four periods of the potential oscillation. (b) Left: Potential oscillation obtained during the electrodeposition of Ag–In alloy. Right: A top view of the spatiotemporal pattern observed at the surface of the electrodeposit. (c) Left: Potential oscillation observed during the deposition of Cu/Cu$_2$O layered nanostructures. Right: Cross-sectional STM image of the film grown under the oscillatory condition. (a) Right: Reprinted from Ref. [105]; (c) Left: Reprinted from Ref. [118]; (c) Right: Reprinted from Ref. [115].)

surface of deposits during the oscillation (Figure 6.2b) [108]. Later, Switzer et al. reported that the electrodeposition of Cu in basic aqueous solutions caused a potential oscillation, producing alternate Cu and Cu_2O multilayers on a nanometer scale (Figure 6.2c) [115]. These results show clearly that the oscillatory electrodeposition is an attractive target for the self-organized formation of ordered structures.

6.1.3
The Important Role of Negative Differential Resistance (NDR) in Electrochemical Oscillations

For reference in later discussions, a brief explanation should first be provided of a general model for electrochemical oscillations, that was established during the last decade [134–136]. According to the literature, all reported electrochemical oscillations can be classified into four [136] (or five [137]) classes, depending on the roles of the true electrode potential (or the Helmholtz-layer potential, E).

Electrochemical oscillations in which the E plays no essential role and is kept essentially constant are called "strictly potentiostatic" type (or Class I). This type can be regarded as chemical oscillators that contain electrochemical reactions. Electrochemical oscillations in which the E is involved as the essential variable, but is still not the autocatalytic variable, are called S-NDR type (or Class II), because oscillations in this case arise from an S-shaped NDR (S-NDR) in the current density (j) versus E curve. When the E is the autocatalytic variable, the oscillations arise from an N-shaped NDR (N-NDR), and these are called N-NDR type. Furthermore, of the N-NDR oscillations, the oscillations in which the N-NDR is hidden by a current increase from another process are called hidden N-NDR or HN-NDR types (or Class IV). It is known that the N-NDR type shows only current oscillations, whereas the HN-NDR type shows both current and potential oscillations. The HN-NDR oscillators are further divided into three or four subcategories, depending on how the NDR is hidden.

As almost all electrochemical oscillations reported to date can be classified into N-NDR or HN-NDR types, the origin of NDR in electrochemical systems should be explained at this point. In general, the electrochemical reactions current, I_{reac}, can be expressed as

$$I_{reac} = nFAkC \tag{1}$$

where n is the number of transferred electrons, F is the Faraday constant, A is the available electrode area, k is the rate constant, and C is the surface concentration of the reacting species. The differential resistance, dI_{reac}/dE (E; the electrode potential), can thus be expressed as follows:

$$dI_{reac}/dE = nF(dA/dE)kC + nFA(dk/dE)C + nFAk(dC/dE) \tag{2}$$

From this equation, three origins for the NDR can be considered [138]:

- Type-I: $dA/dE < 0$: A case where the available electrode area decreases with increasing the electrode polarization. An example is the potential-dependent formation of a passivation layer at the electrode surface.
- Type-II: $dk/dE < 0$: A case where the electron transfer rate constant decreases with increasing the electrode polarization. An example is the potential-dependent desorption of an adsorbed species acting as a promoter for electrochemical reaction.
- Type-III: $dC/dE < 0$: A case where the surface concentration of the electroactive species decreases with increasing the electrode polarization. An example is a decrease in the surface concentration of the electroactive species by Frumkin effect.

As these conditions are easily realized in electrochemical reactions, the electrochemical oscillations appear frequently in a large number of electrochemical reactions.

6.1.4
Outline of the Present Chapter

Previously, the present author has studied the oscillatory electrodeposition of various metals and alloys, with a focus placed on the self-organized formation of ordered structures [92,95–97,100,105,106,127,128,130]. Although the oscillatory electrodeposition studied included NDRs of quite different types, it has been revealed (by the present author) that the layered structures are formed spontaneously in synchronization with the oscillation, independent of the type of NDR.

Initially, details are provided in Section 6.2 of an electrochemical oscillation observed in the hydrogen peroxide (H_2O_2) reduction on a platinum (Pt) electrode as a representative example of an N-NDR oscillator. Here, the H_2O_2 reduction current oscillates spontaneously in synchronization with an adsorption and desorption of under-potential deposited hydrogen (upd-H). As the electrochemical reactions relevant to the current oscillation are very simple, and the electrode functions simply as a catalyst for the reactions, it is very helpful to show the oscillation mechanism first in order to provide a better understanding of the formations of layered structures by oscillatory electrodeposition (as detailed later in the chapter). Two examples of layered structure formation – that is, the formation of Cu–Sn alloy and Ni–P alloy layered structures – are described in Sections 6.3 and 6.4. The former is shown as a representative example of a current oscillation with an NDR, and the latter as an example of a potential oscillation with a hidden NDR. The oscillation mechanisms can be explained in essentially the same framework as that for the oscillation in H_2O_2 reduction. In Section 6.5, a variety of other examples of the formation of ordered deposits by oscillatory electrodeposition are briefly described. Although the mechanisms of the oscillations treated in this section have not yet been fully clarified, the illustrations of the deposits are beautiful enough to show the potential of the oscillatory electrodeposition for producing ordered structures with unique shapes.

6.2
Current Oscillation Observed in H_2O_2 Reduction on a Pt Electrode

H_2O_2 reduction on a Pt electrode is a very interesting reaction in that various oscillations appear only by making slight changes in the experimental conditions [59–62]. Among the various oscillations, the oscillation caused by competition between the H_2O_2 reduction and the formation of upd-H has been most extensively studied. Figure 6.3a shows a current density (j) versus potential (U) curve obtained under a potential-controlled condition in 0.2 M H_2O_2 + 0.3 M H_2SO_4 solution. Although the current oscillation is not observed in this condition, a NDR can be seen at about −0.3 V. When the H_2O_2 concentration is increased, a current oscillation appears in the potential region of the NDR (Figure 6.3b). Figure 6.3c shows a j versus time (t) curve at −0.30 V, with the conditions kept the same as in Figures 6.3a and b. The periodic oscillation is sustained for about 1 h.

A notable point here is that the NDR region shown in Figure 6.3a corresponds to the potential region of the formation of upd-H. Figure 6.4a shows, in schematic form, the potential dependence of the coverage of the upd-H (θ_{upd-H}). The θ_{upd-H} increases with the negative shift of U (or the true electrode potential, E). As formation of the upd-H suppresses the dissociative adsorption of H_2O_2, which is the first step for the H_2O_2 reduction, the H_2O_2 reduction decreases with the negative potential shift – that is, an NDR will appear.

Another important factor for the appearance of an oscillation is the role of ohmic drop in an electrolyte. Figure 6.4b is a schematic illustration of the potential profile in the region between the Pt electrode surface and the position of a reference electrode. The applied potential, U, is kept constant externally by a potentiostat.

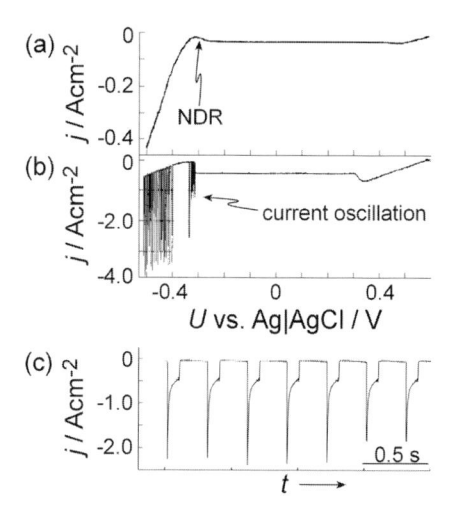

Figure 6.3 (a, b): j versus U curves in 0.3 M H_2SO_4 containing (a) 0.2 M H_2O_2 and (b) 0.7 M H_2O_2, obtained under potential-controlled conditions. (c) A time course of the oscillation in 0.3 M H_2SO_4 + 0.7 M H_2O_2 at −0.30 V versus Ag|AgCl.

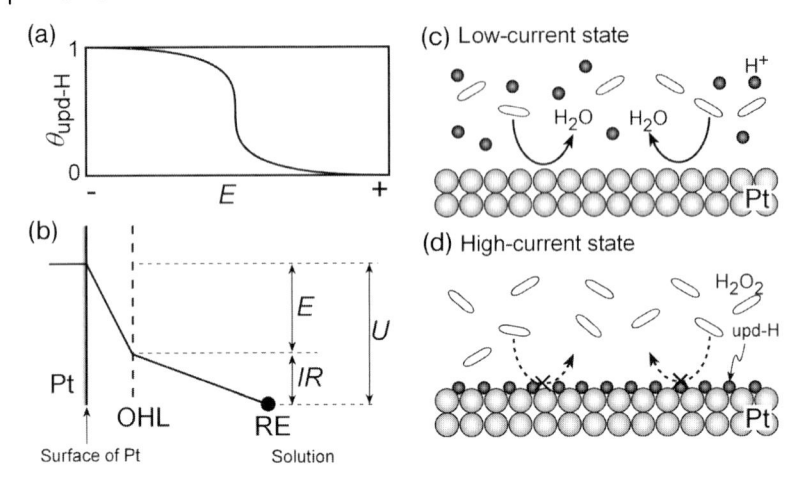

Figure 6.4 Schematic illustrations of (a) $\theta_{upd\text{-}H}$ versus E; (b) potential profile in the region between the electrode (Pt) surface and the position of RE. (c and d) Schematic representations of surface reactions in the low- and high-current states, respectively. OHL and RE in (b) denote the outer Helmholtz layer and the reference electrode, respectively.

Therefore, when a cathodic current flows in the system, the true electrode potential (or Helmholtz double layer potential), E, is more positive than U because E is give by $E = U - jAR$, where A is the electrode area, R is the resistance of the solution between the electrode surface and the reference electrode, and j is taken as negative for the reduction current.

On the basis of this argument, the mechanism for the current oscillation can be explained as follows. In the high-current stage of the current oscillation (Figure 6.4c), E is much more positive than U because of the ohmic drop. This implies that, even if U is kept constant in the region of the NDR, E is much more positive than U, and hence the $\theta_{upd\text{-}H}$ in this stage is small. Thus, the H_2O_2 reduction reaction occurs effectively without retardation by the upd-H. The active H_2O_2 reduction, however, causes decreases in the surface concentrations of the H_2O_2 (hereafter denoted as C) owing to their slow diffusion from the solution bulk. This leads to a gradual decrease in j (in the absolute), and thus to a decrease in the ohmic drop and a negative shift in E. The negative shift in E, in turn, leads to an increase in the $\theta_{upd\text{-}H}$ (Figure 6.4a). Hence, the j decreases owing to a decrease in the effective surface area, and the system goes to a low-current stage (Figure 6.4d). In the low-current stage, only slow reduction occurs at vacant sites (atomic pinholes); thus, C gradually increases by diffusion from the solution bulk. This increase in C induces an increase in j and in turn causes a positive shift in E (and a decrease in the $\theta_{upd\text{-}H}$). When E is shifted to the positive, the j increases owing to an increase in the effective surface area, and the high-current stage is restored again. Thus, the H_2O_2 reduction current oscillates in synchronization with the formation and detachment of the upd-H.

6.3
Nanoperiod Cu–Sn Alloy Multilayers

The oscillatory electrodeposition of Cu–Sn alloy from an acidic solution of Cu^{2+} and Sn^{2+} produces nano-period layered deposits [127,128]. One remarkable point of this system is that a macroscopically uniform nano-period multilayer is formed in synchronization with a current oscillation.

Figure 6.5a compares j versus U curves for metal deposition obtained in various electrolytes. Curve 1 (dashed line) shows j versus U in 0.15 M Cu^{2+} + 0.6 M H_2SO_4, where the current for Cu deposition starts to flow at about +0.04 V. The diffusion-limited, potential-independent current for the Cu^{2+} reduction is reached at ca. −0.3 V, and hydrogen evolution starts at about −0.7 V. Curve 2 (dotted line) is j versus U in 0.15 M Sn^{2+} + 0.6 M H_2SO_4, where Sn deposition starts at −0.43 V, which is more negative than for the Cu deposition. Curve 3 (solid line) is j versus U in 0.15 M Cu^{2+} + 0.15 M Sn^{2+} + 0.6 M H_2SO_4. It should be noted that the curve is deviated from a simple sum of curves 1 and 2. In particular, the current in −0.30 to −0.45 V in curve 3 is higher (in the absolute value) than the diffusion-limited current for the Cu deposition in curve 1, although the Sn deposition does not yet occur at these potentials (curve 2). The results of X-ray diffraction (XRD) and Auger electron spectroscopy (AES) analyses revealed that the current increase in this potential range was due to the formation of Cu–Sn alloy.

Figure 6.5b shows j versus U in Cu^{2+} + Sn^{2+} + H_2SO_4 with (curve 4, solid line) and without (curve 3, dashed line) a cationic surfactant (Amiet-320). Addition of the surfactant causes a drastic change in the j versus U; namely, an NDR appears in a narrow potential region of about 5 mV in the width near −0.42 V, where the Cu–Sn alloy is electrodeposited. Another notable point in the surfactant-added solution is

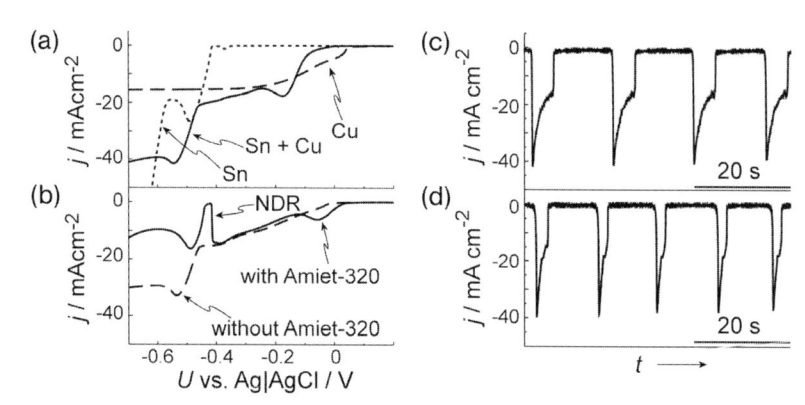

Figure 6.5 (a) j versus U obtained in 0.6 M H_2SO_4 containing (dashed curve) 0.15 M Cu^{2+}, (dotted curve) 0.15 M Sn^{2+}, and (solid curve) 0.15 M Cu^{2+} + 0.15 M Sn^{2+}. The scan rate was 10 mV s^{-1}. (b) j versus U obtained in 0.6 M H_2SO_4 + 0.15 M Cu^{2+} + 0.15 M Sn^{2+} with (solid curve) and without (dashed curve) 0.5 mM Amiet-320. The scan rate was 10 mV s^{-1}. (c, d) Time courses of current oscillations in (c) 0.15 M $CuSO_4$ 0.15 M $SnSO_4$, and (d) 0.1 M $CuSO_4$ 0.1 M $SnSO_4$, both containing 0.6 M H_2SO_4 + 0.5 M citric acid + 0.5 mM Amiet-320. (Reprinted from Ref. [127].)

that a current oscillation appears when the is U kept constant in (and near) the potential region of this NDR. Figures 6.5c and d shows current oscillations observed in (0.15 M Cu^{2+} + 0.15 M Sn^{2+}) and (0.1 M Cu^{2+} + 0.1 M Sn^{2+}), respectively, with both solutions containing commonly 0.6 M H_2SO_4 and 0.5 mM Amiet-320. Thus, the oscillation period and wave form depends on the concentrations of the electroactive species. It should be noted that it is not only the Amiet-320 but also other cationic surfactants (see Figure 6.6) that can cause both the NDR and current oscillation, in almost the same potential region as Amiet-320.

The structure of the deposited Cu–Sn alloy during the current oscillation was investigated using scanning electron microscopy (SEM) and scanning Auger electron microscopy (AEM). Figure 6.7a shows, schematically, the procedure of sample preparation. The deposited film was etched with an Ar^+-ion beam, with the film being rotated. This procedure gave a bowl-shaped hollow of about 1 mm in the diameter at the bottom, together with a slanting cross-section of the deposited film. Figure 6.7b shows an SEM image (top view) of a sample thus prepared. Uniform concentric rings of gray and black colors in the region of the slanting cross-section clearly indicate the formation of a quite uniform layered structure spreading over a macroscopically wide range of 1×1 mm. It was confirmed that the number of a set of the gray and black layers (one period of the multilayer) agreed with the number of the cycles of the current oscillation during which the deposit was formed, indicating that one oscillation cycle produced one layer of the deposit. Figure 6.7c compares the expanded SEM image in the region of the slanting cross-section with the profile (white curve) of the atomic ratio [Cu/(Cu + Sn)] in this region, in which it

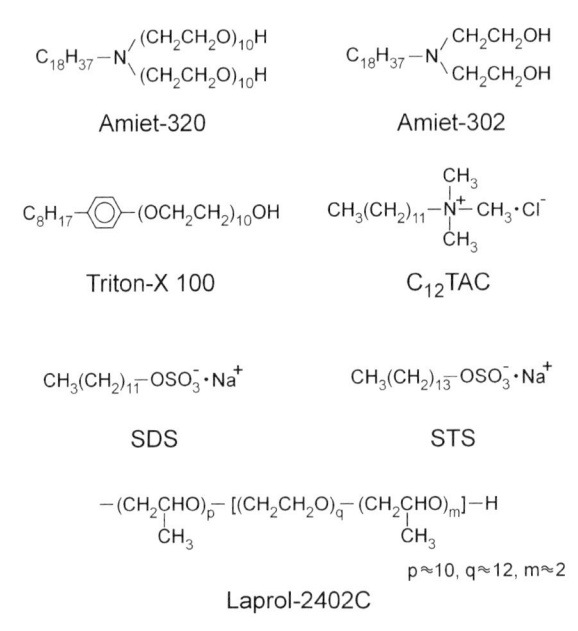

Figure 6.6 Chemical structures of surfactants used in the present investigations. (Reprinted from Ref. [127].)

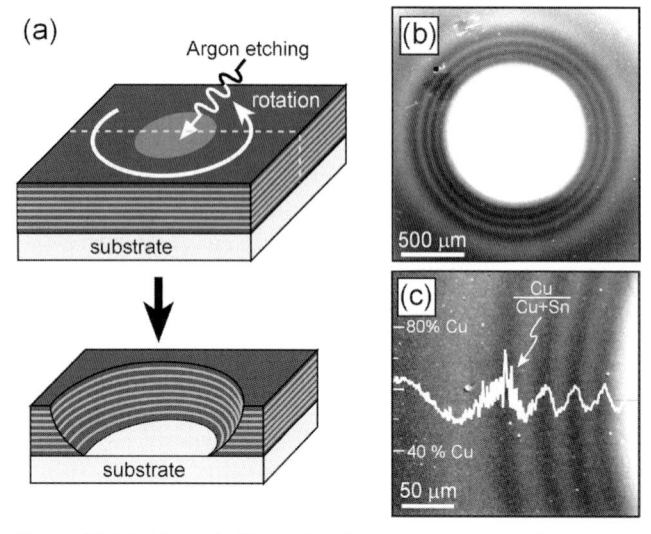

Figure 6.7 (a) Schematic illustration of sample preparation for SEM and AEM analyses. (b) SEM (top view) of a bowl-shaped hollow with a slanting cross-section, prepared in the deposited alloy film by Ar^+ ion etching. (c) Expanded SEM image, compared with the distribution of the atomic ratio [Cu/(Cu + Sn)] obtained with scanning AEM. (Reprinted from Ref. [127].)

can be seen that Cu is rich in the black layer, whilst Sn is rich in the gray layer. This result clearly indicates that the layered structures are formed over a macroscopically wide area.

Auger depth analyses for deposited films obtained during the current oscillation revealed that the thickness of the each layer is several tens of nanometers. Figures 6.8 show Auger depth profiles for deposited films obtained during the oscillations of Figures 6.5c and d, respectively. The thicknesses of one period of the multilayer, calculated from the sputter time in the Ar^+-ion sputtering, are 87 nm and 38 nm, respectively, indicating that the period of layered structure can be tuned by modulation of the electrochemical oscillation.

In this oscillatory system, the NDR and the current oscillation appear only when the cationic surfactant is added to the solution, implying that adsorption of the surfactant causes both the NDR and oscillation. Thus, the oscillation mechanism can be explained in a similar way to that of the oscillation in the H_2O_2 reduction described in Section 6.2. Coverage of the adsorbed cationic surfactant (θ_s) increases with the negative potential shift due the electrostatic force (Figure 6.9a). The adsorption of a cationic surfactant onto a Cu–Sn alloy surface retards the diffusion of electroactive metal ions (Cu^{2+} and Sn^{2+}) to the alloy surface, and thus decreases j (Figure 6.9b). Alternatively, when the surfactant desorbs from the surface, a large j flows (Figure 6.9c). Thus, the adsorbed surfactant plays exactly the same role as the upd-H in the oscillatory system of the H_2O_2 reduction. The atomic ratio of Cu and Sn

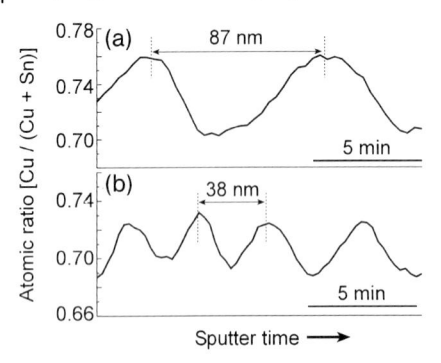

Figure 6.8 Auger depth profiles for two types of deposited film, obtained by combination with the Ar^+ ions sputtering technique. The electrolyte (and externally applied U): (a) 0.15 M $CuSO_4$ + 0.15 M $SnSO_4$ (−432 mV) and (b) 0.1 M $CuSO_4$ + 0.1 M $SnSO_4$ (−418 mV), both containing 0.6 M H_2SO_4, 0.5 M citric acid, and 0.5 mM Amiet-320. (Reprinted from Ref. [127].)

in the electrodeposited Cu–Sn alloy depends on U – that is, the Cu-rich and Sn-rich alloy is formed in more positive and negative potentials, respectively. As the E oscillates in synchronization with the current oscillation via the ohmic drop in the electrolyte, a layered structure with periodic modulation of the Cu and Sn contents is formed as the resultant deposit.

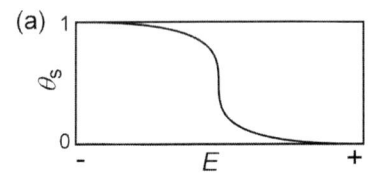

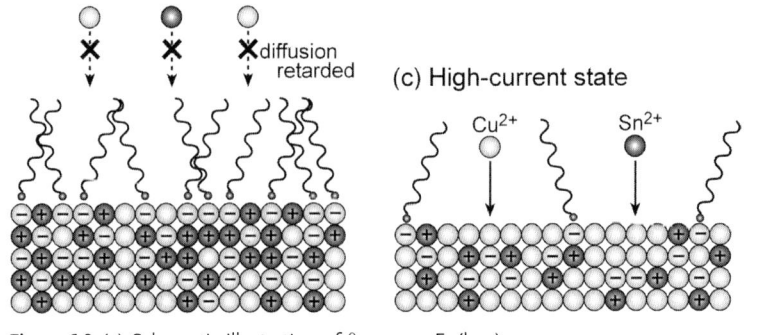

Figure 6.9 (a) Schematic illustration of θ_s versus E. (b, c) Schematic representation of surface reactions in the low- and high-current states, respectively. (Reprinted from Ref. [127].)

6.4
Nano-Scale Layered Structures of Iron-Group Alloys

Iron-group alloys – that is, the alloys of iron-group metals (Fe, Co, and Ni) with other metallic or non-metallic elements (P, Mo, and W), such as Fe–P, Co–W, Ni–P, and Ni–W – are widely used in industry owing to their unique properties [139–144]. Studies conducted to date have also clarified that the incorporated elements (P, Mo, and W) in the alloys play the crucial role in their unique properties. The iron-group alloys are mainly produced either by electroless deposition [139–141,144] or by electrodeposition [142,143]. It is known that the incorporated elements such as P, Mo, and W cannot be electrodeposited by themselves independently. Interestingly, however, the elements can be co-deposited with iron-group metals, resulting in the iron-group alloys. This phenomenon is referred to as "induced co-deposition" [145].

The induced co-deposition of iron-group alloys has another interesting aspect in that it leads to layered structures, in which the iron-group metals and the incorporated elements change their contents periodically. The formation of the layered structures was commonly observed in all reported induced co-deposition systems – a finding which suggests that the induced co-deposition has common general mechanisms. It was also reported that an electrochemical oscillation was observed when the layered Ni–P alloy was co-deposited [129,130]. It has been found that the induced co-deposition commonly has a hidden NDR, which leads to the formation of layered structure as well as to the electrochemical oscillation.

Figure 6.10a shows the j versus U for Ni–P alloy electrodeposition, obtained with an EQCM electrode under a potential-controlled condition. The j in Figure 6.10a starts to appear at about -0.45 V, and monotonously increases (in the absolute value) with a negative potential shift. Concurrently with measurement of the j versus U in Figure 6.10a, the frequency shift (Δf) in the EQCM, caused by a mass change (Δm) in the EQCM electrode, was also measured. Figure 6.10b shows the derivative of the Δm with respect to t, $d\Delta m/dt$, calculated from the Δf. Because the Δm can be attributed to the Ni–P alloy electrodeposition in this electrolyte system, $d\Delta m/dt$ is in proportion to the rate of Ni–P alloy deposition. Thus, it can be seen from Figure 6.10b that the rate of Ni–P alloy deposition increases monotonously, with a negative potential shift in a U range from -0.45 to -0.80 V, whereas it suddenly starts to decrease at around -0.80 V and continues to decrease until it reached about -0.95 V, in contrast to the current density in Figure 6.10a. This implies that the Ni–P alloy deposition current has an NDR in the U range of -0.95 V $< U <$ -0.80 V, which is hereafter called the U_{NDR} region. The j versus U in Figure 6.10a does not show any NDR, indicating that the NDR is hidden by an overlap of a hydrogen evolution current, which increases with a negative potential shift is this U range more steeply than the decrease in the Ni–P alloy deposition current. Figure 6.10c shows a time course of a potential oscillation, which appeared spontaneously when j was kept at a constant value in a range of $-55 < j < -75$ mA cm^{-2}; that is, in a range of j when U was in the U_{NDR} region of Figure 6.10a. It should be noted that the highest and lowest values of the oscillating potential in Figure 6.10c almost coincide with the highest and lowest potentials of the U_{NDR} region, respectively. It also should

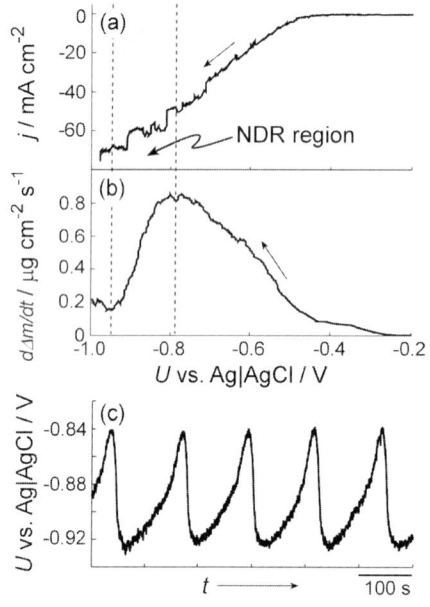

Figure 6.10 (a) j versus U obtained with the EQCM-Au electrode in an electrolyte of 0.65 M $NiSO_4$ + 0.25 M NaH_2PO_2 + 0.5 M H_3PO_4 + 0.3 M H_3BO_3 + 0.35 M NaCl (pH ≅ 1) under potential-controlled conditions, and (b) $d\Delta m/dt$ versus U obtained simultaneously with the measurement of j versus U in (a). The scan rate was 10 mV s^{-1}. (c) U versus t at a constant j of -75 mA cm^{-2}. (Reprinted from Ref. [130].)

be noted here that the Ni–P alloy deposition current in the presence of NaH_2PO_2 starts to flow at more positive potential than in its absence.

Figure 6.11 illustrates an Auger depth profile for a deposited film formed during the potential oscillation. The result clearly shows that the Ni–P alloy with a layered structure is formed in the deposit. Furthermore, the number of layers in the deposit agreed with the number of cycles of potential oscillation during which the deposit was formed, indicating that one oscillation cycle produced one layer of the deposit. The thickness of one layer was estimated as a few hundreds of nanometers, by

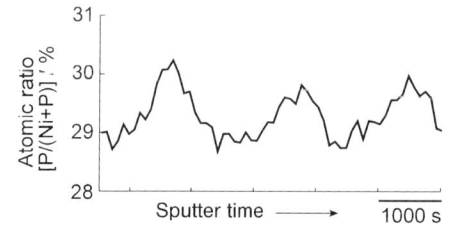

Figure 6.11 An Auger depth profile for a deposit produced under the potential oscillation. (Reprinted from Ref. [130].)

dividing the thickness of the deposit (which was measured with an optical microscope) by the number of the oscillation cycles.

In order to investigate the generality of the mechanism of formation of iron-group alloys, several experiments were conducted on other induced co-deposition systems. Figure 6.12 shows the results for the electrodeposition of Co–W and Ni–W alloys, where essentially the same behavior as for the Ni–P deposition was observed in these co-deposition systems. Namely, the deposition current in the presence of Na_2WO_4, which is the W-source for the Co–W and Ni–W alloys, began to flow at a more positive potential than in the absence of Na_2WO_4. In addition, $d\Delta m/dt$ versus U in the presence of Na_2WO_4 showed a clear NDR, although $d\Delta m/dt$ in the absence of Na_2WO_4 showed no NDR. These results indicated that the electrodeposition of the Co–W and Ni–W alloys occurs by essentially the same mechanism as that of the Ni–P alloy, and suggesting the presence of a general mechanism for the induced co-deposition.

As mentioned above, the Ni–P alloy deposition current in the presence of NaH_2PO_2 begins to flow at a more positive potential than in its absence. This fact indicates that NaH_2PO_2 (or related species) acts as a promoter for the Ni–P deposition reaction. In addition, the fact that the Ni–P alloy is formed indicates that the species acting as the promoter is drawn into the deposition reaction itself. This means that the promoter is present in the adsorbed form at the electrode (or deposit) surface. The above considerations strongly suggest that the NDR arises from desorption (detachment) of the adsorbed (anionic) promoter. Thus, the NDR in this oscillatory system can be classified as an HN-NDR type.

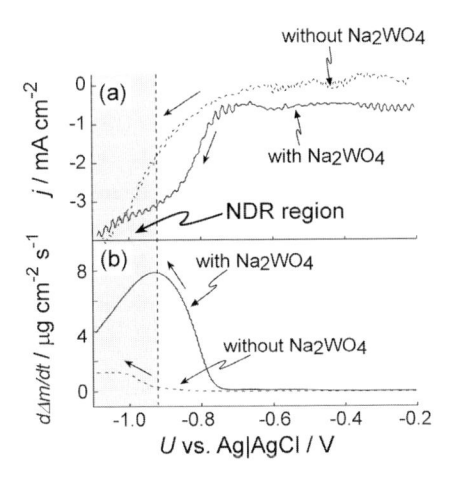

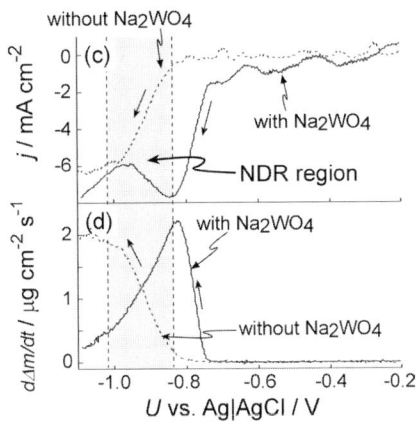

Figure 6.12 (a) The j versus U in 0.1 M $CoSO_4 + 0.25$ M $Na_2SO_4 + 0.1$ M sodium citrate without (dashed curve) and with (solid curve) 50 mM Na_2WO_4, observed under the potential-controlled conditions, and (b) the $d\Delta m/dt$ versus U, obtained simultaneously with the measurement of j versus U in (a). The scan rate was 10 mV s^{-1}. (c) j versus U in 0.1 M $NiSO_4 + 0.25$ M $Na_2SO_4 + 0.1$ M sodium citrate without (dashed curve) and with (solid curve) 50 mM Na_2WO_4, observed under potential-controlled conditions, and (d) the $d\Delta m/dt$ versus U, obtained simultaneously with measurement of j versus U in (c). The scan rate was 10 mVs^{-1}. (Reprinted from Ref. [130].)

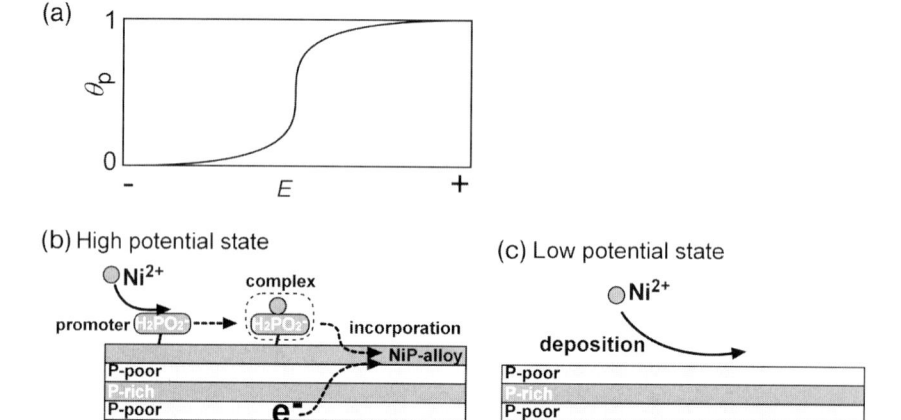

Figure 6.13 (a) Schematic illustration of θ_p versus E. (b, c) Schematic illustration for explaining the promotion effect of adsorbed $H_2PO_2^-$, which works as an origin of the NDR.

The oscillation mechanism in Ni–P alloy electrodeposition can be explained as follows. Coverage of the adsorbed promoter (θ_p), which is an anionic species (most likely $H_2PO_2^-$), decreases with the negative potential shift due the electrostatic force (Figure 6.13a). The adsorption of an anionic species onto a Ni–P alloy surface promotes the Ni–P deposition, and thus increases j (Figure 6.13b). Alternatively, when the promoter desorbs from the surface, the j decreases (Figure 6.13c). Thus, the adsorbed promoter plays an exact opposite role as the upd-H and the cationic surfactant in the oscillatory systems of the H_2O_2 reduction and Cu–Sn alloy deposition, respectively.

Now, let us consider the mechanism for the potential oscillation and formation of a layered structure in the Ni–P alloy deposition, observed under the galvanostatic condition with a constant applied j (see Figure 6.10c). At the positive end of the oscillation, the θ_p is high (Figure 6.13a), which leads to the high Ni–P alloy deposition rate. The occurrence of the high-rate Ni–P deposition, on the other hand, leads to a decrease in the surface Ni^{2+} concentration (Cs) and thus causes a gradual negative U shift to keep the constant j. When U has shifted to the negative and reached the U_{NDR} region, desorption of the adsorbed promoter begins. In this region, the negative U shift leads to a decrease in the θ_p of the adsorbed promoter, and to a decrease in the Ni–P alloy deposition current; this in turn leads to a further negative U shift to keep the constant j. Hence, this is an *autocatalytic process*, and the potential passes rapidly to the negative end of the potential oscillation. At this stage, the adsorbed promoter is almost absent from the electrode surface, and only the slow Ni–P and Ni deposition occurs, with the constant j being maintained by the hydrogen evolution. Thus, Cs gradually increases by diffusion of Ni^{2+} ions from the electrolyte bulk. This increase in Cs induces a gradual increase in the Ni–P alloy (and Ni metal) deposition current, which leads to a gradual positive shift in U to maintain a constant j. When U has shifted to the positive and reached the U_{NDR} region, adsorption of the promoter begins to occur, leading to an increase in θ_p. In the U_{NDR} region, the positive shift in U leads to an increase in θ_p and thus to an increase in the Ni–P alloy deposition current; this in turn leads to a further positive U shift to keep a constant j. This also is an autocatalytic process, and the system rapidly passes back to the positive end of the potential oscillation. The periodic modulation in the P content in the deposit formed under the potential oscillation (Figure 6.11) can also be explained by the decrease in P content at the negative potentials, caused by the additional deposition of Ni metal at the negative potentials.

6.5
Other Systems

6.5.1
Nano-Multilayers of Cu/Cu$_2$O

In 1998, Switzer *et al.* found that a spontaneous potential oscillation appears in an alkaline Cu(II)-lactate solution, which leads to the self-organized formation of layered nanostructures of Cu/Cu$_2$O. Formation of the layered structure was directly evidenced using scanning tunneling microscopy [115], scanning electron microscopy (Figure 6.14a) [115], Auger electron microscopy [116], and second-harmonic

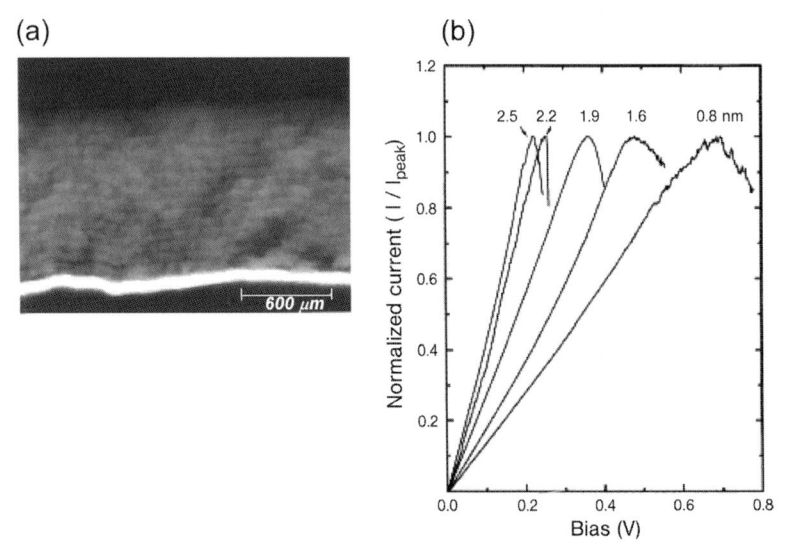

Figure 6.14 (a) Backscattered SEM image of a cross-section of the Cu/Cu$_2$O film grown under the oscillatory condition. (b) NDR curves for layered Cu/Cu$_2$O nanostructures as a function of the Cu$_2$O layer thickness. The NDR maximum shifts to higher applied bias for samples with thinner Cu$_2$O layers. (a) Reprinted from Ref. [116]; (b) Reprinted from Ref. [118].)

generation [117]. Very interestingly, these techniques revealed that the resultant deposits could function as a resonant tunneling device, showing sharp NDR signatures at room temperature in perpendicular transport measurements [118]. Furthermore, it was also shown that the bias for the NDR maximum could be controlled simply by tuning the oscillation period (Figure 6.14b). Since the discovery by Switzer et al., similar oscillations have found in Cu(II)-citrate [119] and Cu(II)-tartrate [120,121] solutions. Leopold and Nyholm and colleagues have conducted extensive studies on the mechanism of oscillation using various techniques, including in-situ electrochemical quartz crystal microbalance [120], local pH measurements [121], and confocal Raman spectroscopy [122]. Subsequently, the appearance of the oscillation was attributed to local pH variations caused by the liberation of lactate (or citrate and tartrate) from the Cu(II) complex.

Alternatively, Wang and colleagues found that periodic nanostructures were formed during the electrodeposition of Cu from "ultra-thin" electrolytes (Figure 6.15)

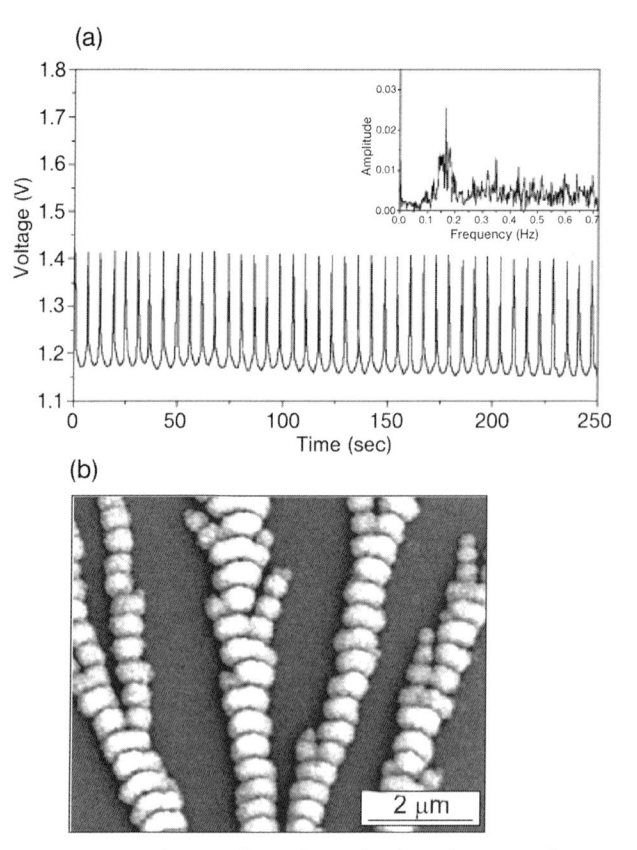

Figure 6.15 (a) Voltage oscillation during the electrodeposition of Cu in ultra-thin electrolyte. Inset: the Fourier transform of the voltage oscillation. (b) SEM view of the filaments of the electrodeposit formed under the oscillation. (Reprinted from Ref. [123].)

[123]. Analyses of the deposits by TEM diffraction [124] and scanning near-field optical microscopy (SNOM) [125], revealed that the periodic nanostructures corresponded to the alternating growth of Cu and Cu_2O. Interestingly, the Cu/Cu_2O nanostructures were formed in synchronization with a spontaneous voltage oscillation, in a similar manner to the system identified by Switzer *et al.* This occurred despite the fact that the $CuSO_4$ aqueous solution did not contain any ligands such as lactate, citrate, and tartrate. A simple mathematical model based on Switzer's model, in which the variation of the local pH plays a critical role, was also proposed.

6.5.2
Ag–Sb Alloy with Periodical Modulation of the Elemental Ratio

During the early 20th century, Raub and colleagues found that the electrodeposition of an Ag–In alloy gives rise to propagating spiral patterns at the deposit's surface [128]. Although this discovery was long forgotten, the same phenomenon was found – after almost 50 years – in Ag–Sb (Figure 6.16a) [108] and Ag–Bi [109] alloy electrodeposition by Krastev *et al.*, and in Ir–Ru alloy [110] electrodeposition by Saltykova *et al.* The appearances of potential oscillations in the same electrolyte were also reported, implying that the origin of the spiral patterns and the oscillation was common. A combined analysis of SEM cross-sections and AES depth profiles of the Ag–Sb alloy deposits revealed that the deposited film had a layered structure of the alloy, but with different composition (Figure 6.16b) [111]. Nagamine *et al.* performed detailed analyses of the Ag–Sb alloy deposits by using *in-situ* optical microscopy and an *ex-situ* electron probe (X-ray) microanalyzer. These authors revealed that: (i) several types of propagating stripe pattern appear, depending on the applied current density [113]; and (ii) not only Ag and Sb but also O plays a part in the spiral pattern formation [114]. Unfortunately, however, little is known at present about the origin of the potential oscillation and the propagating spiral patterns.

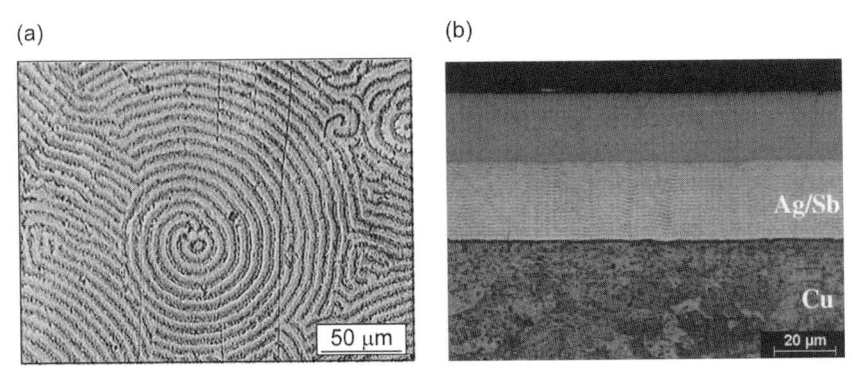

Figure 6.16 (a) Optical microscopic image of a spiral wave on AgSb-alloy deposit. (b) SEM image of a cross-section of the AgSb-alloy deposit. (a) Reprinted from Ref. [111]; © Elsevier Science Ltd.; (b) Reprinted from Ref. [112].)

6.6
Summary

Oscillatory electrodeposition represents an interesting target from the point of view of the self-organized formation of micro- and nano-sized ordered structures, because ever-changing self-organized spatiotemporal patterns during the oscillations are recorded as architectures of electrodeposits. In this chapter, attention was focused on the self-organized formations of layered nanostructure by oscillatory electrodeposition. The oscillatory electrodeposition systems described here led to the formation of layered nanostructures, despite the oscillations having different mechanisms. The results outlined thus far clearly indicate that the NDRs – and thus the oscillations and layered structure formation – can arise from various mechanisms, depending on the reaction systems and electrolyte compositions. This implies that oscillatory electrodeposition systems have a high possibility of producing ordered micro- and nano-structures. This important fact was made clear only after the mechanisms of various oscillatory deposition systems were revealed. A clarification of these mechanisms will prove invaluable for systematic tuning of the layered structures, by designing and modulating the electrochemical oscillations.

References

1 Geissler, M. and Xia, Y. (2004) *Adv. Mater.*, **16**, 1249.
2 Lehn, J.M. (1990) *Angew. Chem. Int. Ed.*, **29**, 1304.
3 Kunitake, T. (1992) *Angew. Chem. Int. Ed.*, **31**, 709.
4 Ringsdorf, H., Schlarb, B. and Venzmer, J. (1988) *Angew. Chem. Int. Ed.*, **27**, 113.
5 Murray, C.B., Kagan, C.R. and Bawendi, M.G. (2000) *Annu. Rev. Mater. Sci.*, **30**, 545.
6 Storhoff, J.J. and Mirkin, C.A. (1999) *Chem. Rev.*, **99**, 1849.
7 Ulman, A. (1996) *Chem. Rev.*, **96**, 1533.
8 Love, J.C., Estroff, L.A., Kribel, J.K., Nuzzo, R.G. and Whitesides, G.M. (2005) *Chem. Rev.*, **105**, 1103.
9 Li, X.M., Huskens, J. and Reinhoudt, D.N. (2004) *J. Mater. Chem.*, **14**, 2954.
10 Johannes, V.B., Giovanni, C. and Kern, K. (2005) *Nature*, **437**, 671.
11 Grzybowski, B.A., Bishop, K.J.M., Campbell, C.J., Fialokowski, M. and Smoukov, S.K. (2005) *Soft Mater.*, **1**, 114.
12 Teichert, C. (2002) *Phys. Rep.*, **365**, 335.
13 Hildebrand, M., Ipsen, M., Mikhailov, A.S. and Ertl, G. (2003) *New J. Phys.*, **5**, 61.
14 Smoukov, S.K., Bitner, A., Campbell, C.J., Kandere-Grzybowkia, K. and Grzyboski, B.A. (2005) *J. Am. Chem. Soc.*, **127**, 17803.
15 Höll, H., Langa, S., Carstensen, S.J., Christophersen, M. and Tiginyanu, I.M. (2003) *Adv. Mater.*, **15**, 183.
16 Nakanishi, S., Tanaka, T., Saji, Y., Tsuji, E., Fukushima, S., Fukami, K., Nagai, T., Nakamura, R., Imanishi, A. and Nakato, Y. (2007) *J. Phys. Chem. C*, **111**, 3934.
17 Karthaus, O., Maruyama, N., Cieren, X., Shimomura, M., Hasegawa, H. and Hashimoto, T. (2000) *Langmuir*, **16**, 6071.
18 Shimomura, M. and Sawadaishi, T. (2001) *Curr. Opin. Colloid Interface Sci.*, **6**, 11.
19 Horany, G. and Visy, C. (1979) *J. Electroanal. Chem.*, **103**, 353.
20 Yamazaki, T. and Kodera, T. (1989) *Electrochim. Acta*, **34**, 969.

21 Eiswirth, M., Lübke, M., Krischer, K., Wolf, W., Hudson, J.L. and Ertl, G. (1992) *Chem. Phys. Lett.*, **192**, 254.
22 Wolf, W., Lübke, L., Koper, M.T.M., Eiswirth, M. and Ertl, G. (1995) *J. Electroanal. Chem.*, **399**, 185.
23 Plenge, F., Varela, M. and Krischer, K. (2005) *Phys. Rev. Lett.*, **94**, 198301.
24 Schell, M., Albahadily, F.N., Safar, J. and Xu, Y. (1989) *J. Phys. Chem.*, **93**, 4806.
25 Karantonis, A., Koutsafits, D. and Kouloumbi, N. (2006) *Chem. Phys. Lett.*, **422**, 78.
26 Triplovic, A., Popovic, K. and Adzic, R.R. (1991) *J. Chim. Phys. Phys. Chim. Biol.*, **88**, 1635.
27 Nakabayashi, S. and Kita, A. (1992) *J. Phys. Chem.*, **96**, 1021.
28 Okamoto, H. (1992) *Electrochim. Acta*, **37**, 969.
29 Raspel, F., Nicholos, R.J. and Kolb, D.M. (1990) *J. Electroanal. Chem.*, **286**, 279.
30 Strasser, P., Lübke, L., Raspel, F., Eiswirth, M. and Ertl, G. (1997) *J. Chem. Phys.*, **107**, 979.
31 Lee, J., Christoph, J., Strasser, P., Eiswirth, M. and Ertl, G. (2001) *J. Chem. Phys.*, **115**, 1485.
32 Mukouyama, Y., Kikuchi, M., Samjeske, G., Osawa, M. and Okamoto, H. (2006) *J. Phys. Chem. B*, **110**, 11912.
33 Koper, M.T.M., Schmidt, T.J., Markovic, N.M. and Ross, P.N. (2001) *J. Phys. Chem. B*, **105**, 8381.
34 Malkahandi, S., Bonnefont, A. and Krischer, K. (2005) *Electrochem. Commun.*, **7**, 710.
35 Miller, B. and Chen, A.C. (2006) *J. Electroanal. Chem.*, **588**, 3141.
36 Chen, A.C. and Miller, B. (2004) *J. Phys. Chem. B*, **108**, 2245.
37 Xu, L.Q., Gao, Q.Y., Feng, J.M. and Wang, J.C. (2004) *Chem. Phys. Lett.*, **397**, 265.
38 Koper, M.T.M. and Gaspard, P. (1991) *J. Phys. Chem.*, **95**, 4945.
39 Koper, M.T.M., Gaspard, P. and Sluyters, J.H. (1992) *J. Phys. Chem.*, **96**, 5674.
40 Fetner, N. and Hudson, J.L. (1990) *J. Phys. Chem.*, **94**, 6505.
41 Strbac, S. and Adzic, R.R. (1992) *J. Electrochem. Chem.*, **337**, 355.
42 Cattarin, S. and Tributsch, H. (1990) *J. Electrochem. Soc.*, **137**, 3475.
43 Catterin, S. and Tributsch, H. (1993) *Electrochim. Acta*, **38**, 115.
44 Nakanishi, S., Mukouyama, Y., Karasumi, K., Imanishi, A., Furuya, N. and Nakato, Y. (2000) *J. Phys. Chem. B*, **104**, 4181.
45 Nakanishi, S., Mukouyama, Y. and Nakato, Y. (2001) *J. Phys. Chem. B*, **105**, 5751.
46 Mukouyama, Y., Nakanishi, S., Chiba, T., Murakoshi, K. and Nakato, Y. (2001) *J. Phys. Chem. B*, **105**, 7246.
47 Mukouyama, Y., Kikuchi, M. and Okamoto, H. (2005) *J. Solid. State Electrochem.*, **9**, 290.
48 Koper, M.T.M. (1996) *Ber. Bunsen-Ges. Phys. Chem. Chem. Phys.*, **100**, 497.
49 Flätgen, G., Krischer, K. and Ertl, G. (1996) *J. Electroanal. Chem.*, **409**, 183.
50 Nakanishi, S., Sakai, S.-I., Hatou, M., Mukouyama, Y. and Nakato, Y. (2002) *J. Phys. Chem. B*, **106**, 2287.
51 Strasser, P., Lübke, M., Eickes, C. and Eiswirth, M. (1999) *J. Electroanal. Chem.*, **462**, 19.
52 Li, Z.L., Ren, B., Xiao, X.M., Chu, X. and Tian, Z.Q. (2002) *J. Phys. Chem. A*, **106**, 6570.
53 Lee, H.P. and Nobe, K. (2000) *J. Phys. Chem. B*, **104**, 4181.
54 Bassett, M.R. and Hudson, J.L. (1990) *J. Phys. Chem. B*, **137**, 1815.
55 Gu, Z.H., Olivier, A. and Fahiday, T.Z. (1990) *Electrochim. Acta*, **35**, 933.
56 Gu, Z.H., Chen, J., Fahiday, T.Z. and Plivier, A. (1994) *J. Electroanal. Chem.*, **367**, 7.
57 Gu, Z.H., Chen, J. and Fahiday, T.Z. (1993) *Electrochim. Acta*, **38**, 2631.
58 Albahadily, F.N., Gingland, J. and Schell, M. (1989) *J. Chem. Phys.*, **90**, 813.
59 Tsitsopoulos, L.T., Tsotsis, T.T. and Webster, I.A. (1987) *Surf. Sci.*, **191**, 225.
60 Dewald, H.D., Parmananda, P. and Rollins, R.W. (1991) *J. Electroanal. Chem.*, **306**, 297.

61 Dewald, H.D., Parmananda, P. and Rollins, R.W. (1993) *J. Electrochem. Soc.*, **140**, (1969).
62 Cooper, J.F., Muller, R.H. and Tobias, C.W. (1980) *J. Electrochem. Soc.*, **127**, 1733.
63 Russell, P. and Newman, J. (1987) *J. Electroanal. Chem.*, **133**, 1051.
64 Diem, C.B. and Hudson, J.L. (1987) *AIChE J.*, **33**, 218.
65 Wang, Y., Hudson, J.L. and Jaeger, N.I. (1990) *J. Electrochem. Soc.*, **137**, 485.
66 Pearlstein, A.J. and Johnson, J.A. (1989) *J. Electrochem. Soc.*, **136**, 1290.
67 Pagitsas, M. and Sazou, D. (1991) *Electrochim. Acta*, **36**, 1301.
68 Li, W., Wang, X. and Nobe, K. (1990) *J. Electrochem. Soc.*, **137**, 1184.
69 Li, W. and Nobe, K. (1993) *J. Electrochem. Soc.*, **140**, 1642.
70 Moina, C. and Posadas, D. (1989) *Electrochim. Acta*, **34**, 789.
71 Shiomi, Y., Karantonis, A. and Nakabayashi, S. (2001) *Phys. Chem. Chem. Phys.*, **3**, 479.
72 Sazou, D., Pagitsas, M. and Kokkinidis, G. (1990) *J. Electroanal. Chem.*, **289**, 217.
73 Sazou, D. and Pagitsas, M. (1992) *J. Electroanal. Chem.*, **323**, 247.
74 Franck, U.F. and Meunier, L. (1953) *Z. Naturforsch B*, **8**, 396.
75 Bell, J., Jaeger, N.I. and Hudson, J.L. (1992) *Phys. Chem.*, **96**, 8671.
76 Lev, O., Wolffberg, A., Sheintuch, M. and Pisman, L.M. (1988) *Chem. Eng. Sci.*, **43**, 1339.
77 Haim, D., Lev, O., Pisman, L.M. and Sheintuch, M. (1992) *J. Phys. Chem.*, **96**, 2676.
78 Lev, O., Sheintuch, M., Piseman, L.M. and Yarnitzky, C. (1988) *Nature*, **336**, 458.
79 Ozanam, F., Chazalviel, J.N., Radi, A. and Etman, M. (1991) *Ber. Bunsen-Ges. Phys. Chem.*, **95**, 98.
80 Chazalviel, J.N. and Ozanam, F. (1992) *J. Electrochem. Soc.*, **139**, 2501.
81 Lewerenz, H.J. and Schlichthoerl, G. (1992) *J. Electroanal. Chem.*, **327**, 85.
82 Corcoran, S.G. and Sieradzki, K. (1992) *J. Electrochem. Soc.*, **139**, 1568.
83 Eidel'berg, M.I., Sandulov, D.B. and Utimenko, V.N. (1991) *Zh. Prikl. Khim.*, **64**, 665.
84 Eidel'berg, M.I., Sandulov, D.B. and Utimenko, V.N. (1990) *Elektrokhimiya*, **26**, 272.
85 Miadokova, M. and Siska, J. (1987) *Collect. Czech. Chem. Commun.*, **52**, 1461.
86 Taveira, L.V., Macak, J.M., Sirotna, K., Dick, L.F.P. and Schmuki, P. (2006) *J. Electrochem. Soc.*, **153**, B137.
87 Marvey, E., Buckley, D.N. and Chu, S.N.G. (2002) *Electrochem. Solid State Lett.*, **5**, G22.
88 Suter, R.M. and Wong, P.Z. (1989) *Phys. Rev. B*, **39**, 4536.
89 Argoul, F. and Arneodo, A. (1990) *J. Phys.*, **51**, 2477.
90 St-Pierre, J. and Piron, D.L. (1990) *J. Electrochem. Soc.*, **137**, 2491.
91 St-Pierre, J. and Piron, D.L. (1987) *J. Electrochem. Soc.*, **134**, 1689.
92 Fukami, K., Nakanishi, S., Tada, T., Yamasaki, H., Sakai, S.-I., Fukushima, S. and Nakato, Y. (2005) *J. Electrochem. Soc.*, **152**, C493.
93 Wang, M., Feng, Y., Yu, G.-W., Gao, W.-T., Zhong, S., Peng, R.-W. and Ming, N.-B. (2004) *Surf. Interf. Anal.*, **36**, 197.
94 Piron, D.L., Nagatsugawa, I. and Fan, C.L. (1991) *J. Electrochem. Soc.*, **138**, 3296.
95 Nakanishi, S., Fukami, K., Tada, T. and Nakato, Y. (2004) *J. Am. Chem. Soc.*, **126**, 9556.
96 Tada, T., Fukami, K., Nakanishi, S., Yamasaki, H., Fukushima, S., Nagai, T., Sakai, S.-I. and Nakato, Y. (2005) *Electrochim. Acta*, **50**, 5050.
97 Fukami, K., Nakanishi, S., Yamasaki, H., Tada, T., Sonoda, K., Kamikawa, N., Tsuji, N., Sakaguchi, H. and Nakato, Y. (2007) *J. Phys. Chem. C*, **111**, 1150.
98 Wen, S.X. and Szpunar, J.A. (2006) *J. Electrochem. Soc.*, **153**, E45.
99 Saliba, R., Mingotaud, C., Argoul, F. and Ravaine, S. (2002) *Electrochem. Commun.*, **4**, 269.
100 Fukami, K., Nakanishi, S., Sawai, Y., Sonoda, K., Murakoshi, K. and Nakato, Y. (2007) *J. Phys. Chem. C*, **111**, 3216.

101 de Almeidi Lima, M.E., Bouteillon, J. and Diard, J.P. (1992) *J. Appl. Electrochem.*, **22**, 577.
102 Doerfler, H.D. (1989) *Nova Acta Leopold*, **61**, 25.
103 Pajdowski, L. and Podiadkly, J. (1977) *Electrochim. Acta*, **22**, 1307.
104 Schlitte, F.W., Eichkorn, G. and Fischer, H. (1968) *Electrochim. Acta*, **13**, 2063.
105 Nakanishi, S., Sakai, S.-I., Nishimura, K. and Nakato, Y. (2005) *J. Phys. Chem. B*, **109**, 18846.
106 Nagai, T., Nakanishi, S., Mukouyama, Y., Ogata, Y.H. and Nakato, Y. (2006) *Chaos*, **16**, 037106.
107 Raub, E. and Schall, A. (1938) *Z. Metallk*, **30**, 149.
108 Krastev, I. and Koper, M.T.M. (1995) *Phys. A*, **213**, 199.
109 Krastev, I., Valkova, T. and Zielonka, A. (2003) *J. Appl. Electrochem.*, **33**, 1199.
110 Saltykova, N.A., Estina, N.O., Baraboshkin, A.N., Pornyagin, O.V. and Pankratov, A.A. (1993) *Abstract Book of the 44th Meeting of the ISE*, Berlin, p. 376.
111 Nakabayashi, S., Krastev, I., Aogaki, R. and Inokuma, K. (1998) *Chem. Phys. Lett.*, **294**, 204.
112 Nakabayashi, S., Inokuma, K., Nakao, A. and Krastev, I. (2000) *Chem. Lett.*, **29**, 88.
113 Nagamine, Y. and Hara, M. (2004) *Physica A*, **327**, 249.
114 Nagamine, Y. and Hara, M. (2005) *Phys. Rev. E*, **72**, 016201.
115 Switzer, J.A., Hung, C.J., Huang, L.Y., Switzer, E.R., Kammler, D.R., Golden, T.D. and Bohannan, E.W. (1998) *J. Am. Chem. Soc.*, **120**, 3530.
116 Bohannan, E.W., Huang, L.Y., Miller, F.S., Shumsky, M.G. and Switzer, J.A. (1999) *Langmuir*, **15**, 813.
117 Mishina, E., Nagai, K., Barsky, D. and Nakabayashi, S. (2002) *Phys. Chem. Chem. Phys.*, **4**, 127.
118 Switzer, J.A., Maune, B.M., Raub, E.R. and Bohannan, E.W. (1999) *J. Phys. Chem. B*, **103**, 395.
119 Eskhult, J., Herranen, M. and Nyholm, L. (2006) *J. Electroanal. Chem.*, **594**, 35.
120 Leopold, S., Herranen, M. and Carlsson, J.-O. (2001) *J. Electrochem. Soc.*, **148**, C513.
121 Leopold, S., Herranen, M., Carlsson, J.O. and Nyholm, L. (2003) *J. Electroanal. Chem.*, **547**, 45.
122 Leopold, S., Arrayet, J.C., Bruneel, J.L., Herranen, M., Carlsson, J.O., Argoul, F. and Sarvant, L. (2003) *J. Electrochem. Soc.*, **150**, C472.
123 Zhong, S., Wang, Y., Wang, M., Zhang, M.-Z., Yin, X.-B., Peng, R.-W. and Ming, N.-B. (2003) *Phys. Rev. E*, **67**, 061601.
124 Zhang, M.-Z., Wang, M., Zhang, Z., Zhu, J.-M., Peng, R.-W. and Ming, N.-B. (2004) *Electrochim. Acta*, **49**, 2379.
125 Wang, Y., Cao, Y., Wang, M., Zhong, S., Zhang, M.-Z., Feng, Y., Peng, R.-W. and Ming, N.-B. (2004) *Phys. Rev. E*, **69**, 021607.
126 Schlitte, F., Eichkorn, G. and Fischer, H. (1968) *Electrochim. Acta*, **13**, 2063.
127 Nakanishi, S., Sakai, S.-I., Nagai, T. and Nakato, Y. (2005) *J. Phys. Chem. B*, **109**, 1750.
128 Nakanishi, S., Sakai, S.-I., Nagai, T. and Nakato, Y. (2006) *J. Surf. Sci. Soc. Jpn.*, **27**, 408.
129 Lee, W.G. (1960) *Plating*, **47**, 288.
130 Sakai, S.-I., Nakanishi, S. and Nakato, Y. (2006) *J. Phys. Chem. B*, **110**, 11944.
131 Krischer, K., Mazouz, N. and Grauel, P. (2001) *Angew. Chem. Int. Ed.*, **40**, 851.
132 Grauel, P., Varela, H. and Krischer, K. (2001) *Faraday Discussions*, **120**, 165.
133 Christoph, J. and Eiswirth, M. (2002) *Chaos*, **12**, 215.
134 Koper, M.T.M. and Sluyters, J.H. (1994) *J. Electroanal. Chem.*, **371**, 149.
135 Krischer, K. (2001) *J. Electroanal. Chem.*, **501**, 1.
136 Strasser, P., Eiswirth, M. and Koper, M.T.M. (1999) *J. Electroanal. Chem.*, **478**, 50.
137 Mukouyama, Y., Nakanishi, S., Konishi, H., Ikeshima, Y. and Nakato, Y. (2001) *J. Phys. Chem. B*, **105**, 10905.
138 Krischer, K. (1995) Principles of temporal and spatial pattern formation in

electrochemical systems, in *Modern Aspects of Electrochemistry*, (eds R.W. White, O.M. Bockris and R.E. Conway), Vol. 32, Plenum, New York.

139 Tulsi, S.S. (1986) *Trans. Inst. Metal. Finish*, **64**, 73.

140 Parker, K. (1992) *Plat. Surf. Finish*, **79**, 29.

141 Williams, J.E. and Davison, C. (1990) *J. Electrochem. Soc.*, **137**, 3260.

142 Paseka, I. and Velicka, J. (1997) *Electrochim. Acta*, **42**, 237.

143 Burchardt, T. (2001) *Int. J. Hydrogen Energy*, **26**, 1193.

144 Flis, J. and Duguatte, D.J. (1984) *J. Electrochem. Soc.*, **40**, 425.

145 Brenner, A. (1963) *Electrodeposition of Alloys*, Academic Press, New York.

7
Electrochemical Corrosion Behaviour of Nanocrystalline Materials
Omar Elkedim

7.1
Introduction

Nanocrystalline materials have received much attention as a result of their unique physical, chemical and mechanical properties, and have been the subjects of intensive research activities both in the scientific and industrial communities.

Nanocrystalline materials – that is, single or multiphase polycrystalline solids with a characteristic grain size of a few nanometers – represent a promising class of new materials. Owing to the extremely small crystallite dimensions (typically 1 to 100 nm), they are characterized structurally by a large volume fraction of interfaces which may lead to improvements in a variety of properties [1].

Current spending on corrosion-related problems accounts for more than 3% of the world's gross domestic product (GDP), and characterization of the corrosion behavior of nanocrystalline materials is important both for prospective engineering applications and for a better understanding of the above fundamental physicochemical properties [2,3]. In many cases the industrial application of novel materials will ultimately depend on their corrosion resistance over extended periods of service. For this reason, corrosion problems must be considered at an appropriate stage of material development.

A limited number of investigations have concentrated on the corrosion of nanocrystalline materials. Rofagha *et al.* [4,5] studied nanocrystalline Ni and Ni–P produced by an electrodeposition technique, whereupon the observed behaviour was considered to be consistent with substantial contributions to the bulk electrochemical behavior from the intercrystalline regions (i.e., grain boundaries and triple junctions) of these materials.

Inturi and Szklarska-Smialowska [6] have observed improved localized corrosion resistance in HCl for sputter-deposited nanocrystalline type 304 stainless steel in comparison with conventional material, and attributed this to the fine grain size and homogeneity of the nanocrystalline materials.

Nanostructured Materials in Electrochemistry. Edited by Ali Eftekhari
Copyright © 2008 WILEY-VCH Verlag GmbH & Co. KGaA, Weinheim
ISBN: 978-3-527-31876-6

Thorpe et al. [7] studied the corrosion behavior of nanocrystalline $Fe_{32}Ni_{36}Cr_{14}P_{12}B_6$ alloy obtained by crystallization of the melt-spun amorphous ribbon. These authors determined that the corrosion resistance of this material was significantly greater than that of its amorphous counterpart, and attributed the improvement to the observed greater Cr enrichment of the electrochemical surface via rapid interphase boundary diffusion [8].

It is difficult to predict the electrochemical behavior of nanocrystalline materials from the known properties of their coarse-grained polycrystalline analogues. Thus, several groups have observed that nanocrystalline materials exhibit enhanced oxidation and corrosion resistance compared to their conventional microcrystalline counterparts [9–12]. In contrast, results obtained in other studies have shown nanocrystalline materials to have higher rates of dissolution and corrosion [13,14].

The aim of this chapter is to outline the aqueous corrosion research activities currently being undertaken on nanocrystalline materials, with key results being extracted and reviewed from publications covering the past few years of research activity in this area. The chapter will include an outline of the electrochemical corrosion behavior of nanocrystalline materials, and this will be followed by a more comprehensive examination of the subject. Ultimately, it is hoped that this chapter will provide help and encouragement to those research groups currently working with nanocrystalline materials, hopefully to achieve consistent development of these products.

7.2
Electrochemical Corrosion Behavior of Nanocrystalline Materials

In general, the corrosion resistance of nanocrystalline materials in aqueous solutions is of major importance when assessing a wide range of potential future applications. Many of these applications require a good understanding of the corrosion properties of the materials as a function of grain size. In contrast to earlier expectations, that the increased density of grain boundary and triple junction defects in nanostructures would have a detrimental effect on the overall corrosion performance of nanocrystalline metals, extensive research conducted over the past 15 years has shown otherwise.

Grain size reduction in nanocrystalline materials has been shown to considerably improve the corrosion performance for a wide range of electrochemical conditions. Many of these studies have shown that this is due mainly to the elimination of localized attack at grain boundaries which, for conventional polycrystalline materials, is one of the most detrimental mechanisms of degradation. Several explanations have been given for this effect, including: (i) the solute dilution effect by grain size refinement; (ii) crystallographic texture changes with decreasing grain size; and (iii) grain size-dependent passive layer formation [49].

Both, beneficial and detrimental effects of nanostructural formation on corrosion performance have been reported in the case of nanocrystalline materials. To illustrate these conflicting results, mention should first be given first to Jiang and

7.2 Electrochemical Corrosion Behavior of Nanocrystalline Materials | 293

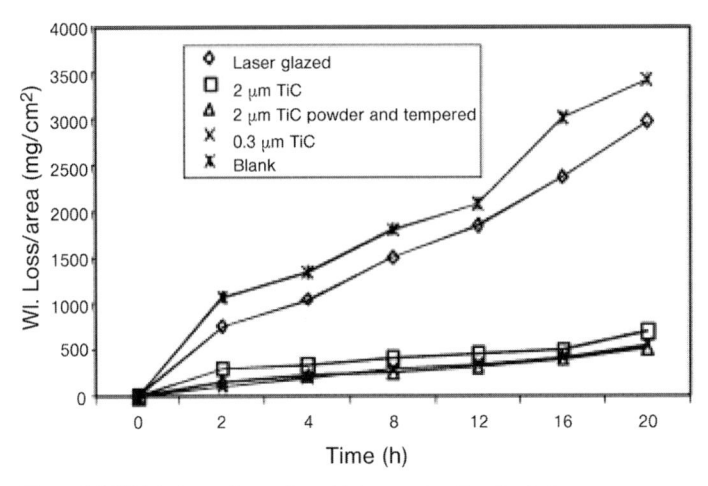

Figure 7.1 Weight loss of samples with immersion time in static A390 melt (weight loss of untreated H13 steel: $3500\,\mathrm{mg\,cm^{-2}}$).

Molian [15], who performed laser glazing and alloying of micro- and nanoparticles of TiC on H-13 steel in order to improve the performance and extend the life of die-casting dies subjected to the harsh environment of repeated heating and cooling cycles (a 1.5-kW CO_2 laser was used to conduct these experiments).

The corrosion/erosion resistance of laser-glazed and alloyed samples was evaluated in simulated metal casting conditions. Figure 7.1 shows the corrosion resistance in terms of weight loss per unit area as a function of time in static melts for samples of as-received, laser-glazed, LSA with 2 μm TiC powder, LSA with 2 μm TiC powder and tempered at 205 °C, and LSA with 300 nm TiC powder. The as-received specimens (blank) had much lower corrosion resistance because H-13 steel can dissolve in an aluminum melt to form multilayer intermetallic compounds of the form Al_4FeSi [16]. Laser glazing had only a small beneficial effect that could be attributed to the uniform dispersion of secondary carbides. A change in surface composition is clearly needed in order to obtain substantial improvements.

Laser-alloyed samples with 300 nm and 2μm powder with tempering provided the best corrosion resistance, the weight loss being decreased from 3500 to $500\,\mathrm{mg\,cm^{-2}}$ by alloying the steel with nanocrystalline powder. This decrease in weight loss corresponds to an increase in corrosion resistance of approximately 85% for the laser-alloyed samples. It is interesting to note that samples laser-alloyed with 2 μm powder (without tempering) exhibited an inferior corrosion resistance compared to samples alloyed with 300 nm powder. The improved corrosion resistance is attributed to the smooth surface, low porosity, and possible ferrite microstructure achieved with the use of powders of nanocrystalline TiC. For the laser-alloyed samples with 2 μm powder and tempering, the improved corrosion resistance may also be explained by the stress reduction and uniform morphology from the tempering process.

Pardo and coworkers [17–20] performed extensive studies on the electrochemical corrosion behavior of nanocrystalline metallic glasses. The influence of Cr concentration on the corrosion resistance of $Fe_{73.5}Si_{13.5}B_9Nb_3Cu_1$ metallic glass in simulated environments contaminated with SO_2 has been studied (0.1 M Na_2SO_4) [17]. The corrosion kinetics has been analyzed using direct current electrochemical techniques.

Figure 7.2 shows anodic polarization curves corresponding to the amorphous, crystalline and nanocrystalline materials without Cr, and with 6 wt.% Cr. In the absence of Cr, the nanocrystalline material presented the highest trend to passivation in the aggressive medium tested, and its anodic polarization curve was at current densities lower than those observed both for amorphous and crystalline

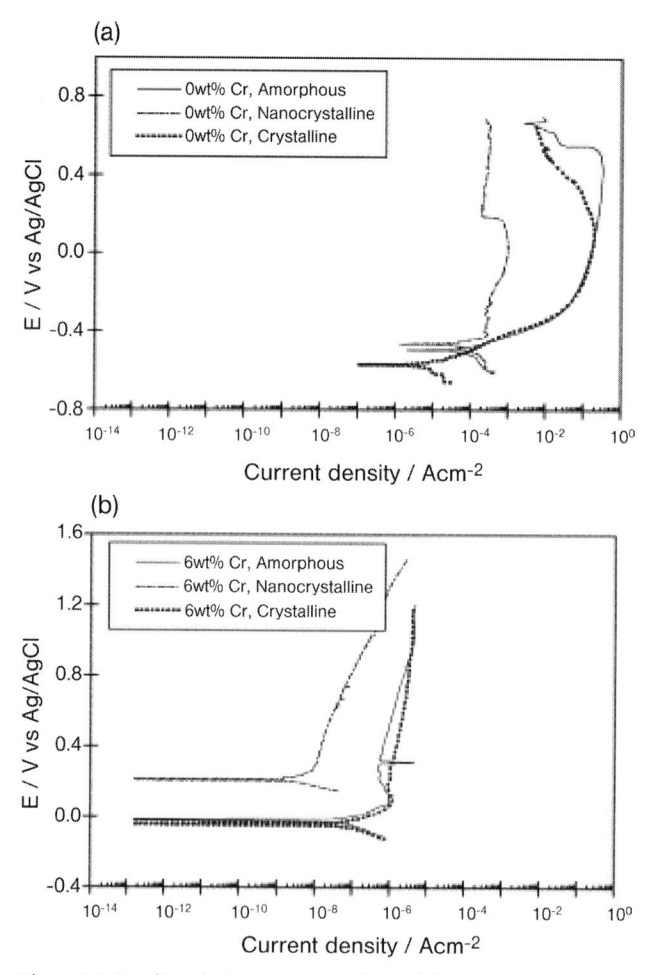

Figure 7.2 Anodic polarization curves obtained for amorphous, nanocrystalline and crystalline samples with (a) 0 wt.% Cr and (b) 6 wt.% Cr in a 0.1 M Na_2SO_4 solution.

materials. When the materials contained up to 6 wt.% Cr, the anodic polarization curves showed a clear trend to passivation, with the anodic current densities lower than those obtained for materials in the absence of Cr. Once more, the material in a nanocrystalline state presented the best anodic behavior because the anodic polarization curve was displaced to the lowest current densities and more noble corrosion potentials. The nanocrystalline material displayed a number of properties that justified its high resistance to corrosion. These properties were based on the equilibrium between a reasonable structural homogeneity, close to that of the material in an amorphous state, and a lower proportion of internal tensions with respect to the material in a vitreous state. The latter property was a result of the preparation process.

The corrosion of amorphous as well as nanocrystalline $Zr_{69.5}Cu_{12}Ni_{11}Al_{7.5}$ ribbons was studied by means of a salt spray test, and also by potentiodynamic polarization [21]. Figure 7.3 shows the polarization curves in alkaline solution, indicating the very similar behavior independent of the microstructure. The salt spray test leads to the formation of a 4 to 6 μm-thick corrosion layer consisting of partially hydrated zirconium oxide (amorphous or t-ZrO_2, $A = 0.512$ nm, $C = 0.525$ nm) and Cu(Ni)-crystals (fcc, $A = 0.37$ nm). Glassy alloys are often expected to exhibit a better corrosion behavior due to the higher homogeneity – that is, the lack of defects or segregation. Nanocrystalline materials, however, are also relatively homogeneous. Alternatively, the driving force for the corrosion process should be smaller for the nanocrystalline material which was formed from an amorphous precursor material. Only minimal evidence was presented for a slightly higher corrosion resistance of the nanocrystalline microstructure in the salt spray test.

Cremaschi et al. [22] analyzed amorphous, nanocrystalline and crystalline FeSiB based alloys by means of the potentiodynamic anodic polarization technique in alkaline and neutral chloride media, studying in particular the influence on their corrosion behavior of both pH and the addition of Sn, Cu, Nb, and Al [22]. The electrochemical behavior of the materials was seen to depend heavily on the pH of the medium, with generalized corrosion occurring at neutral pH values, as opposed

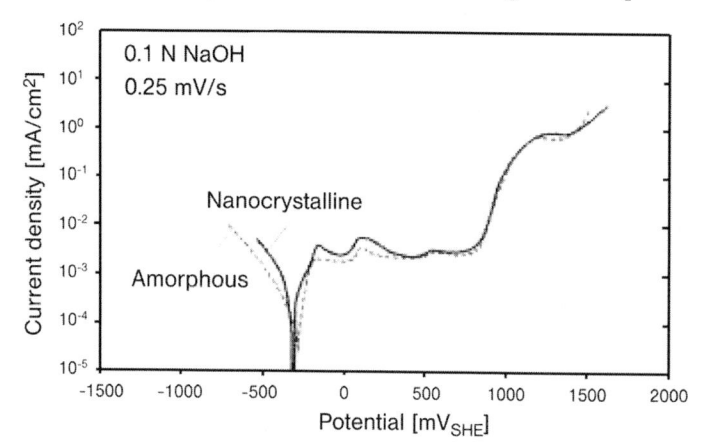

Figure 7.3 Polarization curves of amorphous and nanocrystalline $Zr_{69.5}Cu_{12}Ni_{11}Al_{7.5}$.

to a passive region followed by an abrupt increase in the current in alkaline media. In contrast, lightly adherent deposits were found on the surface of specimens containing Nb, regardless of pH. These deposits were rich in O, Nb, Si, and Na, and were not found on the surface of alloys without Nb. The electrochemical response was basically the same in both groups of alloys. The structural change was not reflected by the U_c value, which decreased as the pH rose. The shape of the electrochemical curve depended on the annealing temperature. A likely explanation for the formation of passive films underlying the flakes is that hydroxide compounds grow easily in alkaline media.

Nie et al. [23] have studied the abrasive wear/corrosion properties of Al_2O_3 coatings fabricated using plasma electrolysis. Figure 7.4 shows the polarization curves of the "thick" (250 μm) alumina-coated alloy sample and the untreated Al alloy substrate. Both types of sample were immersed in 0.5 M NaCl solution for 1 h, and 1 or 2 days before the corrosion tests were conducted. A stainless steel AISI 316L sample was also used in the corrosion test for comparison, and the polarization corrosion curve plotted (see Figure 7.4). The corrosion potentials, corrosion rates and anodic/cathodic Tafel slopes (β_A and β_c) were calculated from these tests after which, based on the approximate linear polarization at the corrosion potential (E_{corr}), the polarization resistance (R_p) values were determined by the relationship [24]:

$$R_p = \beta_A \times \beta_C / 2.3 \times i_{corr}(\beta_A + \beta_C) \tag{1}$$

where i_{corr} is the corrosion current density.

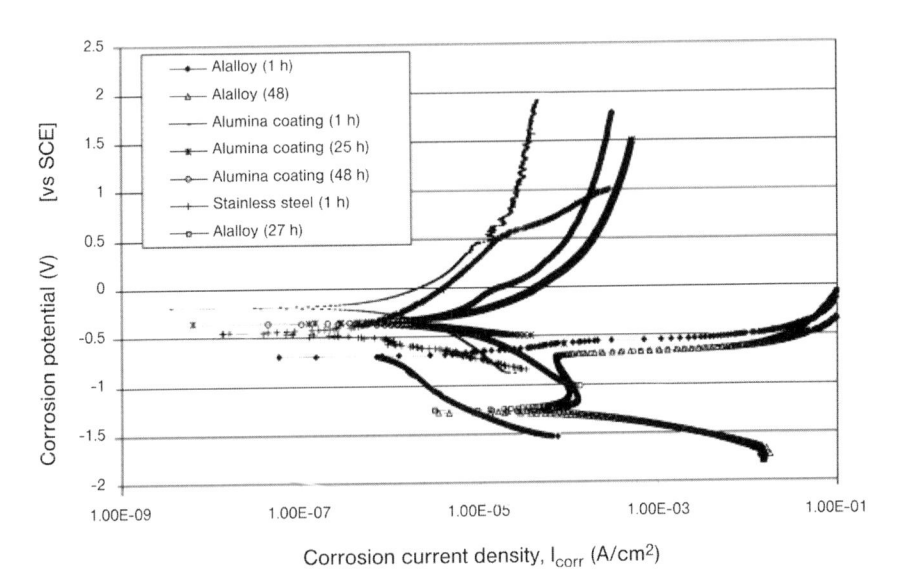

Figure 7.4 Potentiodynamic polarization curves of untreated substrate materials and plasma electrolytic oxidation (PEO) alumina coatings in 0.5 M NaCl solution after different immersion times.

7.2 Electrochemical Corrosion Behavior of Nanocrystalline Materials | 297

The corrosion current of the coated Al alloys increased slightly after immersion in the solution, as the porous surface required more current to fully passivate. However, the E_{corr} and R_p values of the alumina coating showed no significant reduction with increasing immersion time. Although the value of R_p of the Al substrate after 1 h of immersion was of the same order as that of the coatings, it was much lower than the latter after several days of immersion in the corrosion solution. After testing, large corrosion pits were present on the uncoated aluminum surface. These resulted from the fact that the corrosion resistance was considerably decreased after the thin protective oxide film on the uncoated aluminum substrate surface had been broken down by the corrosion processes. The PEO-coated Al alloys possessed excellent corrosion resistance in the solution, and considerably better even than the stainless steel.

El-Moneim et al. [25–27] have investigated the effect of grain size corrosion behavior of nanocrystalline NdFeB magnets in N_2-purged 0.1 M H_2SO_4 electrolyte by in-situ inductively coupled plasma solution analysis, gravimetric and electrochemical techniques, and hot extraction [H]-analysis [25]. Figure 7.5 shows the corrosion rates of the magnets with average grain sizes of 100 and 600 nm as a function of immersion time in N_2-purged acid solution at a sample rotation speed of 720 rpm. The corrosion rates presented represent the sum of partial dissolution rates of magnet components measured instantaneously during immersion in the test solution using on-line inductively coupled plasma (ICP) solution analysis (Figure 7.5).

It should also be noted that the corrosion rate of the annealed NdFeB magnet with a mean grain size of 600 nm was significantly lower than that of hot-pressed sample with 100 nm grain size. This indicates a beneficial effect of grain growth on the corrosion resistance of these magnets. This fact is further confirmed by the data in Figure 7.6, which summarizes the corrosion rates estimated gravimetrically after

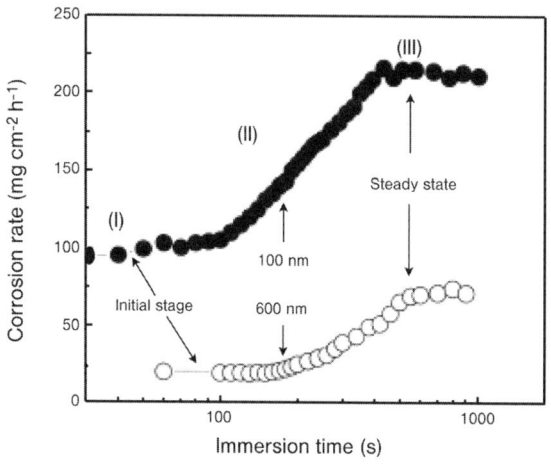

Figure 7.5 Corrosion rates measured by ICP solution analysis for NdFeB magnets with grain sizes of about 100 and 600 nm in N_2-purged H_2SO_4 at 25 °C and 720 rpm as a function of immersion time.

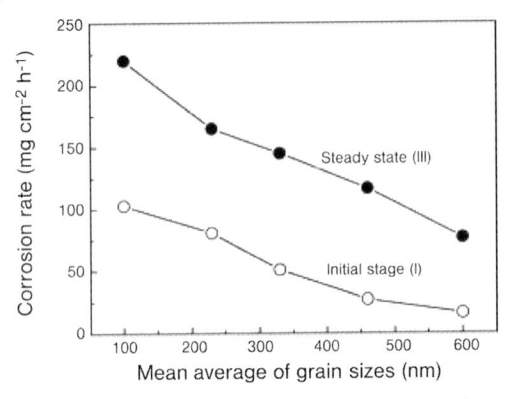

Figure 7.6 Changes in the corrosion rates measured gravimetrically after 1 and 10 min of magnet immersion in N_2-purged H_2SO_4 at 25 °C and 720 rpm as a function of the mean average grain size of ferromagnetic phase.

immersion for 1 and 10 min in 0.1 M H_2SO_4 solution at 720 rpm, as a function of the mean average grain size of the ferromagnetic phase. It is clear that the corrosion rates of NdFeB magnets generally decreases as the grain size of the matrix phase increases. Such corrosion inhibition with grain growth is attributed to the observed change in microstructure upon annealing; that is, the heterogeneity and reduction in volume fraction of the Nd-rich intergranular phase.

Isotropic nanocrystalline $Nd_{14}Fe_{80}B_6$ and $Nd_{12}Dy_2Fe_{73.2}Co_{6.6}Ga_{0.6}B_{5.6}$ magnets with different grain sizes ranging from 60 to 600 nm have been produced from melt-spun materials by hot pressing at 700 °C, and subsequent annealing at 800 °C for 0.5 to 6 h [26]. Partial substitution of Fe with Co and Ga leads to an improvement in corrosion resistance, and also reduces the affinity and binding energy for hydrogen in these materials. A coarsening of the microstructure results in the materials having a better corrosion performance.

Alloying additions (Co and Ga) and annealing have been used to clarify the corrosion property–microstructure relationships of nanocrystalline NdFeB magnets in 0.1 M H_2SO_4 solution, with the following conclusions being drawn:

- Additions of Co and Ga to nanocrystalline NdFeB magnets modify the composition of constituent phases and refine the microstructure. Further, annealing at 800 °C for various times leads to grain growth and heterogeneity of the microstructure.

- Nanocrystalline magnets with Co and Ga additions exhibit a lower absorption of corrosion hydrogen, though their surface activity for hydrogen reduction is increased. This effect provides an additional aspect explaining the beneficial effect of Co and Ga to the corrosion resistance. The beneficial roles of Co and Ga additions can, in general, be attributed to: (i) decreasing the electrochemical potential difference between intergranular phases and matrix phase with a consequence reduction of the strength of galvanic corrosion; (ii) retarding the

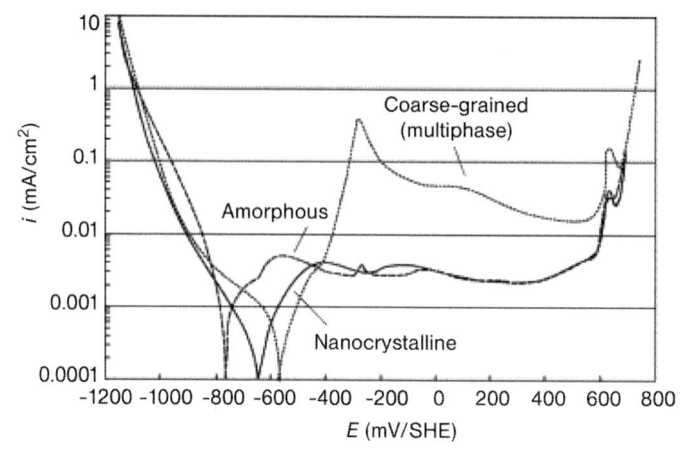

Figure 7.7 Polarization curves of amorphous, nanocrystalline and coarse-grained $(Ni_{70}Mo_{30})_{90}B_{10}$ alloys in 0.8 M KOH; sweep rate 20 mV min^{-1}.

diffusion of hydrogen through Co- and Ga-modified intergranular phase regions; and (iii) enhancing the rate of the adsorbed hydrogen atoms recombination on the magnet surface.

The corrosion behavior of nanocrystalline $(Ni_{70}Mo_{30})_{90}B_{10}$ alloys prepared by crystallization from the amorphous state was studied and compared with that of their amorphous and coarse-grained counterparts by Alves et al. [28]. The potentiodynamic polarization curves of these coarse-grained alloys, measured in a de-aired 0.8 M KOH solution at room temperature, are shown in Figure 7.7.

The $(Ni_{70}Mo_{30})_{90}B_{10}$ alloys achieve their corrosion resistance by the formation of an oxide passive layer, as shown in the polarization curve presented in Figure 7.7. Amorphous and nanocrystalline alloys exhibit very similar performance, characterized by a good passive behavior with low critical passivation current density as well as low passive current density. Thus, in terms of corrosion resistance, no significant change is observed when the melt-spun amorphous material crystallizes into a very fine nanocrystalline microstructure (upon annealing at 600 °C for 1 h) [28].

The high corrosion resistance of amorphous alloys has been partly attributed to the absence of crystal defects, which could act as initiation sites for corrosion. Metallic glasses are well known for their improved corrosion resistance arising from its chemically homogeneous single-phase nature without compositional fluctuations and crystal defects [29,30].

A study conducted by Zander and Köster [31] on the corrosion of nanocrystalline $Zr_{69.5}Cu_{12}Ni_{11}Al_{7.5}$ ribbons involved an investigation similar to that with an amorphous precursor, using a salt spray test in 5% NaCl solution (pH 6.5). Both, scanning electron microscopy (SEM) (Figure 7.8) and transmission electron microscopy (TEM) (Figure 7.9) investigations showed the presence of a nanocrystalline passive layer with grain sizes of about 10 nm and few amorphous areas compared to

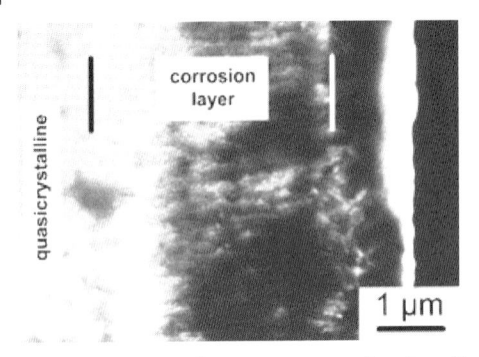

Figure 7.8 Corrosive layer on nanocrystalline $Zr_{69.5}Cu_{12}Ni_{11}Al_{7.5}$ after salt spray test (cross-sectional fracture surface; SEM image) [33].

$Zr_{69.5}Cu_{12}Ni_{11}Al_{7.5}$, which formed mainly an amorphous passive layer. Electron diffraction measurements of the nanocrystalline passive layer also revealed the pattern of the unknown fcc phase, with a lattice parameter of $a = 0.37$ nm (probably Cu(Ni,Al) solid solution) similar to that formed on the amorphous counterpart.

Elkedim et al. [32] compared the mechanically activated field-activated pressure assisted synthesis (MAFAPAS) FeAl corrosion behavior with that of a bulk material obtained by extrusion at 1000 °C of the milled FeAl powders (Figure 7.10). The as-extruded material had a sub-micrometer grain size in the range of 700 to 800 nm (Fe–40Al alloy). Globally, a better corrosion resistance was observed in the case of the MAFAPAS iron aluminide, and in particular the nanocrystalline material exhibited a good passive behavior with a low passive current density.

The enhancement of corrosion resistance has been attributed to the larger fraction of the interphase boundaries and the fast diffusion character of the nanocrystalline materials. It is due to the fast diffusion effect that a large amount of Al will

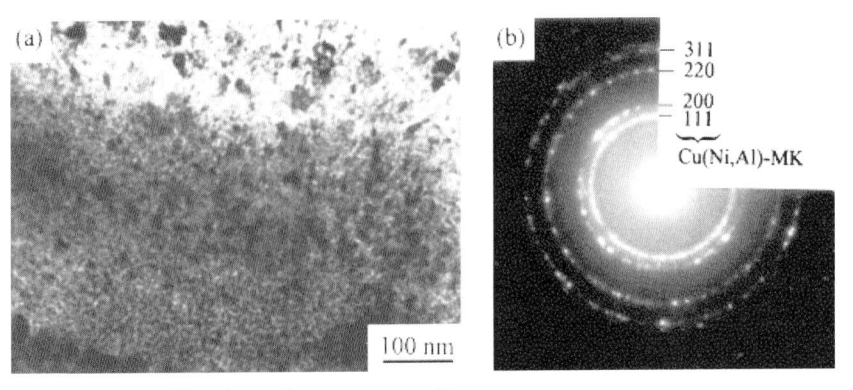

Figure 7.9 Passive film observed on a nanocrystalline $Zr_{69.5}Cu_{12}Ni_{11}Al_{7.5}$ ribbon after salt spray test [33]. (a) TEM image; (b) electron diffraction pattern.

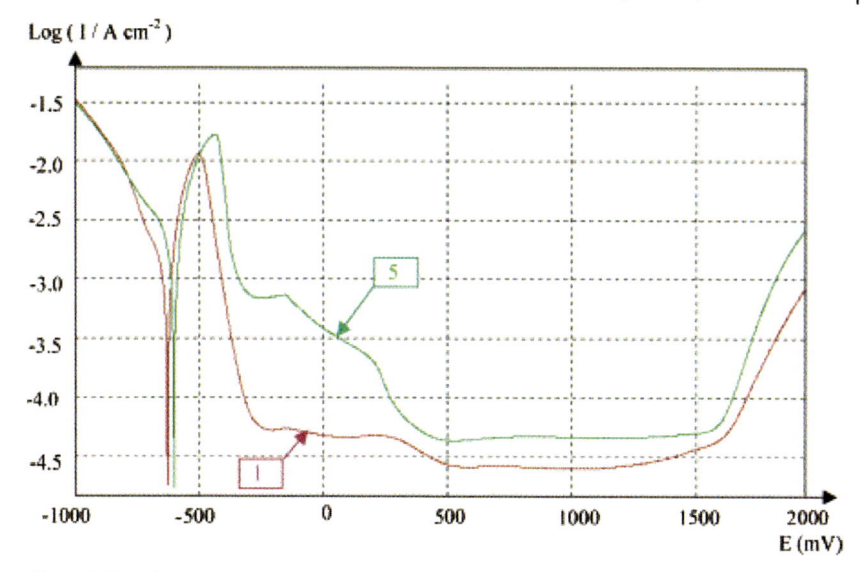

Figure 7.10 Polarization curves comparing the corrosion behavior of nanocrystalline (relative density 99%) MAFAPAS FeAl with a bulk Fe–40Al in 0.5 M H_2SO_4. Curve 1: nanocrystalline MAFAPAS FeAl. Curve 5: as-extruded FeAl.

accumulate on the surface of the samples to form a protective film. Indeed, it has been reported that the presence of aluminum has a significant effect on the corrosion properties of iron by improving its ability to form a passive film in sulfuric acid [33].

Gang et al. [34] compared the potentiodynamic polarization curves of the nanocrystalline high-velocity oxy-fuel (HVOF) FeAl coatings with that of a massive FeAl sample obtained by extrusion of the same milled powders (Figure 7.11). All curves exhibited a typical active–passive–transpassive behavior as the potential increased, with all curves also showing a passive region consisting of two domains.

In terms of corrosion, when compared to the as-extruded material, the nanocrystalline coatings exhibited globally the same active–passive–transpassive behavior, but a poorer corrosion resistance. However, the latter finding, when compared to bulk materials of the same nature, was not very surprising and was consistent with the results from other studies of various metals such as stainless steel, titanium, and nickel-based alloys. The roughness and connecting porosity generate a higher surface area in contact with the solution, and this is responsible for the higher current densities seen for sprayed deposits compared to bulk materials that are free of porosity.

Figure 7.12 shows the potentiodynamic polarization curves for the Fe–10Cr cast alloy and the nanocrystalline coating in 0.05 M H_2SO_4 + 0.25 M Na_2SO_4 solution, as reported by Meng et al. [35]. Fe–10Cr nanocrystalline coatings with a grain size of 20 to 30 nm were synthesized on glass substrates by magnetron sputtering.

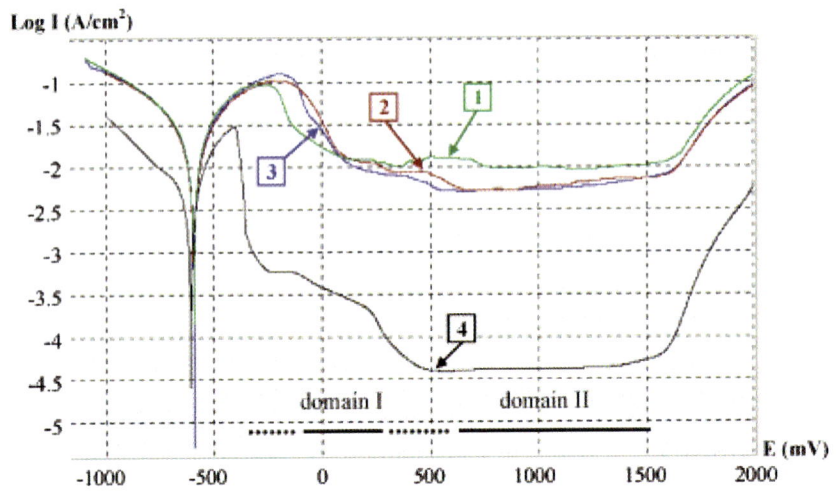

Figure 7.11 Potentiodynamic polarization curves comparing the corrosion behaviors of three HVOF coatings (curves 1, 2, and 3 for the coatings 1, 2, and 3, respectively) with an extruded FeAl sample (curve 4).

The nanocrystalline Fe–10Cr was seen to have a higher i_{corr}, a lower i_{max} and a lower i_p than did the Fe–10Cr cast alloy, yet both materials had similar values of E_{tr}. This suggests that, for nanocrystalline Fe–10Cr, the active dissolution was accelerated, and the passivation ability and chemical stability of the passive film were improved compared to those for Fe–10Cr cast alloy. Compared with the Fe–10Cr

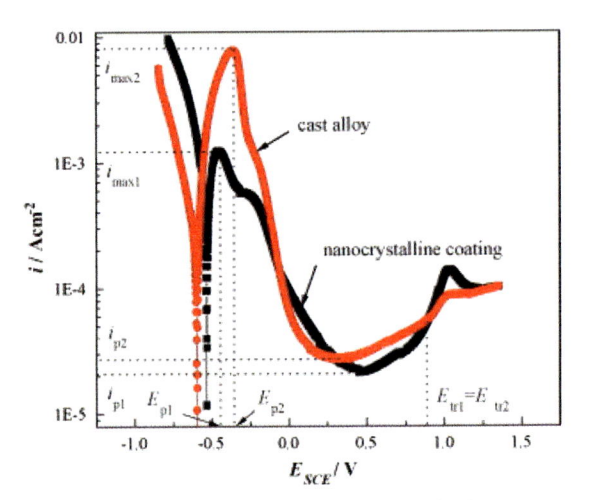

Figure 7.12 Potentiodynamic polarization curves for the Fe–10Cr nanocrystalline coating and the cast alloy in 0.05 M H_2SO_4 + 0.25 M Na_2SO_4 solution.

cast alloy, the active dissolution of the nanocrystalline Fe–10Cr coating was accelerated; Cr was more easily enriched in the passive film due to the large number of metal grain boundaries, which supply diffusion paths for Cr to the surface, and therefore it was more easily passivated. The passive films on both materials were n-type semiconductors in acidic solution without Cl^-, and p-type semiconductors in acidic solution with Cl^- [35].

Wang et al. [36] studied the electrochemical corrosion behavior of nanocrystalline Co compared with coarse-grained (CG) Co coatings in different corrosion media characterized using potentiodynamic polarization testing, electrochemical impedance spectroscopy (EIS), and X-ray photoelectron spectroscopy (XPS).

The typical anodic potentiodynamic polarization of nanocrystalline and CG Co coatings in 10 wt.% NaOH and 10 wt.% HCl solutions are shown in Figure 7.13a and b. Although a typical active–passive–transpassive–active behavior can be clearly observed in nanocrystalline Co in NaOH solution, only active behavior without passivation was displayed in HCl solution.

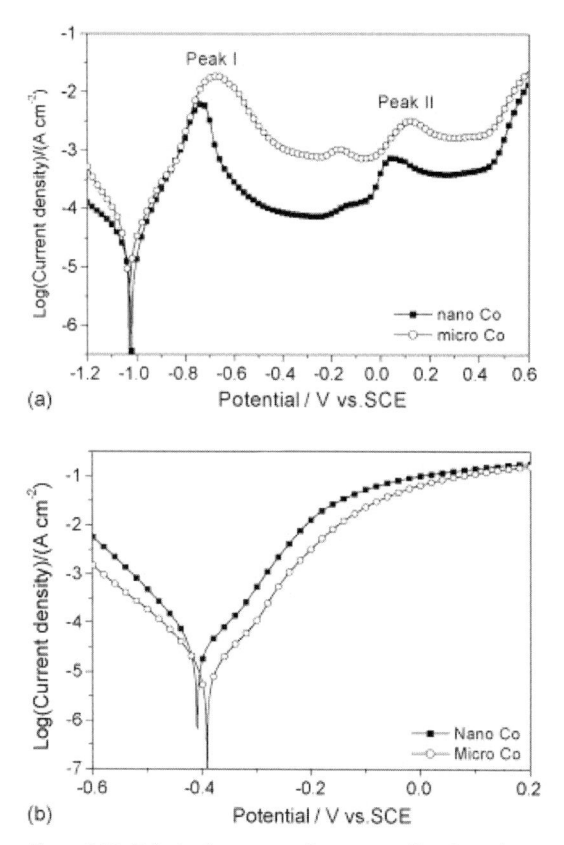

Figure 7.13 Polarization curves of nanocrystalline Co and coarse-grained Co measured in (a) 10 wt.% NaOH and (b) 10 wt.% HCl solutions, respectively.

The results indicated that nanostructure enhances both the formation of passive films and the stability of the passive film on Co coatings, and this was consistent with previous results obtained with nanocrystalline Fe–Cr coatings [37]. Thus, it may be concluded that the corrosion resistance of nanocrystalline Co coatings in NaOH solution was greatly enhanced with a reduction of grain size due to the formation of stable $Co(OH)_2/Co_3O_4$ duplex passive films. However, in HCl solution – where no obvious passive phenomena occurred – nanocrystalline Co coatings exhibited a much higher corrosion current density and lower corrosion resistance compared to CG Co coatings.

The corrosion properties of micro- and nanocrystalline Co and Co–1.1 wt.% P alloys were studied in de-aired 0.1 M H_2SO_4 solution using open-circuit potential measurement, polarization tests, AC impedance measurements and XPS, by Jung and Alfantazi [38]. Figure 7.14 presents the typical potentiodynamic polarization curves of micro Co, nano Co and nano Co–1.1P in de-aired 0.1 M H_2SO_4 solution, with all samples exhibiting active dissolution but without any distinctive transition to passivation up to $-0.1\ V_{SCE}$.

In comparison to that of nanocrystalline (nano) Co, the anodic polarization curve of nano Co–1.1P shifted to a more positive value of potential, and lowered the anodic dissolution rate to $-0.1\ V_{SCE}$. The overpotential for H_2 evolution also decreased, and the cathodic reaction rates increased rapidly upon increasing the cathodic potential.

Thus, it can be concluded that the effects of P for nano Co–1.1P on the hydrogen evolution reaction are more significant than grain size reduction. This is also consistent with the anodic behavior. Considering that the grain size of nano Co–1.1P (10 nm) is smaller than that of nano Co (20 nm), the anodic dissolution kinetics for nano Co–1.1P would be expected to increase. However, an addition of P content led to a positive shift of E_{corr} of some 59 mV, and the anodic dissolution rates were

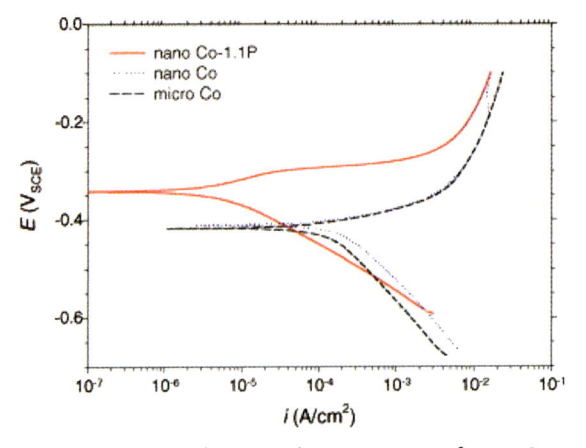

Figure 7.14 Potentiodynamic polarization curves of micro Co, nano Co and nano Co–1.1P alloys in de-aired 0.1 M H_2SO_4 solution (scan rate $= 0.5\ mV\ s^{-1}$).

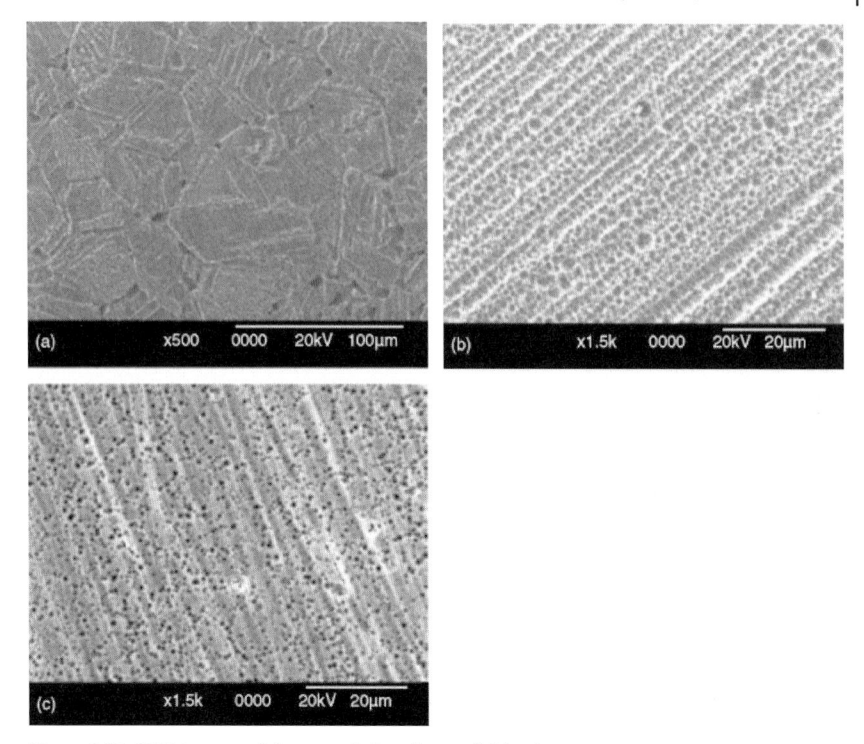

Figure 7.15 SEM images of the corroded surfaces of: (a) micro Co; (b) nano Co; (c) nano Co–1.1P after potentiodynamic polarization scan in de-aired 0.1 M H_2SO_4.

significantly reduced compared to that of nano Co. Therefore, it may be concluded that P content plays a more important role than grain size reduction in the corrosion behavior of nano Co–1.1P.

Figure 7.15 shows SEM images of the corroded surfaces of micro Co, nano Co and nano Co–1.1P after potentiodynamic polarization scans in 0.1 M H_2SO_4. As shown in Figure 7.15a, the micro Co corroded extensively along the grain boundary and triple junctions, whereas nano Co exhibited relatively uniform corrosion and the corroded surface retained the grinding marks. Such examination of corrosion morphologies showed that the detrimental preferential grain boundary attack was largely eliminated by nanoprocessing.

The electrochemical corrosion behavior of the nanocrystalline coating and bulk steel in solutions of 0.25 M $Na_2SO_4 + 0.05$ M H_2SO_4 and 0.5 M $NaCl + 0.05$ M H_2SO_4 was investigated by using potentiodynamic polarization, potentiostatic polarization, and AC impedance techniques [39].

A nanocrystallized 309 stainless steel (309SS) coating has been fabricated on a glass substrate by DC magnetron sputtering. The coating, which had an average grain size of less than 50 nm, had a ferritic (bcc) structure rather than the austenitic (fcc) structure of the bulk steel.

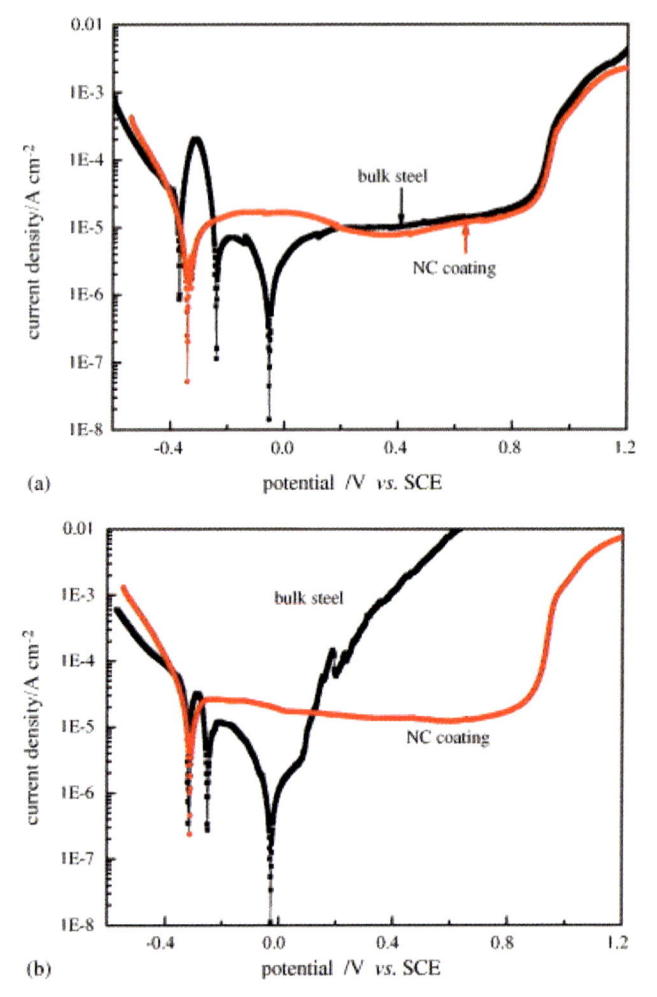

Figure 7.16 Potentiodynamic polarization plots of the nanocrystalline (NC) coating and bulk steel in (a) Na_2SO_4 and (b) NaCl solution.

In Na_2SO_4 solution (Figure 7.16a), three corrosion potentials – at which the anodic current density is equal to the cathodic current density – exist in the active, active–passive and passive regions, respectively. The three corrosion potentials for bulk steel revealed that it was an unstable, passive system, while the nanocrystalline coating, directly entering the passive region under the applied potential, showed excellent passivation ability. The current densities, which increased dramatically with a potential at approximately $0.9\,V_{SCE}$ for the two materials, should be attributed to the oxygen evolution reaction. Thus, the good localized corrosion resistances of these two materials is clear. For the bulk steel, such resistance is attributed to its high resistance to the breakdown of its passivity by sulfate anions [40], and its high Cr content. The nanocrystalline coating showed a slightly better corrosion behavior due to its nanocrystalline microstructure.

A major difference was found between the electrochemical behavior of the nanocrystalline coating and bulk steel in NaCl solution (Figure 7.16b). The nanocrystalline coating not only displayed a much wider passive region than the bulk steel, but also exhibited no pitting corrosion – unlike the bulk steel, which had a pitting potential of approximately $0.08\,V_{SCE}$. This distinct difference in pitting resistance may not be related to the different lattice structure, however, because ferritic stainless steel usually exhibits a lower breakdown potential than austenitic stainless steel of the same Cr content [41]. Consequently, it is believed that the nanocrystallization structure rather than the lattice structure is responsible for this observed difference.

Potentiodynamic polarization, electrochemical impedance spectroscopy, and SEM were each employed to characterize the corrosion behavior of the sputtered nanocrystalline Cu–20Zr films and the corresponding cast Cu–20Zr alloy in HCl aqueous solutions of various concentration (0.1, 0.5, and $1.0\,\mathrm{mol\,L^{-1}}$) [42].

Figure 7.17a–c shows, respectively, the potentiodynamic curves of the nanocrystalline Cu–20Zr films and the corresponding cast alloy in 0.1, 0.5, and 1.0 M HCl solution. As seen from Figure 7.17, this illustrates qualitatively that the anodic dissolution process of Cu–20Zr alloy was clearly retarded by nanocrystallization. In other words, the nanocrystalline Cu–20Zr film has an improved corrosion resistance compared with the cast Cu–20Zr alloy.

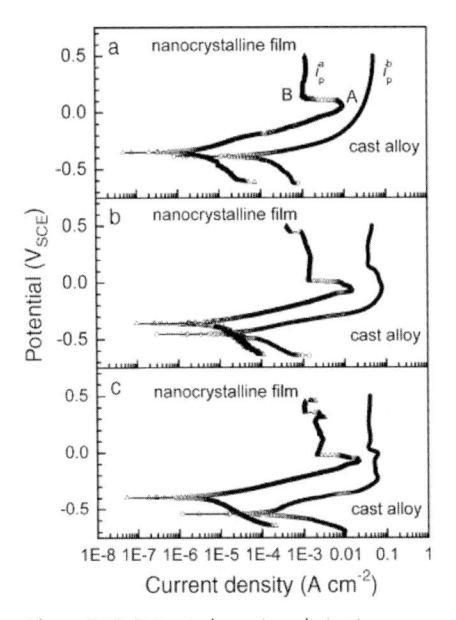

Figure 7.17 Potentiodynamic polarization curves of the nanocrystalline Cu–20Zr films and the cast Cu–20Zr alloy in (a) 0.1 M, (b) 0.5 M, and (c) 1.0 M HCl solution.

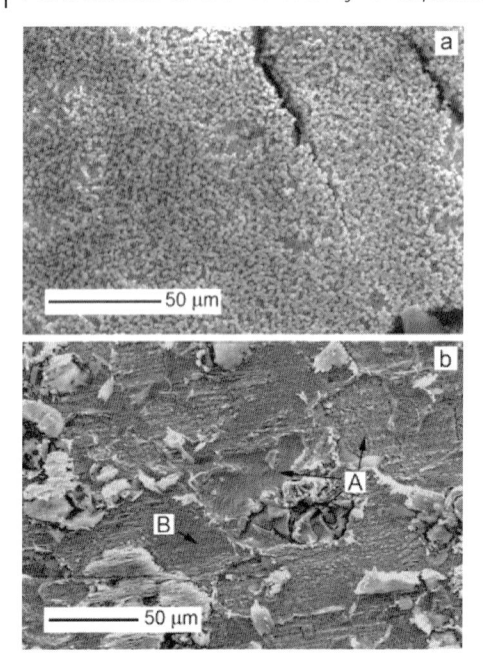

Figure 7.18 SEM images of (a) nanocrystalline Cu–20Zr film and (b) cast alloy after potentiodynamic polarization in 0.1 M HCl solution.

Figure 7.18 shows the SEM images of the corroded surfaces of nanocrystalline Cu–20Zr films and cast Cu–20Zr alloy in the potential range for simultaneous dissolution of Cu and Zr. In the case of the nanocrystalline Cu–20Zr films, a corrosion product layer covers its surface. Energy dispersive X-ray (EDX) analysis of the corroded surface of the nanocrystalline Cu–20Zr films indicated that the corrosion product was a copper–chloride complex.

The results showed the corrosion resistance of the sputtered nanocrystalline Cu–20Zr films to be superior to that of the cast Cu–20Zr alloy. This improvement in corrosion resistance through nanocrystallization is explained by: (i) the formation of a continuous Cu-rich layer at the corrosion potential; and (ii) the formation of a copper–chloride complex layer in the potential range for simultaneous dissolution of Cu and Zr [42].

Further investigations were dedicated the corrosion of nanocrystalline nickel and its alloys [43–48]. For example, Wang *et al.* [43] studied the effects of grain size reduction on the electrochemical corrosion behavior of nanocrystalline Ni produced by pulse electrodeposition. Potentiodynamic polarization testing, electrochemical impedance spectroscopy and X-ray photoelectron spectroscopy were used to confirm the electrochemical measurements and the suggested mechanisms.

The grain size distribution was determined from TEM dark-field images by measuring approximately 500 grains, yielding average grain sizes of 250 nm, 54 nm, and 16 nm. Figure 7.19 typically presents the bright-field image, dark-field image,

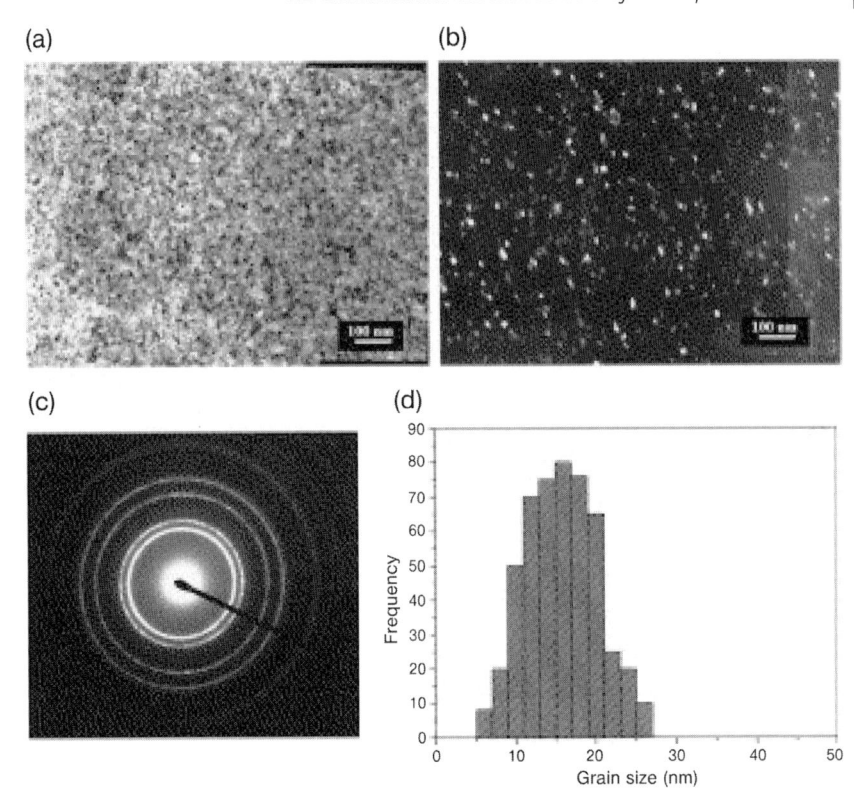

Figure 7.19 Bright-field (a) and dark-field (b) TEM images with electron diffraction pattern (c) and grain size distribution (d) of nanocrystalline nickel coating (with 16 nm average grain size).

electron diffraction pattern and grain size distribution of a nanocrystalline Ni coating with an average grain size of 16 nm. The first four rings in Figure 7.19c represent (1 1 1), (2 0 0), (2 2 0) and (3 1 1) of the nanocrystalline Ni coating, respectively. An analysis of the pattern verified a single-phase, face-centered cubic Ni structure.

Typical anodic potentiodynamic polarization curves of electrodeposited Ni coatings with different grain size, measured in 10 wt.% NaOH solution, are presented in Figure 7.20a, together with magnifications of anodic Tafel curves in the passive range for nanocrystalline Ni coatings (Figure 7.20b). Based on the data in Figure 7.20b, a typical passivation behavior can be clearly observed in nanocrystalline Ni coatings in which the electrode is anodically polarized to a more positive potential, whereas the value of the corresponding current remains limited. Originally, it was considered that passivation of the electrode in the potential range of -200 mV to 300 mV, involved water molecules which were adsorbed onto the electrode surface, and the formation of $Ni(OH)_2$ passive films.

The XPS analysis (Figure 7.21) indicated that the passive film formed on the nanocrystalline Ni was composed of stable and continuous Ni hydroxides.

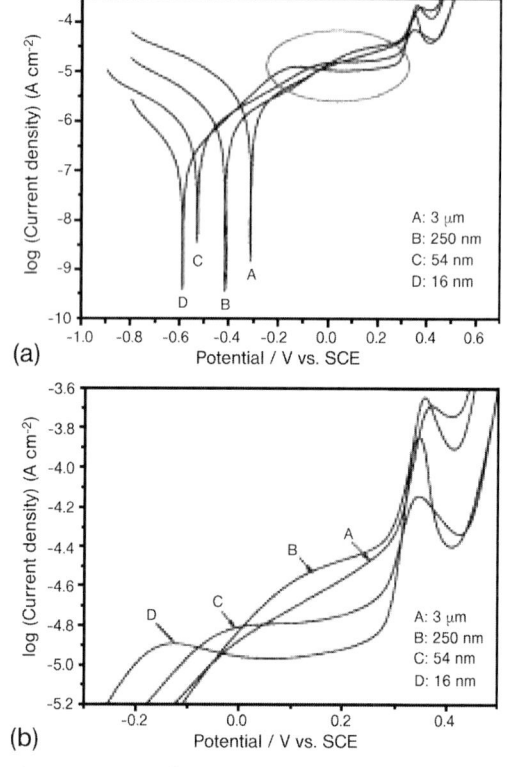

Figure 7.20 (a) Polarization curves of Ni coatings with different grain size measured in 10 wt.% NaOH solution. The ellipse highlights (b) a magnification of the anodic Tafel curves in the passive range for nanocrystalline Ni coatings (16 mm).

The corrosion resistance of NC Ni is believed to be improved by the rapid formation of continuous Ni hydroxide passive films at surface crystalline defects, and the relatively higher integrity of passive films as a result of the smooth and protective nature of the passive films formed on nanocrystalline Ni coatings [43].

Liu et al. [46] reported the electrochemical corrosion behaviors of Ni-based superalloy nanocrystalline coating fabricated by a magnetron sputtering technique. The potentiodynamic polarization curves for the Ni-based superalloy nanocrystalline coating and cast superalloy in 0.25 M Na_2SO_4 + 0.05 M H_2SO_4 and 0.5 M NaCl + 0.05 M H_2SO_4 solutions are shown in Figure 7.22. In Na_2SO_4 acidic solution, the nanocrystalline coating and cast superalloy were able to passivate under a certain potential. Although the E_{corr} and the breakdown potential of the nanocrystalline coating and the cast superalloy were similar, the minimum passive current density (i_{pass}) of the coating was only slightly larger than that of the cast alloy. Subsequently, the passive film formed on the nanocrystalline coating proved to be less stable than that of the cast alloy.

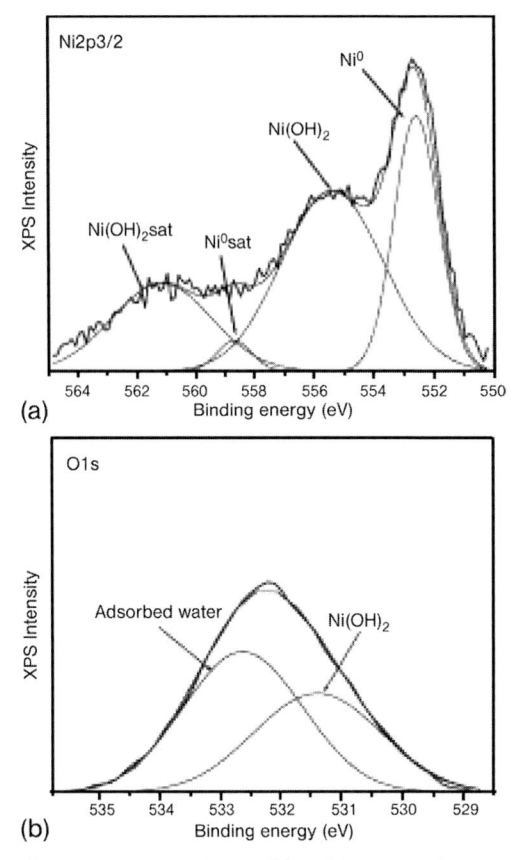

Figure 7.21 Deconvolution of the XPS spectra of Ni 2p3/2 and O1s acquired from the nanocrystalline Ni after anodic potentiodynamic polarization followed by additional 15 min potentiostatic polarization in 10% NaOH solution.

A major difference was identified between the electrochemical behavior of the nanocrystalline coating and the cast alloy when in NaCl acidic solution. Here, the nanocrystalline coating showed a much more positive breakdown potential and a wider passive region compared to the cast alloy, while the minimum passive current density (i_{pass}) of the cast alloy was one magnitude larger than that of the nanocrystalline coating. The results indicated that the nanocrystalline coating was less susceptible to chloride ion corrosion than was the cast alloy.

The corrosion behavior of electrodeposited nanocrystalline, ultra-fine grained and polycrystalline copper foils was studied in 0.1 M NaOH solution by Yu et al. [49]. Figure 7.23 shows the typical potentiodynamic polarization curves for the electrodeposited and commercially available electronic-grade cold-rolled annealed polycrystalline (EG-CRA). All samples displayed a typical active–passive–transpassive behavior, with the formation of a duplex passive film.

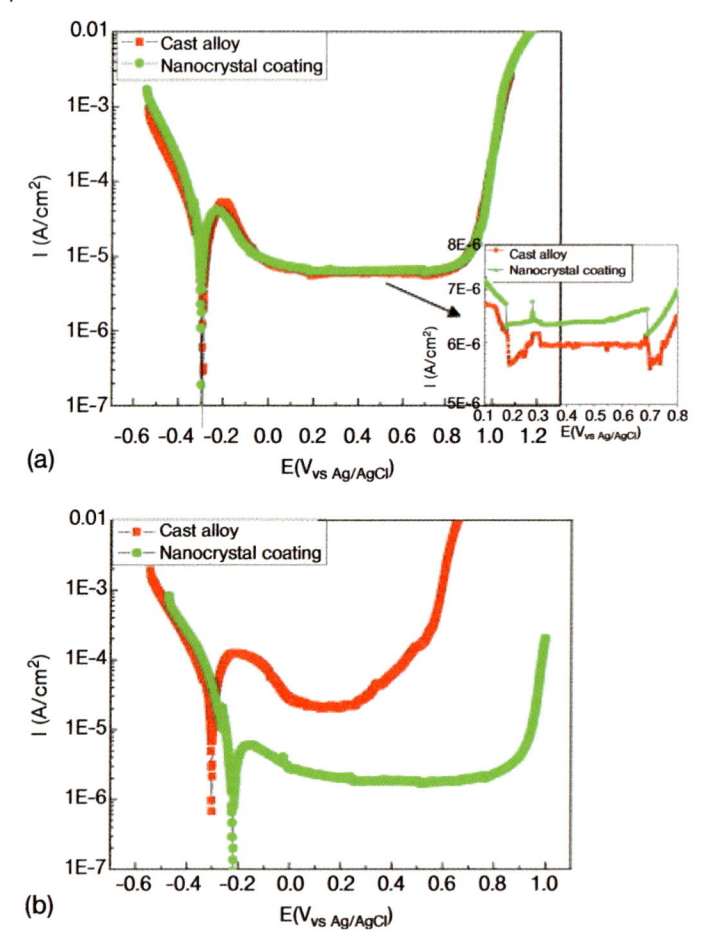

Figure 7.22 Potentiodynamic polarization plots of the cast alloy (■) and the nanocrystalline coating (●). (a) 0.25 M Na_2SO_4 + 0.05 M H_2SO_4; (b) 0.5 M NaCl + 0.05 M H_2SO_4.

The results showed that both, electrodeposited copper foils with grain sizes ranging from 45 nm to 1 μm and conventional polycrystalline cold-rolled annealed copper foil, displayed an active–passive–transpassive corrosion behavior in 0.1 M NaOH (pH 13). The general shape of the potentiodynamic polarization curves was not affected by grain size reduction to the nanocrystalline range. A similarly fine, needle-like corrosion product morphology was observed for all copper foils.

Later, Luo et al. [50] studied the corrosion behavior of nanocrystalline copper bulk prepared by inert gas condensation and *in-situ* warm compress (IGCWC). Figures 7.24 and 7.25 show typical anodic polarization curves obtained from coarse-grain, nanocrystalline and annealed nanocrystalline copper samples, respectively. The electrochemical data measured in these investigations are listed in

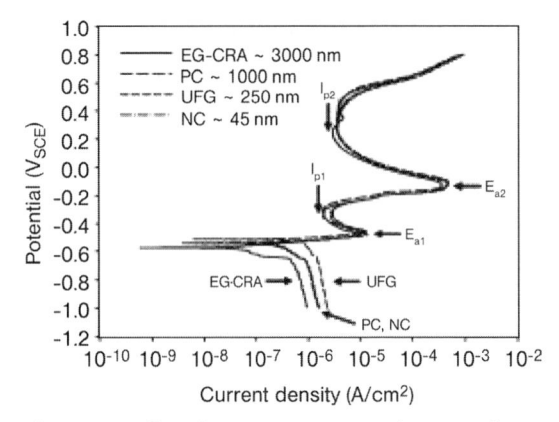

Figure 7.23 Effect of grain size on potentiodynamic polarization curves obtained in de-aired 0.1 M NaOH at pH 13.

Table 7.1. The curves were seen to demonstrate qualitatively similar behavior, but with different values of their electrochemical data. In comparison with coarse-grain copper, nanocrystalline copper exhibited a lower corrosion resistance. The primary passive potential E_{cr} of nanocrystalline copper ($E_{cr}^{nc} = 1.00$ V) was more negative than that of coarse-grain copper ($E_{cr}^{c} = 1.26$ V), which indicated that the former was easier to passivate than the latter. It is evident that nanocrystalline copper exhibits a lower activation energy for passivation than does coarse-grain copper. The critical current density for passivity I_{cr} of nanocrystalline copper ($I_{cr}^{nc} = 158$ mA cm^{-2}) was much greater than that of coarse-grain copper ($I_{cr}^{c} = 104$ mA cm^{-2}), indicating that the nanocrystalline structure enhanced the kinetics of anodic dissolution that resulted in a greater dissolution rate for nanocrystalline copper. The passive current

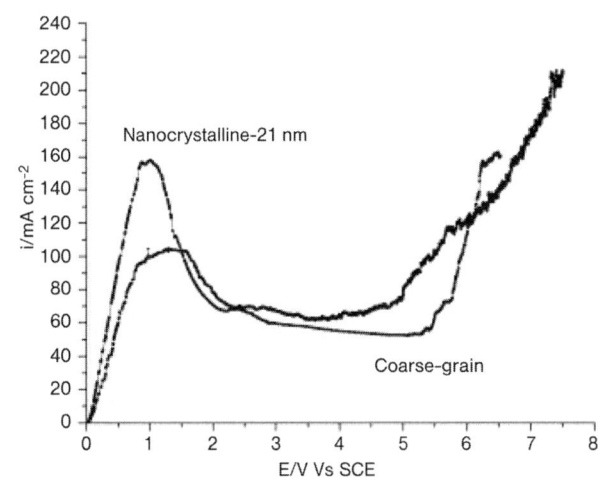

Figure 7.24 Anodic polarization curves of coarse-grain copper and nanocrystalline copper.

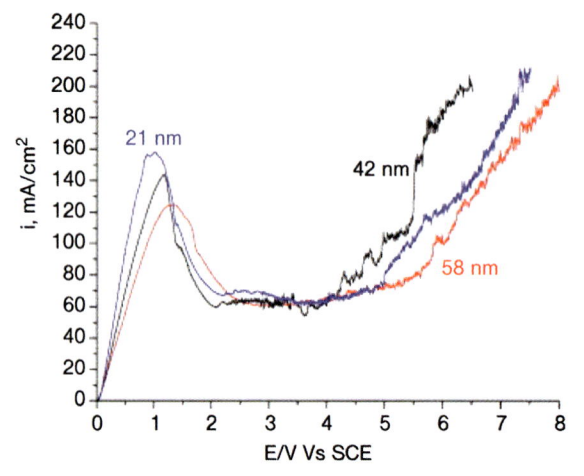

Figure 7.25 Anodic polarization curves of nanocrystalline copper with different grain size.

density I_p of nanocrystalline copper ($I_p^{nc} = 70\,\text{mA}\,\text{cm}^{-2}$) was greater than that of coarse-grain copper ($I_p^c = 50\text{–}60\,\text{mA}\,\text{cm}^{-2}$). It can be deduced that the passivation film formed on the surface of nanocrystalline copper had lower protection characteristics than that of coarse-grain copper. The transpassive potential E_{TP} of nanocrystalline copper ($E_{TP}^{nc} = 5.0\,\text{V}$) was more negative than that of coarse-grain copper ($E_{TP}^c = 5.7\,\text{V}$), indicating that the passivation film formed on nanocrystalline copper was of lower stability than that formed on coarse-grain copper. In addition to that for nanocrystalline copper, the corrosion behavior is seen to be greatly affected by reducing the average grain size to the nanocrystalline range [51] and by defects, such as micro-gaps, produced in the preparation of nanocrystalline copper bulk. With the grain size increasing, the E_{cr} and E_{TP} values of nanocrystalline copper also rise, while the I_{cr} value decreases, as studied in 0.1 M $CuSO_4$ + 0.05 M H_2SO_4 solution.

Table 7.1 Anodic polarization experimental results for copper samples.

Sample	Critical current density for passivity, I_{cr} (mA cm^{-2})	Passive current density, I_p (mA cm^{-2})	Transpassive potential, E_{TP} (V)	Primary passive potential, E_{cr} (V)	Passive potential range, W_p (V)
Nanocrystalline copper - 21 nm	158	70	5.0	1.00	2.0–5.0
Nanocrystalline copper - 42 nm	144	60	4.3	1.17	2.0–4.3
Nanocrystalline copper - 58 nm	125	60–70	5.6	1.24	2.5–5.6
Coarse-grain copper	104	50–60	5.7	1.26	2.5–5.7

Finally, compared to coarse-grain copper, nanocrystalline copper showed a decreased resistance to corrosion which was mainly attributed to the high activity of surface atoms and intergranular atoms. The high activity of surface atoms and intergranular atoms, resulting from the reduction in grain size, led to an enhancement of passivation ability and an increase in dissolution of the passive film. In addition to defects such as micro-gaps, produced during the fabrication of nanocrystalline samples, there were major effects on the overall corrosion performance of the nanocrystalline sample.

7.3
Conclusions

This chapter has highlighted the electrochemical corrosion of nanocrystalline materials, with some authors having reported that nanocrystalline materials exhibited a relatively improved corrosion resistance. Alternatively, results obtained in other studies suggested that nanocrystalline materials had relatively higher dissolution and corrosion rates than their conventional, coarse-grained counterparts. A numbers of reports had also demonstrated that there were no significant differences in corrosion behavior between nanocrystalline materials and their conventional counterparts. Clearly, a better understanding of the corrosion mechanism of nanocrystalline materials, within various environments, is required to explain these seemingly contradictory results.

References

1 Gleiter, H. (1989) *Prog. Mater. Sci.*, **33**, 223.
2 Koch, G.H., Brongers, M.P.H., Thompson, N.G., Virmani, Y.P. and Payer, J.H. (2001) Corrosion Cost and Preventive Strategies in the United States. Report FHWA-RD-01-156(Report by CC Technologies Laboratories Inc. to Federal Highway Administration (FHWA), Office of Infrastructure Research and Development, McLean.
3 Renner, F.U., Stierle, A., Dosch, H., Kolb, D.M., Lee, T.-L. and Zegenhagen, J. (2006) *Nature*, **439**, 707.
4 Rofagha, R., Langer, R., El-Sherik, A.M., Erb, U., Palumbo, G. and Aust, K.J. (1991) *Scr. Metall. Mater.*, **25**, 2867.
5 Rofagha, R., Erb, U., Ostrander, D., Palumbo, G. and Aust, K.T. (1993) *J. Nano Mater.*, **2**, 1.
6 Inturi, R.B. and Szklarska-Smialowska, Z. (1992) *Corrosion*, **48**, 398.
7 Thorpe, S.J., Ramaswami, B. and Aust, K. T. (1988) *J. Electrochem. Soc.*, **135**, 2162.
8 Tong, H.Y., Shi, F.G. and Lavernia, E.J. (1995) *Scripta Metall.*, **21**, 511.
9 Elkedim, O. and Gaffet, E. (1997) Organic and Inorganic Coatings for Corrosion Prevention, in *European Federation of Corrosion, No. 20* (eds P. Fedrizzi and L. Bonora), The institute of Materials, London, pp. 267–275.
10 Aita, C.R. and Tait, W.S. (1992) *Nanostruct. Mater.*, **1**, 269–292.
11 Heim, U. and Schwitzgebel, G. (1999) *Nanostruct. Mater.*, **12**, 19–22.
12 Kirchheim, R., Huang, X.Y., Cui, P., Birringer, R. and Gleiter, H. (1992) *Nanostruct. Mater.*, **1**, 167.

13 Lopez-Hirata, V.M. and Arce-Estrada, E.M. (1997) *Electrochim. Acta*, **42**, 61–65.

14 Vinogradov, A., Mimaki, T., Hashimoto, S. and Valiev, R. (1999) *Scripta Mater.*, **3**, 319–326.

15 Jiang, W. and Molian, P. (2001) *Surface and Coatings Technology*, **135**, 139–149.

16 Yu, M., Shivpuri, R. and Rapp, R.A. (1995) *J. Mater. Eng. Perform.*, **42**, 175–181.

17 Pardo, A., Otero, E., Merino, M.C., López, M.D., Vázquez, M. and Agudo, P. (2001) *Corrosion Sci.*, **43**, 689–705.

18 Pardo, A., Otero, E., Merino, M.C., López, M.D., Vázquez, M. and Agudo, P. (2001) *J. Non-Crystall. Solids*, **287**, 421–427.

19 Pardo, A., Otero, E., Merino, M.C., López, M.D., Vázquez, M. and Agudo, P. (2002) *Corrosion Sci.*, **44**, 1193–1211.

20 Pardo, A., Merino, M.C., Otero, E., López, M.D. and M'hich, A. (2006) *J. Non-Crystall. Solids*, **352**, 3179–3190.

21 Köster, U., Zander, D., Triwikantoro, Rüdiger, A. and Jastrow, L. (2001) *Scripta Mater.*, **44**, 1649–1654.

22 Cremaschi, V., Avram, I., Pérez, T. and Sirkin, H. (2002) *Scripta Mater. Surface Coatings Technol.*, **46**, 95–100.

23 Nie, X., Meletis, E.I., Jiang, J.C., Leyland, A., Yerokhin, A.L. and Matthews, A. (2002) *Surface Coatings Technol.*, **149**, 245–251.

24 Jones, D.A. (1996) *Principles and Prevention of Corrosion*, 2nd edn Prentice-Hall, UK.

25 El-Moneim, A.A., Gebert, A., Schneider, F., Gutfleisch, O. and Schultz, L. (2002) *Corrosion Sci.*, **44**, 1097–1112.

26 El-Moneim, A.A., Gebert, A., Uhlemann, M., Gutfleisch, O. and Schultz, L. (2002) *Corrosion Sci.*, **44**, 1857–1874.

27 El-Moneim, A.A., Gutfleisch, O., Plotnikov, A and Gebert, A. (2002) *J. Magn. Magn. Mater.*, **248**, 121–133.

28 Alves, H., Ferreira, M.G.S. and Köster, U. (2003) *Corrosion Sci.*, **45**, 1833–1845.

29 Hashimoto, K. and Masumoto, T. (1981) Corrosion behavior of amorphous alloys in *Treatise on Material Science and Technology* (ed. H. Herman), Vol. 20, Academic Press, New York. p. 291.

30 Köster, U. and Alves, H. (1997) Electrochemical properties of rapidly solidified alloys. In *Proceedings of the 9th International Conference on Rapidly Quenched Material* (eds P. Duhaj P. Mrafko, and P. Svec), Elsevier, Amsterdam, p. 368.

31 Zander, D. and Köster, U. (2004) *Mater. Sci. Eng. A*, **375–377**, 53–59.

32 ElKedim, O., Paris, S., Phigini, C., Bernard, F., Gaffet, E. and Munir, Z.A. (2004) *Mater. Sci. Eng. A*, **369**, 49–55.

33 Frangini, S., De Cristofaro, N.B., Mignone, A., Lascovitch, J. and Giorgi, R. (1997) *Corrosion Sci.*, **398**, 431.

34 Ji, G., Elkedim, O. and Grosdidier, T. (2005) *Surf. Coat. Technol.*, **190**, 406–416.

35 Meng, G., Li, Y. and Wang, F. (2006) *Electrochim. Acta*, **51**, 4277–4284.

36 Wang, L., Lin, Y., Zeng, Z., Liu, W., Xue, Q., Hu, L. and Zhang, J. (2007) *Electrochim. Acta*, **52**, 4342–4350.

37 Meng, G.Z., Li, Y. and Wang, F.H. (2006) *Electrochim. Acta*, **51**, 4277.

38 Jung, H. and Alfantazi, A. (2006) *Electrochim. Acta*, **51**, 1806–1814.

39 Ye, W., Li, Y. and Wang, F. (2006) *Electrochim. Acta*, **51**, 4426–4432.

40 Leinartas, K., Samuleviciene, M., Bagdonas, A., Juskenas, R. and Juzeliunas, E. (2003) *Surf. Coat. Technol.*, **168**, 70.

41 Fujimoto, S., Hayashida, H. and Shibata, T. (1999) *Mater. Sci. Eng. A.*, **267**, 314.

42 Lu, H.B., Li, Y. and Wang, F.H. (2006) *Thin Solid Films*, **510**, 197–202.

43 Wang, L., Zhang, J., Gao, Y., Xue, Q., Hu, L. and Xu, T. (2006) *Scripta Mater.*, **55**, 657–660.

44 Balaraju, J.N., Selvi, V.E., William Grips, V. K. and Rajam, K.S. (2006) *Electrochim. Acta*, **52**, 1064–1074.

45 Ghosh, S.K., Dey, G.K., Dusane, R.O. and Grover, A.K. (2006) *J. Alloys Compounds*, **426**, 235–243.

46 Liu, L., Li, Y. and Wang, F. (2007) *Electrochim. Acta*, **52**, 2392–2400.

47 Vara, G., Pierna, A.R., Garcia, J.A., Jimenez, J.A. and Delamar, M. (2007) *J. Non-Crystall. Solids*, **353**, 1008–1010.

48 Sriraman, K.R., Ganesh Sundara Raman, S. and Seshadari, S.K. (2007) *Mater. Sci. Eng. A*, **460–461**, 39–45.

49 Yu, B., Woo, P. and Erb, U. (2007) *Scripta Mater.*, **56**, 353–356.

50 Luo, W., Qian, C., Wu, X.J. and Yan, M. (2007) *Mater. Sci. Eng. A*, **452–453**, 524–528.

51 Kim, S.H., Aust, K.T., Erb, U., Gonzalez, F. and Palumbo, G. (2003) *Scripta Mater.*, **48**, 1379–1384.

8
Nanoscale Engineering for the Mechanical Integrity of Li-Ion Electrode Materials

Katerina E. Aifantis and Stephen A. Hackney

8.1
Introduction

Portable power technology plays a key role in the advancement of electronic devices required by modern civilization. Lap-top computers, cellular phones and most portable electronic memory devices require the use of rechargeable batteries. Such batteries have also found use in biomedical devices, including pacemakers and implantable defibrillators. Thus, research focused on the development of rechargeable, high-energy density power sources continues to be driven by technological and commercial applications.

Although various types of battery chemistries exist – such as nickel–metal hydride (Ni-MH) and nickel–cadmium (Ni-Cd) – lithium (Li) ion batteries are the dominant force as they have significant advantages in energy density. The large volumetric and gravimetric energy densities exhibited by Li-ion batteries allow their volume and mass to be reduced by 20% and 50%, respectively, as compared to other battery chemistries; in fact, a Li battery can provide three times the voltage of a Ni-Cd or Ni-MH battery. Furthermore, the self-discharge rate of Li batteries is very small over a long period of time, which makes them extremely reliable, while their operating voltage allows a reduction in the number of batteries required to operate a device. All of the aforementioned properties contribute to the miniaturization of electronic devices. Finally, it should be mentioned that – unlike other battery base materials (Cd, Ni) – Li is non-toxic and, as opposed to Ni-Cd batteries, Li-batteries have no memory effect. Moreover, the charging capacity (total time integrated battery current per mass) is not reduced by repeatedly charging and discharging to insufficient levels, and therefore partial charging is possible (the discharging process takes place during the operation of the respective electronic device).

One major technological barrier to improvements in energy density and reliability of the Li-ion battery systems is related to the stability of the anode and cathode materials. One of the reasons that Li-ion chemistries exhibit high energy densities is because of the relatively high cell voltage. This means that there is a large

Nanostructured Materials in Electrochemistry. Edited by Ali Eftekhari
Copyright © 2008 WILEY-VCH Verlag GmbH & Co. KGaA, Weinheim
ISBN: 978-3-527-31876-6

thermodynamic driving force which enables the transport of Li ions from the negative to the positive electrode on discharge, and that relatively large voltages must be applied to fully recharge the battery. The thermodynamic instability of the Li-ion battery chemistries results in materials reliability issues that continue to be addressed by multiple investigators. The goals of this type of research are to improve the storage life and cycle life of Li-ion batteries, while reducing cost and maintaining or improving the energy density.

The reliability issues of electrode materials used in Li-ion systems appear to be related to both mechanical and chemical instabilities. In this chapter, experimental evidence of mechanical instabilities and the development of nanocomposite structures that mitigate this effect will be examined. Moreover, the limiting behavior of the mechanical integrity of the nanocomposite structures will be examined using a fracture mechanics approach. It is proposed that such an approach can provide additional design criteria for nanocomposite structures, and allow a preliminary study of the relationship between capacity and mechanical stability as a materials design criteria of nanocomposite structures.

8.2
Electrochemical Cycling and Damage of Electrodes

The loss of capacity during the electrochemical cycling of secondary Li-ion batteries has been attributed to a variety of materials instabilities (an electrochemical cycle consists of a complete charge and discharge process). There has been much consideration of the chemical instabilities between the electrolyte and the electrode materials, but in these investigations the mechanical integrity of the electrode materials shall be the main consideration. The mechanical forces which are at issue are associated with the volume change of the electrode material as it reacts with Li (lithiation), and as Li is removed from the electrode material (delithiation). Therefore, before examining the particulate material chemistries that are used in Li anodes and cathodes, the mechanical response of metals upon the formation of Li alloys will be illustrated.

8.2.1
Fracture Process of Planar Electrodes

The structural damage of Li-ion battery electrodes is considered to play a significant role in their electrochemical properties (also known as "cyclability" and "capacity"). Therefore, we will examine the available literature with regards to the structural effects of stress during electrochemical cycling. Here, the coupling between electrochemical cycling and mechanical damage (or fracturing) in a simple planar geometry will be examined. Beaulieu *et al.* [1] electrochemically cycled anodes comprised of a SiSn thin film (active material) deposited on an inactive Cu film, each of which was sputtered in turn on a stainless steel substrate. These thin-film electrodes were cycled versus Li, and therefore act as the positive electrode.

In Figure 8.1, the damage effects of electrochemical cycling SiSn/Cu anodes are shown [1]. Figure 8.1a depicts the film prior to cycling (the scratches were made by a razor as markers to monitor mechanical deformation), while Figure 8.1b shows the film after complete lithiation (i.e., when Li has been added); the SiSn alloy reacts with Li, while the Cu is inert. The authors reported no difference between the two micrographs, which leads to the conclusion that the film expands solely at the out-of-plane direction during the first discharge. Figure 8.1c is taken after 1 h during the first charge (i.e., Li de-insertion) when, as the Li is removed, the film shrinks and cracks; the crack width continues to expand as Li is fully removed (Figure 8.1d). In order to visualize the microstructure better, Figure 8.1e is taken at higher magnification to show that the active material "islands" created by fracture are in the order of 100 μm. Then, upon re-charging the anode (i.e., Li is inserted into the SiSn alloy), Figure 8.1f is observed. Surprisingly, Figures 8.1f and b appear almost identical,

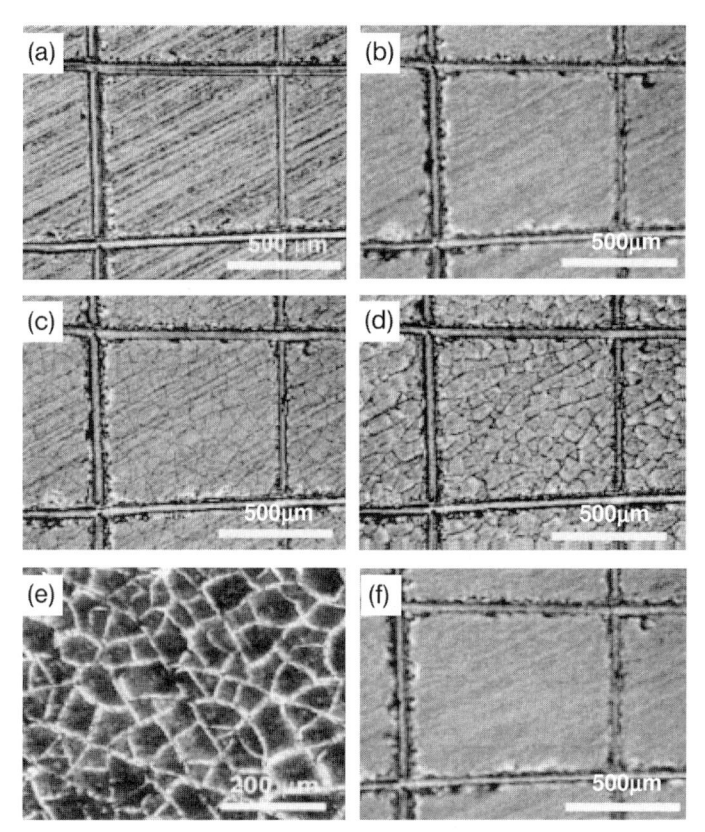

Figure 8.1 Experimental evidence on electrochemical cycling a SiSn thin film. (a) Before cycling begins; (b) after first discharge; (c) at 1 h during the first charge; (d) after complete first charge; (e) greater magnification of microstructure after complete discharge; (f) after second discharge. (Reproduced from Ref. [1].)

which implies that as Li is inserted the fractured active particles expand such that the cracks close up. In particular, it is noted that fracture occurs only once, during the first charge (i.e., during the first time Li is removed). The planar electrode can then be continuously cycled (i.e., the active material expands and contracts reversibly), without further cracking. In this connection it should be mentioned that it has been shown in Ref. [1] that 30 μm-wide flakes with 1 to 8 μm thickness can expand and contract up to 100% during cycling, without noticeable damage. In this connection it should be noted that scanning electron microscopy (SEM) images of SnSb and SnAg electrodes taken after 150 electrochemical cycles look similar to those taken after the first few cycles [2]. This also verifies the fact that severe damage takes place initially, and a stable morphological configuration is approached in the long run.

It should be emphasized here that the aforementioned SnSi film, after a discharge/charge cycle, resembles a dry lake bed fracture that occurs as the mud at the bottom of an empty lake bed begins to dry (Figure 8.2) [1]. Numerous experiments and theoretical investigations [3,4] have been conducted on dry lake bed fracture processes, a consideration of which allows us to explain how, even though fracture occurs in the thin film, it remains in electrical contact with the remainder of the electrode – that is, how the SiSn remains in contact with the Cu, and hence the substrate. In Ref. [1], it was reported that during Li de-insertion (from the SnSi films) the same sequence of events that occurs during the drying of "mud" takes place. The initial effect that Li removal has is the formation of a series of cracks, which create separate "flake-like" particles. Subsequent Li removal results in a shrinkage of these flake-like particles. It should be noted that, during shrinking, the center of the particles remains fixed, and hence remains firmly attached to the substrate, whereas the edges move with respect to the substrate. With their center being firmly connected with the substrate, electron transfer is possible and hence the electrochemical cycle can be completed. Therefore, even though fractured, the film is able to expand and contract repeatedly and respond to the applied voltages.

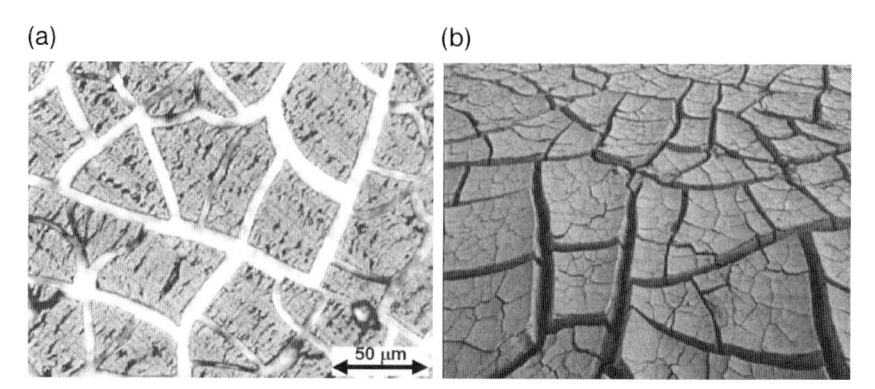

Figure 8.2 (a): Optical micrograph of a Li alloy film after expansion and contraction due to electrochemical cycling. (b): Cracked mud in a dry lake bed bottom. (Reproduced from Ref. [1].)

The volume expansion of the SiSn film upon full lithiation is over 200%. This is expected to produce significant compressive stresses within the SiSn film. The magnitude of these stresses has been predicted in Ref. [5] for an infinitely thick substrate as

$$\sigma = -B\frac{\Delta V}{3V} \tag{1}$$

where σ is the stress, ΔV is the volume change, and B is the biaxial elastic modulus of the film. Based on the experimental results of Beaulieu et al. [1], these compressive stresses are expected to cause plastic deformation of the active material during lithiation. The compressive nature of the stress prohibits crack opening. However, when the Li is removed, the plastically deformed film witnesses a tensile stress which approximates the magnitude of the initial compressive stress. Fracture of the film is expected when this tensile stress exceeds the critical fracture stress [5]:

$$\sigma_{fracture} = \frac{K_{1c}}{\sqrt{\pi h}} \tag{2}$$

where K_{1c} is the fracture toughness of the material and h is the active film thickness.

It should be noted that the resistance to fracture increases as the film thickness decreases; this occurs because, as the volume of the film decreases, there is less strain energy available to do the work of creating a crack. Huggins and Nix [5] make the argument that this result rationalizes the experimental observations of Ref. [6], the authors of which showed that the cycle life of Sn particles increases as the particle size decreases. This rationalization involves the assumption that the strain energy in the particles induced by lithiation and delithiation will not exceed the energy required to form a crack in the smaller particles. This argument is examined in a more quantitative manner in the following section.

8.2.2
Electrochemical Cycling of Particulate Electrodes

Many commercial electrodes consist of a porous structure formed by an agglomeration of powders consisting of a hard, active material powder and a soft, conductive binding material. These types of electrodes may have a significant thickness while maintaining a high surface area for contact with the electrolyte, thus providing advantages for the kinetics of the charge and discharge process. Upon electrochemical cycling both the anode and the cathode react to form compounds or alloys with the diffusing Li ions [7–10], and thus undergo a volume change during charge and discharge, as was discussed for the planar thin-film anodes in Section 8.2.1. For a pure metal anode material in a porous electrode, such as Sn and Si, the volume change can be as much as 300%. Huggins and Nix [5] have suggested that this volume change variation through the thickness of the particle results in stresses large enough to fracture the particle, leading to the observed change in particle morphology during the first few cycles [2]. Volume changes are also known to occur in cathode porous electrode materials during charging and discharging, and this has recently been the topic of an extensive theoretical study related to the situation in which the porous cathode consists of individual $LiMn_2O_4$ particles.

Thus, it is apparent that stress-induced decrepitation of active material particulates is a problem which influences the electrochemical properties of both the anode and the cathode in Li-ion batteries. However, the most detailed investigations of the role of stress and dimensional changes have been carried out for $LiMn_2O_4$ materials, and these will be used as an example of particulate active material that degrades as a result of stress-induced fracture.

The compound, $LiMn_2O_4$, is a cubic spinel oxide that can be used in the positive electrode and can be charged to cubic Mn_2O_4. The full discharge of Mn_2O_4 to a composition $Li_2Mn_2O_4$ creates a tetragonal structure. The insertion of Li in Mn_2O_4 to full discharge gives a 14% volume increase, while the charge from $Li_2Mn_2O_4$ to Mn_2O_4 gives a 14% volume decrease [8]. Thus, the particles which are partially charged such that Mn_2O_4 is present on the oxide particle surface and $Li_2Mn_2O_4$ is present in the particle interior, will exhibit a significant volume misfit between the interior and exterior.

To the present authors' knowledge, the $LiMn_2O_4$ system is the only Li-ion cathode material in which the dimensional changes during electrochemical cycling and the associated topology have been studied experimentally at the nanoscale [12–16]. Figure 8.3 shows crystal morphologies of parent and electrochemically cycled $LiMn_2O_4$ particles (which can be used as the Li insertion material, in the cathode),

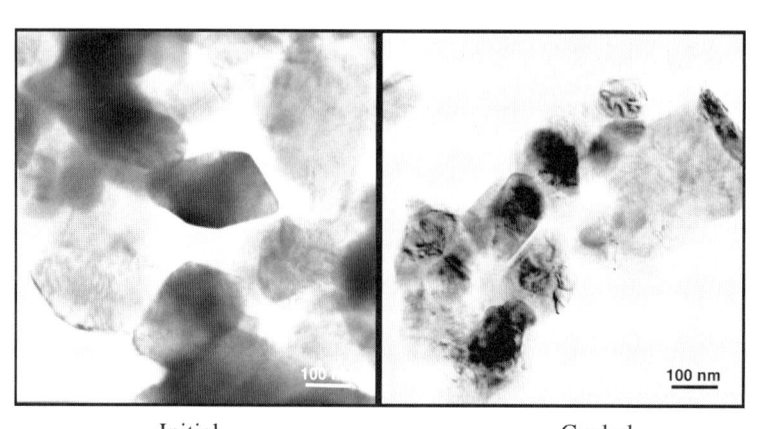

Initial   Cycled

Figure 8.3 Crystal morphologies of the parent and cycled $LiMn_2O_4$ (4.2 V–3.3 V) samples. The initial morphology for the uncycled powder particles consists of single or bi-crystals of sizes 50 nm to 500 nm. After multiple charge/discharge cycles between $Mn_2O_4 \leftrightarrow LiMn_2O_4$, it is evident from the high-frequency spatial variation of contrast in the TEM image that there is strain within the crystals. (Reproduced from Ref. [1].)

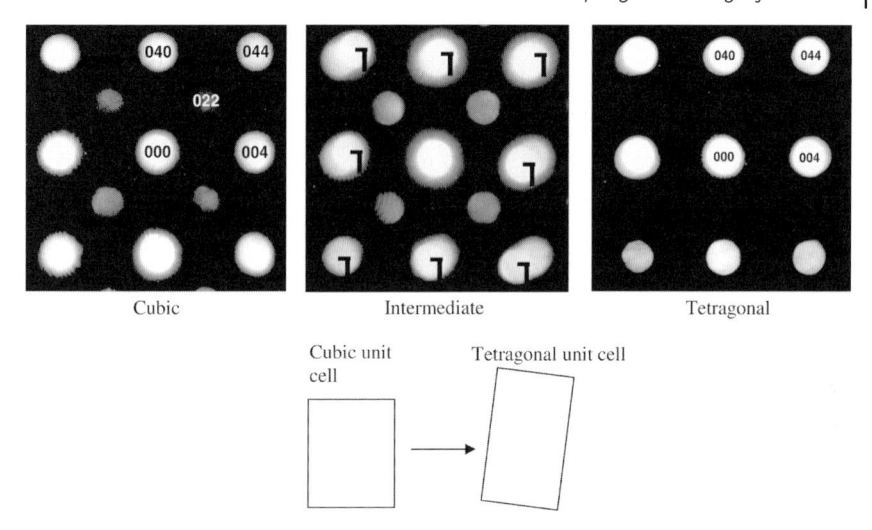

Figure 8.4 When deep discharge occurs below 3.3 V, the structure transforms from cubic to tetragonal with a 14% increase in unit cell volume. The reciprocal space representation from the convergent beam electron diffraction pattern shows how the lattice both expands and rotates. Note the fourfold symmetry of the cubic [1 0 0] pattern. The pattern intermediate between the cubic and tetragonal structures shows a two-phase structure in which the volume/symmetry change is accommodated by a lattice rotation. The mismatch between the cubic and tetragonal spinel structures results in a near-5° rotation of the two unit cells. (Reproduced from Ref. [11].)

indicating the accumulation of strain as a result of the insertion/de-insertion/re-insertion process of Li ions within them. In fact, as the battery is cycled, the Li ions are inserted or removed from the host material; this solid-state diffusion process results in Li-concentration gradients across individual particles, which, in turn, result in gradients in unit cell volume and symmetry.

The structural changes that Li insertion/de-insertion has in the cathode material are demonstrated in Figure 8.4 for the reciprocal lattice of the $LiMn_2O_4$ material, which as a result of the charge/discharge of the battery undergoes the transformation $Mn_2O_4 \leftrightarrow Li_2Mn_2O_4$. This type of structural/volume change, and the misfit strain associated with it, leads to the development of internal stresses due to the difference in the molar volume and elastic moduli of the interior and exterior of the particle. It is proposed that a direct result of the elastic stresses is the development of structural damage near the particle surface at the nanoscale, leading to fracture in deep discharge cells, as illustrated in Figure 8.5. More detail on the experimental procedures, and further interpretation of similar electron micrographs, can be found in reports published by Hackney and coworkers [12–16].

If the electrochemically active particles are approximated as spherical in shape, then a variety of elasticity-based solutions apply to this situation, the simplest being the misfitting spherical core within a spherical shell, as shown in Figure 8.6.

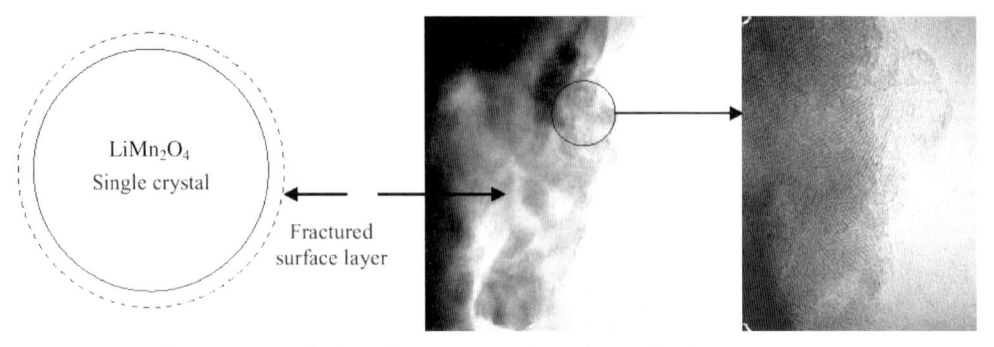

Figure 8.5 Deep discharged $LiMn_2O_4$ particles with associated fractured surface layers. Chemo-mechanical stresses develop as a result of Li-insertion and de-insertion. (Reproduced from Ref. [11].)

The three-dimensional (3D) elasticity problem must be solved in order to predict the magnitude of the stresses that would result from the volume misfit between the inner and outer sphere, which in turn is an approximation of the geometry of an electrochemically active particle that is partially delithiated, with the interior of the particle having a larger molar volume than the exterior of the particle. This was recently established by Christensen and Newman [17], using a very rigorous approach. Here, the approach of Christensen *et al.* is extended to include the ideas of brittle fracture theory introduced by Huggins and Nix [5].

The stress analysis for the volume change of the interior (or exterior) of the sphere can be carried out using the Eshelby type [18,19] experiment, where it is considered that a spherical section is removed from the center of the sphere, creating a spherical cavity. After the spherical section is increased in size, it is then re-inserted into the cavity. The resulting stresses and strains are assumed to simulate the physical situation of a misfitting sphere inside a spherical particle.

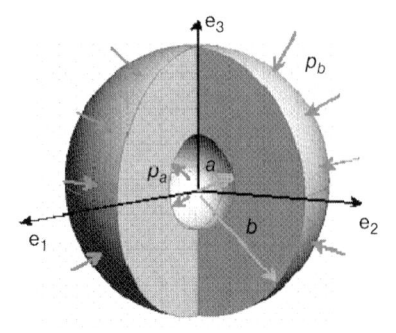

Figure 8.6 Misfitting sphere of radius a with a concentric spherical shell of radius b. The stress, Pa, results from the constraint of the outer sphere on the inner sphere.

This problem is greatly simplified for the case of isotropic elastic constants and spherical symmetry.

The percentage volume expansion of the interior sphere can be expressed as a change in the radius of a sphere, Δ, from an initial radius, r_i,

$$\frac{\Delta V}{V} \cdot 100 = \frac{\left[\frac{4}{3}\pi(r_i + \Delta)^3 - \frac{4}{3}\pi(r_i)^3\right]}{\frac{4}{3}\pi(r_i)^3} \cdot 100 \qquad (3)$$

where it may be seen that Δ is linearly dependent on r_i for a given value of $\Delta V/V$. The radial stress, σ_{rr} resulting from the misfitting center portion of the sphere is obtained using the spherically symmetric isotropic solution of the equilibrium elasticity equation [20]:

$$\sigma_{rr} = \frac{E}{(1+v)(1-2v)}\left[(1+v)A - 2(1-2v)B/r^3\right] \qquad (4)$$

and the radial displacement is given as

$$u_r = Ar + B/r^2 \qquad (5)$$

The constants A and B are integration constants determined by the boundary conditions, while E and v are the elastic modulus and Poisson's ratio, respectively. These equations must be solved for both the interior sphere and exterior spherical shell; thus, there are four integration constants to be determined by the boundary conditions. These boundary conditions are: (i) the solution must be bounded (less than infinity) at the center of the interior sphere; (ii) the stress must be balanced at the interface between the inner sphere and the outer spherical shell; (iii) the radial stress is zero at the outer surface of the spherical shell (free surface); and (iv) the displacements of the inner sphere and external spherical shell must be continuous across their interface. The solution for the special case of the interior and exterior materials having the same modulus and Poisson's ratio is

$$\sigma_{rr} = 2\Delta a^2(b^3 - r^3)(5b^3 + 2a^3)\frac{E}{r^3(4a^3v - 2a^3 - b^3v - b^3)(3b^3 + 4a^3)} \qquad (6)$$

The solution shows that the radial stresses are linear with Δ and are plotted for the outer shell of the particle for two extreme cases in Figure 8.7.

The compressive radial stresses will *not* lead to the radial crack growth in the particle that would be analogous to the thin film fracture observed during several delithiation studies and predicted by Huggins and Nix. However [5], the tangential stresses – the so-called "hoop stresses" – are tensile in character and are expected to produce radial cracks within the delithiated surface layer. The elasticity solution for the hoop stress, $\sigma_{\theta\theta}$, for spherical symmetry and for the special case where the

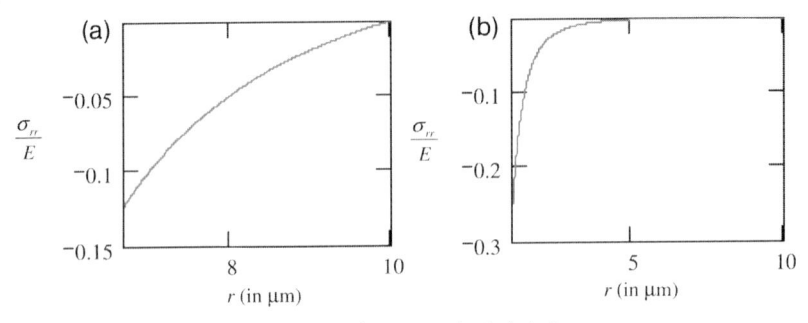

Figure 8.7 Compressive radial stress in the outer spherical shell normalized to the elastic modulus when the outer spherical shell is delithiated. (a) $v=0.33$, $b=10\,\mu m$, $a=6.67\,\mu m$, $\Delta=0.66\,\mu m$. (b) $v=0.33$, $b=10\,\mu m$, $a=1.1\,\mu m$, $\Delta=0.11\,m$.

inner sphere has the same elastic modulus as the outer spherical shell is determined as

$$\sigma_{\theta\theta} = \sigma_{\phi\phi} = a^2(5b^3 + 2a^3)\Delta \frac{2r^3(2v + 4v^2 - 1) + b^3(2v^2 + v - 1)}{r^3(4a^3v - 2a^3 - b^3v - b^3)(3b^3 + 4a^3)} \tag{7}$$

As observed for the radial stress, the hoop stress increases linearly with Δ. These tensile stresses act parallel to the spherical surface to cause crack opening and Mode I fracture at the surface of partially delithiated particles.

As may be observed in Figure 8.8, the hoop stresses are not constrained to be zero at the free surface, and thus may initiate the growth of cracks at the particle surface as is observed experimentally (see Figure 8.5). Moreover, it is apparent from Figure 8.8 that the hoop stresses at the surface are greater when the delithiated

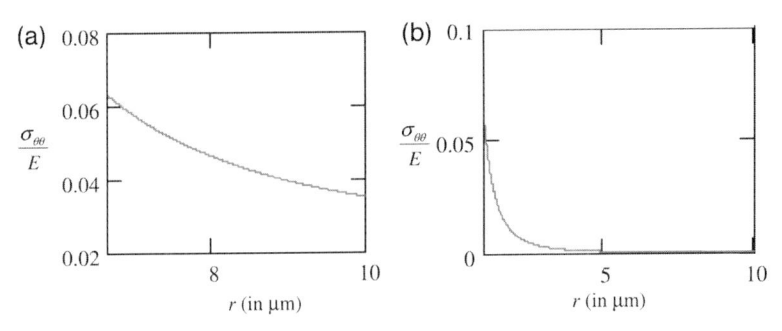

Figure 8.8 Tangential tensile stress (hoop stress) in the outer spherical shell normalized to the elastic modulus when the outer spherical shell is delithiated. (a) $v=0.33$, $b=10\,\mu m$, $a=6.67\,\mu m$, $\Delta=0.66\,\mu m$. (b) $v=0.33$, $b=10\,\mu m$, $a=1.1\,\mu m$, $\Delta=0.11\,\mu m$.

surface layer is thinner ($b - a$ is relatively small) for a given value of $\Delta V/V$ of the inner sphere.

The application of the Griffith brittle crack theory [20] to the misfitting sphere geometry can be carried out in a manner similar to the Huggins and Nix approach [5] for the planar, thin-film geometry by integrating the local strain energy of the sphere over the volume of the sphere and comparing this to the energy required to form the crack. For perfectly brittle fracture, the energy to form a crack is the surface area of the crack multiplied by the surface energy per unit area (γ). When the surface energy required to form the crack is greater than the available strain energy, then fracture by crack growth is not energetically favorable. However, when the available strain energy is greater than the surface energy work required to form the crack, then the particle may fracture and loss of electrochemical activity may occur. The simplifying assumption is made here that a single crack across the center of the particle will result in the complete relaxation of the misfit stress. Using this assumption, the comparison of the strain energy term and the surface energy term as a function of the particle diameter is shown in Figure 8.9 for some specific ancillary parameters. The plot indicates that the surface energy term is larger than the strain energy term until the particle diameter approaches ~5 nm. Thus, the plot suggests for these assumptions that particles less than ~10 nm in diameter are unlikely to fracture, while fracture may initiate when particles are larger than this value. These propositions are supported by experimental observations that the capacity fade is reduced in $LiMn_2O_4$-based cathodes if the electrode consists of submicron particles [21,22] (more details on the aforementioned mechanical analysis can be found in Aifantis and Hackney [23]).

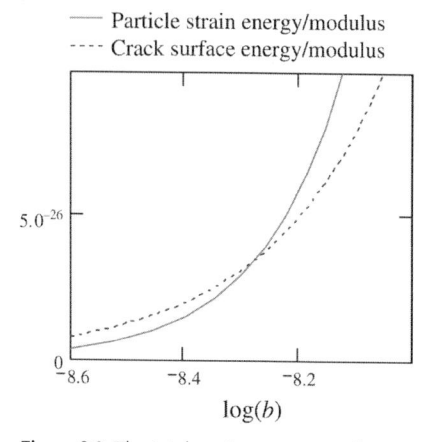

Figure 8.9 The total strain energy as a function of b normalized with respect to modulus compared to crack surface energy as a function of b normalized with respect to modulus. This plot is for the ratio of surface energy to modulus of 10^{-10} m, $a = 0.67b$ and $\Delta = 0.1a$. The cross-over point where the total particle strain energy becomes greater than the crack surface energy ($4\pi\gamma b^2/E$) approximates the particle size at which fracture may initiate.

The mechanisms of deformation during electrochemical cycling that were shown in the previous sections produce fracture in the electrode materials. It has been argued that this fracture process is the result of a volume mismatch between the lithiated and delithiated materials. Efforts to avoid such loss of mechanical integrity have focused on reducing the size of the electrochemically active particles. This avenue of investigation appears to have originated with the studies of Yang et al. [6], who showed that an increase in cycle life could be correlated with reduced particle size. Huggins and Nix [5] have rationalized this result by showing that, for a given volume expansion, there is a critical thickness of a planar, thin-film electrode below which fracture will not occur. In the previous section, experimental results on the fracture of electrochemically active powder particles in porous electrodes were presented. Moreover, elasticity theory and the Griffith brittle crack approach was applied to the spherical geometry approximating active particles in porous electrodes in order to develop a quantitative framework for particle size dependence on fracture during electrochemical cycling. It was shown that the physical basis for this particle size effect is that the particle diameter is less than the critical crack length for the maximum strain present during the electrochemical cycle.

8.3
Electrochemical Properties for Nanostructured Anodes

The overall conclusion that can be drawn from the experimental and theoretical observations of the previous section is that the smaller the dimensions of the materials that react with Li, the more optimum the electrochemical properties of the electrode. In this section, therefore, nanostructured anodes, comprising of Sn, Si, and Bi will be examined.

The driving force for developing nanostructured materials for high cycle life electrodes is that current commercial Li-ion batteries use graphite as the active negative electrode material. The problem is that graphite has a very low Li atomic density at full Li capacity in the carbon intercalation compound (LiC_6) [24–28]; hence, the volumetric Li capacity is rather low. It has been known for at least 20 years that certain metals such as Si, Sn and Al [29–34] have capacities between 900 and 4000 mA·h g^{-1} [35] upon the formation of lithium-rich alloys, as they have a very rich Li intercalation density per host atom (i.e., $Li_{4.4}Si$, $Li_{4.4}Sn$), as opposed to 372 mA·h g^{-1} that result from the formation of LiC_6 [36]. The drawback, however, is that during maximum Li insertion Si suffers a 310% volume increase [35]; in fact, all metals that can act as active sites (i.e., Sn, Bi, Si Al) exhibit over a 100% volume increase after Li alloy formation. As has been discussed, the large values of $\Delta V/V$ due to the lithiation and delithiation process lead to significant internal stresses within the individual active particles of a porous electrode, and can result in fracture of the electrochemically active material when certain energy criteria are satisfied (as shown in Figure 8.9). As a result of the loss of mechanical integrity, the capacity is significantly reduced after the first few charge/discharge cycles. For example, 800 nm Si thin films prepared by deposition at room temperature fail after the third electrochemical cycle [37]. The fracturing of individual particles is believed to degrade the electrochemical properties of the electrode because cracking produces an active material that is no longer in electrical contact with the remainder of the electrode, and is therefore unable to respond to the applied voltages necessary to recharge or control the discharge of the battery [36]. Moreover, corrosive agents which are present in the battery (HF and residual H_2O) are believed to attack the surfaces of the active material and the fracture of individual particles increases the surface area available to chemical attack. Of course, the problems associated with the high surface area of active particles also raises concerns about using nanoparticulate active materials. Therefore, the most effective way to minimize fracture and damage is not only the use of nanostructured metal anodes [38–40], but also the subsequent embedment/encapsulation of the nanostructured metals in carbon or other inert materials [41–46] in order to protect the active site surface and further constrain the active site expansion.

8.3.1
Nanostructured Metal Anodes

8.3.1.1 Sn and Sn-Sb Anodes at the Nanoscale

Lithiation of Sn to form $Li_{4.4}Sn$ produces a theoretical capacity of $990\,mA\cdot h\,g^{-1}$, while the expansion that the Sn undergoes upon Li insertion is 290%. In order to minimize the effects of this expansion, anodes comprising of Sn nanoparticles (also referred to as ultra-fine particles) have been constructed, with diameters ranging from 100 to 300 nm [47]. The capacity of these anodes during the first cycle approaches the theoretical capacity of $990\,mA\cdot h\,g^{-1}$, but this decays rapidly and after 20 cycles drops to $220\,mA\cdot h\,g^{-1}$ [47]. In attempts to reduce this capacity decay, Sb (which gives a capacity of $660\,mA\cdot h\,g^{-1}$ upon the formation of Li_3Sb [48]) has been added to the Sn, forming Sn–Sb alloys. The initial capacity of these Sn–Sb nanoparticles is not as high as that for pure Sn, but the capacity retention is significantly higher. Among the various such alloys examined (Sn–30.7%Sb, Sn–46.5%Sb, Sn–47.2%Sb, Sn–58.5%Sb, Sn–80.8% Sb), the one with a 46.5%Sb content was the most efficient, with a starting capacity of $701\,mA\cdot h\,g^{-1}$, and a final capacity, after 20 cycles, of $566\,mA\cdot h\,g^{-1}$. Transmission electron microscopy (TEM) images of the aforementioned pure Sn nanoparticles with mean diameter 185 nm, as well as of the Sn-46.5at%Sb alloy, with mean diameter 138 nm, are shown in Figure 8.10.

8.3.1.2 Si Anodes at the Nanoscale

The theoretical capacity produced upon the formation of $Si_{4.4}Li$ is $4200\,mA\cdot h\,g^{-1}$. However, it has been observed by Li *et al.* [49] that, after the fifth electrochemical cycle, the capacity of bulk Si is reduced by 90%. The main reason for this capacity loss is the pulverization that the Si undergoes due to the 300% expansion it experiences upon the formation of Li alloy. Si at the nanoscale, on the other hand, has shown to maintain a high capacity after several electrochemical cycles, since the deformation mechanisms are less severe. Si thin-film electrodes prepared by

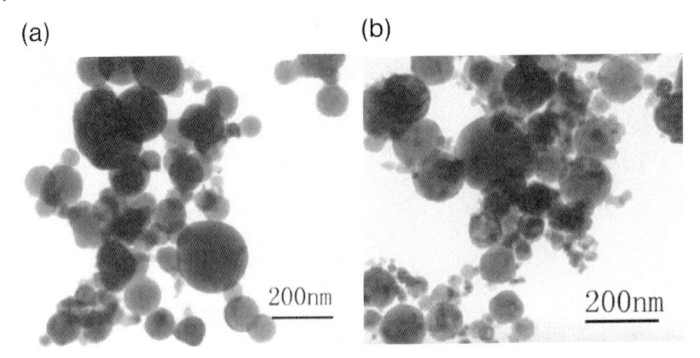

Figure 8.10 TEM images of pure Sn nanoparticles (a) and Sn–46.5%Sb alloy nanoparticles (b). (Reproduced from Ref. [47].)

chemical vapor deposition, were able to maintain a 4000 mA·hg^{-1} discharge capacity for over 10 electrochemical cycles [50]. Furthermore, evaporated Si thin films (40 nm) have been proven to give stable capacities up to 3000 mA·hg^{-1} for over 25 cycles [51], while amorphous thin films (100 nm) gave starting capacities of 3500 mA·hg^{-1} with a stable capacity of 2000 mA·hg^{-1} after 50 cycles [36] (Figure 8.11b). These results are consistent with the predictions reported elsewhere [5], which suggested that thin films are more resistant to fracture than bulk materials. According to Ref. [52], smaller capacities than those achieved for thin films, but still much higher than those given by any other active materials, have been achieved

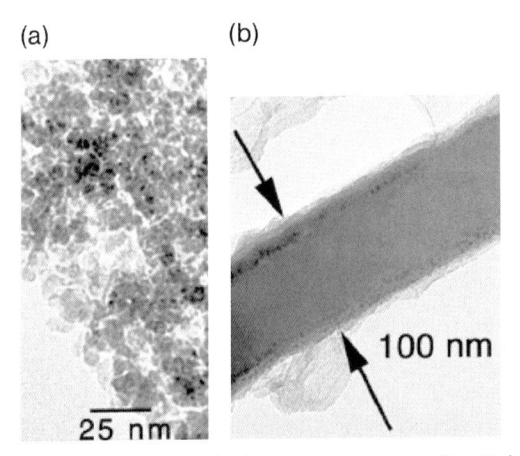

Figure 8.11 Micrographs showing: (a) nanocrystalline Si clusters (with 12 nm average diameter) prepared by gas-phase condensation and ballistic consolidation on planar Cu current collectors; and (b) 100 nm-thick nanostructured thin films of Si prepared by evaporation onto planar Ni current collectors. (Reproduced from Ref. [36].)

for Si nanoparticles of 80 nm diameter. These nanoparticles gave a capacity of 1700 mA·h g^{-1} on the tenth cycle [52]. Elsewhere [53], capacities of 1100 mA·h g^{-1} for nanocrystalline particles of 12 nm average diameter prepared in thin-film form were achieved, with 50% capacity retention being observed after the 50th cycle. Figure 8.11a shows nanoclusters of Si, the starting capacity of which was 2400 mA h g^{-1}, although after the 50th cycle their capacity had fallen to 525 mA·h g^{-1}.

8.3.1.3 Bi Anodes at the Nanoscale

The last alternative anode material candidate that will be examined is Bi. The theoretical capacity upon the formation of Li$_3$Bi is 385 mA·h g^{-1} [53,54]. Although this cannot be compared to the high capacities produced upon the formation of Li$_{4.4}$Si or Li$_{4.4}$Sn, it is of interest to examine the electrochemical cycling properties of Bi as it allows some general conclusions to be drawn about particle size. Bulk Bi, and even microscale Bi particles, have a very low cyclability that is attributed to the large volume expansion of Bi (Bi expands 210% upon maximum Li insertion [54]). Nanoscale Bi particles have therefore been produced, with an average diameter of 300 nm [54]. In order to constrain their expansion, they were embedded inhomogeneously in a graphite-PVDF/HFP SOLEF copolymer (Solvay) which acts as a binder (Figure 8.12), thus creating a porous electrode structure [54]. The relative capacity of an anode containing 12% Bi nano particles is reduced to 50% after 10 cycles. This occurs because the Bi agglomerates combine to form dendrites (Figure 8.12b) [54], which results in a loss of electrochemical activity.

The authors in Ref. [54] observed a morphology change of the Bi particles, where the agglomerates of Bi particles were transformed into a dendritic structure. The lack of morphological stability is the result of direct contact between the active metal and the electrolyte. Such dendrite formation has been also noted for other high-surface-area electrode materials [47]. In the sequel, therefore, techniques that prevent direct contact of the active site with the electrolyte – and hence result in improved capacity retention – will be elaborated.

Figure 8.12 Micrographs showing Bi nanoparticles of 300 nm diameter (a), and dendrite formation after 10 cycles (b). (Reproduced from Ref. [54].)

8.3.2
Embedding/Encapsulating Active Materials in Less-Active Materials

The previous examples suggest that the capacity loss in cycled electrodes is due to mechanical and electrochemical instabilities (expansion of active sites and reaction between electrolyte and active site). Researchers have tried to deal with the large volume expansions of the high-energy density materials by reducing the diameters of the active particles below the critical crack length. However, this leads to large surface areas between the electrolyte and the active material, thereby increasing the probability of deleterious interactions between the electrolyte and the solid. The issue of harmful interactions (as shown in Section 8.3.1.3 for Bi) between the surface of the electrochemically active material and the electrolyte raises serious concerns about using active particles having nanoscale diameters. This has led to the idea of developing a nanocomposite approach where the high-energy density material (Sn, Si, etc.) has a nanometer length scale, but is surrounded by a matrix of less-active material (Li_2O, FeC, C, etc.) with respect to Li. The composite materials may have a micron scale diameter, and thus a smaller surface area than a nanoscale particle, while at the same time nanometer scale active sites may undergo lithiation/delithiation at large $\Delta V/V$, with the possibility of avoiding fracture. This would seem to provide an ideal compromise between having a small surface area for the composite particles to prevent harmful interactions with the electrolyte, while at the same time presenting a nanoscale active particle size within the protective matrix. Thus, the nanocomposite electrode approach not only reduces the surface area available to chemical attack, but the nanometer length scale of the active material also increases the probability that the critical crack length is larger than the particle size, precluding the possibility for fracture [5]. The electrochemical benefit from dealing with fracture and electrolyte interaction in this way is that the initial capacity of the active material is retained after continuous electrochemical cycling.

It should be noted that Fuji was the first to patent nanocomposite materials with an inert matrix containing a nanoscale active material [55]. Investigators at Fujifilm Celtec [56] developed composite structures based on Li_2O–Sn nanocomposites formed from the initial lithiation of SnO_2. The chemical reaction to form these materials was carried out in an electrochemical cell, where the SnO_2 particle in the positive electrode was cycled versus Li metal, with the first discharge giving rise to the irreversible conversion of tin oxide (SnO_2) to metallic tin (Sn) and lithium oxide (Li_2O) via the reaction [6,56,57]:

$$4Li + SnO_2 \rightarrow Sn + 2Li_2O \tag{8}$$

This chemical reaction results in micron-scale particles containing nanoscale Sn active sites embedded in a relatively inert Li_2O matrix. The nanocomposite Sn/Li_2O can then be cycled with the reversible alloying/dealloying of the Sn with Li [18,19]:

$$x\,Li^+ + xe^- + Sn \leftrightarrow Li_xSn, \quad 0 \leq x \leq 4.4 \tag{9}$$

This initial study led to several alternative chemistries and configurations, but almost all approaches relied upon a composite structure consisting of active material, such as Sn, with a less-active material. Some examples of this approach are tin oxide glass composites (e.g., $SnO_2-B_2O_3-P_2O_5$) [56,58–60], tin intermetallic compounds (e.g., Sn–Fe [61], Cu–Sn [32], Sn–Sb [31], Ni–Sn [62], Sn–Ca [63]), and tin oxide metal composites (SnO_2–Mo). It should be noted here that Si reacts with Li in the same way as Sn, and therefore Eqs. (8) and (9) can easily be re-written for Si.

8.3.2.1 Sn-Based Anodes

Anodes comprised of SnO_2 nanofibers [64] are synthesized via the template method [65,66] under which SnO_2 is precipitated within the pores of a microporous membrane. The template membrane is then burned and subsequently heated so as to obtain a crystalline structure with the SnO_2 nanofibers protruding from the underlying current collector surface, like the bristles of a brush (Figure 8.13; [64]). For comparative purposes, Li, Martin and Scorsati [64] synthesized a SnO_2 thin-film electrode with the same SnO_2 content as the nanostructured one (this is how they referred to the nanofiber electrode) shown in Figure 8.13. The thin-film anode was synthesized in the same manner as the nanostructured anode, but without using the template membrane on the Pt current collector, such that the resulting film thickness was 550 nm. In order to compare the two types of anode, they were continuously cycled (separately). The capacity of the thin-film anode decreased with increasing numbers of electrochemical cycles; the initial capacity was approximately $675\,mA\cdot h\,g^{-1}$, but after 50 cycles it had fallen to $420\,mA\cdot h\,g^{-1}$. As shown previously, this is the usual trend observed with the electrochemical cycling of active materials. Cycling, however, of the nanostructured anode of Figure 8.13 did not result in a capacity loss but rather a capacity increase. In Figure 8.14, it can be seen that the initial capacity of the nanostructured anode was $700\,mA\cdot h\,g^{-1}$ (much higher than the respective value for the thin film), and

Figure 8.13 (a): SEM image of a nanostructured anode, with SnO_2 nanofibers protruding from a Pt collector. (b): TEM image of a single uncycled SnO_2 nanofiber. (Reproduced from Ref. [64].)

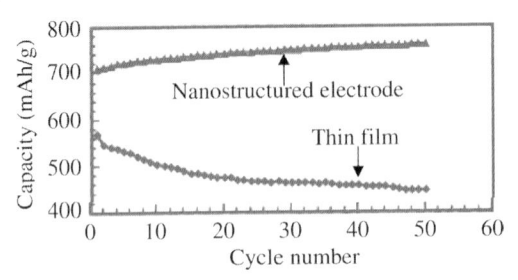

Figure 8.14 Comparison of capacity retention between Sn nanostructured and Sn thin-film electrodes, at a charge/discharge rate of 8 C over the potential window of 0.2–0.9 V. (Reproduced from Ref. [64].)

after 50 cycles it had reached a value of 760 mA·h g^{-1}. The capacity eventually stabilized and did not increase further. In particular, it has been possible to perform 800 cycles on nanostructured SnO_2.

It should be noted that similar nanostructured electrodes to those shown in Figure 8.13 have been synthesized for V_2O_5 using a 50-nm pore diameter polycarbonate template membrane [67]. Again, better performance was achieved for the nanostructured V_2O_5 electrode compared to the respective thin-film configuration. Further information on the use of V_2O_5 as a cathodic material can be found in Ref. [68]. It should be noted that nanostructured anodes, fabricated via the template method, have improved cyclability because the absolute volume expansions of the nanofibers are small, and the brush-like configuration accommodates the expansion of each fiber [2,69].

Another method that has been shown to protect the active sites from fracture, and hence allows for improved cyclability, is illustrated in Ref. [70], where 200 nm Sn particles were embedded in micron-sized carbon particles. These studies have demonstrated significant improvements of cycle life relative to pure Sn. A more recent such study is that by Wang et al. [30], who encapsulated Sn_2Sb alloy in C microspheres, as shown in Figure 8.15. The initial reversible capacity of Sn_2Sb alloy powder was 689 mA·h g^{-1}, and after 60 cycles only 20.3% capacity retention was noted. The initial reversible capacity of Sn_2Sb encapsulated in C microspheres was slightly lower at 649 mA·h g^{-1}, but after 60 cycles a retention of 87.7% was observed (Figure 8.16).

The advanced cyclability of these Sn_2Sb encapsulated particles can be attributed to the following:

- Aggregation between the metal active particles and the electrolyte (to form hazardous dendrites, as illustrated for Bi nanoparticles in Figure 8.12) is avoided as the C microspheres act as barriers.

- The carbon microsphere acts as a buffering matrix (i.e., as a cushion) which relieves the volume changes of the active metal during cycling.

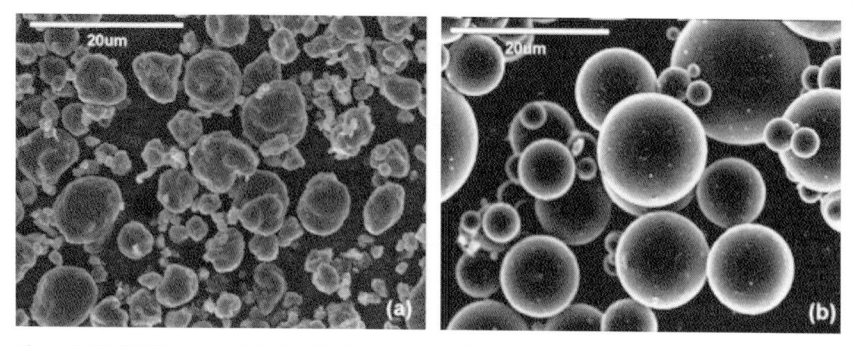

Figure 8.15 SEM images of: (a) Sn_2Sb alloy particles; (b) Sn_2Sb/C microspheres (CM/Sn_2Sb) synthesized through carbonization at 1000 °C. (Reproduced from Ref. [30].)

- The nanoscale diameters of the Sn particles inhibit crack formation by limiting the available strain energy for crack formation.
- The C microsphere itself is an active material for additional Li^+ ion storage; in particular, the Li is initially attracted to the active site as it is more chemically active with respect to Li, and after maximum Li has been stored there, the C can store additional Li.

8.3.2.2 Si-Based Anodes

In addition to the aforementioned advantages that C encapsulation of Sn_2Sb alloy offers, Chen *et al.* [71] have illustrated an additional advantage for spherical nanostructured Si/C composite particles coated with C. (The Si/C nanocomposite is designed to contain 20 wt.% silicon, 30 wt.% graphite, and 50 wt.% PF-pyrolyzed carbon.) As seen in Figure 8.17b, after heat-treating the carbon-coated Si/C spherical particle, hard carbon forms on surface. This hard carbon surface offers additional electrochemical stability for the following reason. During cycling, the electrolyte

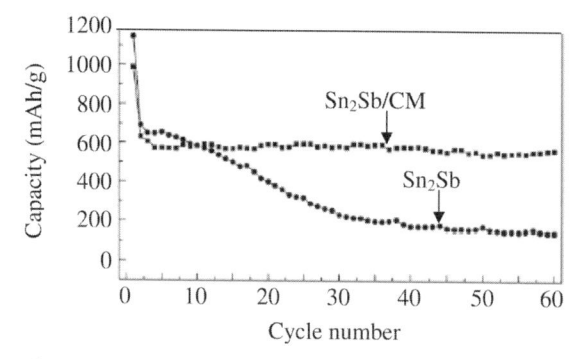

Figure 8.16 Comparison of the cyclability of Sn_2Sb powder with Sn_2Sb enclosed in C microspheres (Sn_2Sb/CM). (Reproduced from Ref. [30].)

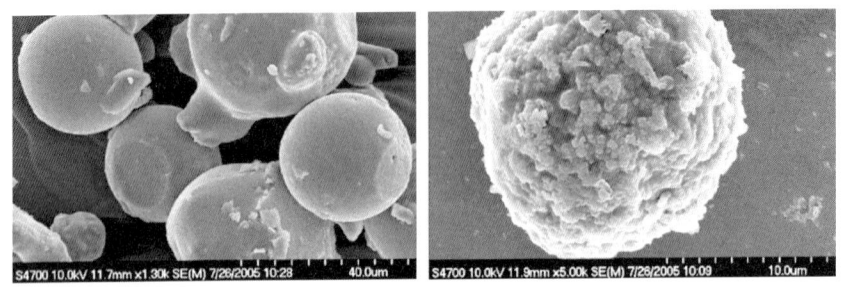

Figure 8.17 Nanostructured Si/C encapsulated in C. Left: Before heat treatment. Right: After heat treatment, the carbon coating is transformed to hard carbon. (Reproduced from Ref. [71].)

decomposes on the electrode, forming a solid electrolyte interphase (SEI) passivation layer on the electrode surface; this layer covers the Si active site, thus affecting its ability to host Li ions. However, coating the Si/C with a hard carbon shell protects the Si, as the SEI passivation layer forms on the hard carbon shell. Moreover, the SEI layer on hard carbon is stable and therefore the degradation of the capacity during cycling is reduced.

The capacity retention of Si/C and carbon-coated Si/C is compared in Figure 8.18. It can be seen that, after the second cycle the capacity of both electrodes increases. This implies that a few cycles are required in order to activate the Si/C composite. Further information on how initial cycling allows the electrochemical kinetics in Si/C to reach an optimal state can be found in Ref. [72]. The cyclability of nanostructured Si/C is better than that of pure nano Si [73], due to the fact that C buffers the expansion of Si and therefore minimizes mechanical damage of the Si active sites, allowing for better electrochemical cycling. The cyclability of carbon-coated Si/C is even better, however, as the carbon coating offers additional stability, from both mechanical and electrochemical points of view as was described at the end of Section 8.3.2.1. The expansion of Si is further minimized, and the Si does not come in direct contact with the electrolyte.

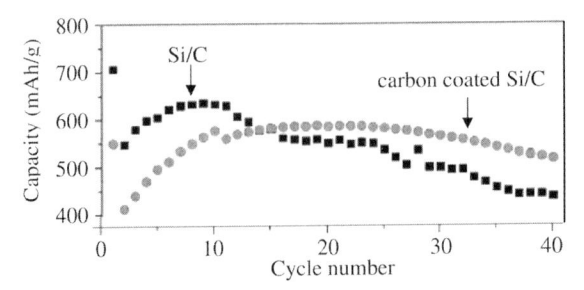

Figure 8.18 Comparison of the discharge capacity of spherical nanostructured Si/C and carbon-coated Si/C. (Reproduced from Ref. [71].)

In spite of the continued experimental development of the nanocomposite approach for active material in porous electrodes, there is evidently little or no progress concerning the limits of the mechanical integrity of such structures. In the following, the specific problem of volume expansion of the active particle within the inactive (or less-active) matrix is considered. Specifically, fracture of the matrix associated with stresses induced by the active material volume change will be examined. This is one of the many issues related to the mechanics of the nanocomposite electrode material concept.

8.4
Modeling Internal Stresses and Fracture of Li-anodes

8.4.1
Stresses Inside the Matrix

As was illustrated in the previous sections, nanosized active sites embedded in a less-active matrix have improved electrochemical properties during cycling. From a mechanical point of view, this is believed to be due to the fact that: (i) deformation mechanisms at the nanoscale are less severe and therefore electrochemical connectivity throughout the electrode is better; and (ii) the surrounding matrix minimizes the expansion of the active sites. In this section, we will show a theoretical analysis which examines some limits for mechanical integrity of the matrix of nanocomposite materials. The framework employed is that of linear elastic fracture mechanics, as proposed by Dempsey *et al.* [74]. In order to model these active/inactive composite electrodes, it is assumed that active spherical (or cylindrical) sites are distributed periodically in an inert matrix, as shown in Figure 8.19. It should be noted that the analysis that follows is contained in Refs. [75–77].

In Section 8.2 it was shown that, after the first few electrochemical cycles, crumbling occurs at the active site surface. If the active sites are surrounded by a matrix it

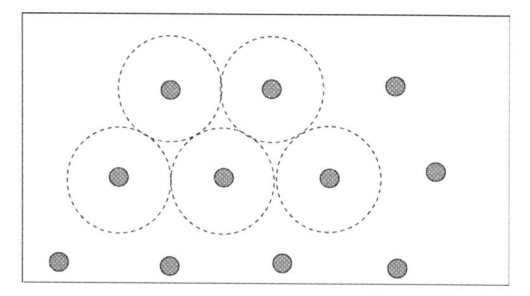

Figure 8.19 Active sites (shaded) embedded in an inert matrix. A unit cell can be thought of as an active surrounded by a circular area of inert material. This is a two-dimensional analogue to the actual three-dimensional problem.

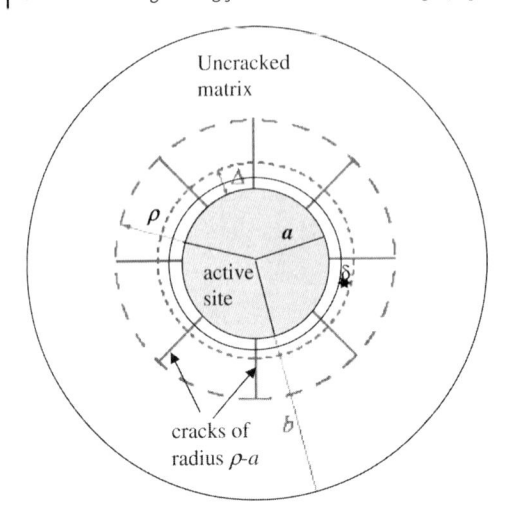

Figure 8.20 Configuration of unit cell used in analysis, where a and b are the radii of the active site and matrix, Δ is the free expansion of the active site if it were not constrained (by the matrix), δ is the radial distance that the matrix pushes back as it opposes this expansion, and ρ is the crack radius.

can be assumed that fracture occurs at the active site/matrix interface after continuous cycling due to the large active site volume expansions. As the matrix is more brittle than the active sites, it is assumed that the crumbling occurs inside the matrix, thereby forming a damage zone (Figure 8.20). As this region is severely damaged it is assumed to support only radial stresses (all other stress components vanish in that region), and can therefore be approximated by a number of radial cracks with length $\rho - a$. Furthermore, it should be noted that in Figure 8.20, a and b are the radii of the Si and matrix; Δ is the radial displacement to which the active site would expand to if it were not surrounded by the matrix, while δ is the radial distance the matrix pushes back as it opposes the active site expansion.

As mentioned in Section 8.2.1, internal stresses are produced in the electrodes during charging. The general solution for the stress distribution inside a cylindrical shell that contains a misfitting core is [78]

$$\sigma_r = \frac{A}{r^2} + 2C \tag{10}$$

and the corresponding radial displacement inside the shell is

$$u_r(r) = \frac{1}{E_g}\left[-\frac{A(1+\nu_g)}{r} + 2C(1-\nu_g)r\right] \tag{11}$$

where E_g and ν_g are the elastic modulus and Poisson's ratio of the shell, while the constants A and C are found from the boundary conditions. If one assumes that the battery system is tightly constrained (i.e., $u_r(b) = 0$), and that the displacement at the core/shell interface is $u_r(a) = \Delta - \delta$, then

$$A = -\frac{E_g a b^2 (\Delta - \delta)}{(b^2 - a^2)(1 - v_g)}; C = -\frac{E_g a (\Delta - \delta)}{2(b^2 - a^2)(1 - v_g)} \quad (12)$$

Δ is a known parameter (i.e., it corresponds to a 300% increase for Si) while δ is found in Aifantis-Hackney [75] (by considering the strain energy of the system) as

$$\delta = -\frac{E_g \Delta [a^2(1+v_g) + b^2(1-v_g)](a+\Delta)^2(1-v_s)}{E_s a^2 (a^2 - b^2)(1-v_g^2) - E_g [a^2(1+v_g) + b^2(1-v_g)](a+\Delta)^2(1-v_s)} \quad (13)$$

Therefore, by inserting Eqs. (8.13) and (8.12), as well as the appropriate material parameters into Eq. (8.10), one can obtain the elastic radial stress that the matrix experiences, prior cracking.

Now, we shall examine the energy of the battery system once cracks develop. Upon maximum Li-insertion, the stress that the Li ions exert inside the active site is constant, and therefore the pressure that the active site exerts onto the matrix is also constant and is set equal to p. Similarly, the pressure that is exerted onto the unit cell under examination by a neighboring cell is constant and set equal to q. It has been shown [76] that the hoop stress right in front of the crack tip, which is responsible for crack growth can be expressed as

$$\sigma_{\theta\theta}(\rho^+) = \frac{pa^2}{b^2} \left[\frac{1 - 3S(\rho/b)^2 + 2(\rho/b)^3}{2(\rho/b)^2 (1 - (\rho/b)^3)} \right] \quad (14)$$

where $S = qb^2/(pa^2)$. Based on Ref. [77] the energy released during crack growth can be written as

$$G(\rho) = \frac{2(1-v_g)\rho \sigma_{\theta\theta}^2(\rho^+)}{nE_g} \quad (15)$$

which in turn allows the stability index to be defined as

$$\kappa = \frac{b}{G}\frac{dG}{d\rho} \quad (16)$$

8.4.2
Stable Crack Growth

As long as crack growth is stable, it is possible to continue charging and discharging the battery since, as explained in Section 8.2.1, the particles that form by cracking

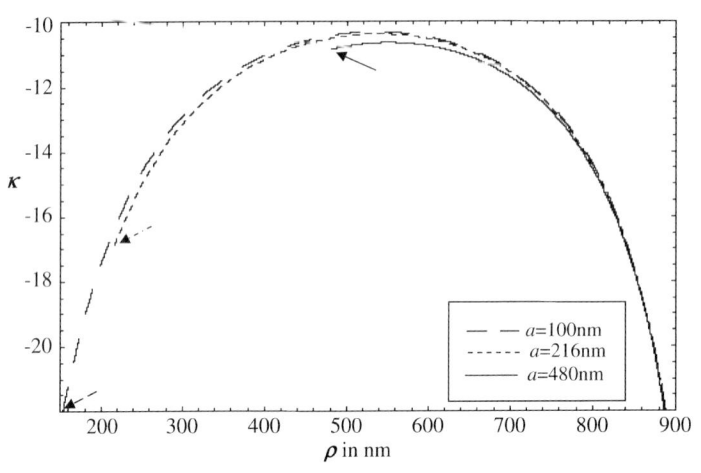

Figure 8.21 Stability plots for systems with different active site volume fractions; the arrows indicate the start of each plot. (Reproduced from Ref. [77].)

remain in electrical contact with the electrode and electrochemical cycling can be performed. Unstable cracking cannot be controlled, however, and this can result in complete fracture of the anode. So, it is of interest to determine the size of the active sites that will result in a more stable battery system.

To do this the stability index κ is plotted with respect to the crack radius ρ, for $b = 1\,\mu m$ and varying a; in other words, the active site volume fraction is varied. The various stability behaviors obtained are shown in Figure 8.21. The active sites are taken to comprise of Si, while the matrix is of soda glass [77].

The more negative the values at which the stability curves start out, the more stable the system, as a greater energy difference is required for crack growth to initiate. It can therefore be predicted from Figure 8.21 that smaller active sites (i.e., smaller active site volume fractions f, where $f = a^3/(b)^3$), result in more stable anodes, as crack growth is initiated with greater difficulty for such systems.

8.4.3
Griffith's Criterion

Based on Griffith's theory, a crack will continue growing as long as the energy released (G) during its propagation is greater than the energy (G_c) required to create the new crack surface. Therefore, by plotting Eq. (15) together with the fracture energy (G_c) of the matrix material (which is a material constant), the crack radius at which the two energies intersect can be determined and hence an estimate can be made of the distance at which crack growth will stop. Thus, in Figure 8.22 G is plotted for various active site volume fractions together with G_c. The active sites are taken to be Si, whereas the matrix is Y_2O_3. It can be seen that, the smaller the volume fraction of the active sites, the smaller the distance at which crack growth stops.

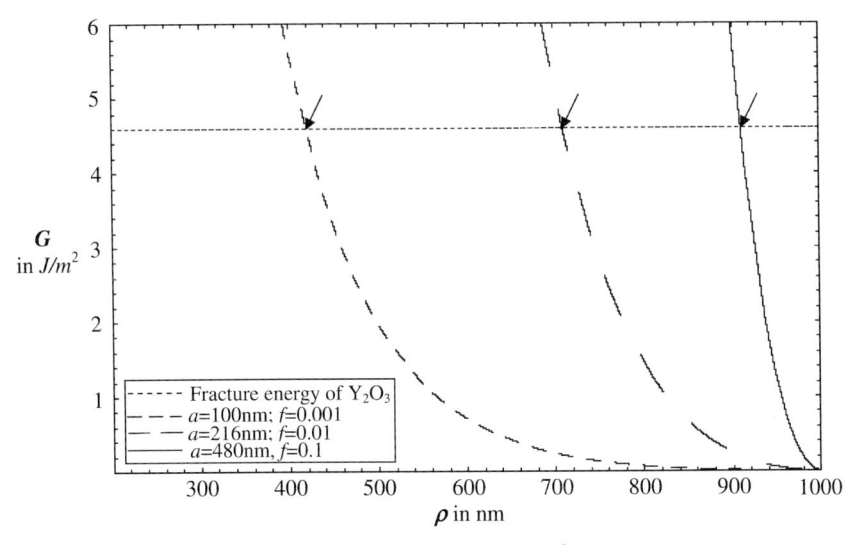

8.4 Modeling Internal Stresses and Fracture of Li-anodes

Figure 8.22 Griffith's criterion for an electrode that comprises of Si active sites and Y_2O_3 matrix; various active site volume fractions are considered. The arrows indicate the distance at which crack growth stops. (Reproduced from Ref. [77].)

Hence, it is again predicted that smaller volume fractions of the active sites are more stable as they allow for less cracking. Furthermore, by plotting G and G_c, for the same volume fractions (i.e., $a = 100$ nm; $b = 1$ μm) and same active site materials, but with different matrix materials, it can be predicted which matrix material allows for the smallest crack propagation distance before cracking ceases; the results are shown in Table 8.1, where it can be seen that the most preferable material, based on these considerations, is Y_2O_3 (that is why it was used in Figure 8.22). It should be noted that in the construction of Figure 8.22, the number of radial cracks, n, present had to be assumed, as Eq. (15) suggests. In order to obtain accurate values, respective experiments must be performed, as n not only depends on the matrix material but also changes according to ρ. Due to a lack of such experimental data, however, and for illustration purposes, it was assumed here that $n = 20$.

Table 8.1 Distance at which cracking stops.

Matrix material	G_c (J m^{-2})	Critical crack radius ρ at which cracking stops (nm)
SrF$_2$	0.36	725
ThO$_2$	2.5	450
Y$_2$O$_3$	4.6	425
KCl	0.14	840

From both Sections 8.4.2 and 8.4.3 it was seen that nanosized active sites are mechanically more stable from a theoretical point of view. This is in qualitative agreement with the experimental evidence of Section 8.3, according to which nanosized active sites have not only a higher capacity but also greater capacity retention during cycling.

8.4.4
No Cracking

Finally, it is of interest to examine why the SnO_2 nanofibers (see Section 8.3.2.1) comprised of nanosized active Sn and inactive Li_2O (Eq. (8.8)), have no capacity loss, whatsoever, during cycling. It is believed that this may be due to the fact that mechanical damage is minimal – that is, no cracking occurs. From Eq. (14), the size of the active sites that will result in no cracking can be estimated. When the crack radius (ρ) equals the radius of the active site (a), it implies that the crack length ($\rho - a$) is zero and hence no cracks are present. Therefore, by defining the outer radius b, and the pressure p in Eq. (14), and setting $\rho = a$, it is possible to solve for the a; that is, the active site radius that will theoretically result in no cracking. It should be noted that, upon fracture, $\sigma_{\theta\theta}$ would be equal to the ultimate tensile strength (UTS) of the matrix, and hence the UTS of various materials is employed for comparison in Table 8.2.

In Ref. [77], the internal pressure p under the self-equilibrated loading case was found to be

$$p = \Delta \left\{ \frac{a^2}{\rho E_g} \left[\frac{\rho}{a} - \frac{(1-v_g)}{2} - \frac{3(1-v_g)\rho^2}{2(b^2+b\rho+\rho^2)} \right] + \frac{(a+\Delta)1-2v_s}{E_s} \right\}^{-1} \quad (17)$$

It can be seen from the data in Table 8.2 that no matter the matrix material, no cracking will result (in theory) when the active site radius is about 75% of the matrix radius. It is possible therefore that the size relationship between the Sn and Li_2O particles in the nanofibers may have been about this order, and therefore cracking was limited, allowing for 100% capacity retention. Further details on the aforementioned analysis can be found in Ref. [77].

Table 8.2 Size of Si sites that results in no cracking.

Matrix material	Radius a of active site that results in no cracking when $b = 1\ \mu m$
Al_2O_3	757
B_4C	781
BeO	746
WC	745
ZrO_2	758

8.5
Conclusions and Future Outlook

By employing experimental evidence, the advanced electrochemical properties of active/inactive nanocomposites – which represent the most promising anode materials for the next generation of Li-ion batteries – were identified. The major drawback in using materials (e.g., Sn, Si, Bi) that produce very high capacities during the formation of Li compounds is that they experience over a 200% volume expansion; as a result, after continuous electrochemical cycling, fracture and mechanical degradation of the anode takes place. As greater mechanical stability corresponds to better electrochemical contact within the anode, and hence allows for a greater lifetime, the following two fabrication methods have been developed to prevent fracture:

- A general trend noted for all materials that can be employed as active sites is that their capacity retention during electrochemical cycling increases significantly when their dimensions are at the nanoscale. This is attributed to the fact that deformation mechanisms at the nanoscale are less severe.

- Encasing the nanosized active materials in an inert or less-active with respect to Li matrix buffers the expansion of the active sites, thus minimizing mechanical damage and also protecting the active site surface from reacting with the electrolyte.

By performing a mechanical analysis based on linear elastic fracture mechanics, it was proven – from a theoretical point of view – that nanoscale materials are in fact more stable. It was also illustrated that theoretical considerations can lead to design criteria, as the crack stability and fracture of any material system can be predicted. One drawback, however, in trying to relate theory and experiment, is that the mechanical properties of the active sites are not known as a function of Li concentration. Also, the fracture properties of matrix materials are not known. Although as a first approximation it is possible to use the bulk elastic modulus, and Poisson's ratio, and some hypothetical matrix materials with known fracture energies, it is important to have accurate values of the parameters required for the mechanical formulation at the nanoscale in order to be able to develop a comprehensive study relating theory and experiment. The theoretical formulation illustrated will then allow for the development of complete design criteria.

References

1 Beaulieu, L.Y., Eberman, K.W., Turner, R. L., Krause, L.J. and Dahn, J.R. (2001) *Electrochem. Solid-State Lett.*, **4**, A137.
2 Besenhard, J.O., Yang, J. and Winter, M. (1997) *J. Power Sources*, **68**, 87.
3 Groisman, A. and Kaplan, E. (1994) *Europhys. Lett.*, **25**, 415.
4 Kitsunezaki, S. (1999) *Phys. Rev. E*, **60**, 6449.
5 Huggins, R. A. and Nix, W.D. (2000) *Solid State Ionics*, **6**, 5.
6 Yang, J., Winter, M. and Besenhard, J.O. (1996) *Solid State Ionics*, **90**, 281.
7 Huggins, R.A. (1989) *J. Power Sources*, **26**, 109.

8 Huggins, R.A. (1992) in *Fast Ion Transport in Solids* (eds B. Scrosati, A. Magistris, C.M. Mari and G. Mariotto), Kluwer Academic Publishers, p. 143.

9 Huggins, R.A. (1999) *J Power Sources*, **81–82**, 13.

10 Huggins, R.A. (1999) in *Handbook of Battery Materials* (ed. J.O. Besenhard), Wiley-VCH, p. 359.

11 Aifantis, K.E. and Hackney, S.A. (2003) *J. Mech. Behav. Mater.*, **14**, 403.

12 Thackeray, M.M., Johnson, C.S., Kahaian, A.J., Kepler, K.D., Vaughey, J.T., Shao-Horn, Y. and Hackney, S.A. (1999) *ITE Battery Lett.*, **1**, 26.

13 Shao-Horn, Y., Hackney, S.A., Armstrong, A.R., Bruce, P.G., Johnson, C.S. and Thackeray, M.M. (1999) *J. Electrochem. Soc.*, **146**, 2404.

14 Mao, O., Turner, O.R.L., Courtney, I.A., Fredericksen, B.D., Buckett, M.I., Krause, L.J. and Dahn, J.R. (1999) *Electrochem. Solid State Lett.*, **2**, 3.

15 Thackeray, M.M., Johnson, C.S., Kahaian, A.J., Kepler, K.D., Skinner, E., Vaughey, J.T., Shao-Horn, Y. and Hackney, S.A. (1999) *J. Power Sources*, **81–82**, 60.

16 Shao-Horn, Y., Hackney, A., Kahaian, A.J., Kepler, K.D., Vaughey, J.T. and Thackeray, M.M. (1999) *J. Power Sources*, **81–82**, 496.

17 Christensen, J. and Newman, J. (2006) *J. Electrochem. Soc.*, **153**, A1019.

18 Eshelby, J.D. (1957) *Proc. R. Soc. London*, **376**, A241.

19 Eshelby, J.D. (1959) *Proc. R. Soc. London*, **561**, A252.

20 Dieter, G.E. (1990) *Mechanical Mettalurgy*, 3rd edn. McGraw-Hill.

21 Kang, S.H. and Goodenough, J.B. (2000) *J. Electrochem. Soc.*, **147**, 3621.

22 Kang, S.H., Goodenough, J.B. and Rabenberg, L.K. (2001) *Electrochem. Solid-State Lett.*, **4**, A49.

23 Aifantis, K.E. and Hackney, S.A. submitted).

24 Guerard, D. and Herold, A. (1975) *Carbon*, **13**, 337.

25 Endo, M., Kim, C., Nishimura, K., Fujino, T. and Miyashita, T. (2000) *Carbon*, **38**, 183.

26 Kambe, N., Dresselhaus, M.S., Dresselhaus, G., Basu, S., McGhie, A.R. and Fischer, J.E. (1979) *Mater. Sci. Eng.*, **40**, 1.

27 Sato, K., Noguchi, M., Demachi, A., Oki, N. and Endo, M. (1994) *Science*, **64**, 556.

28 Dahn, J.R., Zheng, T., Liu, Y. and Xue, J.S. (1995) *Science*, **270**, 590.

29 Lindsay, M.J., Wang, G.X. and Liu, H.K. (2003) *J. Power Sources*, **119**, 84.

30 Wang, K., He, X.M., Ren, J.G., Jiang, C.Y. and Wan, C.R. (2006) *Electrochem. Solid State Lett.*, **9**, A320.

31 Yang, J., Takeda, Y., Imanishi, N. and Yamamoto, O. (1999) *J. Electrochem. Soc.*, **146**, 4009.

32 Kepler, K.D., Vaughey, J. and Thackeray, M.M. (1999) *Electrochem. Solid State Lett.*, **2**, 309.

33 Mao, O., Dunlap, R.A. and Dahn, J.R. (1999) *J. Electrochem. Soc.*, **146**, 405.

34 Vaughey, J.T., O'Hara, J. and Thackeray, M.M. (2000) *Electrochem. Solid State Lett.*, **3**, 13.

35 Beaulieu, L.Y., Eberman, K.W., Turner, R.L., Krause, L.J. and Dahn, J.R. (2001) *Electrochem. Solid-State Lett.*, **4**, A137.

36 Graetz, J., Ahn, C.C., Yazami, R. and Fultz, B. (2003) *Electrochem. Solid-State Lett.*, **6**, A194.

37 Moon, T., Kim, C. and Park, B. *J Power Sources* (in print).

38 Trifonova, A., Wachtler, M., Wagner, M.R., Schroettner, H., Mitterbauer, Ch., Hofer, F., Möller, K.C., Winter, M. and Besenhard, J.O. (2004) *Solid State Ionics*, **168**, 51.

39 Mukaibo, H., Osaka, T., Reale, P., Panero, S., Scrosati, B. and Wachtler, M. (2004) *J. Power Sources*, **132**, 225.

40 Yin, J.T., Wada, M., Tanase, S. and Sakai, T. (2004) *J. Electrochem. Soc.*, **151**, A583.

41 Wang, G.X., Yao, J. and Liu, H.K. (2004) *Electrochem. Solid-State Lett.*, **7**, A250.

42 Dimov, N., Kugino, S. and Yoshio, M. (2004) *J. Power Sources*, **136**, 108.

43 Hwang, S.-M., Lee, H.-Y., Jang, S.-W., Lee, S.-M., Lee, S.-L., Baik, H.-K. and Lee, J.-Y. (2001) *Electrochem. Solid-State Lett.*, **4**, A97.

44 Patel, P., Kim, I.-S. and Kumta, P.N. (2005) *Mater. Sci. Eng. B*, **116**, 347.
45 Kim, I.-S., Blomgren, G.E. and Kumta, P.N. (2003) *Electrochem. Solid-State Lett.*, **6**, A157.
46 Kim, I.-S., Kumta, P.N. and Blomgren, G.E. (2000) *Electrochem. Solid-State Lett.*, **3**, 493.
47 Wang, Z., Tian, W. and Li, X. (2006) *J. Alloys Compounds*, 247doi:10.1016/j.jallcom.2006.08.247.
48 Wang, K., He, X., Ren, J., Wang, L., Jiang, C. and Wan, Chunron. (2006) *Electrochimica Acta*, **52**, 1221.
49 Li, H., Huang, X. and Chen, L. (1999) *J. Power Sources*, **81**, 340.
50 Sayama, K., Yagi, H., Kato, Y., Matsuta, S., Tarui, H. and Fujitani, S. (2002) Abstract 52, The 11th International Meeting on Lithium Batteries, Monterey, CA, June 23–28.
51 Takamura, T., Ohara, S., Suzuki, J. and Sekine, K. (2002) Abstract 257, The 11th International Meeting on Lithium Batteries, Monterey, CA, June 23–28.
52 Li, H., Huang, X., Chen, L., Wu, Z. and Liang, Y. (1999) *Electrochem. Solid-State Lett.*, **2**, 547.
53 Crosnier, O., Brousse, T., Devaux, X., Fragnaud, P. and Schleich, D.M. (2001) *J. Power Sources*, **94**, 169.
54 Crosnier, O., Devaux, X., Brousse, T., Fragnaud, P. and Schleich, D.M. (2001) *J. Power Sources*, **97**, 188.
55 Idota, Y., Mishima, M., Miyiaki, Y., Kubota, T. and Miyasaka, T. (1997) US Patent 5,618,641.
56 Idota, Y., Kubota, T., Matsufuji, A., Maekawa, Y. and Miyasaka, T. (1997) *Science*, **276**, 1395.
57 Courtney, I.A. and Dahn, J.R. (1997) *J. Electrochem. Soc.*, **144**, 2943.
58 Morimoto, H., Nakai, M., Tatsumisago, M. and Minami, T. (1999) *J. Electrochem. Soc.*, **146**, 3970.
59 Kim, J.Y., King, D.E., Kumta, P.N. and Blomgren, G.E. (2000) *J. Electrochem. Soc.*, **147**, 4411.
60 Lee, J.Y., Xiao, Y. and Liu, Z. (2000) *Solid State Ionics*, **133**, 25.
61 Mao, O. and Dahn, J.R. (1999) *J. Electrochem. Soc.*, **146**, 414.
62 Ehrlich, G.M., Durand, C., Chen, X., Hugener, T.A., Spiess, F. and Suib, S.L. (2000) *J. Electrochem. Soc.*, **147**, 886.
63 Fang, L. and Chowdari, B.V.R. (2001) *J. Power Sources*, **97–98**, 181.
64 Li, N., Martin, C.R. and Scorsati, B. (2001) *J. Power Sources*, **97**, 240.
65 Che, G., Lakshmi, B.B., Martin, C.R., Fisher, E.R. and Ruoff, R.A. (1998) *Chem. Mater.*, **10**, 260.
66 Che, G., Fisher, E.R. and Martin, C.R. (1998) *Nature*, **393**, 346.
67 Sides, C.R., Li, N., Patrissi, C.J., Scrosati, B. and Martin, C.R.August (2002) *MRS Bulletin*.
68 Pistoia, G., Pasquali, M., Geronov, Y., Manev, V. and Moshtev, R.V. (1989) *J. Power Sources*, **27**, 35.
69 Brousse, T., Retoux, R. and Schleich, D. (1998) *J. Electrochem. Soc.*, **145**, 1.
70 Kim, Il-seok, Blomgren, G.E. and Kumta, P.N. (2004) *Electrochem. Solid-State Lett.*, **7**, A44.
71 Chen, L., Xie, X., Wang, B., Wang, K. and Xie, J. (2006) *Mater. Sci. Eng. B*, **131**, 186.
72 Yang, J., Wang, B.F., Wang, K., Liu, Y., Xie, J.Y. and Wen, Z.S. (2006) *Electrochem. Solid-State Lett.*, **6**, A154.
73 Li, H., Huang, X.J., Chen, L.Q., Zhou, G.W., Zhang, Z., Yu, D.P., Mo, Y.J. and Pei, N. (2000) *Solid-State Ionics*, **135**, 181.
74 Dempsey, J.P., Slepyan, L.I. and Shekhtman, I.I. (1995) *Int. J. Fract.*, **73**, 223.
75 Aifantis, K.E. and Hackney, S.A. (2003) *J. Mech. Behav. Mater.*, **14**, 403.
76 Dempsey, J.P., Palmer, A.C. and Sodhi, D.S. (2001) *Eng. Fract. Mech.*, **68**, 1961.
77 Aifantis, K.E., Hackney, S.A. and Dempsey, J.P. (2007) *J. Power Sources*, **165**, 874.
78 Little, R.W. (1973) *Theory of Elasticity*, Prentice-Hall, Englewood Cliffs, NJ.
79 Aifantis, K.E. and Dempsey, S.A. (2005) *J. Power Sources*, **143**, 203.

9
Nanostructured Hydrogen Storage Materials Synthesized by Mechanical Alloying

Mieczyslaw Jurczyk and Marek Nowak

9.1
Introduction

9.1.1
The Aim of the Research

Today, increases in the world's population and its economic growth are raising the energy demand at a dramatic pace [1]. Under the International Energy Agency scenario of a world population of 10 billion people by 2050, the energy demand will increase from 13.6 TW in 2002 to 33.2 TW – an additional demand of 20 TW compared to today. If only 50% of this is to be supplied by nuclear energy, there would be a call for 10 000 new 1-GW power plants to be built by 2050. The available uranium resources, however, would only be sufficient to feed this number of conventional fission reactors for 13 to 16 years.

Another possibility is to change towards a sustainable energy future, with a larger utilization of renewable and carbon-free energy sources. Besides the political will to make the change, the necessary technologies must be both reliable and available. In aiming to ensure this situation, materials research must provide the basis for innovative technical solutions capable of supplying the required energy.

In 1959, Professor Feynman presented his famous idea of nanostructured materials production [2]. He suggested that a combination of single atoms could lead to the formation of new, specialized materials with unconventional properties. Today, it is possible to prepare metal/alloy nanocrystals with nearly monodispersive size distribution. Nanostructures represent key building blocks for nanoscale science and technology. They are needed to implement the "bottom-up" approach to nanoscale fabrication, whereby well-defined nanostructures with unique properties are assembled into functional as well as mechanical properties [3].

During the past few years, interest in the study of nanostructured materials has been accelerating, stimulated by recent advances in materials synthesis and characterization techniques, and also the realization that these materials exhibit many interesting and unexpected properties with a number of potential technological applications. For example, hydrogen storage nanomaterials are the key to the future of the storage and batteries/cells industries [4–6].

Since 1996 a research program has been initiated at Institute of Materials Science and Engineering, Poznan University of Technology, in which fine grained, intermetallic compounds were produced by mechanical alloying (MA), high-energy ball milling (HEBM), hydrogenation–disproportionation–desorption–recombination (HDDR), or mechanochemical processing (MCP) [6–9]. The mechanical synthesis of nanopowders and their subsequent consolidation is an example of how this idea can be realized in metals by the so-called bottom-up approach. Alternatively, other methods have been developed which are based on the concept of the production of nanomaterials from conventional bulk materials via a "top-down" approach. The investigations by severe plastic deformation [e.g., cyclic extrusion compression method (CEC) or equal channel angular extrusion (ECAE)] [10–13] show that such a transformation is indeed possible.

Metal hydrides (MH) which reversibly absorb and desorb hydrogen at ambient temperature and pressure – the so-called "hydrogen storage alloys" – are regarded as important materials for solving energy and environmental issues. Studies in metal–hydrogen grew rapidly during the 1970s [14–21], as a result of which many hydride alloys (TiNi, TiFe, LaNi$_5$) were studied and developed. During the early 1980s, AB$_2$-type Ti/Zr–V–Ni–M systems and rare-earth-based AB$_5$-type Ln–Ni–Co–Mn–Al alloys were investigated. These systems obtained long-life hydrogen storage electrode alloys, but the Ti/Zr system has a higher capacity. During the late 1980s and early 1990s, the non-stoichiometric AB$_2$ and AB$_5$ systems with improved performance, such as higher capacity and long cycle life, were developed [18].

The Ni–hydrogen battery came to the market in 1990. The nickel–metal hydride (Ni-MH) battery, using hydrogen storage alloy as the negative electrode material, has recently undergone extensive investigation and development because of its advantages such as high energy density and freedom from charge in electrolyte concentration during charge–discharge cycling over the nickel–cadmium (NiCd) battery [18]. Compared to the NiCd battery, a Ni-MH cell has several advantages, including almost twice the capacity, no memory effect, and no environmental pollution problems.

Conventionally, the microcrystalline hydride materials have been prepared by arc or induction melting and annealing. However, either a low storage capacity by weight or poor absorption–desorption kinetics in addition to a complicated activation procedure have limited the practical use of metal hydrides. Substantial improvements in the hydriding–dehydriding properties of metal hydrides could possibly be achieved by the formation of nanocrystalline structures by non-equilibrium processing techniques such as mechanical alloying or HEBM [5,6,17,18]. In MA, elemental powdered materials are used as the starting materials, whilst in HEBM the starting material is an alloy with a desired composition [6].

Mechanical alloying was developed during the 1970s at the International Nickel Company as a technique for dispersing nanosized inclusions into nickel-based alloys [22]. During recent years, however, the MA process has been used successfully

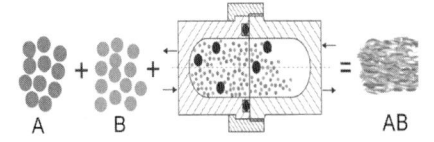

Figure 9.1 A schematic cross-sectional representation of the mechanical alloying process for synthesizing nanometer-sized powders (SPEX 8000 mixer mill).

to prepare a variety of alloy powders, including powders exhibiting supersaturated solid solutions, quasicrystals, amorphous phases and nano-intermetallic compounds [22,23]. Indeed, the MA technique has been proved to be a novel and promising method for alloy formation (Figure 9.1).

The raw materials used for MA are commercially available high-purity powders that have particle sizes in the range of 1 to 100 µm. During the MA process, the powder particles are periodically trapped between colliding balls and are plastically deformed. Such a feature occurs by the generation of a wide number of dislocations, as well as other lattice defects. Furthermore, the ball collisions cause fracturing and cold welding of the elementary particles, forming clean interfaces at the atomic scale. Further milling leads to an increase of the interface number, and the sizes of the elementary component area decrease from millimeter to submicrometer lengths. Concurrent to this decrease in the elementary distribution, some nanocrystalline intermediate phases are produced inside the particles, or at their surfaces. As the milling duration develops, the content fraction of such intermediate compounds increases, leading to a final product the properties of which are a function of the milling conditions. It has been shown, that MA has produced amorphous phases in metals, although differentiation between a "truly" amorphous, extremely fine-grained material, or a material in which very small crystals are embedded in an amorphous matrix in so-produced materials, has not been straightforward on the basis of diffraction basis [23]. Only supplementary investigations with neutron diffraction can confirm unambiguously that the phases produced by MA are truly amorphous. The milled powder is finally heat-treated to obtain the desired microstructure and properties. Annealing leads to grain growth and release of microstrain.

Nanocrystalline materials exhibit quite different properties from both crystalline and amorphous materials. This is due to the structure, in which extremely fine grains are separated by what some investigators have characterized as "glass-like" disordered grain boundaries. The generation of new metastable phases or materials with an amorphous grain boundary phase offers a wider distribution of available sites for hydrogen, and thus totally different hydrogenation behavior. The mechanism of amorphous phase formation by MA is due to a chemical solid-state reaction, which is believed to be caused by the formation of a multilayer structure during milling [23].

Recently, it has been shown that MA of TiFe, ZrV_2, $LaNi_5$ or Mg_2Ni is effective for improving the initial hydrogen absorption rate, due to a reduction in particle size and to the creation of new clean surfaces [5,6]. The proper engineering of microstructures by using unconventional processing techniques will lead to advanced nanocrystalline intermetallics representing a new generation of metal hydride materials.

In order to optimize the choice of intermetallic compounds for a battery application, a better understanding of the role of each alloy constituent on the electronic properties of the material is crucial. Semi-empirical models showed that the energy of the metal–hydrogen interaction depends both on geometric as well as electronic factors [6,24].

The aim of this chapter is to review the advantages of some nanostructured materials, and their applications in electrochemistry. The influence of chemical composition on the structural properties of nanocrystalline TiFe-, ZrV_2-, $LaNi_5$- and Mg_2Ni-type alloys was analyzed. The electrochemical behavior of nanocrystalline hydrogen storage compounds, under the conditions of their performance in rechargeable nickel hydride (Ni-MH) batteries, is presented. Finally, the electronic properties of nanocrystalline alloys are compared to those of microcrystalline samples.

9.1.2
Types of Hydride

Hydrogen reacts with many elements to form hydrides. The binary hydride, a reaction product, can be of different types:

- ionic hydrides (LiH)
- covalent molecular hydrides (CH_4)
- metallic hydrides.

All of the binary hydrides which can be formed by a direct reaction of hydrogen with the metal are illustrated in Figure 9.2. The densities of the hydrogen in most of

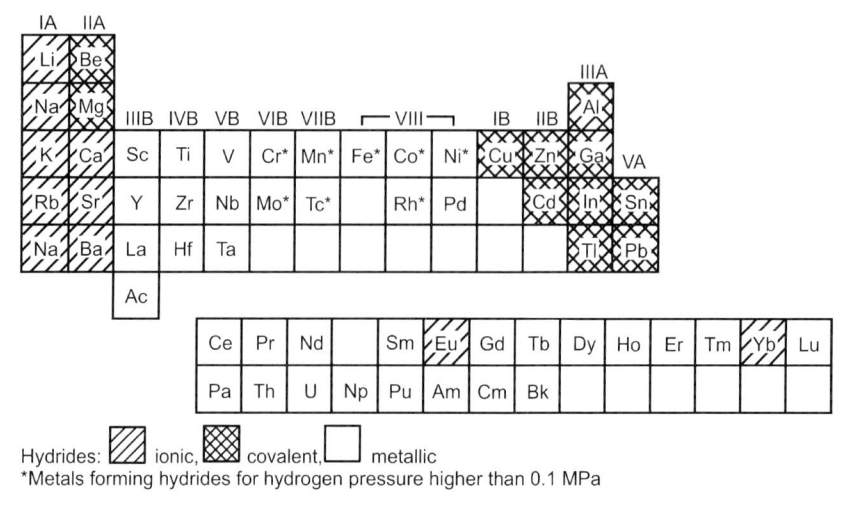

Hydrides: ▨ ionic, ▩ covalent, ☐ metallic
*Metals forming hydrides for hydrogen pressure higher than 0.1 MPa

Figure 9.2 Metals which form binary hydrides.

Table 9.1 Hydrogen densities of some metal hydrides compared with liquid hydrogen.

Compounds	PdH$_{0.8}$	MgH$_2$	TiH$_2$	VH$_2$	Liquid hydrogen
Hydrogen density [atom × 10^{-22} cm^{-3}]	4.7	6.6	9.2	10.4	4.2

these metallic hydrides are greater than that of liquid hydrogen (Table 9.1). Most metallic hydrides are formed with pure hydrogen at room temperature.

9.1.3
The Absorption–Desorption Process

Of primary importance in the hydrogen storage by metal hydrides is the pressure "plateau" at which the material reversibly absorbs/desorbs large quantity of hydrogen. Figure 9.3 represents the hydride formation process as a plot of hydrogen pressure against the ratio of hydrogen to metal atoms. With increasing hydrogen pressure, hydrogen dissolves in the metal to form a solid solution, represented by point α. A hydride phase then precipitates at constant pressure "p", as determined by the phase role, until the metal is fully converted to hydride at point β. Finally, the hydrogen pressure rises further. When the pressure is reduced, there is usually some hysteresis, such that the plateau pressure on decomposing the hydride is lower than that during its formation.

For many applications, the plateau pressure should be close to ambient pressure because this allows the use of lightweight storage containers. The plateau pressure can be tailored to a specific application through alloying. For example, the plateau pressure of LaNi$_5$ can be lowered by the addition of a few percent of tin [18].

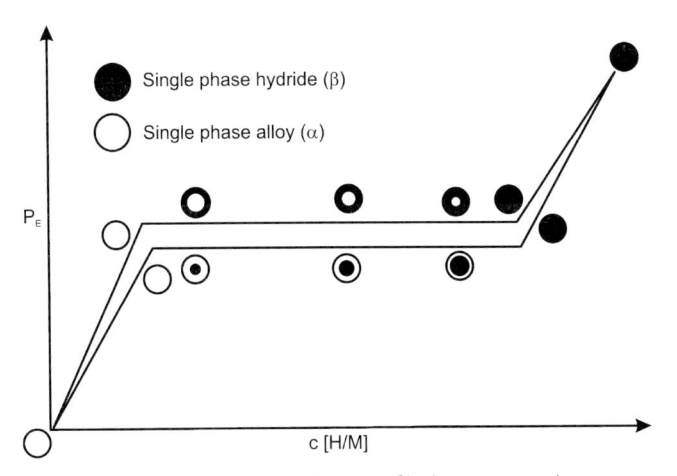

Figure 9.3 Pressure composition diagram of hydrogen in metal.

Table 9.2 Binary intermetallic compounds and their hydrides.

System	Hydrides
AB	$TiFeH_2$
AB_2	$ZrV_2H_{5.5}$
AB_5	$LaNi_5H_6$
A_2B	$Mg_2NiH_{3.6}$

9.1.4
Hydrides Based on Intermetallic Compounds of Transition Metals

The group of metallic hydrides is increased in number if, in addition to binary systems, one also includes hydrides based on intermetallic compounds of transition metals. Currently, a wide set of alloys composed of rare earth (RE) with nickel (AB_5-type) and alloys of zirconium and vanadium with nickel (AB_2-type), as well as titanium and magnesium with nickel (AB- or A_2B-type) are offered for use as hydrogen storage materials (Tables 9.2 and 9.3).

In all of these alloys, component A is the one which forms the stable hydride. Component B performs several additional functions:

- it can play a catalytic role in enhancing the hydriding–dehydriding kinetic characteristics
- it can alter the equilibrium pressure of the hydrogen absorption–desorption process to desired level
- it can increase the stability of the alloy, preventing dissolution or formation of a compact oxide layer of component A.

The TiFe, ZrV_2 and $LaNi_5$ phases are familiar materials which absorb large quantities of hydrogen under mild conditions of temperature and pressure. These types of hydrogen-forming compound have recently proven to be very attractive as negative electrode material in rechargeable Ni-MH batteries [17–19]. Magnesium-based hydrogen storage alloys have also been considered as possible candidates for electrodes in Ni-MH batteries [6,25,26].

Table 9.3 The properties of alloys offered for use as hydrogen storage materials.

Type of alloy	Structure	Density [g/cm^2]	p_d [atm]	(H/M) max [wt %]
AB_5 - $LaNi_5$	$CaCu_5$	6.6 ÷ 8.6	0.024 ÷ 23	1.1 ÷ 1.9
AB_2 - ZrV_2	$MgCu_2$/$MgZn_2$	5.8 ÷ 7.6	10^{-3} ÷ 182	1.5 ÷ 2.4
AB – TiFe	CsCl	6.5	0.1 ÷ 4.1	1.3 ÷ 1.9
A_2B - Mg_2Ni	Mg_2Ni	3.46	10^{-5}	3.6

p_d – equilibrium pressure.

TiFe and ZrV$_2$ alloys crystallize in the cubic CsCl and MgCu$_2$ structures, and at room temperature absorb up to 2 H per formula unit (f.u.) and 5.5 H f.u.$^{-1}$, respectively. In contrast, LaNi$_5$ alloy crystallizes in the hexagonal CaCu$_5$ structure, and at room temperature can absorb up to 6 H f.u.$^{-1}$ [18,20]. Nevertheless, the application of these types of material in batteries has been limited due to slow absorption–desorption kinetics, in addition to a complicated activation procedure. The properties of hydrogen host materials can be modified substantially by alloying them to obtain the desired storage characteristics, such as the correct capacity at a favorable hydrogen pressure. TiFe alloy is lighter and cheaper than the LaNi$_5$-type material, and several approaches have been adopted to improve the activation of this alloy. For example, the replacement of Fe by some amount of transition metals to form a secondary phase may improve the activation properties of TiFe. On the other hand, the electrochemical activity of ZrV$_2$-type materials can be stimulated by substitution, in which Zr is partially replaced by Ti and V is partially replaced by other transition metals (Cr, Mn, Ni) [18]. Independently, it was found that the respective replacement of La and Ni in LaNi$_5$ with small amounts of Zr and Al resulted in a prominent increase in the cycle life time, without causing much decrease in capacity [18].

Magnesium-based alloys have also been extensively studied during the past few years [25,26]. The microcrystalline Mg$_2$Ni alloy can reversibly absorb and desorb hydrogen only at high temperatures. Upon hydrogenation at 250°C, Mg$_2$Ni transforms into the hydride phase Mg$_2$NiH$_4$. Substantial improvements in the hydriding–dehydriding properties of Mg$_2$Ni metal hydrides might possibly be achieved by the formation of nanocrystalline structures [5]. The hydrogen content in Mg$_2$NiH$_4$ is also relatively high, at 3.6 wt.%, but is only 1.5 wt.% in LaNi$_5$H$_6$.

In order to provide storage for hydrogen, the hydride should:

- be able to store large quantities of hydrogen
- be readily formed and decomposed, and have reaction kinetics satisfying the charge–discharge requirements of the system
- have the capability of being cycled without alteration in pressure–temperature characteristics during the life of the system
- have low hysteresis
- have good corrosion stability
- have low cost
- be at least as safe as other energy carriers.

9.1.5
Prospects for Nanostructured Metal Hydrides

A few currently available technologies permit hydrogen to be stored directly by modifying its physical state in gaseous or liquid form. However, these methods are unable to meet all of the following proposed criteria [15,16]:

- high hydrogen content per unit mass and unit volume
- limited energy loss during operation

- fast kinetics during charging
- high stability with cycling
- cost of recycling and charging infrastructures
- safety concerns in regular service or during accidents.

Currently, much effort is being expended in the development of hydrogen storage systems, whether nanostructured metals or alloys hydrides, and carbon nanostructures. For vehicle applications, depending on the temperature of hydrogen absorption/desorption below or above 150 °C, the alloy hydrides can be distinguished as either high- or low-temperature materials. The principal disadvantages of alloy hydrides – other than cost – are the low hydrogen content at low temperature (e.g., La-based alloys) and the difficulty of reducing the desorption temperature and pressure of alloy hydrides with high hydrogen storage capacity and fast rated kinetics? (e.g., Mg-based materials). In order to solve the above-mentioned problems, the use of composite materials, starting from La- or Mg-alloys, and of novel catalyzed metal hydrides, has been systematically investigated. Among the latter group, alanates and other hydrides of light elements (Li, B, Na) have provided very interesting results, with hydrogen contents reaching 5–6 mass% for $NaBH_4$ and up to 18 mass% for $LiBH_2$ [15].

Alternatively, the high hydrogen uptake capacity supposed for fullerenes has stimulated hydrogen scientists to investigate more closely the synthesis of the new form of carbon [27]. In theory, the hydrogen storage capacity of these new materials seems rather hopeful, as when one hydrogen atom is joined to each carbon atom (which is quite probable), there appears the possibility of producing a sorbing matrix based on these materials which allows a storage of up to 7.7 wt.% H_2. This value both meets and exceeds all of the requirements for this class of material. Thus, seeking the means to conduct the reversible and complete reaction:

$$C_x + x/2 H_2 \leftrightarrow C_x H_x \qquad (1)$$

and considering features of structures and properties would allow the application of materials as systems for hydrogen storage in many fields of engineering and technologies. Moreover, the absence of materials from this class restrains the wide application of hydrogen as fuel and energy carrier.

Today, nanostructured metal hydrides represent a new class of materials in which outstanding hydrogen sorption may be achieved by correct engineering of the microstructure and surface [4–6]. The case for these materials is extremely important, because it can be applied to hydrogen storage systems and Ni-MH batteries. The techniques of MA and HEBM have been successfully used to improve the hydrogen sorption properties of various metal hydrides. Nanocrystalline powder alloys, in contrast to conventional hydrides, readily absorb hydrogen, with no need for prior activation. These materials show substantially enhanced absorption and desorption kinetics even at relatively low temperatures. Thus, nanostructured materials will clearly play an important role in the field of Ni-MH batteries.

9.2
The Fundamental Concept of the Hydride Electrode and the Ni-MH Battery

9.2.1
The Hydride Electrode

The interaction of hydrogen and metal M is represented as [21]:

$$M + x/2 H_2(g) \leftrightarrow MH_x(s) \tag{2}$$

where MH_x is the hydride of metal M.

Some alloys can be charged and discharged electrochemically. Equation (3) demonstrates the electrochemical charging and discharging reactions:

$$M + xH_2O + xe^- \leftrightarrow MH_x + xOH^- \tag{3}$$

9.2.2
The Ni-MH Battery

The electrochemical reaction of a Ni-MH cell can be represented by the following half-cell reactions [19].

9.2.2.1 Normal Charge–Discharge Reactions
During charging, the nickel hydroxide $Ni(OH)_2$ positive electrode is oxidized to nickel oxyhydroxide NiOOH, while the alloy M negative electrode forms MH by water electrolysis. The reactions on each electrode proceed via solid-state transitions of hydrogen. The overall reaction is expressed only by a transfer of hydrogen between alloy M and $Ni(OH)_2$:

- Nickel positive electrode

$$Ni(OH)_2 + OH^- = NiOOH + H_2O + e^- \tag{4}$$

- Hydride negative electrode

$$M + H_2O + e^- = MH + OH^- \tag{5}$$

- Overall reaction

$$Ni(OH)_2 + M = NiOOH + MH \tag{6}$$

9.2.2.2 Overcharge Reactions
In sealed Ni-MH cell, the M electrode has a higher capacity than the $Ni(OH)_2$ electrode, thus facilitating a gas recombination reaction. During an overcharge

situation, the MH electrode is charged continuously, forming hydride, while the Ni electrode starts to evolve oxygen gas according to Eq. (7):

- Nickel positive electrode

$$2OH^- = H_2O + \tfrac{1}{2}O_2 + 2e^- \tag{7}$$

- Hydride negative electrode

$$2M + 2H_2O + 2e^- = 2MH + 2OH^- \tag{8}$$

$$2MH + \tfrac{1}{2}O_2 = 2M + H_2O \tag{9}$$

- Actual (non-ideal) case

$$H_2O + e^- = \tfrac{1}{2}H_2 + OH^- \tag{10}$$

$$2MH + \tfrac{1}{2}O_2 = 2M + H_2O \tag{11}$$

The oxygen diffuses through the separator to the MH electrode, where it reacts chemically, producing water [Eq. (11)] and preventing a pressure rise in the cell.

9.2.2.3 Over-Discharge Reaction

During the over-discharge process, hydrogen gas starts to evolve at the Ni electrode [Eq. (12)]. The hydrogen diffuses through the separator to the MH electrode, where it dissociates to atomic hydrogen by a chemical reaction [Eq. (13)], followed by a charge transfer reaction [Eq. (14)], ideally causing no pressure rise in the cell:

- Nickel positive electrode

$$2H_2O + 2e^- = H_2 + 2OH^- \tag{12}$$

- Hydride negative electrode

$$H_2 + 2M = 2MH \tag{13}$$

$$2OH^- + 2MH = 2H_2O + 2e^- + 2M \tag{14}$$

9.3
An Overview of Hydrogen Storage Systems

In view of the promising features of the Ni-MH batteries, a large number of hydrogen storage systems have been characterized to date at the present authors' [6,8,24,28,29]. In addition, MA has been used recently to prepare a nanocrystalline TiFe-, ZrV$_2$- LaNi$_5$- and Mg$_2$Ni-type alloys (Figure 9.4).

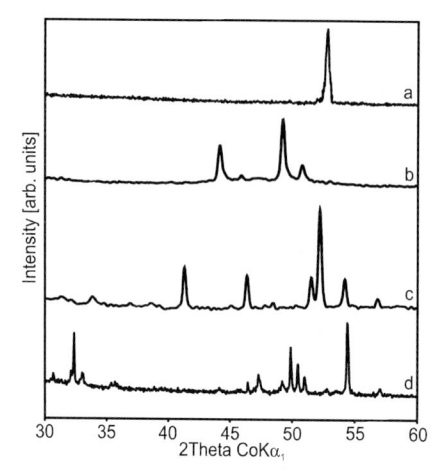

Figure 9.4 X-ray diffraction spectra of nanocrystalline TiFe (a), ZrV$_2$ (b), LaNi$_5$ (c) and Mg$_2$Ni (d) alloys produced by mechanical alloying followed by annealing (TiFe MA for 20 h and heat-treated at 700°C for 30 min; ZrV$_2$ MA for 40 h and heat-treated at 800°C for 30 min; LaNi$_5$ MA 30 h and heat-treated at 700°C for 30 min; Mg$_2$Ni MA 90 h and heat-treated at 450°C for 30 min).

9.3.1
The TiFe-Type System

Figure 9.5 illustrates a series of X-ray diffraction (XRD) spectra of a mechanically alloyed Ti–Fe powder mixture (53.85 wt.% Ti + 46.15 wt.% Fe) which has been

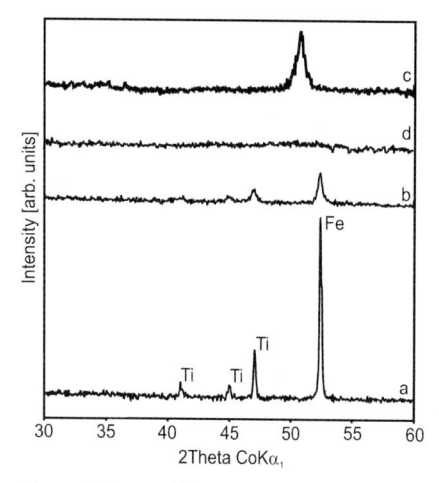

Figure 9.5 X-ray diffraction spectra of a mixture of Ti and Fe powders MA for different times in an argon atmosphere. (a) Initial state (elemental powder mixture); (b) after MA for 2 h; (c) after MA for 20 h; and (d) heat-treated at 700°C for 30 min.

Table 9.4 Discharge capacities of TiFe alloy prepared by different methods on 3rd cycle (current density of charging and discharging was 4 mA g^{-1}).

Microstructure	Processing method	Lattice constant a [Å]	Discharge capacity [mA·h g^{-1}]
Microcrystalline	arc melting and annealing*	2.977	0.00
Amorphous	MA	–	5.32
Nanocrystalline	MA and annealing	2.973	7.50

*900 °C/3 days.

subjected to milling for increasing times. The originally sharp diffraction lines of Ti and Fe gradually become broader (Figure 9.5, spectrum b), and their intensity decreases with milling time. The powder mixture milled for more than 20 h has been transformed completely to the amorphous phase, without the formation of other phases (spectrum c). During the MA process, the crystalline size of the Ti decreases with MA time, and reaches a steady value of 20 nm after 15 h of milling. This size of crystallites appears to be favorable for the formation of an amorphous phase, which develops at the Ti–Fe interfaces. Formation of the nanocrystalline alloy TiFe was achieved by annealing the amorphous material in a high-purity argon atmosphere at 700 °C for 30 min (see Table 9.4; Figures 9.4a and 9.5d). All diffraction peaks were assigned to those of CsCl-type structure with cell parameter $a = 2.973$ Å. When nickel is added to TiFe$_{1-x}$Ni$_x$, the lattice constant a increases.

The microstructure and possible local ordering in the TiNi samples was studied using transmission electron microscopy (TEM). The sample milled for 5 h was mostly amorphous, as was apparent from the high-resolution image (Figure 9.6a). The selected area electron diffraction (SAED) pattern (see inset of Figure 9.6a) contains

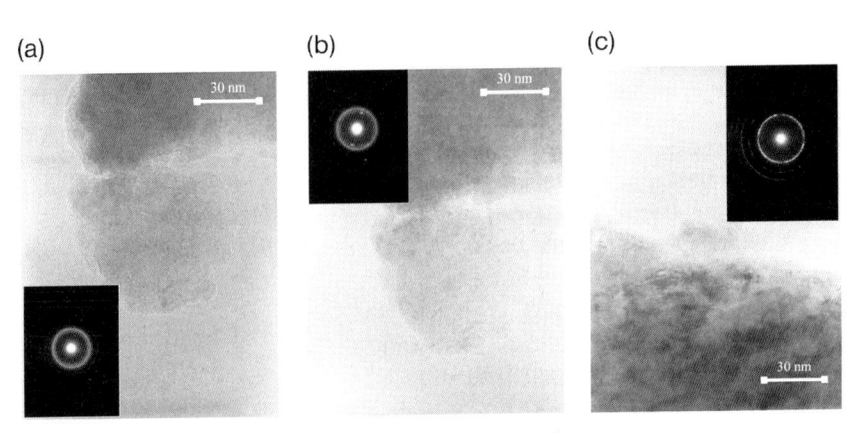

Figure 9.6 TEM images and electron diffraction patterns (insets) of the milled TiNi sample. (a) Typical amorphous fragment; (b) the same region after 25-min exposure to an electron beam; (c) crystalline grain.

broad rings at positions expected for TiNi with CsCl structure. There are, however, additional weakly diffuse rings, most probably from TiO_2. The amorphous alloy was found to be unstable upon exposure to an electron beam and underwent some crystallization. In the image acquired after a 25-min exposure to the electron beam (Figure 9.6b), the formation of ordered regions (defined as patches with parallel lattice fringes) may be seen. Accordingly, additional sharp reflections appear in the SAED pattern (inset to Figure 9.6b). Apart from the prevailing amorphous phase, the milled sample contained a small amount of crystalline alloy with a CsCl structure (Figure 9.6c). The lack of any sharp reflections in the XRD pattern suggests that the amount of crystalline phase is very low, and/or it is formed during the TEM observations.

The microstructure of the annealed sample is shown in Figure 9.7. An analysis of high-resolution images (Figure 9.7a and b) revealed the presence of well-developed crystallites with broad range of sizes from 4 nm up to more than 30 nm. The SAED pattern obtained from the large area (200 μm) (Figure 9.7c) contained sharp rings which corresponded to a TiNi alloy with CsCl structure.

The amorphization process of the studied materials was also examined using differential scanning calorimetry (DSC) measurements. XRD studies of MA samples after heating in DSC showed that the observed large exothermic calorimetric effects must be attributed to the formation of the ordered compounds as a crystallization product.

The electrochemical pressure-composition (e.p.c.) isotherms for absorption and desorption of hydrogen were obtained from the equilibrium potential values of the electrodes, measured during intermittent charge and/or discharge cycles at constant current density, by using the Nernst equation [30]. Due to the amorphous nature of the studied alloys prior to annealing, the hydrogen absorption–desorption characteristics proved unsatisfactory as, compared to nanocrystalline TiFe, the storage capacity was considerably smaller (Figure 9.8). Annealing causes transformation

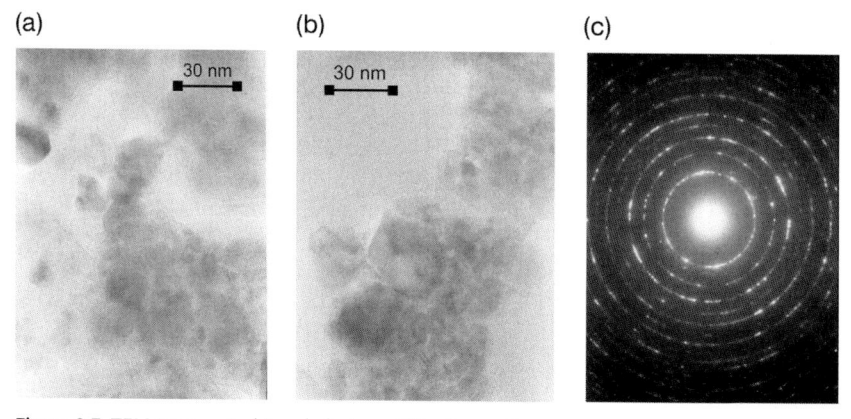

Figure 9.7 TEM images (a, b) and electron diffraction patterns (c) of the annealed TiNi sample. Nanocrystallites of an alloy are clearly visible in (a) and (b).

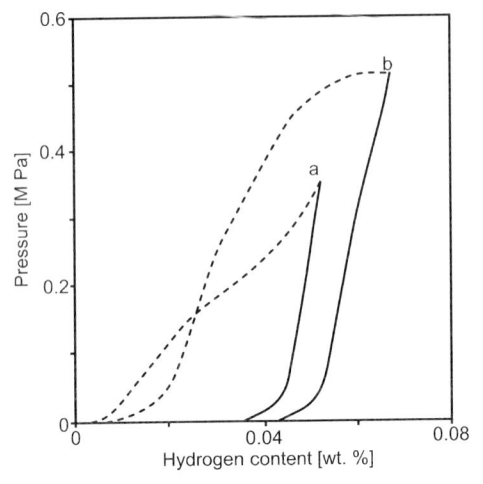

Figure 9.8 Electrochemical isotherms pressure-composition for absorption (dashed line) and desorption (solid line) of hydrogen on: (a) amorphous and (b) nanocrystalline TiFe alloys.

from the amorphous to the crystalline structure, and produces grain boundaries. Anani et al. [17] noted that grain boundaries are necessary for the migration of the hydrogen into the alloy. At this point it is worth noting, that although the characteristics for microcrystalline and nanocrystalline materials are very similar in respect to hydrogen contents, there are small differences in the plateau pressures. When the amount of Ni in TiFe$_{1-x}$Ni$_x$ was increased, the pressure in the plateau region continued to decrease and the hydrogen storage capacity was increased. The hydrogenation behavior of the amorphous structure was different from that of the thermodynamically stable, crystalline material. Notably, for amorphous TiFe material, the plateau totally disappears.

The discharge capacities of the studied microcrystalline, amorphous and nanocrystalline TiFe materials are listed in Table 9.4. The discharge capacity of electrodes prepared from TiFe alloy powder by the application of MA and annealing displayed a very low capacity (7.50 mA·h g^{-1} at 4 mA g^{-1} discharge current), whereas the arc-melted electrodes had no capacity [31]. The reduction in powder size, and the creation of new surfaces, is effective for the improvement of the hydrogen absorption rate.

Materials obtained when Ni was substituted for Fe in TiFe led to a great improvement in activation behavior of the electrodes. It was found that the increasing nickel content in TiFe$_{1-x}$Ni$_x$ alloys led initially to an increase in discharge capacity, giving a maximum at $x = 0.75$ [32]. In the annealed nanocrystalline TiFe$_{0.25}$Ni$_{0.75}$ powder, a discharge capacity of up to 155 mA·h g^{-1} (at 40 mA g^{-1} discharge current) was measured (Table 9.5). The electrodes which were mechanically alloyed and annealed from the elemental powders displayed maximum capacities at around the third cycle but, especially for $x = 0.5$ and 0.75 in TiFe$_{1-x}$Ni$_x$ alloy, degraded slightly with cycling. However, this may have been due to the easy formation of the oxide layer (TiO$_2$) during cycling.

Table 9.5 Structural parameters and discharge capacities for nanocrystalline TiFe$_{1-x}$Ni$_x$ materials (current density of charging and discharging was 40 mA g^{-1}).

x	a [Å]	V [Å^3]	Discharge capacity on 3-rd cycle [mA·h g^{-1}]
0.0	2.973	26.28	0.7
0.25	2.991	26.76	55
0.5	3.001	27.03	125
0.75	3.010	27.27	155
1.0	3.018	27.49	67

The discharge capacities of the materials studied are listed in Table 9.5. The discharge capacity of electrodes prepared by the application of MA TiFe alloy powder is very low (Figure 9.9). It is important to note, that the mechanically alloyed TiFe alloys showed a higher discharge capacity (0.7 mA·h g^{-1}) than their arc-melted counterparts (0.0 mA·h g^{-1}). The reduction in powder size, and the creation of new surfaces, proved effective for the improvement of the hydrogen absorption rate.

Materials obtained when Ni was substituted for Fe in TiFe$_{1-x}$Ni$_x$ led to a great improvement in the activation behavior of the electrodes. Again, increasing the nickel content in TiFe$_{1-x}$Ni$_x$ alloys led initially to an increase in discharge capacity, with a maximum at $x = 0.75$ [33].

On the other hand, the discharge capacity of nanocrystalline TiNi$_{0.6}$Fe$_{0.1}$Mo$_{0.1}$Cr$_{0.1}$Co$_{0.1}$ powder was largely unchanged during cycling (Figure 9.9). The

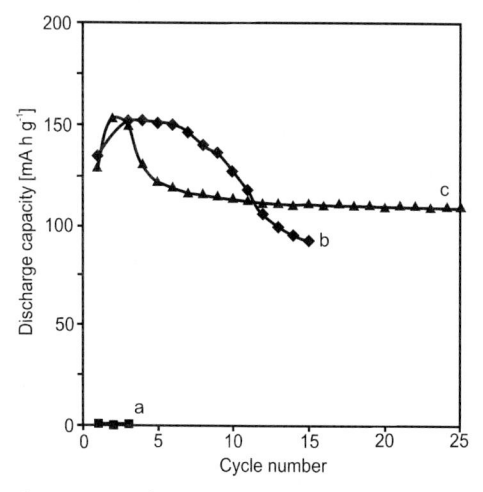

Figure 9.9 Discharge capacity as a function of cycle number of electrode prepared with nanocrystalline TiFe (a), TiFe$_{0.25}$Ni$_{0.75}$ (b), and TiNi$_{0.6}$Fe$_{0.1}$Mo$_{0.1}$Cr$_{0.1}$Co$_{0.1}$ (c) (solution, 6 M KOH; temperature, 20°C). The charge conditions were: 40 mA g^{-1}; the cut-off potential versus Hg/HgO/6 M KOH was −0.7 V.

alloying elements Mo, Cr and Co, when substituted simultaneously for iron atoms in the nanocrystalline Ti(Fe-Ni) master alloy, prevented oxidation of this electrode material.

9.3.2
The ZrV$_2$-Type System

The nanocrystalline ZrV$_2$ and Zr$_{0.35}$Ti$_{0.65}$V$_{0.85}$Cr$_{0.26}$Ni$_{1.30}$ alloys were synthesized by MA, followed by annealing [34,35]. The electrochemical properties of nanocrystalline powders were measured and compared with those of amorphous material. The behavior of the MA process has been studied using XRD, whereupon the powder mixture milled for more than 25 h was seen to transform absolutely to the amorphous phase. The formation of ordered alloys was achieved by annealing the amorphous materials in a high-purity argon atmosphere at 800 °C for 30 min (see Figure 9.4, spectrum b).

Figure 9.10 shows the discharge capacities of the amorphous and nanocrystalline electrodes as a function of charge/discharge cycling number. The electrode prepared with nanocrystalline Zr$_{0.35}$Ti$_{0.65}$V$_{0.85}$Cr$_{0.26}$Ni$_{1.30}$ material showed better activation and higher discharge capacities. This improvement was due to a well-established diffusion path for hydrogen atoms along the numerous grain boundaries. The discharge capacities of the studied ZrV$_2$-type materials are listed in Table 9.6. The electrochemical results showed very little difference between the nanocrystalline and microcrystalline powders, as compared to the substantial differences between these and the

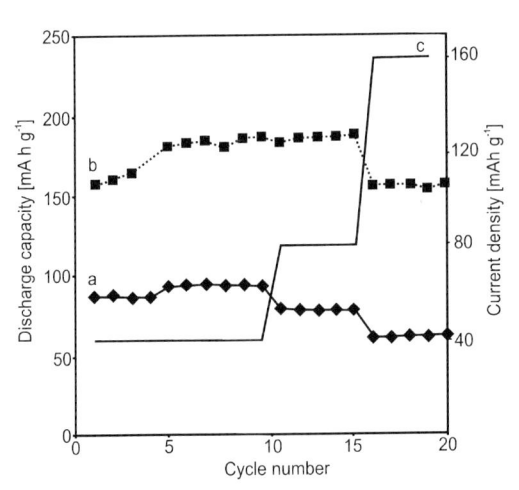

Figure 9.10 Discharge capacity as a function of cycle numbers of electrode prepared with (a) amorphous and (b) nanocrystalline Zr$_{0.35}$Ti$_{0.65}$V$_{0.85}$Cr$_{0.26}$Ni$_{1.30}$ (solution, 6 M KOH; temperature, 20 °C). The charge conditions were: 40 mAg^{-1}; the discharge conditions were plotted on; (c) cut-off potential versus Hg/HgO/ 6 M KOH was −0.7 V.

Table 9.6 Discharge capacities of microcrystalline, amorphous and nanocrystalline $Zr_{0.35}Ti_{0.65}V_{0.85}Cr_{0.26}Ni_{1.30}$ materials (current density of charging and discharging was 160 mA g^{-1}).

Preparation method	Structure type	Discharge capacity at 18th cycle [mA·h g^{-1}]
arc melting and annealing*	Microcrystalline ($MgZn_2$)	135
MA	Amorphous	65
MA and annealing	Nanocrystalline ($MgZn_2$)	150

*1000°C/7 days.

amorphous powder. In the annealed nanocrystalline $Zr_{0.35}Ti_{0.65}V_{0.85}Cr_{0.26}Ni_{1.30}$ powders prepared by MA and annealing, discharge capacities of up to 150 mA·h g^{-1} (at 160 mA g^{-1} discharge current) have been measured [35].

Independently, it has been shown, that the electrochemical properties of hydrogen storage alloys, which do not contain nickel, can be stimulated by HEBM of the precursor alloys with a small amount of nickel powders [34]. The ZrV_2 and $Zr_{0.5}Ti_{0.5}V_{0.8}Mn_{0.8}Cr_{0.4}$ alloy powders have been prepared in this way, after which it was confirmed that the discharge capacity of electrodes prepared by the application of ZrV_2 and $Zr_{0.5}Ti_{0.5}V_{0.8}Mn_{0.8}Cr_{0.4}$ alloy powders with 10 wt.% nickel powder addition was impossible to estimate because of the extremely high polarization. In electrodes prepared by application of high-energy ball-milled ZrV_2/Ni and $Zr_{0.5}Ti_{0.5}V_{0.8}Mn_{0.8}Cr_{0.4}$/Ni alloy powders with 10 wt.% nickel powder, the discharge capacities were considerably improved, increasing from 0 to 110 mA·h g^{-1} and 214 mA·h g^{-1}, respectively (Table 9.7).

The alloy elements such as Ti substituted for Zr and Mn, and Cr substituted for V in ZrV_2/Ni-based materials, greatly improved the activation behavior of the electrodes. It is worthy of note that annealed nanocrystalline ZrV_2/Ni-based powders have greater capacities (about 2.2-fold) than the amorphous parent alloy powders. In general, the electrochemical properties are closely linked to the size and crystallographic perfection of the constituent grains, which in turn are a function of the processing or grain refinement method used to prepare the hydrogen storage alloys.

Table 9.7 Discharge capacities of nanocrystalline ZrV_2-type materials without and with 10 wt.% of Ni powder (current density of charging and discharging was 4 mA g^{-1}).

Composition	Discharge capacity [mA·h g^{-1}]
ZrV_2	0
ZrV_2/Ni	110
$Zr_{0.5}Ti_{0.5}V_{0.8}Mn_{0.8}Cr_{0.4}$	0
$Zr_{0.5}Ti_{0.5}V_{0.8}Mn_{0.8}Cr_{0.4}$/Ni	214

9.3.3
The LaNi$_5$-Type System

The properties of hydrogen host LaNi$_5$ materials can be modified substantially by alloying. It was found that the substitution of Ni in LaNi$_5$ by small amounts of Al, Mn, Si, Zn, Cr, Fe, Cu or Co altered the hydrogen storage capacity, the stability of the hydride phase, and also corrosion resistance [18,20,36–41]. In general, in the transition metal sublattice of LaNi$_5$-type compounds, substitution by Mn, Al and Co has been found to offer the best compromise between high hydrogen capacity and good resistance to corrosion [6,18]. During the MA process, the originally sharp diffraction lines of La and Ni gradually become broader, and their intensity decreases with milling time. The powder mixture milled for more than 30 h was transformed completely to the amorphous phase. Formation of the nanocrystalline alloy was achieved by annealing the amorphous material in a high-purity argon atmosphere at 700 °C for 30 min (see Figure 9.4, spectrum c). According to atomic force microscopy (AFM) studies, the average size of the amorphous La-Ni powders was of order of 25 nm (Figure 9.11).

The discharge capacity of an electrode prepared by the application of nanocrystalline LaNi$_5$ alloy powder is low (Figure 9.12) [37,38]. It was found that the substitution of Ni by Al or Mn in La(Ni,M)$_5$ alloy leads to an increase in discharge capacity. The LaNi$_4$Mn electrode, when mechanically alloyed and annealed, displayed the maximum capacity at the first cycle, but the discharge capacity degraded strongly with cycling. On the other hand, alloying elements such as Al, Mn and Co substituting nickel greatly improved the cycle life of LaNi$_5$-type material (Figure 9.12). With the increase of cobalt content in LaNi$_{4-x}$Mn$_{0.75}$Al$_{0.25}$Co$_x$, the material showed an increase in discharge capacity which passed through a wide maximum for

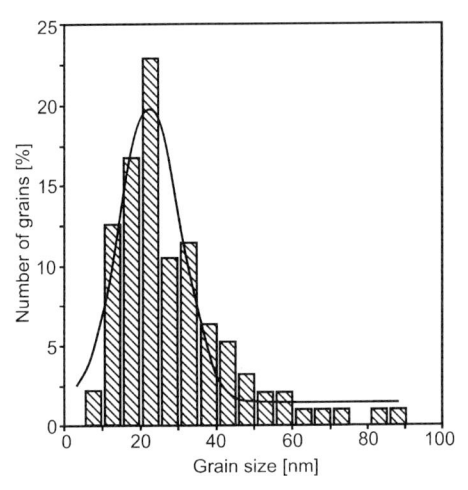

Figure 9.11 Histogram of a mixture of La and Ni powders mechanically alloyed for 30 h under an argon atmosphere.

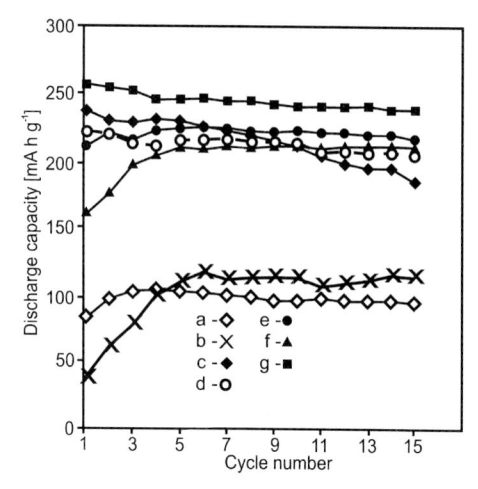

Figure 9.12 Discharge capacities as a function of cycle number of LaNi$_5$-type negative electrodes made from nanocrystalline powders prepared by MA followed by annealing: (a) LaNi$_5$; (b) LaNi$_4$Co; (c) LaNi$_4$Mn; (d) LaNi$_4$Al; (e) LaNi$_{3.75}$CoMn$_{0.25}$; (f) LaNi$_{3.75}$CoAl$_{0.25}$; (g) LaNi$_{3.75}$Mn$_{0.75}$Al$_{0.25}$Co$_{0.25}$ (solution, 6 M KOH; temperature, 20 °C). The charge conditions were 40 mA g^{-1}; the cut-off potential versus Hg/HgO/6 M KOH was −0.7 V.

$x = 0.25$ [40]. In nanocrystalline LaNi$_{3.75}$Mn$_{0.75}$Al$_{0.25}$Co$_{0.25}$, discharge capacities of up to 258 mA·h g^{-1} (at 40 mA g^{-1} discharge current) were measured, which compared well with results reported by Iwakura *et al.* for microcrystalline MmNi$_{4-x}$Mn$_{0.75}$Al$_{0.25}$Co$_x$ alloys (mishmetal) [41].

The cleanliness of the surface of microcrystalline and nanocrystalline LaNi$_5$-type alloys was studied using X-ray photoelectron spectroscopy (XPS) and Auger electron spectroscopy (AES) [37,42]. The element-specific Auger intensities of the microcrystalline LaNi$_{4.2}$Al$_{0.8}$ sample, as a function of the sputtering time and converted to depth, are shown graphically in Figure 9.13a.

As can be seen, there is relatively high concentration of carbon and oxygen immediately on the surface, which may be due to carbonates or to adsorbed atmospheric CO_2. The carbon concentration decreases strongly towards the interior of the sample. At the metal interface itself only oxygen is present, which makes it very likely that only an oxide layer is formed, and no other compounds, the latter growing apparently with a smaller probability. It was also found that, at the oxide–metal interface, mainly lanthanum and nickel atoms were present. Taking into account that the escape-depth of the Auger electron from nickel and aluminum atoms is about 2 nm, the concentration of these elements on the metallic surface is significantly lower compared to the average bulk composition. Therefore, the lanthanum atoms which segregate to the surface form a La-based oxide layer under atmospheric conditions. The oxidation process is depth-limited such that an oxide-covering layer with a well-defined thickness is formed, and by which the lower-lying metal is prevented from further oxidation. In this way it is possible to obtain a self-stabilized oxide–metal structure. Very similar behavior was observed for the microcrystalline

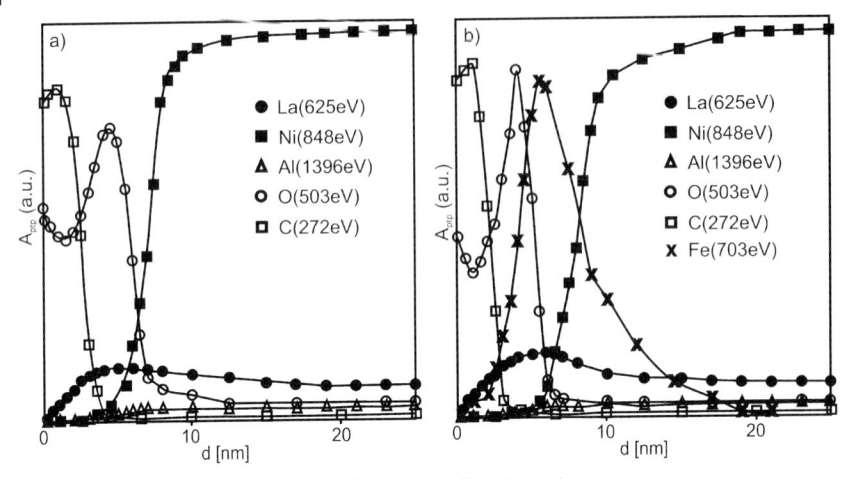

Figure 9.13 Auger electron spectrum of microcrystalline (a) and nanocrystalline (b) LaNi$_{4.2}$Al$_{0.8}$ alloy versus sputtering time, as converted to depth. The sample surface is located on the left-hand side.

Co thin films oxidized under atmospheric conditions. From the peak-to-peak amplitude it is possible to calculate the maximum atomic concentration of oxygen inside the sample as 2 atom%. The concentration of carbon impurities inside the sample was below 0.5 atom%. Figure 9.13b shows the element-specific Auger intensities of the nanocrystalline LaNi$_{4.2}$Al$_{0.8}$ sample as a function of the sputtering time, converted to depth. Similar to results obtained for the microcrystalline sample, there is a relatively high concentration of carbon and oxygen immediately on the surface, but the carbon concentration is greatly decreased towards the interior of the sample. At the oxide–metal interface, only iron impurities and lanthanum atoms were found to be present. As the escape-depth of the Auger electrons from nickel and aluminum atoms is about 2 nm, it is concluded that these elements are virtually absent from the metallic surface. In other words, lanthanum atoms and iron impurities strongly segregate to the surface, where they form an oxide layer under atmospheric conditions. The lower-lying Ni atoms form a metallic subsurface layer, and are responsible for the observed high hydrogenation rate in accordance with earlier findings. The above segregation process is stronger compared to that observed for the microcrystalline sample. The presence of a significant amount of iron atoms in the surface layer of the nanocrystalline LaNi$_{4.2}$Al$_{0.8}$ alloy may be explained by Fe impurities becoming trapped in the MA powders as a result of erosion of the milling media. The amount of Fe impurities was considerably decreased in the subsurface layer of the sample. From the peak-to-peak amplitude, the maximum atomic concentration of oxygen was estimated as ∼2 atom% in the interior of the sample. Similar to results reported for the polycrystalline LaNi$_{4.2}$Al$_{0.8}$ alloy, the concentration of carbon impurities inside the sample was below 0.5 atom%.

9.3.4
The Mg$_2$Ni-Type System

The magnesium–nickel phase diagram shows two compounds, Mg$_2$Ni and MgNi$_2$. The first of these reacts with hydrogen slowly at room temperature to form the ternary hydride Mg$_2$NiH$_4$. At higher temperatures and pressure (e.g., 200°C, 1.4 MPa [25]), the reaction is sufficiently rapid for useful absorption–desorption reactions to occur.

Mechanical alloying is one of main methods used to produce the Mg–Ni alloys which show great expectation as hydrogen storage materials [25,43–46]. Subsequently, Ling et al. [44] noted that HEBM, which gives rise to the creation of fresh surfaces and cracks, is highly effective for the kinetic improvements in the initial hydriding properties.

In a series of earlier studies conducted by the present authors, the nanocrystalline Mg$_2$Ni-type alloys were prepared by MA followed by annealing, whereupon the powder mixture milled for more than 90 h was transformed completely to the amorphous phase, without the formation of any other phase. Formation of the nanocrystalline alloy was achieved by annealing the amorphous material in a high-purity argon atmosphere at 450°C for 30 min. All diffraction peaks were assigned to those of the hexagonal crystal structure, with cell parameters $a = 5.216$ Å, $c = 13.246$ Å (see Figure 9.4, spectrum d) [25]. According to AFM studies, the average size of amorphous Mg–Ni powders was of the order of 30 nm.

Although, at room temperature, the nanocrystalline Mg$_2$Ni alloy absorbs hydrogen, the desorbtion of hydrogen is minimal. At temperatures above 250°C the kinetics of the absorption–desorption process improves considerably, and for nanocrystalline Mg$_2$Ni alloy the reaction with hydrogen is indeed reversible. The hydrogen content in this material at 300°C is 3.25 wt.%. However, upon hydrogenation, Mg$_2$Ni transforms into the hydride Mg$_2$NiH$_x$ phase. It is important to note, that between 210 and 245°C the hydride Mg$_2$NiH$_x$ phase transforms from a high-temperature cubic structure to a low-temperature monoclinic phase. When hydrogen is absorbed by Mg$_2$Ni beyond 0.3 H per f.u., the system undergoes a structural rearrangement to the stoichiometric complex Mg$_2$NiH$_x$ hydride, with an accompanying 32% increase in volume. The electrochemical properties of the alloy are improved after the substitution of some amounts of magnesium by manganese. The results show that the maximum absorption capacity reaches 3.25 wt.% for pure nanocrystalline Mg$_2$Ni alloy, this being lower than the microcrystalline Mg$_2$Ni alloy (3.6 wt.%) due to a significant amount of strain, chemical disorder, and defects introduced into the material during the mechanical alloying process. At the same time, increasing the manganese substitution causes the unit cell to decrease. The concentration of hydrogen in the produced nanocrystalline Mg$_2$Ni alloys decreases greatly with increasing Mn content. The hydrogen content at 300°C in nanocrystalline Mg$_{1.5}$Mn$_{0.5}$NiH was only 0.65 wt.% (Table 9.8; Figure 9.14).

The Mg$_2$Ni electrode, when mechanically alloyed and annealed, displayed the maximum discharge capacity (100 mA · h g^{-1}) at the first cycle but degraded strongly with cycling. The poor cyclic behavior of Mg$_2$Ni electrodes is attributed to the

Table 9.8 Structure, lattice parameters, discharge capacities and hydrogen contents for nanocrystalline Mg$_2$Ni-type materials; data for parent microcrystalline Mg$_2$Ni alloy were also included for comparison [15].

Alloy	Structure and lattice constants	Discharge capacity [mA·hg^{-1}]	Hydrogen content at 300°C [wt.%]
Mg$_2$Ni nanocrystalline	hexagonal $a = 5.216$ Å, $c = 13.246$ Å	100	3.25
Mg$_{1.75}$Mn$_{0.25}$Ni nanocrystalline	hexagonal $a = 5.185$ Å, $c = 13.097$ Å	148	2.50
Mg$_{1.5}$Mn$_{0.5}$Ni nanocrystalline	cubic $a = 3.137$ Å	241	0.65
Mg$_{1.75}$Al$_{0.25}$Ni nanocrystalline	hexagonal $a = 5.193$ Å, $c = 13.173$ Å	105	1.75
Mg$_{1.5}$Al$_{0.5}$Ni nanocrystalline	cubic $a = 3.149$ Å	175	0.26
Mg$_2$Ni microcrystalline	hexagonal $a = 5.223$ Å, $c = 13.30$ Å	–	3.6

formation of Mg(OH)$_2$ on them, and which is considered to arise from the charge–discharge cycles. In trying to avoid the surface oxidation, the effect of magnesium substitution by Mn or Al in Mg$_2$Ni-type material has been examined, and found this alloying to greatly improve the discharge capacities. In nanocrystalline Mg$_{1.5}$Mn$_{0.5}$Ni and Mg$_{1.5}$Al$_{0.5}$Ni alloys, discharge capacities of up to 241 mA·hg^{-1} and 175 mA·hg^{-1} were measured, respectively [25].

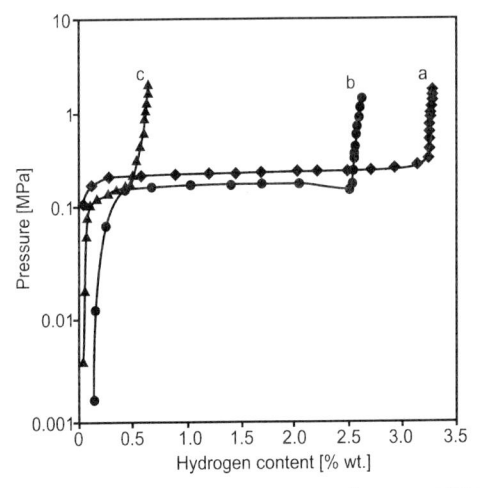

Figure 9.14 Pressure–composition isotherms at 300°C of hydrogen desorption from nanocrystalline Mg$_{2-x}$Mn$_x$Ni-H alloys. (a) $x = 0$; (b) $x = 0.25$; (c) $x = 0.5$.

The surface chemical composition of nanocrystalline Mg_2Ni-type alloy studied with XPS showed a strong surface segregation under ultra-high vacuum (UHV) conditions of Mg atoms in the MA nanocrystalline Mg_2Ni alloy. This phenomenon might also considerably influence the hydrogenation process in such materials.

9.3.5
Nanocomposites

A new class of electrode materials – nanocomposite hydride materials – is proposed for anodes in hydride-based rechargeable batteries [47–50]. These materials are synthesized by the mechanical mixing of two components: a major component having good hydrogen storage properties; and a minor component used as the surface activator. The major component was selected among conventional hydride electrode materials, as alloys of the TiFe-, ZrV_2-, $LaNi_5$- and Mg_2Ni-type type. The minor component was usually nickel, copper, palladium or graphite. Until now, nanocomposite hydride electrodes have shown the following advantages:

- almost complete elimination of the need for initial activation
- an enhancement of the discharge capacity
- a considerable improvement in the stability to charge–discharge at high rates
- an increase in charging efficiency
- a higher resistance to surface degradation during repeated charge/discharge.

In order to improve the electrochemical properties of these as-yet studied nanocrystalline electrode materials, the ball-milling technique was applied to the TiFe-type alloys using the nickel and graphite elements as a surface modifiers [49,50]. The $TiFe_{0.25}Ni_{0.75}$/M-type composite materials, where $M = 10$ wt.% Ni or C, were produced by ball-milling for 1 h. Ball-milling with nickel or graphite of $TiFe_{0.25}Ni_{0.75}$-type materials is sufficient to considerably broaden the diffraction peaks of $TiFe_{0.25}Ni_{0.75}$ (not shown). Additionally, milling with graphite is responsible for a sizeable reduction of the crystallite sizes of $TiFe_{0.25}Ni_{0.75}$/C, from 30 nm to 20 nm.

Figure 9.15 shows, graphically, the discharge capacities as a function of the cycle number for studied nanocomposite materials. When coated with nickel, the discharge capacities of nanocrystalline $TiFe_{0.25}Ni_{0.75}$ powders were increased. The elemental nickel was distributed on the surface of the ball-milled alloy particles homogeneously, the role of these particles being to catalyze the dissociation of molecular hydrogen on the surface of the studied alloy. Mechanical coating with nickel or graphite effectively reduced the degradation rate of the electrode materials. Compared to the uncoated powders, degradation of the coated powders was suppressed. Recently, Raman spectroscopy and XPS investigations indicated that the interaction of graphite with MgNi alloy occurred at the Mg part in the alloy [36]. Graphite inhibits the formation of new oxide layer on the surface of materials once the native oxide layer is broken during the ball-milling process.

A similar behavior was observed recently in the case of Mg_2Cu-based electrode nanomaterials [26]. Mechanically alloyed and annealed nanocrystalline Mg_2Cu alloy, displayed the maximum discharge capacity (26.5 mA $\cdot$ h g^{-1}) at the first cycle, but

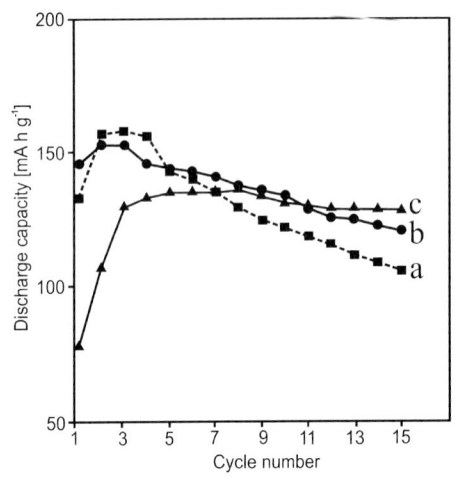

Figure 9.15 Discharge capacity as a function of cycle number for MA and annealed TiFe$_{0.25}$Ni$_{0.75}$ (a), in addition to TiFe$_{0.25}$Ni$_{0.75}$/Ni (b) and TiFe$_{0.25}$Ni$_{0.75}$/C (c) composite electrodes, Solution, 6 M KOH; temperature, 20°C.

degraded strongly with cycling (Table 9.9). The poor cyclic behavior of Mg$_2$Cu electrodes is attributed to the formation of Mg(OH)$_2$ on them, and this is thought to arise from the charge–discharge cycles. In order to avoid the surface oxidation, the effect of palladium coating of the Mg$_2$Cu-type material has been examined. The discharge capacity of coated nanocrystalline Mg$_2$Cu powders with palladium was improved. The elemental Pd was distributed on the surface of ball-milled alloy particles homogeneously, the role of these particles being to catalyze the dissociation

Table 9.9 Structure, lattice parameters, discharge capacities and hydrogen contents for nanocrystalline and nanocomposite Mg$_2$Cu-type materials.

Material	Structure and lattice constants [Å]	Discharge capacity [mA·hg^{-1}] 1st cycle	3rd cycle	Hydrogen content at 300°C [wt.%]
Nanocrystalline Mg$_2$Cu	Orthorhombic $a = 9.119(4)$ $b = 18.343(4)$ $c = 5.271(1)$	26.5	4.7	2.25
Nanocomposite Mg$_2$Cu/Pd	Orthorhombic/f.c.c. $a = 9.046(6)$ $b = 18.463(3)$ $c = 5.274(1)/a = 3.890(7)$	26.3	19.3	1.75

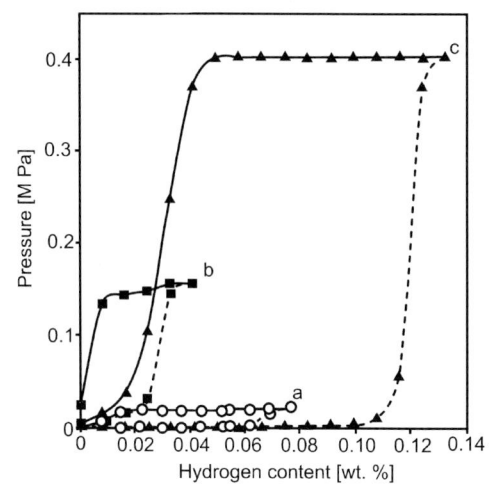

Figure 9.16 Electrochemical pressure–composition isotherms for absorption (solid line) and desorption (dashed line) of hydrogen on microcrystalline Mg_2Cu (a), nanocrystalline Mg_2Cu (b) and nanocomposite Mg_2Cu/Pd (c). 6 M KOH solution; the charge–discharge conditions were 4 mA g^{-1}, and the cut-off potential −0.700 V.

of molecular hydrogen on the surface of the alloy. Mechanical coating with palladium effectively reduced the degradation rate of the electrode material under investigation. Compared to uncoated powders, degradation of the coated powders was suppressed (Table 9.9).

In the case of Mg_2Cu-type alloys, the capacity of the alloy electrode in relation to the amount of absorbed hydrogen (wt.%) was calculated based on the input/output charge. In accordance with the equation: $E_s = -0.9325 - 0.0291 \times \log p(H_2)/p_o$ [18], the charge of one order of magnitude in hydrogen pressure in the alloy results in a change in the electrode potential of 29 mV. The e.p.c. isotherms determined on the studied Mg_2Cu-type materials are illustrated in Figure 9.16. The isotherms show an increase in the equilibrium hydrogen pressure, and an increase in the amount of hydrogen observed for the nanocomposite Mg_2Cu/Pd material (curve c) in comparison with the microcrystalline (curve a) and nanocrystalline (curve b) alloys.

Independently, nanocomposite Mg_2Ni/Pd and $Mg_{2-x}Al_xNi/Pd$-type hydrogen storage materials ($x = 0$, 0.5) have been prepared using MA. The effect of MA processing on Mg-based alloys was studied in detail (Figures 9.17 and 9.18). The cell parameters of all studied materials are listed in Table 9.10. For $Mg_{1.5}Al_{0.5}Ni$, the crystalline phase of a Ti_2Ni-type cubic structure is formed ($a = 3.149$ Å). The effect of palladium coating was examined on Mg_2Ni- and $Mg_{1.5}Al_{0.5}Ni$-type materials, and the discharge capacity of coated nanocrystalline Mg_2Ni- and $Mg_{1.5}Al_{0.5}Ni$-powders with palladium was improved. The elemental Pd was distributed on the surface of ball-milled alloy particles homogeneously, the role of these particles being to catalyze the dissociation of molecular hydrogen on the surface of studied alloy [13].

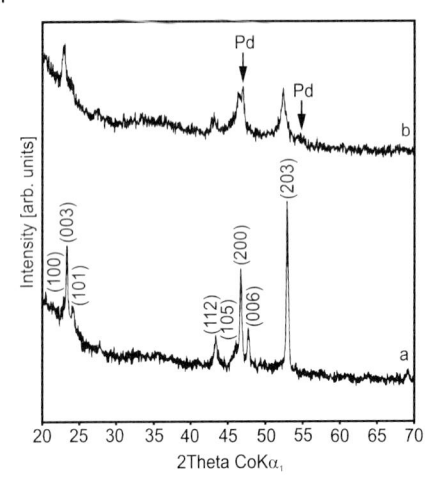

Figure 9.17 X-ray diffraction spectra of nanocrystalline Mg_2Ni- and nanocomposite Mg_2Ni/Pd-type hydrogen storage materials produced by mechanical alloying followed by annealing (see text for details).

Mechanical coating with palladium effectively reduced the degradation rate of the studied electrode material. Compared to the uncoated powders, degradation of the coated powders was suppressed (Figure 9.19).

The results of X-ray fluorescence (XRF) measurements revealed the assumed bulk chemical composition of the polycrystalline and nanocrystalline Mg_2Ni-type alloys.

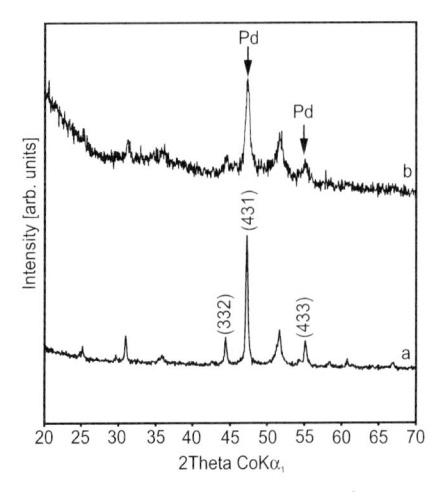

Figure 9.18 X-ray diffraction spectra of nanocrystalline $Mg_{1.5}Al_{0.5}Ni$- and nanocomposite $Mg_{1.5}Al_{0.5}Ni/Pd$-type hydrogen storage materials produced by mechanical alloying followed by annealing (see text for details).

Table 9.10 Structure, lattice parameters and discharge capacities for nanocomposite $Mg_{2-x}Al_xNi/Pd$-type hydrogen storage materials (x = 0, 0.5).

Material	Structure and lattice constants [Å]	Discharge capacity at 1st cycle [mA·h g^{-1}]
Nanocomposite Mg_2Ni/Pd	hexagonal/f.c.c. $a = 5.254$, $c = 13.435/a = 3.8907$	305
Nanocomposite $Mg_{1.5}Al_{0.5}Ni/Pd$	cubic/f.c.c. $a = 3.171/a = 3.8907$	240

On the other hand, core-level XPS measurements showed that the surface segregation of Mg atoms in the MA nanocrystalline samples is stronger compared to that of microcrystalline thin films. In particular, a strong surface segregation of Mg atoms was observed for the Mg_2Ni/Pd composites. Figure 9.20 shows normalized integral intensities of Mg, O, Ni, and Pd XPS peaks versus sputtering time as converted to depth for Mg_2Ni/Pd composites. The XPS Mg-1 s, Ni-$2p_{3/2}$, and Pd-$3d_{5/2}$ peaks were normalized to the intensities of in-situ-prepared pure Mg, Ni, and Pd thin films, respectively. The oxygen 1 s peak was normalized to the O-1 s intensity in the MgO single crystal. Results presented in Figure 9.20 show that Ni and Pd atoms are practically absent on the composite surface. On the other hand, Mg atoms strongly segregate to the surface and form a Mg-based oxide layer under atmospheric conditions. The oxidation process is depth-limited such that an oxide-covering layer with a well-defined thickness is formed by which the lower-lying metal is prevented from

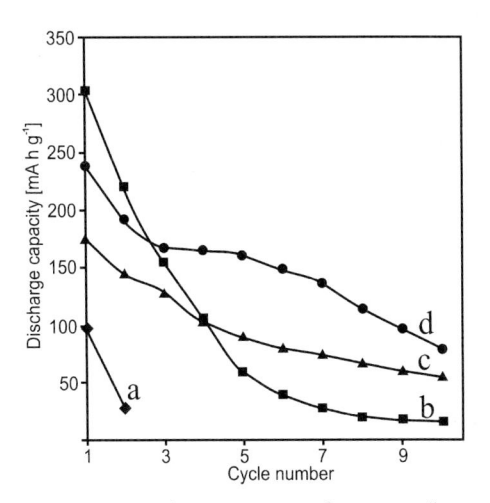

Figure 9.19 Discharge capacities of nanocrystalline Mg_2Ni (a), $Mg_{1.5}Al_{0.5}Ni$ (b) and nanocomposite Mg_2Ni/Pd (c), $Mg_{1.5}Al_{0.5}Ni/Pd$ (d) hydrogen storage materials. The current density of charge–discharge was 4 mA g^{-1}.

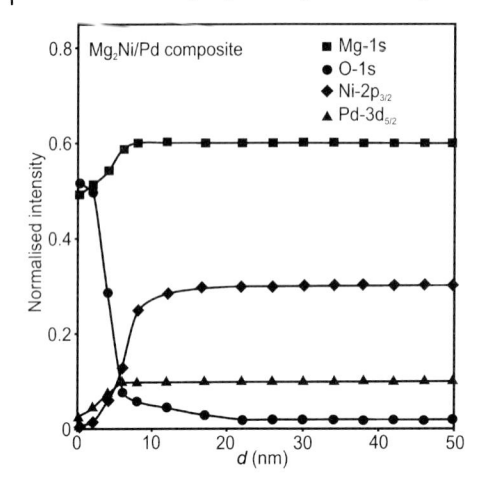

Figure 9.20 Normalized integral intensities of Mg, O, Ni, and Pd XPS peak versus sputtering time as converted to depth for Mg$_2$Ni/Pd composite. The X-ray photoelectron spectroscopy (XPS) Mg-1s, Ni-2p$_{3/2}$, and Pd-3d$_{5/2}$ peaks were normalized to the intensities of *in-situ*-prepared pure Mg, Ni, and Pd thin films, respectively. The oxygen 1s peak was normalized to the O-1s intensity in the MgO single crystal. The XPS measurements were performed immediately after heating under ultra-high vacuum conditions (see text), which allow the removal of adsorbed impurities (mainly carbonates), excluding a stable oxide top layer.

further oxidation. In this way, it is possible to obtain a self-stabilized oxide–metal structure. The lower-lying Ni and Pd atoms form a metallic subsurface layer, and are responsible for the observed relatively high hydrogenation rate. The surface segregation process of Mg atoms in Mg$_2$Ni/Pd composite is stronger compared to that observed for the Mg$_2$Ni nanocrystalline alloy. Furthermore, no segregation effect has been observed for the *in-situ*-prepared microcrystalline Mg$_2$Ni thin films. On the other hand, the Mg$_2$Ni thin films which are naturally oxidized in air for 24 h show a small segregation effect of the Mg atoms to the surface.

9.4
Electronic Properties

The application of hydrogen storage alloys as anode materials has focused attention also on the electronic structure of TiFe, ZrV$_2$, LaNi$_5$ or Mg$_2$Ni, and its modification mainly by Ni atoms, but also by Al, Co, Cr, and Mo impurities [51–60]. Until now, several semi-empirical models [61,62] have been proposed for the heat of formation and heat of solution of metal hydrides, and attempts have been made for justifying the maximum hydrogen absorption capacity of the metallic matrices.

Recently, the electronic structure of Ti-based systems was studied by the tight-binding version of the linear muffin-tin method in the atomic sphere approximation (TB-LMTO ASA) [53]. In the TiFe$_{1-x}$Ni$_x$ alloys, increasing the content of the Ni impurities extended the valence bands and increased the density of states at the Fermi level. Similar effects were observed for the TiNi$_{0.6}$Fe$_{0.1}$Mo$_{0.1}$Cr$_{0.1}$Co$_{0.1}$ system.

Independently, the electronic properties of microcrystalline and nanocrystalline TiFe$_{0.25}$Ni$_{0.75}$ alloys were studied using XPS [54]. In general, the lattice expansion associated with Fe substitution by Ni in TiFe$_{1-x}$Ni$_x$ could cause a narrowing of the Ni-d sub-band due to a decrease in the Ni–Ni interaction. The above effect is manifested as a relatively sharp maximum of the valence band [54,63]. Furthermore, the experimental valence band could also be broader due to the effect of disorder caused by substitution of Fe by Ni. The shape of the XPS valence band of the nanocrystalline TiFe$_{0.25}$Ni$_{0.75}$ alloy is broader compared to that measured for the polycrystalline TiFe$_{0.25}$Ni$_{0.75}$ sample. This is most likely due to a strong deformation of the nanocrystals. Normally, the interior of the nanocrystal is constrained and the distances between atoms located at the grain boundaries are expanded. Furthermore, in the case of MA nanocrystalline TiFe$_{0.25}$Ni$_{0.75}$ alloy the Ni atoms could also occupy metastable positions in the deformed grain. The above behavior could also modify the electronic structure of the valence band.

The electronic structure of Zr(V-Ni)$_2$-type compounds has also been studied [56]. The Ni impurities cause a charge transfer from the Zr and V atoms to the Ni atom, the valence band is wider, and the density of electronic states at the Fermi level decreases by about 30%.

Recently, the effect of substitution at the Ni site on the electronic structure of LaNi$_5$-type compounds was investigated [57,58,65]. Thus, a very good agreement was found between experimental results and *ab-initio* LMTO calculations of the total density of states (DOS) [57]. The occupied part of the conduction band is dominated by the Ni-3d states with a non-negligible bonding contribution of the La-5d states. The main part of the La-5d states is located above the Fermi energy. The XPS signal at E_F is high, and mostly composed of Ni-3d states since the La-5d contribution is practically negligible [57,64].

The XPS valence band spectrum of microcrystalline LaNi$_4$Al is significantly modified compared to that measured for the LaNi$_5$. In general, the lattice expansion associated with nickel substitution by aluminum could cause a narrowing of the Ni-3d sub-band due to a decrease in the Ni–Ni interaction. The above effect is manifested as a relatively sharp maximum of the valence band. On the other hand, the width of the valence band of the LaNi$_4$Al$_1$ alloy is greater in comparison with LaNi$_5$ system. This is due to the contribution of the Al s and p sub-bands, which are located near the bottom of the total valence band [57]. Furthermore, the experimental valence band could be also broader due to the effect of disorder caused by substitution of Ni by Al. Additionally, the XPS valence band of the nanocrystalline LaNi$_4$Al$_1$ alloy is broader compared to that measured for the microcrystalline LaNi$_4$Al$_1$ sample [59].

Binary LaNi$_5$ crystallizes with the CaCu$_5$ structure type in which La occupies site 1 (a) and Ni sites 2(c) and 3(g). The battery electrode material LaNi$_4$Al$_1$ is a substitutional derivative of LaNi$_5$ in which La occupies site 1(a) and Ni and Al sites 2(c) and 3 (g) of space group P6/mmm. Experimental results showed that the La sites do not accommodate Ni and Al atoms. Furthermore, TB LMTO calculation [57] showed that the impurity aluminum atoms prefer the 3 g positions in agreement with experimental data [64]. However, in the case of MA nanocrystalline LaNi$_4$Al$_1$ alloy the Al atoms could also occupy metastable (2c) positions in the deformed grain. The above behavior could also modify the electronic structure of the valence band.

A large number of experimental investigations on Mg–Ni compounds have been performed to date in relation to their electrochemical properties [25,26,66–68]. The band structure calculations were performed for ideal hexagonal Mg$_2$Ni-type structures with P6$_2$22 space group. In this structure, magnesium and nickel atoms each occupied two crystallographic positions: Mg(6i), Mg(6f), Ni(3d), and Ni(3b). The total energy calculations showed that, in both cases, Mg$_{11/6}$Al$_{1/6}$Ni and Mg$_{11/6}$Mn$_{1/6}$Ni, the impurity atoms, Al and Mn, preferred the (6i) position. Al atoms modify the bottom of the valence band, which is by about 0.5 eV wider than for the Mg$_2$Ni system. In the case of Mn atoms, 4d electrons modify the valence band in the range of 3 eV below the Fermi level (E_F), and the value of DOS for $E = E_F$ is higher [25].

The experimental XPS valence bands for nanocrystalline Mg$_2$Ni and Mg$_{1.5}$Mn$_{0.5}$Ni were studied [24,69]. Similar to the effect of the band broadening observed for the nanocrystalline TiFe- and LaNi$_5$-based alloys, such a modification has also been observed in the case of the Mg$_2$Ni system. The reasons responsible for band broadening of the nanocrystalline Mg$_2$Ni are as described above for the nanocrystalline FeTi- and LaNi$_5$-type alloys. It is believed that, also in the case of the nanocrystalline Mg$_{1.5}$Mn$_{0.5}$Ni system, its experimental valence band is broadened compared to that for a microcrystalline alloy.

The experimental XPS valence bands measured for nanocrystalline Mg$_2$Ni and Mg$_{1.5}$Mn$_{0.5}$Ni are shown in Figure 9.21. The experimental XPS valence bands

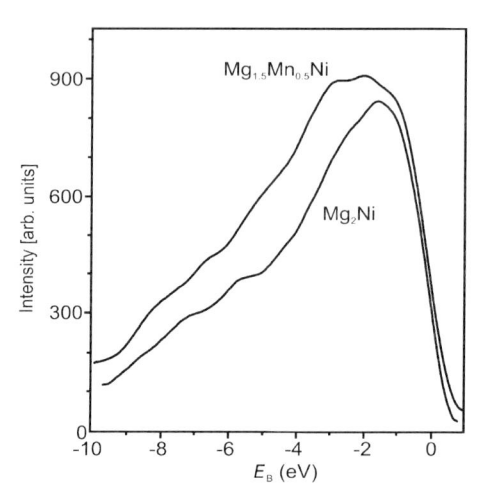

Figure 9.21 X-ray photoelectron spectroscopy (XPS) valence band (Al-K$_\alpha$) spectra for nanocrystalline Mg$_2$Ni and Mg$_{1.5}$Mn$_{0.5}$Ni alloys. The XPS measurements were performed immediately after heating under ultra-high vacuum conditions, followed by removal of a native oxide and possible impurities layer using an ion gun etching system (see text).

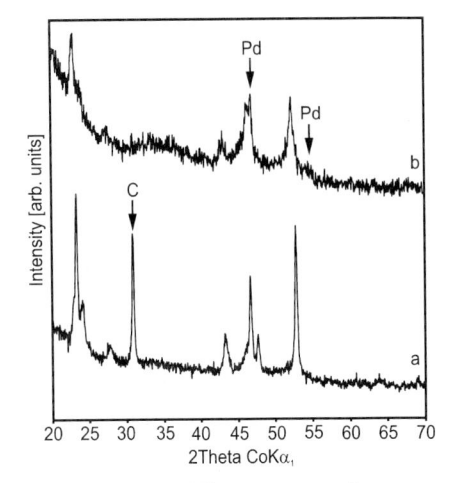

Figure 9.22 X-ray diffraction spectra of nanocomposite Mg_2Ni/C (a) and Mg_2Ni/Pd (b) materials (see text).

measured for MA nanocrystalline alloys showed a significant broadening compared to those obtained by theoretical band calculations. Especially, a clear broadening of the band is visible when comparing the experimental and theoretical (not shown here) XPS valence bands for the Mg_2Ni alloy. The reasons for this band broadening of the nanocrystalline Mg_2Ni-type alloys are probably associated with a strong deformation of the nanocrystals in the MA samples [69]. Normally, the interior of the nanocrystal is constrained and the distances between atoms located at the grain boundaries are expanded. Furthermore, the Al and Mn atoms may also occupy metastable positions in the nanocrystals.

In order to improve the electrochemical properties of the studied nanocrystalline electrode materials, the ball-milling technique was applied to Mg-based alloys using graphite and palladium as surface modifiers (Figure 9.22). When coated with graphite or palladium, the discharge capacity of nanocrystalline $Mg_{1.5}Mn_{0.5}Ni$ and $Mg_{1.5}Al_{0.5}Ni$ powders was increased [66,68]. The elemental graphite was distributed on the surface of ball-milled alloy particles homogeneously (again, the role of these particles was to catalyze the dissociation of molecular hydrogen on the surface of the studied alloy). Mechanical coating with graphite and palladium effectively reduced the degradation rate of the electrode materials. Compared to that of the uncoated powders, degradation of the coated powders was suppressed. Recently, Iwakura et al. [70] have shown that the modification of graphite on the MgNi alloy in the MgNi-graphite composite is mainly surficial in nature. Both, Raman spectroscopy and XPS investigations have shown that the interaction of graphite with MgNi alloy occurred at the Mg region of the alloy. Graphite inhibits the formation of a new oxide layer on the surface of materials when the native oxide layer is broken during the ball-milling process.

The experimental XPS valence bands measured for the Mg_2Ni/Pd and Mg_2Cu/Pd composites are shown in Figure 9.23. Those measured for nanocomposite alloys

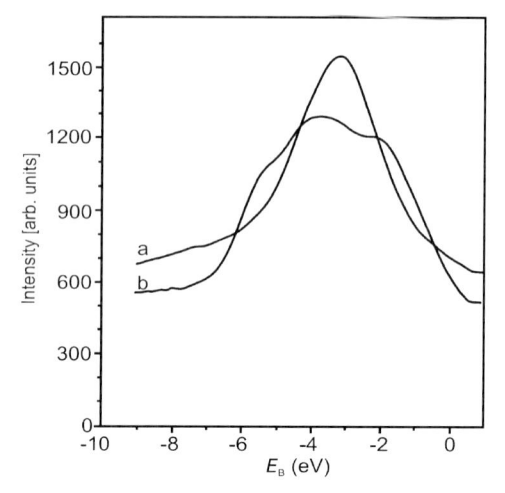

Figure 9.23 X-ray photoelectron spectroscopy (XPS) valence band (Al-Kα) spectra for Mg_2Cu/Pd (a) and Mg_2Ni/Pd (b) nanocomposites. The XPS measurements were performed immediately after heating under ultra-high vacuum conditions, followed by removal of a native oxide and possible impurities layer using an ion gun etching system (see text).

showed a significant broadening compared to those obtained for microcrystalline Mg_2Cu or Mg_2Ni alloys [68]. The maximum of the nanocrystalline Mg_2Ni valence band spectrum was located about 1.78 eV closer to the Fermi level than that measured for nanocomposite Mg_2Ni/Pd. The results also showed a significant broadening of the valence bands of the studied nanocomposites compared to those obtained by theoretical band calculations. Notably, a clear broadening of the band was visible when compared to the experimental XPS valence band for the nanocrystalline Mg_2Cu alloy and nanocomposite Mg_2Cu/Pd material. The reasons for the band broadening of the nanocrystalline Mg_2Ni and Mg_2Cu alloys were probably associated with a strong deformation of the nanocrystals in the MA samples [42]. Normally, the interior of the nanocrystal is constrained and the distances between atoms located at the grain boundaries are expanded. The valence band spectra of the MA samples may also broadened due to an additional disorder introduced during formation of the nanocrystalline structure.

The strong modifications of the electronic structure of the nanocrystalline Mg_2Ni-type alloy may have significant influences on its hydrogenation properties [24], similar to the behavior observed earlier for the nanocrystalline FeTi- and $LaNi_5$-type alloys. It is also believed that, in the case of the nanocrystalline $Mg_{1.5}Mn_{0.5}Ni$ system, the experimental valence band was broadened compared to that measured or calculated for a polycrystalline alloy (not shown here).

The present theoretical studies will, hopefully, stimulate further experimental investigations that may lead to a full understanding of the electronic properties of these technologically important hydrogen storage materials.

9.5
Sealed Ni-MH Batteries

The cyclic behavior of the some nanostructured alloy anodes was examined in a sealed HB 116/054 cell (according to the International standard IEC no. 61808, related to the hydride button rechargeable single cell) [71]. The mass of the active material was 0.33 g. In order to prepare MH negative electrodes, alloy powders were mixed with 5 wt.% tetracarbonylnickel, and the mixture was pressed into tablets that were placed in a small basket made from nickel nets (as the current collector). The diameter of each tested button cell was 6.6 mm, and the thickness 2.25 mm. The sealed Ni-MH cell was constructed by pressing the negative and positive electrodes, polyamide separator and KOH ($\rho = 1.20 \times 10^{-3}$ kg m^{-3}) as the electrolyte solution. The battery with electrodes fabricated from nanocrystalline materials was charged at a current density of $i = 3$ mA g^{-1} for 15 h and, after a 1-h pause, discharged at current density of $i = 7$ mA g^{-1} down to 1.0 V. All electrochemical measurements were performed at $20 \pm 1°C$.

In order to study the quality of the active TiFe$_{0.25}$Ni$_{0.75}$ and TiNi as electrode materials in the Ni-MH battery, the overpotential dependence on the current density ($i = 10, 20, 40$ and 80 mA g^{-1}) was recorded at 15 s of anodic and cathodic galvanostatic pulses (Figure 9.24). It can be seen that the anodic and cathodic parts of the nanocrystalline electrodes are almost symmetrical with respect to the resting potential of the electrode. It may also be concluded that fast rates can be achieved for all of the studied electrodes.

Figure 9.25 shows the discharge capacities of sealed button cells with electrodes prepared from nanocrystalline Ti-based alloys as a function of discharge cycle number. Among TiNi-type materials, the highest discharge capacities were found for

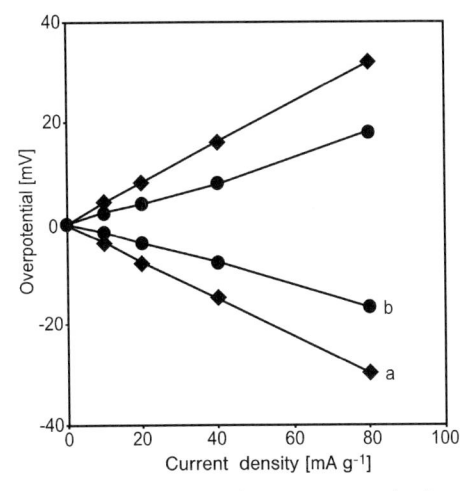

Figure 9.24 Overpotential against current density on activated nanocrystalline: (a) TiFe$_{0.25}$Ni$_{0.75}$; and (b) TiNi electrodes, at 15 s of anodic and cathodic galvanostatic pulses.

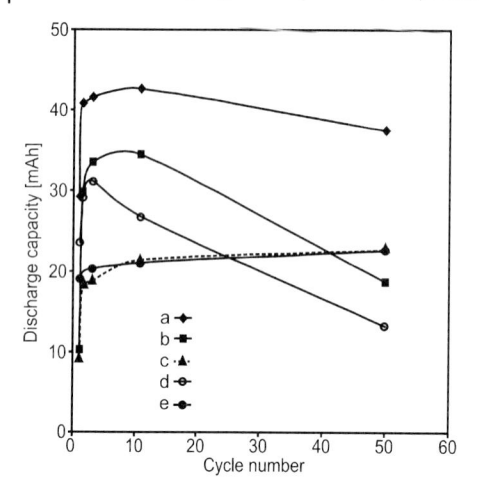

Figure 9.25 Durability of the sealed button cells with negative electrodes made from nanocrystalline Ti-based alloys: (a) $TiFe_{0.25}Ni_{0.75}$; (b) TiNi; (c) $TiNi_{0.875}Zr_{0.125}$; (d) $TiNi_{0.6}Fe_{0.1}Mo_{0.1}Cr_{0.1}Co_{0.1}$; (e) NiCd alloys. The mass of the active material was 0.33 g.

$TiFe_{0.25}Ni_{0.75}$ alloy. It is of interest to note that that the sealed battery using the nanocrystalline $TiFe_{0.25}Ni_{0.75}$ alloy had almost the capacity of the polycrystalline $(Zr_{0.35}Ti_{0.65})(V_{0.93}Cr_{0.28}Fe_{0.19}Ni_{1.0})$ counterpart [72].

Independently, it was found that the discharge capacities of sealed button batteries with electrodes prepared from the nanocrystalline $La(Ni,Mn,Al,Co)_5$ powders had slightly higher discharge capacities than the negative electrodes prepared from microcrystalline powders.

9.6
Conclusions

In conclusion, nanocrystalline TiFe-, ZrV_2-, $LaNi_5$- and Mg_2Ni-type alloys synthesized by MA can store hydrogen reversibly to form hydride, and then release hydrogen electrochemically. Mechanical alloying represents a suitable procedure for obtaining nanocrystalline hydrogen storage materials with high capacities and better hydrogen sorption properties. Currently, there are two main advantages of MA in the synthesis of hydrogen storage powders:

- It can be used to alloy elements with vastly different melting temperatures (e.g., Mg, Ni, or Cu), which is not easily achieved by conventional techniques such as arc or induction melting.
- It is a mature powder synthesis technique that can easily be scaled up from the laboratory to industry (e.g., from a few grams to several tons of powder).

The hydrogen storage properties of nanocrystalline ZrV_2- and $LaNi_5$-type powders prepared by MA and annealing show no major differences from those of melt casting (microcrystalline) alloys. On the other hand, the nanocrystalline TiFe- and Mg_2Ni-type hydrides show substantially enhanced absorption characteristics which are superior to those of conventionally prepared materials. The properties of nanocrystalline electrodes were attributed to the structural characteristics of the compound as a result of MA. At present, the Ni-MH battery represents a key component for advanced information and telecommunication systems. The input materials for these high-tech Ni-MH batteries would be nanostructured hydrogen storage alloys.

References

1 http://www.emrs.c-strasbourg.fr.
2 http://www.zywex.com/nanotech/feynman.html.
3 Gleiter, H. (1990) *Progr. Mater. Sci.*, **33**, 223.
4 Wang, Z.-L., Liu, Y. and Zhang, Z. (eds) (2003) *Handbook of nanophase and nanostructured materials*, Kluwer Academic/Plenum Publishers, New York.
5 Zaluski, L., Zaluska, A. and Ström-Olsen, J.O. (1997) *J. Alloys Comp.*, **253–254**, 70.
6 Jurczyk, M. (2004) *Bull. Pol. Ac.: Tech.*, **52** (1), 67.
7 Jurczyk, M. (1996) *J. Alloys Comp.*, **235**, 232.
8 Jurczyk, M., Smardz, L., Smardz, K., Nowak, M. and Jankowska, E. (2003) *J. Solid State Chem.*, **171**, 30.
9 Jurczyk, M. (1995) *J. Alloys Comp.*, **228**, 172.
10 Wojciechowski, S. (2000) *J. Mater. Proc. Techn.*, **106**, 230.
11 Richert, J. and Richert, M. (1986) *Aluminium*, **62**, 604.
12 Valiev, R.Z., Korrznikov, A.V. and Mulyukov, R.R. (1993) *Mater. Sci. Eng. A*, **168**, 141.
13 Sun, P.I., Kao, P.W. and Chang, C.P. (2000) *Mater. Sci. Eng. A*, **283**, 82.
14 Bradhurst, D.H. (1983) *Metals Forum*, **6**, 139.
15 Conte, M., Prosini, P.P. and Passerini, S. (2004) *Mater. Sci. Eng. B*, **108**, 2.
16 Sandrock, G. (1999) *J. Alloys Comp.*, **293–295**, 877.
17 Anani, A., Visintin, A., Petrov, K., Srinivasan, S., Reilly, J.J., Johnson, J.R., Schwarz, R.B. and Desch, P.B. (1994) *J. Power Sources*, **47**, 261.
18 Kleparis, J., Wojcik, G., Czerwinski, A., Skowronski, J., Kopczyk, M. and Beltowska-Brzezinska, M. (2001) *J. Solid State Electrochem.*, **5**, 229.
19 Sakai, T., Matsuoka, M. and Iwakura, C. (1995) In: (eds K.A. GschneiderJr. and L. Eyring) *Handbook of the Physics and Chemistry of Rare Earth*, Vol.21 Elsevier Science B.V p. 135. Chapter 142
20 Buschow, K.H.J., Bouten, P.C.P. and Miedema, A.R. (1982) *Rep. Prog. Phys.*, **45**, 937.
21 Hong, K. (2001) *J. Alloys Comp.*, **321**, 307.
22 Benjamin, J.S. (1976) *Sci. Am.*, **234**, 40.
23 Suryanarayna, C. (2001) *Progr. Mater. Sci.*, **46**, 1.
24 Jurczyk, M., Smardz, L. and Szajek, A. (2004) *Mater. Sci. Eng. B*, **108**, 67.
25 Gasiorowski, A., Iwasieczko, W., Skoryna, D., Drulis, H. and Jurczyk, M. (2004) *J. Alloys Comp.*, **364**, 283.
26 Jurczyk, M., Okonska, I., Iwasieczko, W., Jankowska, E. and Drulis, H. (2007) *J. Alloys Comp.*, **429**, 316.
27 Tarasov, B.P., Fokin, V.N., Moravsky, A.P., Shul'ga, Yu.M. and Yartys, V.A. (1997) *J. Alloys Comp.*, **253–254**, 25.
28 Jurczyk, M., Rajewski, W., Wojcik, G. and Majchrzycki, W. (1999) *J. Alloys Comp.*, **285**, 250.

29 Jankowska, E. and Jurczyk, M. (2002) *J. Alloys Comp.*, **346**, L1.
30 Kopczyk, M., Wojcik, G., Mlynarek, G., Sierczynska, A. and Beltowska-Brzezinska, M. (1996) *J. Appl. Electrochem.*, **26**, 639.
31 Jurczyk, M., Jankowska, E., Nowak, M. and Wieczorek, I. (2003) *J. Alloys Comp.*, **354**, L1.
32 Jurczyk, M., Jankowska, E., Nowak, M. and Jakubowicz, J. (2002) *J. Alloys Comp.*, **336**, 265.
33 Jurczyk, M., Smardz, L., Makowiecka, M., Jankowska, E. and Smardz, K. (2004) *J. Phys. Chem. Sol.*, **65**, 545.
34 Jurczyk, M., Rajewski, W., Majchrzycki, W. and Wojcik, G. (1998) *J. Alloys Comp.*, **274**, 299.
35 Majchrzycki, W. and Jurczyk, M. (2001) *J. Power Sources*, **93**, 77.
36 Jurczyk, M. (2003) *Curr. Top. Electrochem.*, **9**, 105.
37 Jurczyk, M., Smardz, K., Rajewski, W. and Smardz, L. (2001) *Mater. Sci. Eng. A*, **303**, 70.
38 Jurczyk, M., Nowak, M., Jankowska, E. and Jakubowicz, J. (2002) *J. Alloys Comp.*, **339**, 339.
39 Jurczyk, M., Smardz, L., Smardz, K., Nowak, M. and Jankowska, E. (2003) *J. Solid State Chem.*, **171**, 30.
40 Jurczyk, M., Nowak, M. and Jankowska, E. (2002) *J. Alloys Comp.*, **340**, 281.
41 Iwakura, C., Fukuda, K., Senoh, H., Inoue, H., Matsuoka, M. and Yamamoto, Y. (1998) *Electrochim. Acta*, **43**, 2041.
42 Smardz, L., Smardz, K., Nowak, M. and Jurczyk, M. (2001) *Cryst. Res. Techn.*, **36**, 1385.
43 Aymard, L., Ichitsubo, M., Uchida, K., Sekreta, E. and Ikazaki, F. (1997) *J. Alloys Comp.*, **259**, L5.
44 Ling, G., Boily, S., Huot, J., Van Neste, A. and Schultz, R. (1998) *J. Alloys Comp.*, **268**, 302.
45 Orimo, S., Züttel, A., Ikeda, K., Saruki, S., Fukunaga, T., Fujii, H. and Schlapbach, L. (1999) *J. Alloys Comp.*, **293–295**, 437.
46 Mu, D., Hatano, Y., Abe, T. and Watanabe, K. (2002) *J. Alloys Comp.*, **334**, 232.
47 Bouaricha, S., Dodelet, J.P., Guay, D., Huot, J. and Schultz, R. (2001) *J. Alloys Comp.*, **325**, 245.
48 Chen, J., Bradhurst, D.H., Don, S.X. and Liu, H.K. (1998) *J. Alloys Comp.*, **280**, 290.
49 Jurczyk, M. (2004) *J. Mater. Sci.*, **39**, 5271.
50 Iwakura, C., Inoue, H., Zhang, S.G. and Nohara, S. (1999) *J. Alloys Comp.*, **293–295**, 653.
51 Gupta, M. (1982) *J Phys. F: Metal Phys.*, **12**, L57.
52 Garcia, G.N., Abriata, J.P. and Sofo, J.O. (1999) *Phys. Rev. B*, **59**, 11746.
53 Szajek, A., Jurczyk, M. and Jankowska, E. (2003) *J. Alloys Comp.*, **348**, 285.
54 Smardz, K., Smardz, L., Jurczyk, M. and Jankowska, E. (2003) *Phys. Stat. Sol. (a)*, **196**, 263.
55 Szajek, A., Jurczyk, M. and Jankowska, E. (2003) *Phys. Stat. Sol.*, **196**, 256.
56 Szajek, A., Jurczyk, M. and Rajewski, W. (2000) *J. Alloys Comp.*, **302**, 299.
57 Szajek, A., Jurczyk, M. and Rajewski, W. (2000) *J. Alloys Comp.*, **307**, 290.
58 Smardz, L., Smardz, K., Nowak, M. and Jurczyk, M. (2001) *Cryst. Res. Techn.*, **36**, 1385.
59 Szajek, A., Jurczyk, M., Nowak, M. and Makowiecka, M. (2003) *Phys. Stat. Sol. (a)*, **196**, 252.
60 Szajek, A., Makowiecka, M., Jankowska, E. and Jurczyk, M. (2005) *J. Alloys Comp.*, **403**, 323.
61 Griessen, R. (1988) *Phys. Rev.*, **B38**, 3690.
62 Bouten, P.C. and Miedema, A.R. (1980) *J. Less Common Metals*, **71**, 147.
63 Jankowska, E., Makowiecka, M. and Jurczyk, M. (2006) *Polish J. Chem. Techn.*, **8**, 59.
64 Joubert, J.M., Latroche, M., Percheron-Guégan, A. and Bourée-Vigueron, F.B. (1988) *J. Alloys Comp.*, **275–277**, 118.
65 Jurczyk, M. (2006) *J. Optoelectron. Adv. Mater.*, **8**, 418.

66 Jurczyk, M., Smardz, L., Okonska, I., Jankowska, E., Nowak, M. and Smardz, K. Int. J. Hydrogen Energy (2007), doi: 10.1016/j.ijhydene.2007.07.022.

67 Szajek, A., Jurczyk, M., Smardz, L., Okonska, I. and Jankowska, E. (2007) *J. Alloys Comp.*, **436**, 345.

68 Smardz, K., Smardz, L., Okonska, I., Nowak, M. and Jurczyk, M. Int. J. Hydrogen Energy (2007), doi: 10.1016/j.ijhydene.2007.07.032.

69 Smardz, K., Szajek, A., Smardz, L. and Jurczyk, M. (2004) *Mol. Phys. Rep.*, **40**, 131.

70 Iwakura, C., Inoue, H., Zhang, S.G. and Nohara, S. (1999) *J. Alloys Comp.*, **293–295**, 653.

71 Jankowska, E. and Jurczyk, M. (2004) *J. Alloys Comp.*, **372**, L9.

72 Skowronski, J.M., Sierczynska, A. and Kopczyk, M. (2002) *J. Solid State Electrochem.*, **7**, 11.

10
Nanosized Titanium Oxides for Energy Storage and Conversion
Aurelien Du Pasquier

10.1
Introduction

Titanium dioxide (TiO_2) is a mass-produced ceramic (4×10^6 tons per year [1]) used in a variety of applications in our daily lives, from toothpaste to paint pigments. It is also used, for example, as a catalyst, a photocatalyst, and in gas sensors and optical coatings. For a more complete review of its applications, the reader is referred to Diebold's review of TiO_2 surface science [2]. Here, we shall only focus on the electrochemical applications that involve nanosized TiO_2. The three polymorphs of TiO_2 are rutile, anatase, and brookite. Rutile and anatase (Figure 10.1) are the two phases of industrial interest. Phase transformation from anatase to rutile normally occurs above 450 °C [3]. Reaction of TiO_2 with lithium sources such as LiOH, Li_2CO_3, or $LiNO_3$ leads to the lithium titanate spinel $Li_4Ti_5O_{12}$.

A common starting material for all applications described in this chapter is the nanosized TiO_2 anatase. Hence, the nano-TiO_2 syntheses of industrial significance will first be described, followed by its conversion to $Li_4Ti_5O_{12}$, and finally its applications in batteries and solar cells.

10.2
Preparation of Nanosized Titanium Oxide Powders

10.2.1
Wet Chemistry Routes

The hydrolysis of $TiCl_4$ in water yields anatase nanoparticles with traces of rutile. The $TiCL_4/H_2O$ volume ratio controls the crystallinity and particle size. Powders with up to $250\,m^2\,g^{-1}$ BET surface area and 12 nm crystallite size can be obtained with 1:50 $TiCL_4/H_2O$ volume ratio [4]. Nanosized TiO_2 can also be prepared simply by the hydrolysis of sol–gel precursors such as titanium tetraisopropoxide (TTIP) in

Nanostructured Materials in Electrochemistry. Edited by Ali Eftekhari
Copyright © 2008 WILEY-VCH Verlag GmbH & Co. KGaA, Weinheim
ISBN: 978-3-527-31876-6

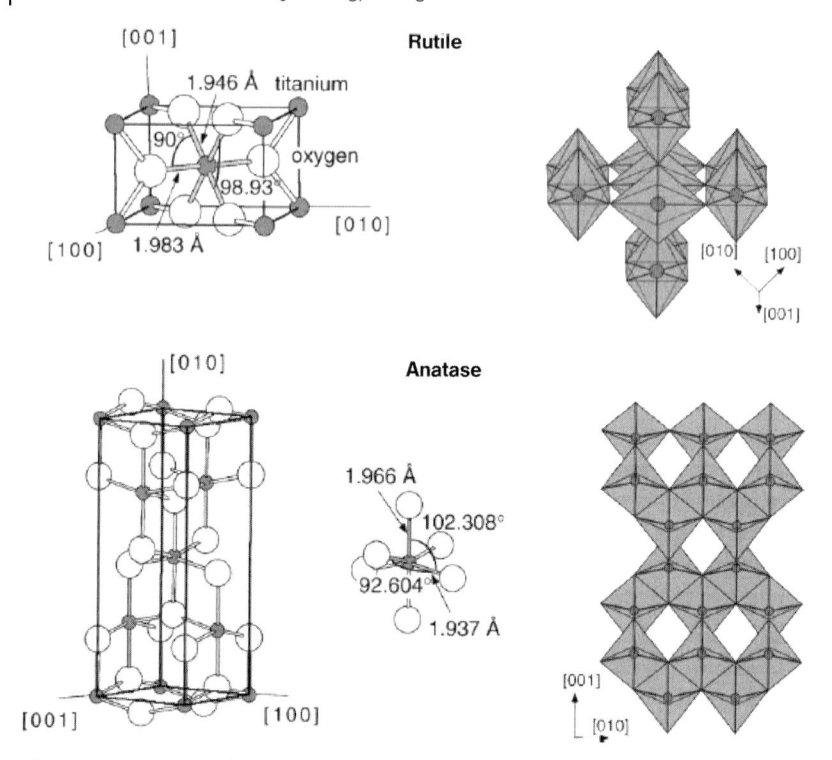

Figure 10.1 Anatase and Rutile TiO$_2$ structures.

pure water at 70 °C, followed by drying at 100 °C. Powders of very high BET surface area (750 m^2 g^{-1}) can be obtained, but they are a mixture of brookite and anatase phases [5]. The phase mixture is not an issue for application such as photocatalytic decomposition of organic pollutants in water, but does limit the efficiency of these powders in dye-sensitized solar cells, where pure anatase phase is preferred.

Altair Nanotechnologies has developed a hydrometallurgical process for the production of ultrafine or nanosized titanium dioxide from titanium-containing solutions, particularly titanium chloride solutions [6]. The process involves total evaporation of the solution, above the boiling point of the solution and below the temperature where there is significant crystal growth. Chemical control additives may be added to control the particle size, with nanosized elemental particles being formed after calcination. The titanium dioxide can be either anatase or rutile. Following calcination, the titanium dioxide is milled to liberate the elemental particles and provide a high-quality nanosized TiO$_2$ with a narrow particle size distribution.

The aqueous titanium chloride solution is generally comprised of water, hydrochloric acid, titanium oxychlorides, and titanium chlorides. The solutions may vary widely in composition with respect to the hydrochloric acid and titanium contents. The solution is further converted to a titanium oxide solid in a process involving total, controlled evaporation of the solution and the formation of a thin film of titanium dioxide. This process is conducted above the boiling point of the solution and below the temperature where there is significant crystal growth. The water and hydrochloric acid are vaporized and the hydrochloric acid may be recovered.

The titanium oxide is next calcined at an elevated temperature to induce and control crystallization. The concentration and type of chemical control agent, as well as the calcination conditions, determine the desired crystalline form and crystal size of the ultrafine titanium dioxide.

Following calcination, the titanium dioxide is milled or dispersed (e.g., in a spray-dryer) to yield a final nanosized or ultrafine titanium dioxide having a narrow particle size distribution. The advantages of the process according to the invention include a superior high-quality ultrafine titanium dioxide due to the narrow particle size distribution, readily controlled physical and chemical characteristics, and low cost processing.

10.2.2
Chemical Vapor Deposition

The oxidation of $TiCl_4$ vapor – which is also known as the "chloride" process in the titania industry [7] – is a gas-phase preparative method or a chemical vapor deposition (CVD) method in which all reactants undergo gas- or vapor-phase chemical reactions to form powders. $TiCl_4$ is a low-cost inorganic precursor which can be oxidized or hydrolyzed to prepare TiO_2 powders, according to:

$$TiCl_4(g) + O_2(g) \rightarrow TiO_2(s) + 2Cl_2(g) \tag{1}$$

The production of nanoparticles is favored by the increase in the equilibrium constant that occurs at a higher reaction temperature. The reaction temperature is typically in the range of 1400 to 1500 °C, which requires a preparation temperature of 900 °C. Under these conditions, anatase TiO_2 is obtained.

10.2.3
Vapor-Phase Hydrolysis

One drawback of the CVD "chloride" route is the high preparation temperature, which may cause rapid reactor corrosion. Vapor-phase hydrolysis occurs naturally in the chloride process, according to the reaction:

$$TiCl_4(g) + 2H_2O(g) \rightarrow TiO_2(s) + 4HCl(g) \tag{2}$$

Although the hydrolysis of $TiCl_4$ is sometimes attached to the oxidation route, it mainly acts as a nucleation agent [8]. However, it is known that $TiCl_4$ hydrolysis yields finer TiO_2 particles at lower temperature. Degussa has developed a fumed TiO_2 synthesis process which is a variant of vapor-phase hydrolysis using a hydrogen flame.

Degussa P25 is the reference TiO_2 photocatalyst against which all other TiO_2 nanomaterials are compared. It consists of mixtures of about 80% anatase/20% rutile TiO_2 nanoparticles of 20 nm average crystallite size and $50\,m^2\,g^{-1}$ BET surface area.

10.2.4
Physical Vapor Deposition

Nanophase Technologies has patented an arc plasma method known as physical vapor deposition (PVD) for the synthesis of nanocrystalline materials, including TiO_2 [9]. The system includes a chamber, a non-consumable cathode shielded against chemical reaction by a working gas (not including an oxidizing gas, but including an inert gas), a consumable anode vaporizable by an arc formed between the cathode and the anode, and a nozzle for injecting at least one of a quench and reaction gas in the boundaries of the arc. In this process, nanosized TiO_2 is prepared from a Ti anode and oxygen as the reaction gas.

10.3
Other TiO_2 Nanostructures

Besides TiO_2 nanospheres, other TiO_2 nanostructures of potential interest for energy storage and conversion have been prepared. Most of these structures can be converted to titanium spinel by lithium treatment. There are several routes for accessing specific TiO_2 morphologies: TiO_2 nanotube arrays can be obtained by Ti anodization in presence of a fluoride-based solution [10–13], in a method similar to that for aluminum anodization (see Chapter 1). Another common method of mesoporous nanostructures preparation is the sol–gel self-assembly in the presence of block copolymer templates [14–16]. When spheres are used as template, honeycomb-like structures can be obtained [17–21] (Table 10.1). Hydrothermal synthesis conditions can also be tuned towards the synthesis of nanowires [22,23], nanoflakes [24], or nanobelts [25] (Table 10.2).

10.4
Preparation of Nano-$Li_4Ti_5O_{12}$

Nanosized $Li_4Ti_5O_{12}$ (n-LTO) has been prepared, using solid-state chemistry, by Amatucci et al. [26] from flash annealing of a nanosized (32 nm) TiO_2 anatase precursor in the presence of $LiNO_3$. These authors obtained particle diameters of less than 100 nm for annealing times less than 1000 s (Figure 10.2). Guerfi et al. [27] also prepared n-LTO by solid-state reaction of nanosized TiO_2 anatase with lithium carbonate. Ball milling and jar milling in the presence of carbon were investigated for mixing the precursors. Zaghib et al. prepared nanocrystalline $Li_4Ti_5O_{12}$ by high-energy grinding of the microcrystalline spinel using a ball mill, obtaining particles of

Table 10.1 TiO$_2$ nanostructures obtained by Ti anodization, and templated sol–gel synthesis.

Description	SEM	Preparation	References
Nanotubes array		Ti Anodization in presence of HF	[10–13]
Mesoporous array		Self-assembly of sol-gel precursor and block copolymer	[14–16]
Honeycomb array		Templated sol–gel synthesis in presence of polystyrene spheres	[17–21]

600 nm in size. However, the electrochemical performance was not significantly different among their materials [28].

In a sol–gel route developed by Graetzel *et al.* [29], nanocrystalline Li$_4$Ti$_5$O$_{12}$ (spinel) was prepared from lithium ethoxide and Ti(IV) alkoxides as the starting reagents. The optimized materials contained less than 1% of anatase as the main impurity and, depending on the synthetic conditions, they exhibited Brunauer–Emmett–Teller (BET) surface areas of 53 to 183 m^2 g^{-1}. Another sol–gel route has

Table 10.2 TiO$_2$ nanostructures obtained by hydrothermal reactions.

Description	SEM	Preparation	Reference(s)
Nanowires		Hydrothermal reaction between NaOH and TiO$_2$, followed by acid washing and heating at 400 °C.	[22,23]
Nanoflakes array		Sol–gel synthesis from titanium tetraisopropoxide with hydrothermally induced phase separation	[24]
Nanowires and nanobelts		TiCl$_4$ hydrolysis in microemulsion	[25]

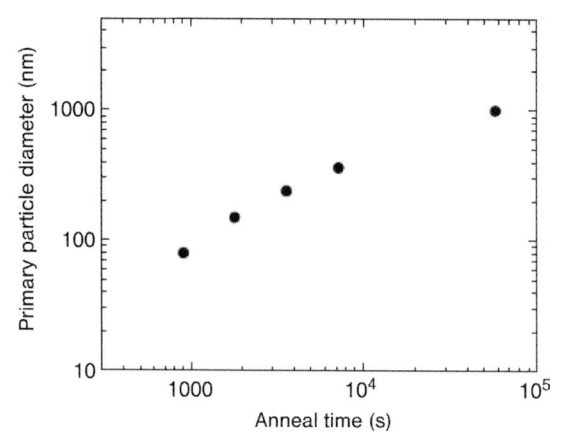

Figure 10.2 Primary particle size as determined by field emission scanning electron microscopy for Li4Ti5O12 fabricated by flash annealing for various times before quenching.

been proposed by Shen et al. [30], starting from tetrabutyl titanate and lithium acetate in isopropyl alcohol. The particle size of 100 nm was calculated from X-ray diffraction (XRD) spectra. A polyol-mediated preparation of n-LTO from titanium tetraisopropoxide and lithium oxide in ethylene glycol was proposed by Kim et al. [31]. Remarkably, small particle diameters of 5 nm were obtained when a temperature of 320 °C was used for annealing.

10.5
Nano-Li$_4$Ti$_5$O$_{12}$ Spinel Applications in Energy Storage Devices

An important report detailing the spinel oxide Li$_{1+x}$Ti$_{2-x}$O$_4$ ($0 < x < 1/3$) was published by Deschanvers et al. in 1971 [32]. Murphy et al. published the initial reports on the use of Li$_4$Ti$_5$O$_{12}$ as lithium intercalation material in 1983 [33]

Since the early 1990s, this material has been electrochemically characterized by Colbow et al. [34], Ferg et al. [35], and Ohzuku et al. [36]. Among the various members of the series, Li$_4$Ti$_5$O$_{12}$ is semi-conducting, and exhibits Li-insertion electrochemistry. The formal potential of Li insertion is 1.55–1.56 V for Li$_4$Ti$_5$O$_{12}$ [37,38]. This material accommodates one mole of Li$^+$ with a theoretical capacity of 175 mAh g^{-1}, based on the mass of the starting host material. It is one of the few Li intercalation materials that accommodates lithium ions without any lattice expansion. This results in an excellent cycle-life in batteries because no electrochemical grinding occurs during the charge and discharge reactions. Furthermore, it was demonstrated that a reduction in LTO crystallite size resulted in a significant increase in discharge rate capability (Figure 10.3). A novel spinel Li$_4$Ti$_5$O$_{12}$ with nanotubes/nanowires morphology and high surface area has been prepared using a low-temperature hydrothermal lithium ion-exchange processing from hydrogen

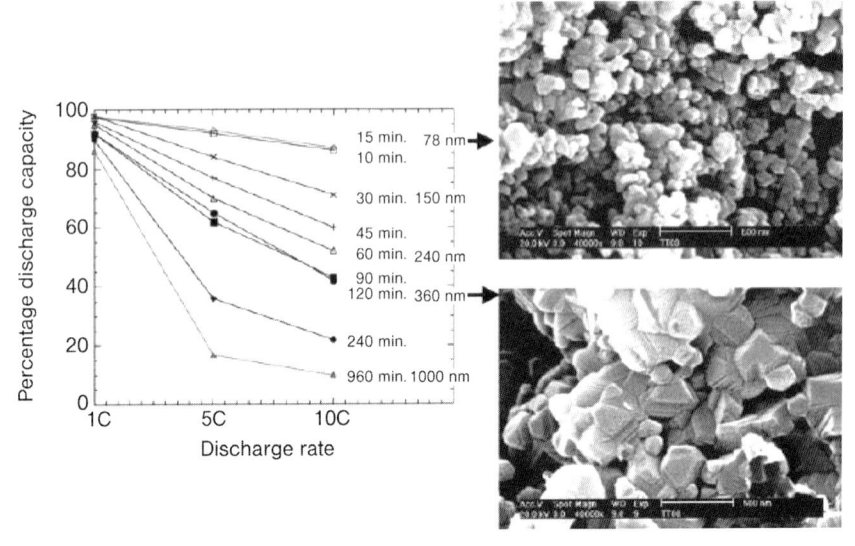

Figure 10.3 Percentage capacity retention at various rates for Li$_4$Ti$_5$O$_{12}$ of various sizes as fabricated by flash annealing at various times, tested versus Li metal anodes.

titanate nanotubes/nanowires precursors. A superior rate capability over Li$_4$Ti$_5$O$_{12}$ nano powders is claimed, although the report [39] did not provide a comparison of both materials.

10.5.1
Asymmetric Hybrid Supercapacitors

In 2001, Amatucci *et al.* reported a nanocrystalline Li$_4$Ti$_5$O$_{12}$ exhibiting a very promising charging rate and stability in a hybrid cell with a supercapacitor-like activated carbon counter electrode [40]. This new type of device was named a non-aqueous asymmetric hybrid supercapacitor (NAH) [41], as it truly combines a battery faradaic reaction on the anode with a capacitive double-layer adsorption on the cathode (Figure 10.4). The two advantages of this approach were: (i) a higher energy density than supercapacitors because the n-LTO constant discharge voltage raises the device average voltage (Figure 10.5); and (ii) a greater cycle-life than conventional batteries because of the absence of lattice expansion on the anode and an absence of intercalation reaction on the cathode (Figure 10.6). The resulting devices had an energy density of 11 W·h kg^{-1} packaged in 500F prototypes, and a cycle-life of over 100 000 cycles [42]. This device structure has been patented and is owned by Rutgers University.

Variants of this device have been investigated for increasing specific energy density. One variant involves the use of conjugated polymer pseudocapacitive cathodes such as poly(flurorphenylthiophene) [43] or poly(methylthiophene) [44]. However, in both cases, the better energy density is traded for lower

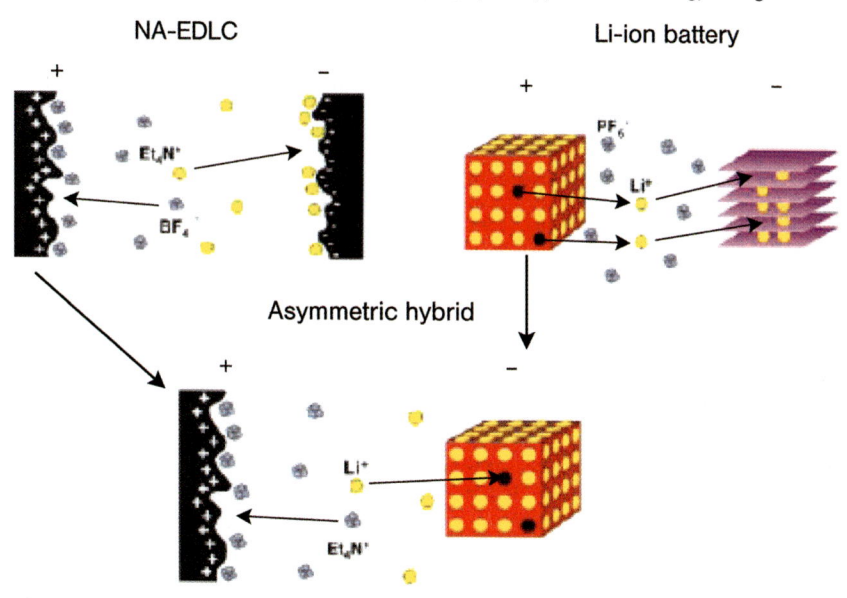

Figure 10.4 Scheme showing the configuration of a NAH cell composed of an activated carbon positive electrode and an intercalation electrode traditionally utilized in non-aqueous EDLC and Li-ion battery chemistry, respectively.

cycle-life because the cathode charge and discharge pseudocapacitive mechanism involves the intercalation–deintercalation of counter-ions in the conjugated polymer structures, a charge storage mechanism which is less reversible than pure double-layer capacitance.

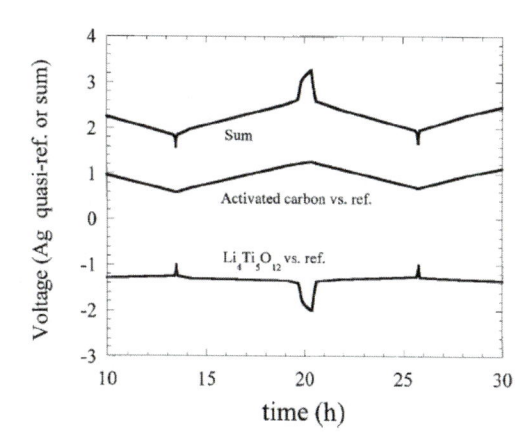

Figure 10.5 Three electrode voltage profiles of anode, cathode and device for n-LTO/activated carbon asymmetric hybrid supercapacitor.

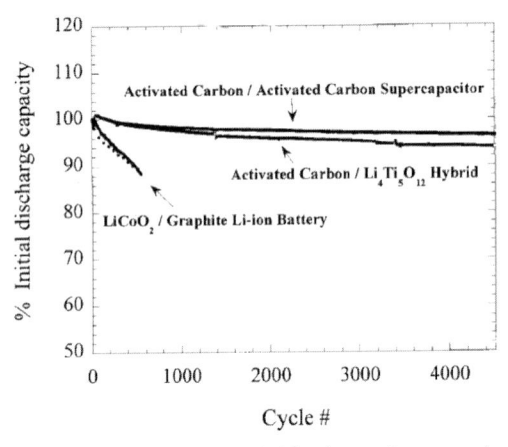

Figure 10.6 Comparative cycle-life of Li-ion battery, carbon/carbon supercapacitor and n-LTO/activated carbon asymmetric hybrid supercapacitor.

10.5.2
High-Power Li-Ion Batteries

The coupling of micron-sized $Li_4Ti_5O_{12}$ anode with lithium $LiMn_2O_4$ cathodes was initially proposed by Abraham et al. [45] in a poly(acrylonitrile)-based gel electrolyte. More recently, Kavan and Graetzel reported on the Li-insertion activity of a nano-sized spinel $Li_4Ti_5O_{12}$ prepared via a sol–gel route [29]. A thin-film electrode (2–4 μm) prepared from nanocrystalline $Li_4Ti_5O_{12}$ exhibited excellent activity toward Li insertion, even at a charging rate as high as 250 °C. The relationship between BET specific surface area and charge rate capability was well established (Figure 10.7).

Plastic "Bellcore-type" batteries have been built with plasticized PVDF-HFP n-LTO anodes and Celgard microporous polyolefin separators, with acetonitrile, $LiBF_4$ 2M electrolyte. When coupled with standard Li-ion battery cathodes, such as $LiCoO_2$ or $LiMn_2O_4$, the n-LTO anodes produce Li-ion batteries of $\sim 50\,W \cdot h\,kg^{-1}$, very fast charging rates (20C or 3 min), high specific power ($>2000\,W\,kg^{-1}$) (Figure 10.8), and excellent cycle-life ($>10\,000$ cycles) [46] (Figure 10.9).

Such high-power Li-ion batteries have other advantages of safety (no risk of Li-plating) and better rate capability than graphite at low temperature [an absence of any solid electrolyte interface (SEI) passivation layer] that make them attractive for use in electric and hybrid-electric vehicles. Currently, these applications are under development by companies such as Altair Nanotechnologies [47]. This new-type of Li-ion battery differs from those using graphite anodes in several aspects, notably that nitrile solvents can be used instead of carbonates because they are stable at the 1.5 V versus Li^+/Li anode voltage. For instance, high-conductivity and low-viscosity solvents such as acetonitrile [48], methoxypropionitrile [49] or methoxypropionitrile [50] enable high power capability and better rate capability than carbonates at low

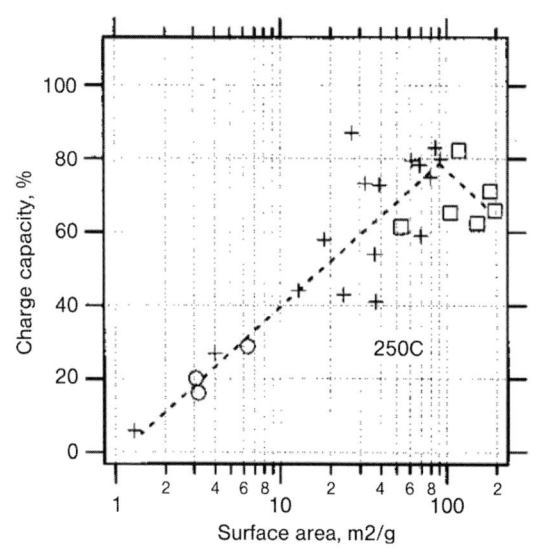

10.5 Nano-Li$_4$Ti$_5$O$_{12}$ Spinel Applications in Energy Storage Devices | 397

Figure 10.7 Charge capacity of Li$_4$Ti$_5$O$_{12}$ materials with varying surface areas at a 250C charging rate. The charge capacity was determined from galvanostatic chronopotentiometry with cut-off voltages of 3 and 1 V. The nominal charge capacity was determined from slow cyclic voltammetry at scan rates of 1 mV s^{-1}, and/or from charging at 2 °C. Electrolyte solution: 1 M LiN (CF$_3$SO$_2$)$_2$ 1 EC/DME ~1:1 (v:v). (From Ref. [29].)

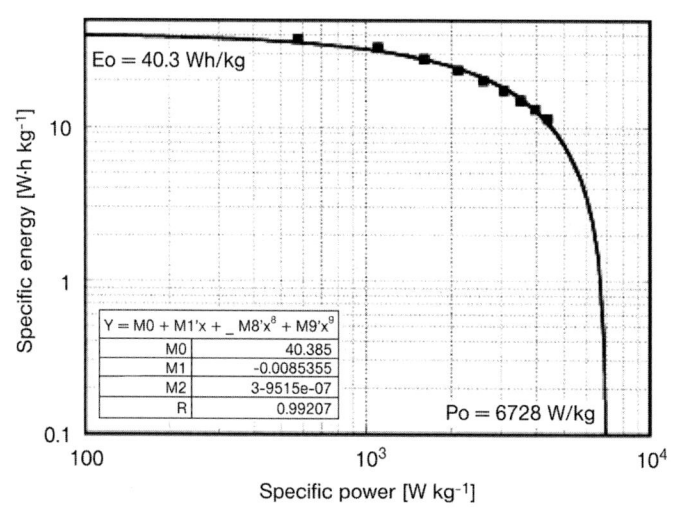

Figure 10.8 Ragone plot of a n-LTO/65%//LiCoO$_2$-10% activated SuperP cell of matching ratio 1.2.

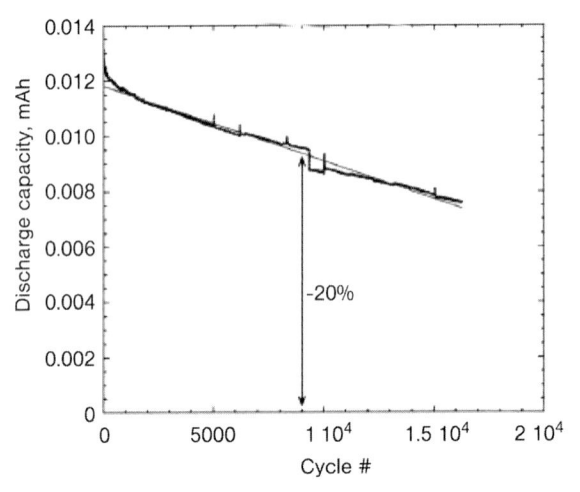

Figure 10.9 Discharge capacity versus cycle number for n-LTO/ 55% LiCoO$_2$-20% activated SuperP cell of matching ratio 1.67 (galvanostatic cycling at a 20C charging rate; charge-discharge rate between 1.6 and 3.0 V).

temperature. A second benefit is that an absence of Li-plating risk enables the design of Li-ion batteries with excess cathode capacity, which is an easy way of improving the device's cycle-life. Finally, the absence of passivation reactions on the anode can suppress any redox-shuttle aging mechanisms responsible for cathode passivation or dissolution. Hence, green and low-cost cathode materials such as LiMn$_2$O$_4$ can achieve better cycle-lives by using n-LTO anodes rather than graphite anodes.

10.6
Nano-TiO$_2$ Anatase for Solar Energy Conversion

10.6.1
TiO$_2$ Role in Dye-Sensitized Solar Cells

Nanosized TiO$_2$ has been used with great success in dye-sensitized solar cells (DSSCs) [51], and is also used in photoelectrolysis electrodes [52]. Unlike the case of semiconductors used for solid-sate solar cells, the functions of light harvesting and charge transport are separated. This considerably relaxes the purity requirements for the semiconductor, resulting in large saving costs for the solar cell manufacturing process. TiO$_2$ is a wide bandgap semiconductor (3.2 eV) which does not absorb any visible light. In order to become photoactive in the visible spectrum, it must be sensitized by a monolayer of dyes. TiO$_2$ sensitization by N-methylphenazinium ion for solar energy conversion was proposed in 1978 [53], but the power conversion efficiency was low and the dye unstable. The breakthrough studies of Graetzel and O'Reagan in 1991 [54] combined the use of large surface area, nanosized anatase TiO$_2$ with more stable Ru-based dyes [55]. The larger surface area of TiO$_2$ considerably increased the power conversion efficiency compared with micron-sized TiO$_2$.

The operating principle of DSSCs is summarized in Eq. (10.3–10.6). The dye transitions to the excited state upon photon absorption [Eq. (10.3)]; the electron is then injected from the dye in the conduction band of TiO$_2$ [Eq. (10.4)]. The dye is then regenerated by reduction from iodide ions which form triodide [Eq. (10.5)], and the triodide is reduced back to iodide at a platinum counter electrode [Eq. (10.6)]:

$$Dye + h\nu \rightarrow Dye^* \tag{3}$$

$$Dye^* + TiO_2 \rightarrow Dye^+ TiO_2 + e^- \tag{4}$$

$$Dye^+ + I^- \rightarrow Dye + 1/3 I_3^- \tag{5}$$

$$I_3^- + 2e^- (Pt) \rightarrow 3I^- \tag{6}$$

Later, Graetzel and colleagues demonstrated a promising 10.4% power conversion efficiency (η) by using a panchromatic black dye [56]. A comparative study of anatase- and rutile-based DSSCs has shown that anatase is the most photoactive phase [57]. The morphology of the rutile phase was found to lower dye coverage and decrease interparticle connectivity, thus slowing electron transport. It has also been found that the addition of micron-sized TiO$_2$ crystals to the nano-TiO$_2$ electrodes increases the haze factor, which results in a better light harvesting and increased power conversion efficiency. Using this concept, DSSCs with η-values up to 11.1% have been demonstrated [58].

As in all types of solar cell, η is expressed as:

$$\eta = \frac{Isc * Voc * FF}{Pin} \tag{7}$$

This is a function of four parameters, where I_{sc} is the short-circuit current, V_{oc} is the open-circuit voltage, FF is the fill factor, and P_{in} is the incident light power. The nature of electron transport in TiO$_2$ has a direct impact on the short-circuit current, as will be reviewed below.

10.6.2
Trap-Limited Electron Transport in Nanosized TiO$_2$

Electron transport in the TiO$_2$ electrode proceeds via diffusion and a trapping–detrapping mechanism [59]. Trap location has recently been demonstrated to occur mostly at the surface of the TiO$_2$ nanoparticles, by studying the relationship between the photoinduced electron density and the TiO$_2$ electrode roughness [60]. The electron chemical diffusion coefficient in mesoporous TiO$_2$ is linked to the density of traps by the relationship [61]:

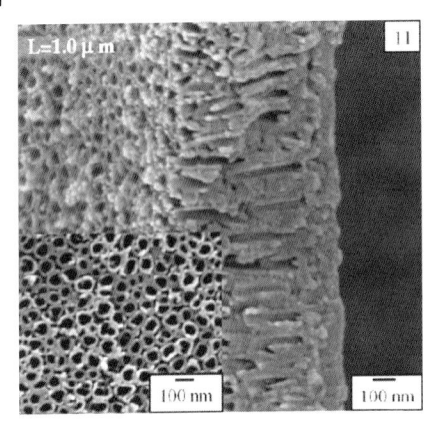

Figure 10.10 Lateral and top views of TiO$_2$ nanotube array formed by anodization. (From Ref. [63].)

$$D = C1 N_{tot}^{-1/\alpha} n^{((1/\alpha)-1)} \qquad (8)$$

where N_{tot} is the total trap density, n is the photoinduced electron density, C1 is a constant independent from particle size, and α is a dispersive parameter ($0 < \alpha < 1$) which appears in the distribution of waiting times that an electron spends in a trap [62]. Therefore, improved electron transport in TiO$_2$ electrodes is achieved by annealing at 450 °C, which causes necking of the nanoparticles and better percolation. Other TiO$_2$ morphologies such as TiO$_2$ nanotube arrays [63] have also been investigated to improve electron transport (Figure 10.10). However, the maximum power conversion efficiency of 4.7% has been limited by a lower dye adsorption than TiO$_2$ nanoparticles films. Furthermore, it was found that faster electron transport also causes faster electron back-transfer, for example, electrical losses by recombination with the dye [64].

10.6.3
Electron Recombination in Dye-Sensitized Solar Cells

In TiO$_2$ DSSCs, electron recombination with the dye cation is several orders of magnitude slower than the electron injection rate, which results in an electron injection yield close to unity. However, electron back-transfer to the dye is only one order of magnitude slower than dye cation regeneration by reduction with I^- from the electrolyte. Under a high applied bias voltage or high illumination conditions, the accumulation of thermalized electrons in the TiO$_2$ conduction intraband states results in a faster recombination with the dye cation (Figure 10.11). This in turn prevents dye regeneration because the dye is reduced faster by TiO$_2$ electrons than it is by I^- ions. This results in a decrease in photocurrent and fill factor, causing a loss in power conversion efficiency.

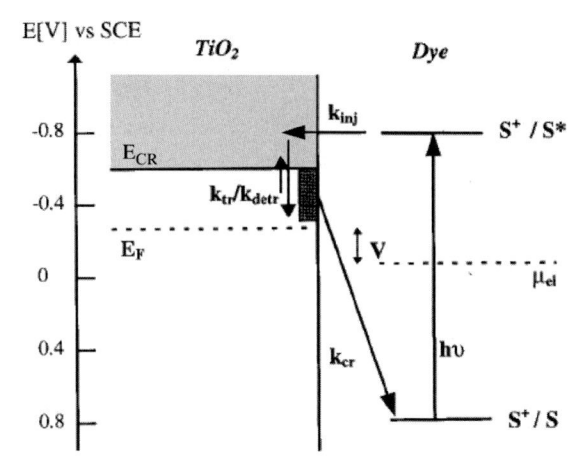

Figure 10.11 Schematics of electron transfer pathways in TiO$_2$ dye-sensitized solar cells. Under illumination, the excited state of the dye (S*) causes ultrafast (k_{inj}) electron injection into the conduction band states of TiO$_2$. The injected electron subsequently thermalizes with electrons accumulated in conduction band/intraband states of the TiO$_2$ (k_{tr}/k_{detr}). The oxidized dye cation is re-reduced by back-transfer of electrons accumulated or photoinjected in the TiO$_2$ conduction band/intraband states (k_{cr} = charge recombination). (From Ref. [65].)

The kinetics of dye electron injection and recombination in TiO$_2$ can be measured using nanosecond transient absorption spectroscopy [65]. In this method, the decay of transient absorption of the dye under pulsed laser excitation is attributed to a recombination of the electrons injected into the metal oxide conduction band with the dye cation when no I$^-$/I$_3^-$ is present in the electrolyte [66]. It has also been shown that surface lithium intercalation in TiO$_2$ electrodes reduces the electron diffusion rate at the surface, simultaneously reducing the electron recombination rate [67].

10.6.4
Preparation of Flexible TiO$_2$ Photoanodes

The low-temperature preparation of TiO$_2$ photoanodes for dye-sensitized solar cells represents a challenge of major industrial significance from a low-cost manufacturing perspective. Not only does it result in energy savings during the deposition process but, more importantly, it enables the use of flexible substrates such as indium tin oxide coated poly (ethylene terephthalate) (ITO/PET) which are required for high-throughput roll-to-roll manufacturing process. Furthermore, such substrates add the advantages of lighter weight and impact resistance. In the case of ITO/PET, the maximum annealing temperature is limited to 150 °C by the substrate melting. This is not sufficient to cause necking of the TiO$_2$ nanoparticles, which results in poor adhesion to the substrate and poor interparticle connection. Low-temperature annealing of TiO$_2$ electrodes without additives has yielded devices with $\eta = 1.22\%$ [68]. Several strategies have been devised to overcome this problem, and the most successful will be reviewed below.

10.6.4.1 Sol–Gel Additives

Konarka Technologies (Lowell, MA, USA) has developed a low-temperature roll-to-roll coating process for TiO_2 dye-sensitized solar cells. An examination of the patent literature [69,70] suggests the use of poly(n-butyl) titanate as a linear polymeric linking agent for TiO_2 nanoparticles reacting at temperatures ranging from 100 to 200 °C. This material has the advantages of acting both as a binder and a surfactant during the film coating and drying steps; subsequently it converts to pure TiO_2 which interconnects the TiO_2 nanoparticles.

In the present authors' laboratory, titanium tetraisopropoxide (TTIP) has been used as an efficient cross-linking additive to Degussa P25 TiO_2 colloidal suspensions in methanol. In this way, a η-value of 3.55% under 48 mW cm^{-2} was obtained after doctor-blade deposition on ITO/PET substrates and annealing at 130 °C for 30 min (Figure 10.12; Table 10.3). A similar approach, combined with preheating of the TiO_2 nanoparticles and ultra-violet (UV)-ozone treatment of the TiO_2 electrodes, was recently reported by Zhang et al. [71]. The UV-ozone treatment favored the elimination of any organic byproducts which remained in the electrode. Solar-to-electric energy conversion efficiencies of 4.0% and 3.27% have been achieved for cells with conductive glass and plastic film substrates, respectively. Scanning electron microscopy (SEM) observations of the TiO_2 films in the presence of TTIP demonstrate necking of the nanocrystals which appear fused together when large amounts of TTIP are used (Figure 10.13).

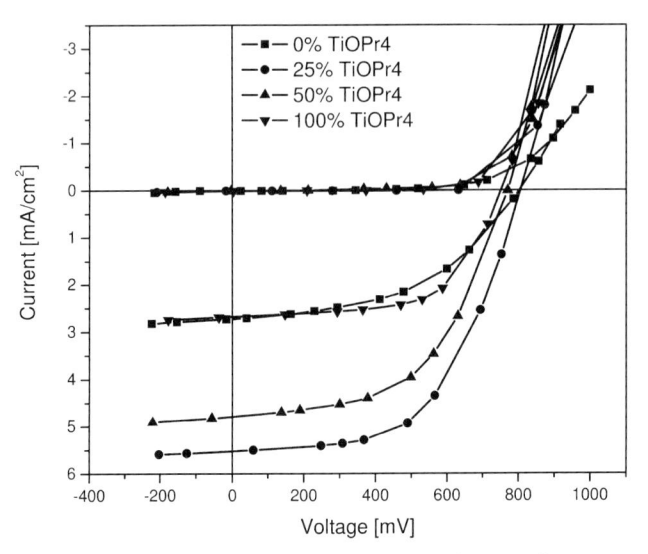

Figure 10.12 Current–voltage (I–V) characteristics of DSSC cells on FTO substrates with various amounts of titanium isopropoxide additive, in the dark and under 101 mW cm^2 AM 1.5G illumination.

Table 10.3 Photovoltaic parameters of DSSC cells on FTO and ITO/PET substrates with various annealing temperatures and 48.7 mW cm^{-2} illumination.

Sample	η [%]	FF	I_{sc} [mA cm^{-2}]	V_{oc} [mV]	R_s [Ω·cm^2]	R_p [Ω·cm^2]
FTO, 450 °C	4.01	0.56	4.53	767	50	519
FTO, 130 °C	4.10	0.63	4.37	723	52	782
PET, 130 °C	3.55	0.48	4.84	747	84	693

10.6.4.2 Mechanical Compression

Hagfeldt and coworkers from the Ångström Solar Center in Uppsala, Sweden [72] originally proposed a pressing technique that consisted of statically or continuously pressing powder films of TiO$_2$ (Degussa P-25) onto flexible electrodes (ITO/PET). A typical pressure for preparing efficient solar cells is 1000 kg cm^{-2} for a few seconds. When this method is used, further annealing of the photoanodes does not produce any improvement in power conversion efficiency. With ITO/PET substrates, the overall cell efficiency (active area 0.32 cm^2) was 4.9% at 10 mW cm^{-2}. Because of the series resistance losses in the conducting plastic layer (ITO/PET, sheet resistance: 60 Ω sq^{-1}), a η-value of only 2.3% was reached under 100 mW cm^{-2} illumination. Durrant et al. [73] used the same method, but built the devices with a polymer electrolyte based on NaI/I$_2$ in poly-(epichlorohydrin-co-ethylene oxide), and obtained η = 5.3% at 10 mW cm^{-2} and η = 2.5% at 100 mW cm^{-2}.

10.6.4.3 Metallic Foils

In order to use flexible substrates and high-temperature annealing of TiO$_2$ electrodes, a promising alternative strategy consists of using flexible metal foils as photoanode substrates, and illuminating the cells through a transparent Pt-coated ITO/PET counter electrode. The metals should have a work function close to TiO$_2$ (~4.3 eV) to provide ohmic contacts, and good corrosion resistance to the

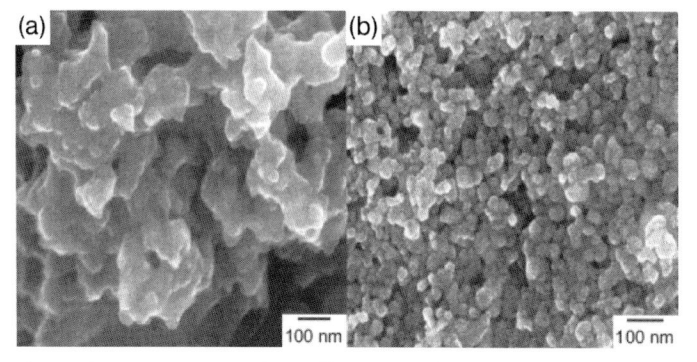

Figure 10.13 SEM images of TiO$_2$ films prepared at room temperature on ITO/PET with molar ratios TTIP:TiO$_2$ of (a) 0.36 and (b) 0.036. (From Ref. [71].)

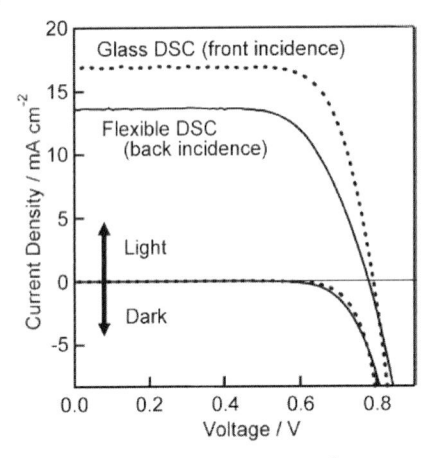

Figure 10.14 Current–voltage (I–V) characteristics of a 7.2% efficient flexible DSSC on Ti substrate and glass. The irradiation was AM 1.5 (100 mW cm^{-2}). (From Ref. [76].)

iodine/triodide redox electrolyte. Promising metals in this respect include Zn, W, Ti, and stainless steel [74]. Another advantage of this approach is that it is better suited for large-area solar cells, where the sheet resistance of FTO can increase the device series resistance. An efficiency of 4.2% under 100 mW cm^{-2} incident power has been reported with nanocrystalline-TiO$_2$ film on ITO/SiO$_2$/stainless steel [75]. More recently, TiO$_2$ electrodes with optimized thickness were prepared on Ti foils, sintered at 500 °C, and flexible dye-sensitized solar cells with 7.2% efficiency under 100 mW cm^{-2} AM 1.5 incident power were reported (Figure 10.14) [76]. (Note: AM 1.5 is the ASTM standard spectrum for solar simulators, mimicking the solar conditions in North America on the ground and corresponding to a solar illumination at a tilt angle of 37° facing the sun.) It should be noted however that, in both cases, the flexible devices were tested with liquid electrolytes.

10.7
Conclusions

Nanosized TiO$_2$ anatase is the starting material for two important applications in energy storage and energy conversion. Currently, several companies are producing this material at low cost, and the concept that it may soon play a major role in new batteries for electric and hybrid vehicles, as well as low-cost solar cells, it indeed exciting. These properties make nanosized TiO$_2$ the ideal candidate for coupling the two functions of energy conversion and energy storage at the materials level. However, such an enterprise is not straightforward, as it appears that lithium ions play opposite roles in the key processes involved in photovoltaic energy conversion and energy storage. In the anatase structure used for DSSCs, Li intercalation reduces the electron diffusion coefficient, but the effect is slow and affects only the surface. However, when the TiO_2 anatase phase is converted to lithium titanium spinel, it becomes inactive in a dye-sensitized solar cell, as the electron diffusion rate becomes too slow in comparison with electron trapping and recombination rates. Conversely, TiO_2 anatase can be used as a reversible lithium intercalation host, although the lower lithium diffusion coefficient makes it unattractive for such an application, and the lithium titanium spinel is preferred. Both types of device are still in need of more stable electrolytes in order to deliver their promise of performance in practical devices. In both, the $Li_4Ti_5O_{12}$-based battery and DSSCs, the record performances that made those devices famous have been obtained with acetonitrile-based electrolytes. Although the very low viscosity and high dielectric constant of acetonitrile are unmatched by any other solvent, the high volatility and potential toxicity mean that it is a poor choice for commercial products. Solvents with lower vapor pressures are required in order to enable the high-temperature operation of the batteries, and solid-state electrolytes are required for DSSCs in order attain 10-year lifetimes. This area of active research into DSSCs [77] will, in time, lead to further developments in the outstanding performance of nanosized TiO_2 and $Li_4Ti_5O_{12}$ for use in consumer products.

Acknowledgments

The authors thank Timothy Spitler, Research Manager at Altair Nanotechnologies for fruitful discussions. Research funding from Altair Nanotechnologies is gratefully acknowledged.

References

1 Kronos International, (1996).
2 Diebold, U. (2003) *Surf. Sci. Rep.*, **48**, 53.
3 Kumar, K.-N.P., Keizer, K. and Bruggraaf, A.J. (1994) *J. Mater. Sci. Lett.*, **13**, 59–61.
4 Addamo, M. et al. (2005) *Colloids and Surfaces A: Physicochem. Eng. Aspects*, **265**, 23.
5 Liu, A.R. et al. (2006) *Mater. Chem. Phys.*, **99**, 131.
6 US Patent. 6,440,383, (2000).
7 Suyama, Y. and Kato, A. (1976) *J. Am. Ceram. Soc.*, **59**, 146.
8 Akhtar, M. Kamal, Vemury, Srinivas and Pratsinis, Sotiris E. (1994) *AIChE J.*, **40**, 1183.
9 US Patent 5,874,684, (1999).
10 Gong, D., Grimes, C.A., Varghese, O.K., Hu, W., Singh, R.S., Chen, Z. and Dickey, E.C. (2001) *J. Mater. Res.*, **16**, 3331.
11 Mor, G.K., Varghese, O.K., Paulose, M., Mukherjee, N. and Grimes, C.A. (2003) *J. Mater. Res.*, **18**, 2588.
12 Cai, Q., Paulose, M., Varghese, O.K. and Grimes, C.A. (2005) *J. Mater. Res.*, **20**, 230.
13 Mor, G.K., Shankar, K., Paulose, M., Varghese, O.K. and Grimes, C.A. (2005) *Nano Lett.*, **5**, 191.
14 Zukalova, M., Zukal, A., Kavan, L., Nazeeruddin, M.K., Liska, P. and Graltzel, M. (2005) *Nano Lett.*, **5**, 1789.
15 Jiu, J., Wang, F., Sakamoto, M., Takao, J. and Adachi, M. (2005) *Solar Energy Mater. Solar Cells*, **87**, 77.

16 Wijnhoven, J.E.G.J. and Vos, W.L. (1998) *Science*, **281**, 802.

17 Jiang, P., Cizeron, J., Bertone, J.F. and Colvin, V.L. (1999) *J. Am. Chem. Soc.*, **121**, 7957.

18 Lai, Qi. and Birnie, D.P., III (2007) *Mater. Lett.*, **61**, 2191.

19 Sakamoto, Y., Kaneda, M., Terasaki, O., Zhao, D.Y., Kim, J.M., Stucky, G.D., Shin, H.J. and Ryoo, R. (2000) *Nature*, **408**, 449.

20 Hwang, Y.K., Lee, K.C. and Kwon, Y.U. (2001) *Chem. Commun.*, 1738.

21 Alberius-Henning, P., Frindell, K.L., Hayward, R.C., Kramer, E.J., Stucky, G.D. and Chmelka, B.F. (2002) *Chem. Mater.*, **14**, 3284.

22 Kasuga, T., Hiramatsu, M. and Hoson, A. (1998) *Langmuir*, **14**, 3160.

23 Armstrong, A.R., Armstrong, G., Canales, J. and Bruce, P.G. (2004) *Angew. Chem.*, **116**, 2336.

24 Ho, W., Yu, J.C. and Yu, J. (2005) *Langmuir*, **21**, 3486.

25 Wang, J., Sunb, J. and Bian, X. (2004) *Mater. Sci. Eng. A*, **379**, 7.

26 Plitz, I., Dupasquier, A. and Badway, F. *et al.* (2006) *Appl. Phys. A: Mater. Sci. Process.*, **82**, 615.

27 Guerfi, A., Sévigny, S., Lagacé, M., Hovington, P., Kinoshita, K. and Zaghib, K. (2003) *J. Power Sources*, **119–121**, 88.

28 Zaghib, K., Simoneau, M., Armand, M. and Gauthier, M. (1999) *J. Power Sources*, **81–82**, 300.

29 Kavan, L. and Graetzel, M. (2002) *Electrochem. Solid-State Lett.*, **5**, A39.

30 Shen, C.-M., Zhang, X.-G., Zhou, Y.-K. and Li, H.-L. (2003) *Mater. Chem. Phys.*, **78**, 437.

31 Kim, D.H., Ahn, Y.S. and Kim, J. (2005) *Electrochem. Commun.*, **7**, 1340.

32 Deschanvers, A., Raveau, B. and Sekkal, Z. (1971) *Mater. Res. Bull.*, **6**, 699.

33 Murphy, D.W., Cava, R.J., Zahurak, S.M. and Santoro, A. (1983) *Solid State Ionics*, **9–10**, 413.

34 Colbow, K.K., Dahn, J.R. and Haering, R.R. (1989) *J. Power Sources*, **26**, 397.

35 Ferg, E., Gummov, R.J., de Kock, A. and Thackeray, M.M. (1994) *J. Electrochem. Soc.*, **141**, L147.

36 Ohzuku, T., Ueda, A. and Yamamoto, N. (1995) *J. Electrochem. Soc.*, **142**, 1431.

37 Harrison, M.R., Edwards, P.P. and Goodenough, J.B. (1985) *Philos. Mag. B*, **52**, 679.

38 Pyun, S.I., Kim, S.W. and Shin, H.C. (1999) *J. Power Sources*, **81–82**, 248.

39 Li, J., Tang, Z. and Zhang, Z. (2005) *Electrochem. Commun.*, **7**, 894.

40 Amatucci, G.G., Badway, F., Du Pasquier, A. and Zheng, T. (2001) *J. Electrochem. Soc.*, **148**, A930.

41 Pell, W.G. and Conway, B.E. (2004) *J. Power Sources*, **136**, 334.

42 Pasquier, A.D., Plitz, I., Gural, J., Menocal, S. and Amatucci, G. (2003) *J. Power Sources*, **113**, 62.

43 Du Pasquier, A., Laforgue, A., Simon, P., Amatucci, G.G. and Fauvarque, J.-F. (2002) *J. Electrochem. Soc.*, **149**, A302.

44 Du Pasquier, A., Laforgue, A. and Simon, P. (2004) *J. Power Sources*, **125**, 95.

45 Peramunage, D. and Abraham, K.M. (1998) *J. Electrochem. Soc.*, **145**, 2615.

46 Pasquier, A.D., Plitz, I., Gural, J., Badway, F. and Amatucci, G.G. (2004) *J. Power Sources*, **136**, 160.

47 http://www.altairnano.com.

48 Du Pasquier, A., Plitz, I., Menocal, S. and Amatucci, G. (2003) *J. Power Sources*, **115**, 171.

49 Wang, Q., Zakeeruddin, S.M., Exnar, I. and Grätzel, M. (2004) *J. Electrochem. Soc.*, **151**, A1598.

50 Wang, Q., Pechy, P., Zakeeruddin, S.M., Exnar, I. and Grätzel, M. (2005) *J. Power Sources*, **146**, 813.

51 Grätzel, M. (2003) *J. Photochem. Photobiology C: Photochem. Rev.*, **4**, 145.

52 Grätzel, M. (2005) *Chem. Lett.*, **34**, 8.

53 Chen, S., Deb, S.K. and Witzke, H. (1978) U.S. Patent 4,080,488.

54 O'Regan, B. and Graetzel, M. (1991) *Nature*, **353**, 737.

55 Kalyanasundaram, K. and Graetzel, M. (1998) *Coord. Chem. Rev.*, **177**, 347.

56 Nazeeruddin, M.K., Kay, A. and Rodicio, I. et al. (1993) *J. Am. Chem. Soc.*, **115**, 6382.
57 Park, N.-G., Van De Lagemaat, J. and Frank, A.J. (2000) *J. Phys. Chem. B*, **104**, 8989.
58 Chiba, Y., Islam, A., Komiya, R., Koide, N. and Han, L. (2006) *Appl. Phys. Lett.*, **88**, 223505.
59 Eppler, A.M., Ballard, I.M. and Nelson, J. (2002) Physica E: Low-Dimensional Systems. *Nanostructures*, **14**, 197.
60 Kopidakis, N., Benkstein, K.D., Van De Lagemaat, J. and Frank, A.J. (2005) *Appl. Phys. Lett.*, **87**, 1.
61 Van de Lagemaat, J. and Frank, A.J. (2001) *J. Phys. Chem. B*, **105**, 11194.
62 Nelson, J. (1999) *Phys. Rev. B - Condensed Matter Mater. Phys.*, **59**, 15374.
63 Mor, G.K., Varghese, O.K., Paulose, M., Shankar, K. and Grimes, C.A. (2006) *Solar Energy Mater. Solar Cells*, **90**, 2011.
64 Frank, A.J., Kopidakis, N. and Lagemaat, J.V.D. (2004) *Coord. Chem. Rev.*, **248**, 1165.
65 Haque, S.A., Tachibana, Y., Klug, D.R. and Durrant, J.R. (1998) *J. Phys. Chem. B*, **102**, 1745.
66 Tachibana, Y., Moser, J.E., Grätzel, M., Klug, D.R. and Durrant, J.R. (1996) *J. Phys. Chem.*, **100**, 20056.
67 Kopidakis, N., Benkstein, K.D., Van De Lagemaat, J. and Frank, A.J. (2003) *J. Phys. Chem. B*, **107**, 11307.
68 Pichot, F., Pitts, J.R. and Gregg, B.A. (2000) *Langmuir*, **16**, 5626.
69 US Patent 6,858,158, (2005).
70 US Patent 7,094,441, (2006).
71 Zhang, D., Yoshida, T., Oekermann, T., Furuta, K. and Minoura, H. (2006) *Adv. Funct. Mater.*, **16**, 1228.
72 Lindström, H., Holmberg, A., Magnusson, E., Malmqvist, L. and Hagfeldt, A. (2001) *J. Photochem. Photobiol. A: Chemistry*, **145**, 107.
73 Haque, S.A., Palomares, E. and Upadhyaya, H.M. et al. (2003) *Chem. Commun.*, **24**, 3008.
74 Man, G.K., Park, N.-G. and Ryu, K. et al. (2005) *Chem. Lett.*, **34**, 804.
75 Kang, M.G., Park, N.-G. and Ryu, K. et al. (2006) *Solar Energy Mater. Solar Cells*, **90**, 574.
76 Ito, S., Ha, N.L.C. and Rothenberger, G. et al. (2006) *Chem. Commun.*, **38**, 4004.
77 Li, B., Wang, L. and Kang, B. et al. (2006) *Solar Energy Mater. Solar Cells*, **90**, 549.

11
DNA Biosensors Based on Nanostructured Materials
Adriana Ferancová and Ján Labuda

11.1
Introduction

Following the discovery of novel materials with unique physical and chemical properties, their introduction into the construction of high-performance biosensors remains the subject of much interest to the scientific community. Previously, low sensor-to-sensor reproducibility and rather poor signal stability have negatively affected the mass production of biosensors and their commercial use [1]. Today, however, nanotechnology plays a major role in the development of biosensors [2], with recent advances having allowed the use of a relatively new group of materials – nanomaterials – as transduction matrices and electrode materials for both chemical sensors and biosensors [3].

Nanoscience and nanotechnology involve the synthesis, characterization, exploration, manipulation and utilization of nanostructured materials, which are characterized by being at least one dimension smaller than 100 nm. Individual nanostructures involve clusters, nanoparticles, nanocrystals, quantum dots, nanowires and nanotubes, whilst collections of nanostructures involve arrays, assemblies, and superlattices of individual nanostructures [4,5].

The construction of electrochemical DNA biosensors is based on the immobilization of single-stranded DNA (ssDNA) or double-stranded DNA (dsDNA) onto the surface of, or in the bulk of, a working electrode. The working principle is based on the detection of specific interactions, such as DNA hybridization and association interactions with low-molecular-weight compounds (drugs, risk chemicals), in addition to the structural damage of DNA. During recent years, nanoparticles and carbon nanomaterials such as carbon nanotubes, nanofibers, fullerenes and diamonds (Figure 11.1) have begun to be used widely in the preparation of DNA biosensors. Indeed, their conjugation with DNA has been the subject of much research interest, such that nanoparticles have now found applications in novel electronic devices, drug delivery systems, biomaterials, and biomedicine [6–8].

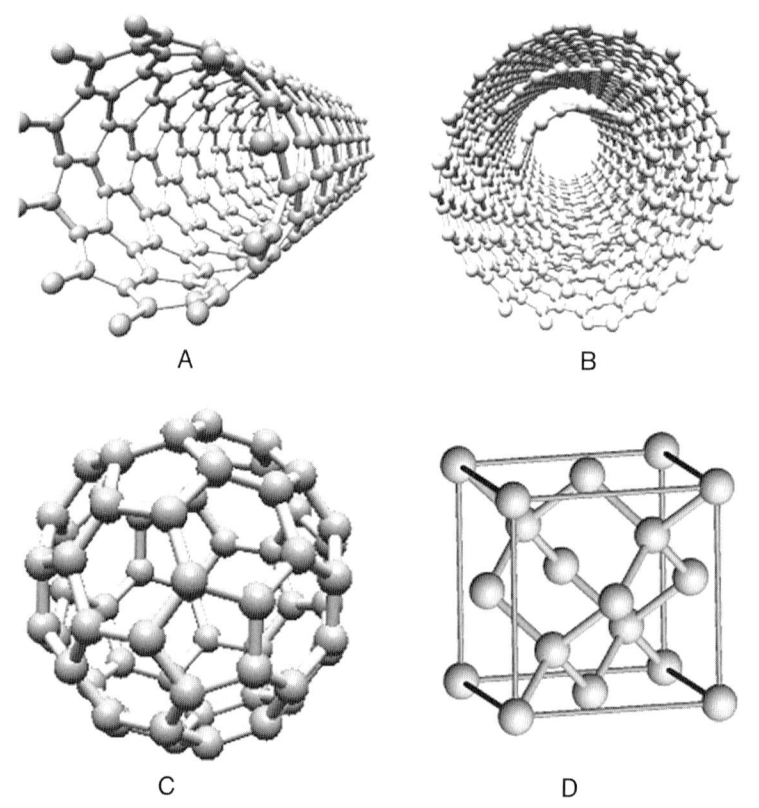

Figure 11.1 Structures of selected nanomaterials. (A) Single-walled carbon nanotube; (B) multi-walled carbon nanotube; (C) fullerene C_{60}; (D) diamond.

The main aim of applying nanomaterials to electrochemical DNA biosensors is to improve the immobilization of DNA molecules, as well as to enhance molecular recognition and signal transduction events [9]. Although the primary advantage of nanostructured materials is their large surface area, the chemical modification of these materials, by enzymes and electroactive molecules (mediators, markers), has led to significant improvements in electrochemical sensing.

11.2
Nanomaterials in DNA Biosensors

11.2.1
Carbon Nanotubes

Carbon nanotubes (CNTs) were discovered in 1991 as multi-walled carbon nanotubes (MWNTs) [10], and in 1993 as single-walled carbon nanotubes (SWNTs) [11].

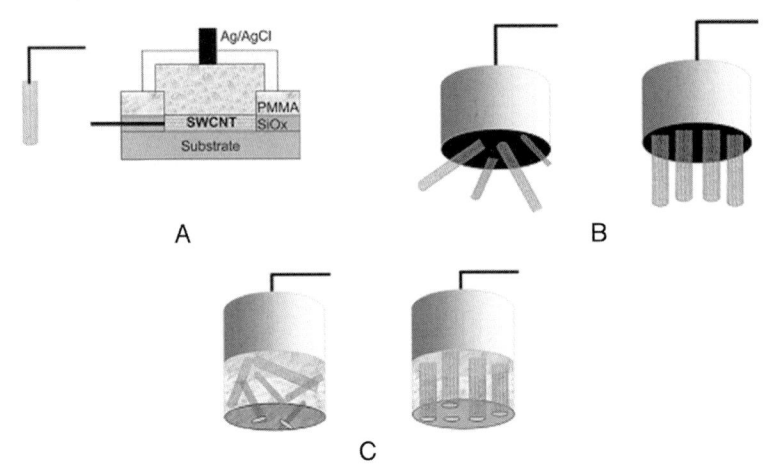

Figure 11.2 Configurations of carbon nanotube (CNT) integration in the electrochemical sensors. (A) Individual single-walled CNT (SCWNT); (B) electrode surface modifiers: non-oriented (left) and oriented (right) CNTs; (C) composites with non-oriented (left) and oriented (right) CNTs. (Reproduced from Ref. [20] with kind permission of Elsevier.)

Since that time, they have undergone intensive investigations and have been identified for many applications in different areas of science and technology [12]. The unique physical-chemical properties of CNTs, such as sorption properties, electron transfer and conductivity, renders them of great interest for applications in analytical sciences (for reviews, see Refs. [13,14]). Typically, CNTs demonstrate an excellent biocompatability [15] and offer an environment which is suitable for the immobilization of biological components. Therefore, they are also widely used in the construction of sensors and biosensors (Figure 11.2) [16–24].

CNTs are formed by rolled-up plates consisting of hexagons of carbon atoms. By comparison, SWNTs (Figure 11.1A) consist of a single CNT with a typical diameter in the range of 0.4 to 2 nm and a length up to few micrometers. The MWNTs (Figure 11.1B), which are formed from several concentric CNTs, have diameters which normally exceed 2 nm, while the lengths may be more than 10 µm [25].

11.2.1.1 Electronic Properties and Reactivity of CNTs

The electronic properties and reactivity of CNTs play important roles in their application as biosensors. The electrochemical properties of porous SWNTs in the form of sheets of papers in aqueous and non-aqueous solutions were studied using electrochemical methods, electrochemical impedance spectroscopy, and an electrochemical quartz crystal microbalance [26–28]. The conductivity of CNTs depends on their structure [29]. For example, MWNTs are regarded as metallic conductors, whereas in the case of SWNTs the problem of conductivity is more difficult. SWNTs may possess different chirality, depending on the angle at which the graphite plate is rolled up. Chiral nanotubes have metallic properties, whereas arm-chair and zig-zag nanotubes have the properties of semi-conductors.

The reactivity of nanotubes increases with a decrease in nanotube diameter, and also depends on their chirality. This property of nanotubes can be affected by their functionalization, doping, or pre-treatment. One widely used method is the purification of the nanotubes with mineral acids such as HNO_3 or H_2SO_4 (or mixtures thereof) [30,31], which causes not only shortening but also an opening of the nanotube ends. Acid pretreatment also enables the CNTs' sidewalls or tips to be functionalized with hydroxyl, carboxyl and carbonyl groups, which can be further modified.

The main problem in the manipulation and application of CNTs is their insolubility in aqueous and also polar media. In such an environment, CNTs have a tendency to coagulate due to hydrophobic interactions and strong attractive van der Waals forces between the nanotubes. However, CNTs can be dispersed by using a variety of methods [32–34]:

- oxidative acid treatments (refluxing in diluted HNO_3)
- non-covalent stabilization in non-polar organic solvents (e.g., dimethylformamide; DMF), using surfactants (sodium dodecyl sulfate; SDS, Nafion) and γ-cyclodextrin
- covalent stabilization (by glucose, DNA, enzymes).

11.2.1.2 CNT–DNA Interaction

In recent years, an investigation of the interactions of various macrobiomolecules with CNTs has attracted much interest. An understanding of these interactions is important for many applications, such as the preparation of the DNA biosensors.

Several groups have investigated the interaction of CNTs and DNA by using a theoretical approach to demonstrate two types of interaction: (i) insertion into the wide nanotube (Figure 11.3A); and (ii) wrapping around the narrow nanotube (Figure 11.3B) [35,36]. Song et al. [37] proposed biomolecule-functionalized SWNTs (base-functionalized SWNTs) which may be used as bio-nanomaterials possessing self-assembly properties due to the hydrogen bond interactions between the purine and pyrimidine bases, as occurs in native DNA. The interaction of zig-zag and arm-chair SWNTs with DNA showed that any local atomic structural distortion on SWNTs caused by sidewall functionalization can alter the electronic structure of

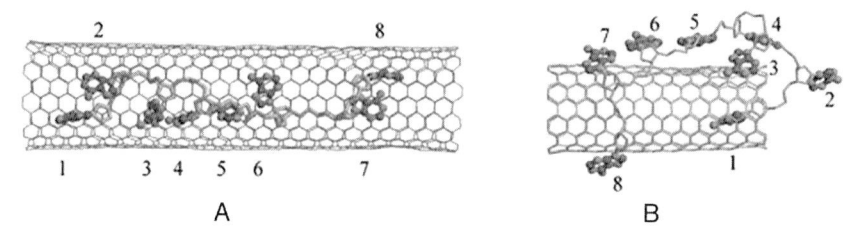

Figure 11.3 Simulations of CNT interactions with a single-stranded DNA oligomer consisting of eight adenine nucleobases (numbered 1 to 8). (A) Insertion of a DNA oligonucleotide into the wide CNT. (B) Wrapping DNA around the narrow CNT. (From Ref. [35] with kind permission from Elsevier.)

the nanotube. By using molecular dynamics simulations it was also shown that, in an aqueous environment, the ssDNA molecule could be inserted spontaneously into the CNTs [38]. Both, the van der Waals and hydrophobic forces were found to be important for this interaction. These results were verified experimentally in a spectroscopic investigation in which Pt-labeled DNA was encapsulated inside MWNTs [39]. According to the theoretical simulations, the nucleosides are able to interact with CNTs in vacuum, and in the presence of an external gate voltage [40]. Fluorescence microscopy measurements showed that single MWNTs incorporated into the membrane may also function as a channel for the transport of DNA of appropriate size [41].

Surface-enhanced infrared adsorption spectroscopy was used to study the interaction of DNA with SWNTs [42,43]. Vibration modes of the DNA–SWNTs complex showed structural changes in DNA which could be interpreted as A–B transition and stabilization of the DNA structure in some DNA fragments. The proposed model of the DNA–SWNT interaction was based on wrapping of the DNA molecule around the CNT. It was also found that DNA may help to disperse CNTs and alter their electrical properties [44–46].

Covalently linked DNA–SWNTs adducts were prepared and studied using X-ray photoelectron spectroscopy (XPS) [47]. These adducts were stable, and bound DNA molecules were accessible for hybridization and showed high specificity towards complementary sequences. The results indicated that the DNA oligonucleotides were chemically bound to the exterior of SWNTs, and neither wrapped around nor inserted into the nanotube. The conclusion was that these adducts might have a role in the development of highly selective and reversible biosensors.

11.2.1.3 CNTs in DNA Biosensors

DNA molecules are widely used to functionalize CNTs, either covalently or noncovalently [48,49]. The CNTs in DNA biosensors may serve as the electrode modifier for an enhanced immobilization of DNA on the electrode surface or, when modified with electroactive markers or enzymes, they may significantly improve the DNA recognition and transduction events. These actions may, in turn, be used as an ultrasensitive method for electric biosensing of DNA [48,50].

Carbon Pastes Carbon nanotubes can be incorporated into the electrode in the form of a paste, in similar fashion to a simple carbon paste electrode (CPE). This approach combines the advantages of the CNT material with the attractivity of the carbon paste, and provides the feasibility to incorporate different substances, low background currents, easy renewal, and composite nature [51]. As a binder, traditional substances such as bromoform [52], mineral oil [51,53] or nujol [54] can be used. Teflon was also reported as a binder in a CNT/Teflon composite [55]. The CNT paste electrodes (CNTPE) exhibit excellent electrochemical catalytic properties towards biologically active materials, and this enables studies to be conducted of the electrochemical behavior of biomolecules such as DNA [56].

DNA can be immobilized on the surface of CNTPEs, prepared by mixing MWNT powder and mineral oil in the ratio 60:40, by the electrically stimulated adsorption from its solution in acetate buffer. It has been shown that the CNTPE is a suitable tool for adsorptive stripping measurements of trace levels of nucleic acids, and it also provides an enhanced signal of guanine oxidation. Moreover, free guanine can be adsorbed at CNTPEs under certain conditions at which no adsorption is observed at conventional CPEs. The interaction between nucleic acids and CNTPEs has mainly a hydrophobic character.

Another means of immobilizing DNA is to mix it with the CNTs and binder [57]. In this way, a DNA biosensor which was very sensitive for dopamine was constructed, with a dopamine detection limit of 2.1×10^{-11} mol L^{-1}. DNA may also be immobilized on the electrode surface by means of entrapment in the conducting polymer, using electrochemical polymerization. Because of its many attractive properties (good ion-exchange capacity, strong adsorptive capability, good conductivity), polypyrrole (ppy) is often used [58]. In this respect, several methods can be used for DNA immobilization, including simple adsorption or covalent binding within amino groups on the ppy surface as well as entrapment in the electropolymerized ppy. A DNA probe is employed as the counter anion in the polymer matrix due to its negative charge. In order to prepare the DNA biosensor with high sensitivity and selectivity, ssDNA was incorporated into the electropolymerized polypyrrole on the surface of a MWNT paste electrode [59]. The biosensor was prepared by covering the MWNT paste electrode with a ppy film which had been electropolymerized during a repetitive cyclic voltammetric scans in a solution containing pyrrole and DNA oligonucleotide. This biosensor was used for the electrochemical detection of DNA hybridization using ethidium bromide as an electrochemical indicator.

Simple adsorption was used to immobilize ssDNA onto an MWNT-modified screen-printed carbon electrode (SPCE) [50]. Cyclic voltammogram of the ssDNA/MWNT/SPCE showed two peaks which could be attributed to the oxidation of guanine and adenine residues of ssDNA (Figure 11.4). A negative shift of both

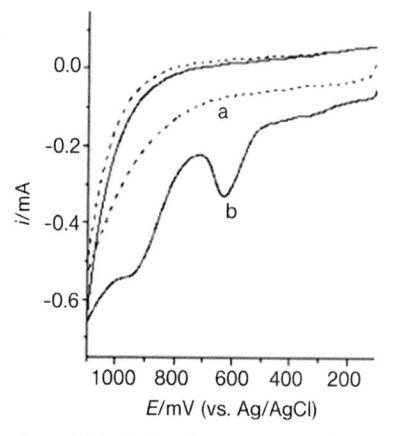

Figure 11.4 Cyclic voltammograms of (a) MWNT/SPCE and (b) ssDNA/MWNT/SPCE in 0.1 M phosphate buffer solution, pH 5.5 at 50 mV s^{-1}. (From Ref. [50] with kind permission of Elsevier.)

peak potentials indicated that ssDNA could be oxidized more easily at MWNT/SPCE than at pre-treated glassy–carbon electrode (GCE) (+0.8 V for adenine, +1.1 V for guanine [60]). This method allowed for an indicator-free detection of ssDNA via signals of guanine.

SPCE were modified with SWNTs dispersed in DMF and covalenty immobilized DNA [61]. The new detection protocol combined the advantages of the recognition ability of protein and the electrochemical activity of DNA was found. Changes in the intrinsic signal of DNA were monitored using binding DNA to single-strand binding protein. Described protocol offers the electrochemical label-free detection of the DNA hybridization. The adsorption of MWNTs – or a mixture of MWNTs and DNA – onto the surface of SPCE was also described [62]. For this, two means of modification were used, namely layer-to-layer coverage and composite (mixed) coverage. For the first approach, the MWNTs suspension was cast onto the electrode surface and, after drying, the DNA layer was added. The basic electrochemical characteristics of the prepared electrodes were obtained by using the signals of a $[Co(phen)_3]^{3+}$ marker, $[Fe(CN)_6]^{3-}$ present in solution and the guanine moiety; these signals were compared to those obtained with electrodes modified by other nanostructured materials (montmorillonite and hydroxyapatite). The MWNTs were shown to greatly enlarge the surface area, and this resulted in higher electrochemical signals than in case of other modifiers and the unmodified electrode (Figure 11.5). Thus, a composite coverage of the electrode surface by a mixture of MWNTs and DNA appears to be much more effective, most likely due to a better access of DNA by the marker particles within the nanostructured film.

Impedance spectroscopic measurements showed good sensitivity for the detection of DNA damage. The biosensor was used successfully to detect DNA damage caused by tin(II) and arsenic(III) compounds [63]. Compared to the simple DNA biosensor, the MWNTs effectively enlarged the signal of the electrochemical DNA marker $[Co(phen)_3]^{3+}$ and provided a suitable detection window for observing the damage to DNA. Both, MWNTs and MWNTs mixed with gold nanoparticles (GNPs) were dispersed in SDS solution and DMF [64]. When the composite and layer-to-layer

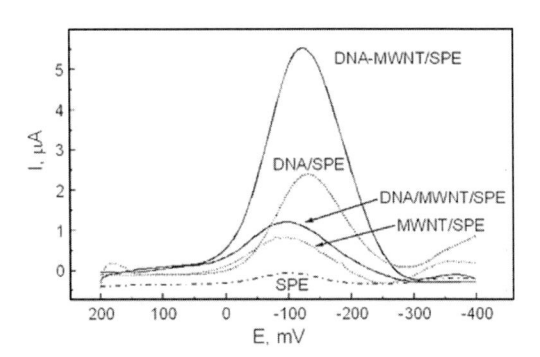

Figure 11.5 Comparison of the differential pulse voltammetric signals of $[Co(phen)_3]^{3+}$ measured at different electrodes. Conditions: 5×10^{-7} M $[Co(phen)_3]^{3+}$ in 5×10^{-3} M phosphate buffer pH 7.0, accumulation 120 s in open circuit.

electrode coverages were tested, the results showed that SDS led to a good dispersion of MWNTs in an aqueous medium. In comparison to simple MWNTs, GNPs–MWNTs nanohybrids showed no significant change in the biosensor properties. The DNA–MWNTs biosensor was successfully applied to the detection of damage to DNA caused by a berberine derivative, thus proving that such DNA biosensors may in general act as effective chemical toxicity sensors.

Solid Electrode Modification CNTs effectively enlarge the transducer surface area and offer a good environment for DNA immobilization on glassy–carbon and metal electrodes. Several methods of CNT and DNA immobilization were utilized. Very often, pre-treated CNTs (using mineral acids or ultrasonication) in the form of dispersion are simply cast onto the electrode surface to form a CNT film. For example, a MWNT dispersion prepared by means of surfactant dihexadecyl-hydrogen-phosphate (DHP) was cast onto the surface of a GCE [65]. This electrode showed strong electrocatalytic activity towards the oxidation of adenine, guanine which was evident from the remarkable enhancement in peak current and lowering of the oxidation potential (Figure 11.6A). A very good adsorptive ability of the MWNT-modified electrode was observed. Moreover, in the case of dsDNA detection, no electrochemical response was observed at the bare GCE; however, at the MWNT-modified GCE, some well-defined oxidation peaks which were attributed to the oxidation of guanine and adenine content of DNA were observed, notably after 2 min accumulation (Figure 11.6B). These authors also highlighted several advantages of the scheme, including direct detection, high sensitivity, rapid response, excellent reproducibility, and extreme simplicity.

Another method of preparing DNA–CNTs biosensors is to use a self-assembly approach. This enables the electrode surface to be easily functionalized by forming a highly organized and well-defined monolayer film. The flexibility for designing

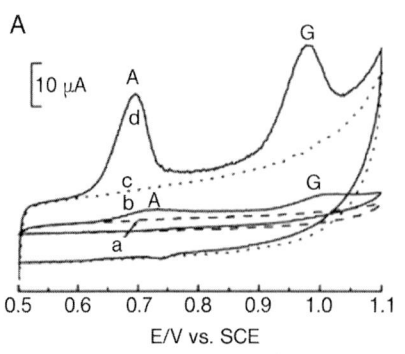

Figure 11.6 (A) Comparison of cyclic voltammograms (CV) of guanine (G) and adenine (A) at: (curve a) DHP/GCE; (curve b) bare GCE; (curve d) MWNT-DHP/GCE. Curve (c) shows CV in blank solution at MWNT-DHP/GCE. (B) Comparison of cyclic voltammograms of dsDNA at: (curve a) bare GCE; (curve b) DHP/GCE; (curve c) MWNT-DHP/GCE; and (curve d) MWNT-DHP/GCE after 2 min accumulation. (From Ref. [65] with kind permission of Springer Science and Business Media.)

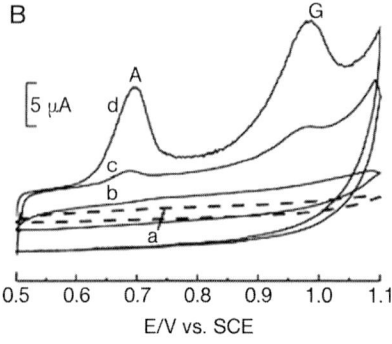

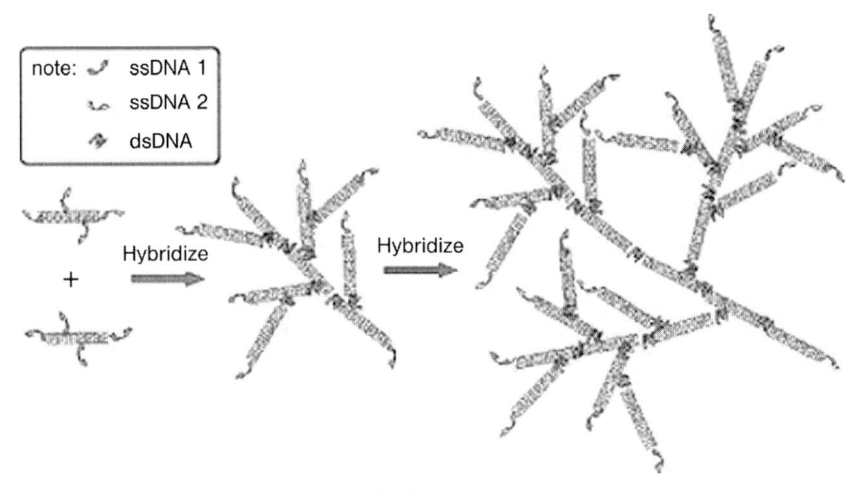

Figure 11.7 Schematic representation of the formation process of the self-assembling of dsDNA-SWNTs. (From Ref. [69] with kind permission of Elsevier.)

different head groups of monolayers by large numbers of electroactive or electroinactive functional groups makes this method especially useful for the preparation of biosensors [66–68]. On finding that DNA may guide the assembling of SWNTs, the process was studied using atomic force microscopy (AFM) [69]. The SWNTs were functionalized with complementary chains, ssDNA1 and ssDNA2, and the hybridization of ssDNA1–SWNTs with ssDNA2–SWNTs was performed such that the hybridized dsDNA–SWNTs product was observed (Figure 11.7). Subsequently, well-defined branched structures were identified, while much looser and more randomly distributed structures were observed in solutions containing unfunctionalized SWNTs or ssDNA–SWNTs. This method can be used in the construction of desired nanoscale architectures of SWNTs for various electrical and molecular sensing applications.

DNA may also play the role of linker to control the self-assembly of SWNTs when preparing electronic devices [70]. Both, the Au contacts and SWNTs are modified with ssDNA by means of thiol-groups, after which hybridization between the complementary ssDNA chains takes place (Figure 11.8). This method permits the simple production of hundreds of devices with high yields, while the measured currents are larger by two orders of magnitude than values reported for direct metal–SWNTs contacts.

Self-assembled MWNTs can be also produced using direct growth on Au substrates by using the chemical vapor method [71]. It was first shown that the nanotubes grew vertically to Au substrate, after which the carboxylic acid groups could be introduced onto the MWNTs surface to enable an immobilization of the DNA probe by forming covalent amide bonds. The prepared biosensor was applied to detect the hybridization process between the DNA probe and target DNA, using methylene blue as an electrochemical indicator. The self-assembled MWNT-based

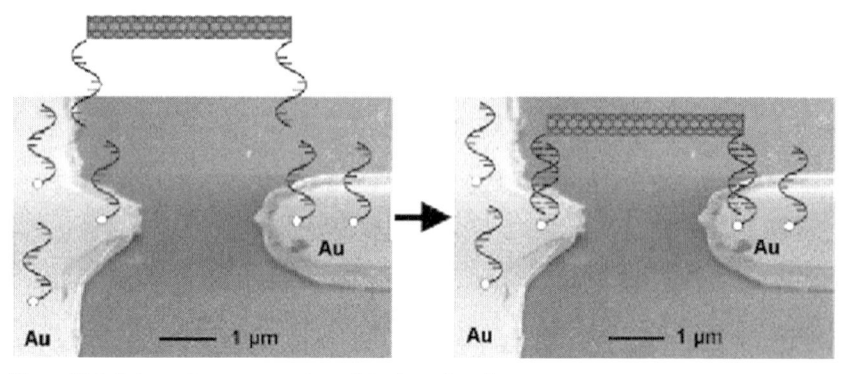

Figure 11.8 Schematic representation of DNA-mediated deposition of SWNTs between two gold electrodes. (From Ref. [70] with kind permission of Elsevier.)

biosensor showed a higher hybridization efficiency in comparison to the random MWNT-based biosensor.

The self-assembly method can be used to immobilize not only CNTs but also the DNA probe [72]. A thiolated ssDNA probe was attached to a Au–CNT hybrid (Figure 11.9), whereupon the hybrid was shown to be a compatible heterostructure for the self-assembly of thiolated DNA where gold nanoparticles served as the anchoring sites. The immobilization process was monitored using electrochemical impedance spectroscopy and voltammetry. The biosensor obtained was able to recognize the complementary and mismatched hybridization events by means of the catalytic oxidation of guanine using $[Ru(bpy)_3]^{3+}$ as the redox-active mediator. The hybridization event was enhanced in the presence of mercaptohexanol, which also displaced non-specifically adsorbed DNA.

The covalent interaction of DNA molecules and CNTs was also employed in the construction of biosensors. Both, gold electrodes [73] and GCEs [74] were modified

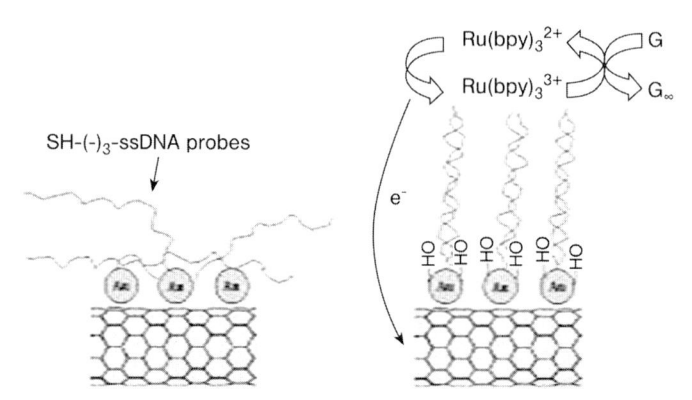

Figure 11.9 Schematic illustration of self-assembly of thiolated oligonucleotides onto Au–CNT hybrid. (From Ref. [72] with kind permission of Elsevier.)

by dropping the MWNT suspension onto the electrode surface, and drying. Calf-thymus DNA was then immobilized on the MWNTs via a diimide-activated amidation between carboxylic acid groups on the MWNTs and amino groups on the DNA bases. An interaction between immobilized DNA and a molecule such as ethidium bromide [73], as well as a hybridization reaction using daunomycin as an intercalative indicator [74], were investigated electrochemically.

Carbon nanotubes are also able to grow directly on graphite electrodes [75]. Following the modification with ethylene diamine, dsDNA can be electrochemically immobilized onto the surface. Here, electrostatic forces can help attach the DNA to the MWNT-modified gold electrode by means of a cationic polyelectrolyte [76]. The DNA biosensors were used to determine the presence of low-molecular-weight molecules such as promethazine hydrochloride [75] and chlorpromazine hydrochloride [76], both of which are able to intercalate into DNA.

A mixture of MWNT dispersion and DNA solution was cast onto the surface of platinum electrode, and allowed to evaporate to dryness [77]. The DNA/MWNT layer obtained in this way was very stable. By incubating the DNA/MWNT-modified electrode in cytochrome c solution, it was immobilized uniformly due to electrostatic interactions between the negatively charged DNA/MWNT layer and the positively charged cytochrome c. Moreover, MWNTs played the key role in promoting the redox process of cytochrome c.

A GCE was modified with a random dispersion of bamboo-type CNT (BCNT), and compared with both modified SWNTs and unmodified GCEs [78]. A dispersion of CNTs was cast onto the electrode surface and allowed to dry by evaporation. DNA was adsorbed from solution using an applied electric potential, after which the oxidation of guanine and adenine bases of DNA was evaluated to compare all three electrodes. The much higher peak currents for both guanine and adenine (Figure 11.10) indicated that BCNTs have a far higher number of electroactive sites than do SWNTs.

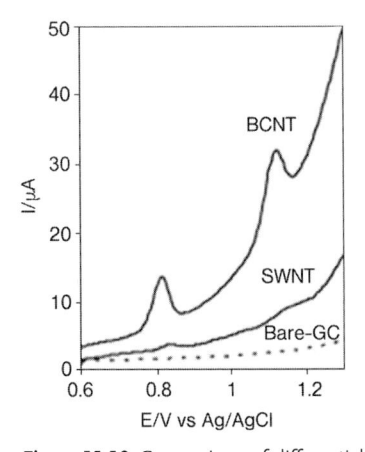

Figure 11.10 Comparison of differential pulse voltammetric signals of adenine and guanine bases at bare GCE, SWNT/GCE, and BCNT/GCE. (From Ref. [78] with kind permission of Elsevier.)

Carbon nanotubes were also immobilized on the surface of electrode with the help of cyclodextrins (CDs). These compounds can effectively enhance the dispersion of CNTs [79,80] and immobilize them by an entrapment within the polymer [81]. The ability of CDs to form supramolecular complexes is well recognized, as they are known to serve as efficient molecular receptors. A composite containing both CDs and CNTs offers the advantages of both materials. Thus, complexes between CDs and CNTs were studied using nuclear magnetic resonance (NMR) or Raman spectroscopy, and evidence of an intermolecular interaction between γ-CD and SWNTs was thus found [79]. It was also shown that SWNTs could be solubilized by threading with large-ring CD (η-CD) [80]. Many applications of the CD/CNT film-coated electrodes for the determination of biological and other organic molecules, including dopamine and epinephrine [82], thymine [83], dopamine [84] or rutine [85], have been reported. In addition, a DNA recognition process was also reported. The modified graphite electrode was prepared by casting the mixture of MWNTs and β-CD onto the surface and evaporating to dryness [86,87]. The sensor was then applied to determine free guanine and adenine levels. In the case of DNA determination, well-defined peaks of guanine and adenine moieties were found after 60 s accumulation at an open circuit. In addition, it was shown that β-CD at the electrode surface could serve as a filter membrane [87].

Conducting polymers have also been used advantageously in the preparation of DNA biosensors, as they can enhance the chemical compatibility of CNTs as well as their solubility. One of the most extensively used polymers is polypyrrole (ppy), which can be generated electrochemically and deposited onto the conducting surfaces. The main advantage of a prepared ppy film is its uniformity and possibility of controlling the film thickness and morphology by an applied current or potential [88]. The ppy film can be deposited from neutral pH, with two of its main features being good conductivity and the contribution to a sufficient stability of the biosensors. However, whilst many applications of ppy in the design of DNA biosensors have been reported, few applications have been determined for CNT-based DNA biosensors.

For these studies, two approaches were used. In the first method, the GCE was modified with a suspension of MWNTs functionalized with carboxylic groups. The ppy film doped with DNA oligonucleotide was then formed using electropolymerization from the solution containing pyrrole and DNA [89,90], with electrochemical impedance being used to detect hybridization. Trapping of the DNA oligonucleotide in the ppy film allowed the ssDNA to hybridize target DNA sequences more easily. In the second method, the ppy film doped with MWNTs was prepared on the surface of the GCE by electropolymerization from the solution containing pyrrole and MWNTs [91]. The film was covered with the DNA probe conjugated with magnetite nanoparticles to form the DNA hybridization biosensor.

An alternative polymer which may be used for DNA–CNTs biosensors is that of chitosan, a cationic polymer which is able to enhance CNTs dispersion [92]. Chitosan also forms a suitable environment for further simple modification and the efficient immobilization of biomolecules [93]. The properties of composite of CNTs and chitosan were studied using microscopic methods and X-ray diffraction (XRD) [94]. DNA is effectively immobilized on the polycationic polymer film of chitosan by means of electrostatic attractions, with non-specific DNA adsorption being avoided by using a high ionic strength [95]. The chitosan–MWNTs solution was spread uniformly on the surface of the graphite electrode and evaporated to dryness [96]. The DNA was then immobilized on the surface of the modified electrode by simple adsorption from solution. As a cationic polymer, chitosan can adsorb anionic chemicals such as CNTs containing carboxylic groups. Therefore, CNTs may be dispersed uniformly in aqueous solutions of chitosan, thus enhancing the stability of the CNTs solution. The formation of DNA/chitosan–CNTs film was confirmed by monitoring the $[Fe(CN)_6]^{3-}/[Fe(CN)_6]^{4-}$ redox couple.

In order to modify the properties of CNTs, they can be doped with various materials, such as metal nanoparticles, redox mediators, and polymers. These conjugates offer a combination of the properties of both materials used, which results in a synergistic effect. DNA was immobilized on the electrode modified with a MWNT/nanoporous ZrO_2/chitosan composite film to form the DNA biosensor [97]. The nanoporous ZrO_2 and MWNTs were shown to have a synergistic effect on the redox behavior of daunomycin. Due to the effectively enlarged surface area and good charge-transport characteristics, an increased ssDNA loading quantity and enhanced detection sensitivity for the DNA hybridization were observed.

Carbon nanotubes can be also used as carriers of nanoparticle tracers, such as CdS [98] and Pt nanoparticles [99,100], for the detection of DNA hybridization. Wang et al. demonstrated a method for amplifying the electrical detection of DNA hybridization [98]. For this purpose, the CNTs side walls were covered with CdS nanocrystals by means of hydrophobic interactions. A major enhancement of the stripping-voltammetric signal of CNT-loaded CdS tags in comparison to single CdS tracer was subsequently observed (Figure 11.11).

The MWNTs suspension was mixed with the Pt solution and the resulted mixture was cast onto the surface of the GCE [100]. After drying, the electrode was incubated in DNA solution to immobilize the DNA probe. When the electrochemical properties of the modified electrode were studied using cyclic voltammetry, the results indicated a larger effective surface of the MWNT/Pt-modified electrode than that of

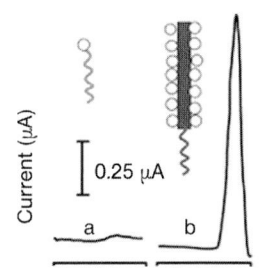

Figure 11.11 Stripping-voltammetric cadmium hybridization signals obtained using a single CdS tracer (a) and with CNT-loaded CdS tags (b). (From Ref. [98] with kind permission of Elsevier.)

the Pt- or MWNT-modified electrodes. A high selectivity of the MWNT/Pt-based hybridization assay was also observed.

11.2.2
Fullerenes

Fullerene and its derivatives represent a class of carbon materials which is beginning to attract much interest in many fields of science. The biological properties and activity of fullerenes and their derivatives, such as DNA cleavage, anti-viral activity, electron-transfer, are well known and have been highlighted [101,102]. Until now, the analytical applications of fullerenes have not been widely recognized [103]. However, their unique structure and properties – which are similar to those of CNTs – make them an attractive material for use as sorbents or chromatographic stationary phases. In addition, in the development of chemically modified sensors and biosensors, the deposition of fullerenes onto various substrates may also be of interest with regards to the decreased resistivity of electrode materials.

Fullerenes (see Figure 11.1C) are closed-cage carbon molecules with pentagonal or hexagonal rings. Their electrocatalytic properties and applications in the preparation of chemically modified electrodes have been widely reviewed [104,105]. Fullerene-modified electrodes demonstrate useful electrocatalytic properties for both chemical and biochemical reactions [104], and consequently they may be used as electron-transfer mediators in biosensors. For example, C_{60}:γ-CD was found to mediate electron transfer to DNA, showing two-way activity towards DNA [106]. In this case, the GCE was modified with complex C_{60}:γ-CD by simple adsorption. As this complex was soluble in water, the modified surface was covered with Nafion in order to prevent leaching of the C_{60}:γ-CD layer. This electrode had good stability and reproducibility; moreover, in the absence of C_{60} no redox wave corresponding to DNA was found.

The interaction between C_{60} derivatives and DNA was investigated by using a DNA-modified gold electrode [107]. As the C_{60} derivative is non-electroactive, the $[Co(phen)_3]^{3+/2+}$ redox pair was used as an electroactive indicator to study this interaction. The kinetics of binding and dissociation were studied using cyclic voltammetry, whereupon the binding targets of the C60 derivatives were found to be the major groove of the double helix and the phosphate backbone of dsDNA.

An ITO electrode was used to study a C_{60}-derivative/porphyrin/DNA complex deposited on the electrode surface by polymerization of 3,4-ethylenedioxythiophene [108]. A porphyrin derivative was shown to bind to DNA as an intercalator, and both the fullerene and porphyrin derivatives were seen to be entrapped by the DNA scaffold. The DNA-modified electrode was used for the electrochemical detection of 16S rDNA extracted from *Escherichia coli* [109]. Here, the DNA probe was immobilized onto a fullerene-impregnated screen-printed electrode which was first activated by exposure to air plasma. A good enhancement of the signal of $[Co(phen)_3]^{3+}$ due to the incorporation of fullerene was observed. In addition, the DNA biosensor was also capable of detecting target oligonucleotides in the presence of mismatching oligonucleotides.

11.2.3
Diamond and Carbon Nanofibers

11.2.3.1 **Diamond**

Although diamond (see Figure 11.1D) is known as a "superhard" material, it has – in similar fashion to the above-mentioned carbon materials – unique electrochemical properties that are of interest in the preparation of modified electrodes. The basic electrochemical properties of diamond films as electrodes, studied by electrochemical and impedance spectroscopy measurements as well as electrode kinetics, have been reviewed by Pleskov [110]. The practical applications of diamond electrodes as detectors for chromatographic and flow-injection methods are also described.

Diamond is usually doped with an acceptor to enhance its electrochemical properties. The electrochemical behavior of three carbon surfaces – boron-doped diamond (BDD), glassy carbon (GC), and pyrolytic graphite (PG) – were compared and demonstrated for the electrochemical oxidation of 4-nitrophenol [111]. An investigation of ferrocyanide redox behavior has shown that the charge-transfer resistance and capacitive currents of BDD are lower than those of GC and PG. In addition, the best detection limit, repeatability and reproducibility were obtained with the BDD electrode.

As diamond has a good biocompatability [112], its surface can be modified with biomolecules. The covalent immobilization of DNA on diamond was investigated using diffuse reflectance infrared spectroscopy [113], while several other investigators have used XPS and impedance spectroscopy to investigate DNA-modified diamond thin films [114–117]. A high stability and sensitivity of the DNA-modified diamond, as well as properties useful for the detection of DNA hybridization events, were also observed. Chips based on diamond produced by chemical vapor deposition (CVD) were covered with DNA using a solidification technique which enables DNA to be bound vertically to the substrate [118]. Subsequently, the amount of oligonucleotide on the CVD diamond chips was shown to be higher than on silicon chips, an outcome that may be beneficial for developing microarray-type DNA chips for DNA diagnostics.

DNA hybridization biosensors based on the BDD film were prepared on a Si substrate and covered, using electropolymerization, with a thin layer of polyaniline/poly(acrylic acid) composite [119]. The DNA probe was immobilized by incubation of the sensor in DNA solution. The carboxylic groups in the polymeric film were found to act as active sites for covalent binding of the DNA probe. Non-specific adsorption of DNA onto polymeric film was not observed, however, and the DNA biosensor showed an appropriate stability and selectivity to DNA sensing. The BDD film electrode with a low background current was used to study the electrochemical behavior of native and thermally denatured fish DNA [120]. Two well-defined peaks corresponding to the oxidation of guanine and adenine residues were observed on cycling voltammogram of the thermally denatured DNA. In addition it was found that, in the presence of cytosine, the guanine peak was decreased, which indicated an interaction of cytosine with denatured DNA via hydrogen bonds. It was also shown that the cationic porphyrins could stabilize the denatured DNA by intercalation and ionic interactions.

An aminophenyl-modified BDD electrode was covered with cross-linker and thiol-modified DNA oligonucleotide probe [121]. The presence of a DNA probe was confirmed by DNA hybridization using fluorescein-labeled complementary/non-complementary target DNA oligonucleotides. Electrochemical AFM measurements were used to characterize the DNA functionalized and hybridized surfaces, and to show that both closed and dense DNA films were obtained. The BDD electrode was also used as a high-performance liquid chromatography (HPLC) detector for the detection of purines and pyrimidines [122]. Well-defined oxidation peaks of cytosine and thymine were observed due to a wide potential window of the BDD electrode, and both low detection limits and high sensitivity and stability were also reported. Subsequently, the electrode was successfully applied to the determination of 5-methylcytosine in a DNA sample after HPLC analysis; acceptable recoveries of approximately 95% was achieved, with good reproducibility.

11.2.3.2 Carbon Nanofibers

Carbon nanofibers represent another form of intensively studied carbon material [123]. For example, they are not only attractive materials for catalysis [124], but their unusual electrochemical properties [125,126] also attracts interest for use as novel electrode materials in electrochemical applications [127]. Vertically aligned carbon nanofibers have been shown to be useful for applications in chemical sensors and biosensors [128], and several uses in enzyme [129] and DNA biosensors [130] have been reported. In the latter situation, the nanofibers surface were modified with amino groups to bind the thiol-modified DNA probe [131], and the biosensor was used to recognize fluorescently labeled target DNA complementary to the immobilized probe.

Vertically aligned carbon nanofibers were also covalently modified with DNA, using photochemical and chemical methods [132]. Both, excellent specificity and reversibility were observed in recognizing complementary and non-complementary sequences. The carbon nanofibers were found to possess an enhanced surface area, in addition to an approximately eight-fold higher amount of hybridized DNA in comparison with GCEs. Vertically aligned carbon nanofibers were also modified with DNA through the carboxylic groups on nanofibers, and used for the direct physical introduction and expression of exogenous genes in mammalian cells [133]. Transcriptional accessibility of DNA was investigated using polymerase chain reaction and *in-vitro* transcription.

11.2.4 Clays

Clays are minerals which belong to the group of phyllosilicates, and their application in chemical sensors has been widely reviewed [134–136]. Clays can act as ion-exchangers, and therefore can serve as useful materials in the preparation of modified electrodes [137]. The ion-exchange properties of different clays incorporated into CPEs were investigated using electrochemical methods [138,139]. Because of their adsorption properties, clays can also serve as a component of biosensors, and many applications of clays in enzyme electrodes have been described [136]; applications in DNA biosensors are still very rare, however.

A modified CPE was prepared using a sodium montmorillonite (MMT) as the modifier [140]. The electrode was used to investigate the electrochemical behavior and determination of guanine. The MMT-modified CPE showed better electron-transfer in comparison to unmodified CPE, resulting in an increase in the guanine peak current as well as lowering the oxidation overpotential. Carbon paste-based screen-printed electrodes were modified with unmodified MMT and modified montmorillonite (MMTmod) in two ways [62]: (i) MMT or MMTmod dispersed in polyvinylalcohol was cast onto the electrode surface and, after drying, covered with a DNA layer (layer-to-layer covering); and (ii) the mixture of MMT or MMTmod dispersion and DNA solution was cast onto the electrode surface and allowed to evaporate to dryness (composite covering). Voltammetric signals of the $[Co(phen)_3]^{3+}$ marker, guanine residue and $[Fe(CN)_6]^{3-}$ were used to evaluate the prepared DNA biosensors. The results showed that the composite covering was more effective because of better access of DNA to the marker particles within the nanostructured films of enhanced active surface area. A comparison of the MMT and MMTmod modifiers showed that the latter exhibit a higher efficiency than the former, due to an increased interlayer distance in the nanomaterial.

11.2.5
Metal Nanoparticles

Metal nanoparticles (NPs) are clusters of a few hundred to a few thousand atoms that are only a few nanometers long [141]. They are known as "dye" compounds and, due to their unique physical properties (which include a high surface-to-bulk ratio based on their small size) they are of major interest for sensing [142]. Among the various NPs, metallic gold and silver are the most often used materials for the preparation of nanoparticles.

Gold nanoparticles (GNPs) can be prepared either electrochemically or non-electrochemically via chemical reduction (usually reduction of Au(III) from $HAuCl_4 \cdot 3H_2O$ to Au(0) using $NaBH_4$), which often produces colloidal gold [143]. Colloidal gold consists of octahedral units of gold homogeneously dispersed in the liquid phase. GNPs can be modified with biomolecules such as oligonucleotides or proteins, and are frequently used to label DNA oligonucleotides and to enhance the sensitivity of the detection of DNA hybridization events [144–146]. By modifying gold with DNA, a significant stabilization of the gold colloid was observed, which represents a prerequisite for further biochemical and/or molecular–biological manipulations [147]. GNPs are hydrophobic, and therefore must be modified in order to become water-soluble (hydrophilic), then they may be attached to biological molecules. The attachment of oligonucleotides to the surface of GNPs can be performed by simple adsorption, via biotin–avidin linkage and via thiol–gold bonds. Such nanoparticles can be used as quantitation tags, as well as encoded electrochemical hosts [146].

Quantum dots (QDs) are also nanostructured materials known as zero-dimensional material [148]. A quantum dot is a location that can contain a single electrical charge, a single electron. The presence or absence of an electron changes the properties of QDs, and they can then be used for several purposes, including information storage or as transducers in sensors. QDs are semi-conductor nanocrystals that are roughly spherical in shape, with diameters between 1 and 12 nm. Because of their reduced size, QDs behave differently from bulk solids due to the quantum-confinement effects that are responsible for their remarkably attractive properties. Due to the availability of precursors and the simplicity of crystallization, CdS and CdSe are normally used for the preparation of QDs. Yet, QDs can be prepared by several methods based on pattern formation (colloidal self-assembled pattern formation by surfactant micellation), organometallic thermolysis or electrochemical deposition. They can also be modified with DNA or proteins by simple adsorption, linkage via thiol groups, electrostatic interaction, and covalent linkage via streptavidin–biotin bonds.

Due to its ultrahigh surface area, colloidal gold can be used advantageously to improve the immobilization of DNA on the electrode surface. Electrochemical DNA hybridization biosensor based on binding events between *Escherichia coli* ssDNA binding protein (SSB) and ssDNA conjugated to GNPs was prepared [149]. The recognition ability of SSB and the electrochemical properties of GNPs, streptavidin-coated beads and biotin-modified CPE were studied, and a specific discrimination of streptavidin-coated beads towards mismatched and non-complementary DNA sequences was observed. The GNPs enabled a lowering of the detection limits to 2.17×10^{-12} mol L^{-1}.

The effect of GNPs on the interaction between the DNA and [Co(phen)$_3$]$^{3+}$ was investigated [150]. For this, GNPs of different size were assembled onto the surface of gold disc electrodes through dithiol molecules, and the DNA biosensor was prepared by casting the DNA solution onto a GNP-modified surface. Subsequently, the concentration of DNA adsorbed onto the electrode surface was found to depend on the GNP size. The DNA biosensor was tested in aqueous solution, as well as a non-aqueous environment containing acetonitrile, and well-developed redox peaks of [Co(phen)$_3$]$^{3+}$ were obtained in both media. By increasing the GNP size, a decrease in not only the concentration of the DNA adsorbed onto the GNP-modified surface but also as in the redox currents of [Co(phen)$_3$]$^{3+}$, was observed.

A DNA hybridization biosensor was prepared by the immobilization of target DNA onto colloid GNPs self-assembled on the cysteamine monolayer-modified gold electrode [151]. The GNPs significantly enlarged both the electrode surface and the amount of immobilized ssDNA. The oligonucleotide probe was modified with silver nanoparticles that were released after the hybridization process. The electrochemical signal of solubilized Ag(I) in the case of one complementary oligonucleotide sequences was much higher than that at the oligonucleotide sequences containing the single-base mismatch. An anodic stripping voltammetric determination resulted in the detection limit of the target nucleotide falling to 5×10^{-12} mol L^{-1}.

The target ssDNA was immobilized on the surface of GCE by the formation of stable electrostatic complex with chitosan [152]. The hybridization reaction was conducted by immersing the prepared biosensor into the medium of the GNP-modified DNA probe. Following hybridization, the electrochemical signal of the GNPs was enhanced by their modification with the silver particles (Figure 11.12A). Moreover, the sensitivity of the biosensor was increased by about two orders of magnitude due to the silver enhancement. This biosensor was successfully applied to the recognition of an oligonucleotide sequence from the target with single nucleotide mismatches (Figure 11.12B).

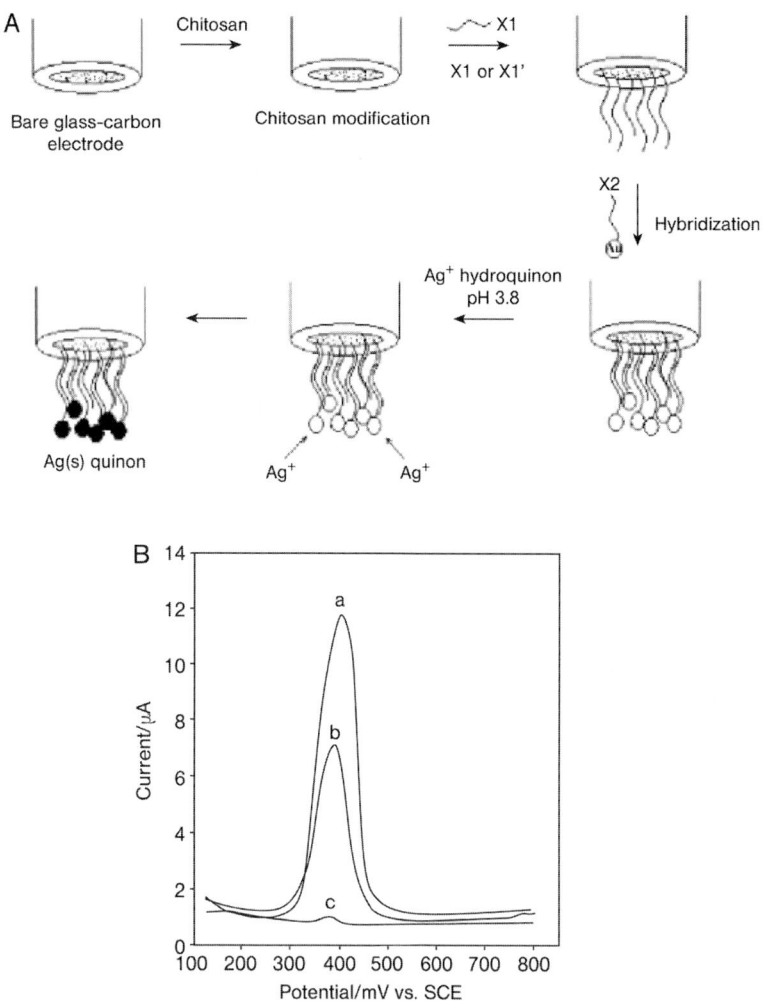

Figure 11.12 (A) Schematic diagram of silver-enhanced colloidal gold electrochemical detection of DNA hybridization. (B) Differential pulse voltammetric response for dsDNA detected by monitoring the silver after hybridization of GNPs-labeled DNA oligonucleotides with complementary oligonucleotides (a), with oligonucleotides containing single-base mismatch (b) and with non-complementary oligonucleotides (c). (From Ref. [152] with kind permission of Elsevier.)

An electrodeposited DNA membrane doped with GNPs was prepared on the surface of a GCE [153]. The catalytic activity of this biosensor towards the oxidation of norepinephrine (NE) was monitored, and the biosensor was used to determine NE concentrations in the presence of ascorbic acid.

A GCE modified with a poly-2,6-pyridinecarboxylic acid film was further modified with GNPs by a combination of electrodeposition and adsorption of nanogold [154]. The ssDNA probe was then immobilized onto the modified electrode by adsorption

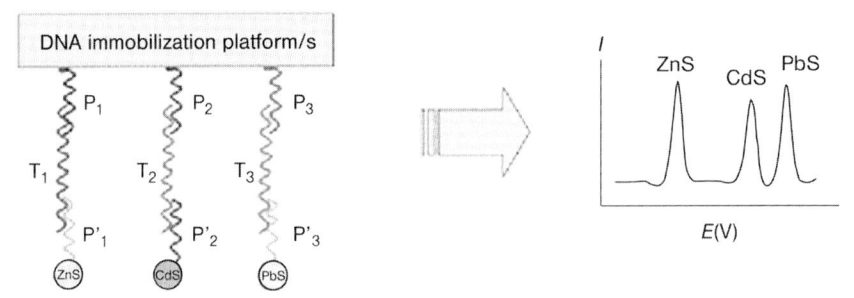

Figure 11.13 Schematic of multiple detection of DNA. P′$_1$, P′$_2$, P′$_3$ are DNA probes connected to QDs; T$_1$, T$_2$, T$_3$ are DNA targets hybridized with corresponding DNA-capturing probes (P$_1$, P$_2$, P$_3$). (From Ref. [146] with kind permission of Elsevier.)

from its solution, and the hybridization process was detected using electrochemical impedance spectroscopy. The difference in surface electron transfer resistance in [Fe(CN)$_6$]$^{3-/4-}$ solution measured on the DNA probe biosensor, and that on the hybridized electrode, was used for the evaluation. Hybridization of the DNA probe with complementary DNA was found rapidly to increase the surface electron transfer resistance, and the prepared biosensor was used to determine the sequence-specific phosphinothricin acetyltransferase gene, with a detection limit of 2.4×10^{-11} mol L^{-1}.

An electrical DNA hybridization device based on GNPs attached to oligonucleotide probes and closely spaced interdigitated electrodes was reported [155]. The oligonucleotide probe was attached in the gap between two microelectrodes, and hybridization of the target DNA and the second nanoparticle-coupled probe brought the GNPs into the gap. Follow-up silver deposition resulted in a conductivity signal only when DNA hybridization took place.

Quantum dots can also act, in similar manner to GNPs, as labels for the detection of DNA hybridization. For example, by attaching PbS, CdS and ZnS to various detection probe sequences, and subsequently stripping the labels at various potentials, the different target sequences can be detected and quantified (Figure 11.13) [146]. Similarly, multiplexed immunoassays of proteins with measurements of different antigens can also be performed.

11.3
Conclusions

In the ongoing development of biosensors, the use of nanotechnology and nanomaterials has raised new possibilities of controlling the properties of transducers, of matrices for the immobilization of biomolecules, and of markers, indicators, and other important building components. Many applications of biosensors for analyte determination have highlighted the use of nanomaterials, with a clear lowering of detection limits (Table 11.1). In particular, the miniscule dimensions of nanomaterials

Table 11.1 Applications of nanomaterials in DNA biosensors.

Nanostructured material	Dispersion agent	Immobilization method	Measured signal/analyte	Electrochemical method	Linear concentration range (M)	Limit of detection (M)	Reference
MWNTs	Mineral oil	Paste	Ethidium bromide	DPV	1×10^{-10}–1.1×10^{-8}	8.5×10^{-11}	[59]
MWNTs	HNO_3	Cast on GCE	α-Naphthol	CPN	(20–120) ppb	2 pg	[61]
MWNTs	DMF	Cast on GCE	Daunomycin	DPV	2.0×10^{-10}–5.0×10^{-8}	1×10^{-10}	[74]
MWNTs	DMF	Cast on GCE	Daunomycin	EIS	1×10^{-10}–1×10^{-6}	5×10^{-11}	[90]
MWNTs	–	Electropolymerization with ppy	Daunomycin	DPV	6.9×10^{-14}–8.6×10^{-13}	2.3×10^{-14}	[91]
MWNTs	Chitosan	Cast on GCE	Daunomycin	DPV	1.49×10^{-10}–9.32×10^{-8}	7.5×10^{-11}	[97]
MWNTs	Nafion	Cast on GCE	Daunomycin	DPV	2.25×10^{-11}–2.25×10^{-7}	1×10^{-11}	[100]
GNPs	–	Self-assembly	Au	SWV	3×10^{-11}–9.55×10^{-9}	2.17×10^{-12}	[149]
GNPs	–	Self-assembly	Dissolved silver	ASV	1×10^{-11}–8×10^{-10}	5×10^{-12}	[151]
CNTs	Mineral oil	Paste	Dopamine	SWV	1×10^{-8}–1.1×10^{-7}	2.1×10^{-11}	[57]
CNTs	–	Deposition on graphite substrate	Promethazine hydrochloride	SWV	2.5×10^{-8}–2.1×10^{-6}	–	[75]

ASV, anodic stripping voltammetry; CPN, chronopotentiometry; DPV, differential pulse voltammetry; EIS, electrochemical impedance spectroscopy; GCE, glassy–carbon electrode; ppy, polypyrrole; SWV, square-wave voltammetry.

enable the miniaturization of biosensors, the preparation of nanoelectrode arrays [156] and offer new environmental, biomedical and *in-vivo* applications [7,140,157,158].

Acknowledgments

The authors are grateful to the Grant Agency VEGA (Grant No. 1/2462/05), the Project of Applied Research of the Ministry of Education of the Slovak Republic, and the Research and Development Assistance Agency (Contract No. APVT-20-015904) for the financial support of these studies.

References

1 Sotiropoulou, S., Gavalas, V., Vamvakaki, V. and Chaniotakis, N.A. (2003) *Biosens. Bioelectronics*, **18**, 211.

2 Jianrong, C., Yuqing, M., Nongyue, H., Xiaohua, W. and Sijiao, L. (2004) *Biotechnol. Adv.*, **22**, 505.

3 Vamvakaki, V. and Chaniotakis, N.A. (2007) *Sens. Actuat. B* **22**, 193.

4 Rao, C.N.R. and Cheetham, A.K. (2001) *J. Mater. Chem.*, **11**, 2887.

5 Kuchibhatla, S.V.N.T., Karakoti, A.S., Bera, D. and Seal, S. (2007) *Prog. Mater. Sci.*, **52**, 699.

6 Singh, K.V., Pandey, R.R., Wang, X., Lake, R., Ozkan, C.S., Wang, K. and Ozkan, M. (2006) *Carbon*, **44**, 1730.

7 Xu, T., Zhang, N., Nichols, H.L., Shi, D. and Wen, X. (2007) *Mater. Sci. Eng. C*, **27**, 579.

8 Sun, Y. and Kiang, C.-H. (2005) DNA-based Artificial Nanostructures: Fabrication Properties, and Applications. in *Handbook of Nanostructured Biomaterials and Their Applications in Nanobiotechnology* (ed. H.S. Nalwa), Vol. 2, American Scientific Publishers, California,. pp. 224.

9 Fortina, P., Kricka, L.J., Surrey, S. and Grodzinski, P. (2005) *Trends Biotechnol.*, **23**, 168.

10 Iijima, S. (1991) *Nature*, **354**, 56.

11 Iijima, S. and Ichihashi, T. (1993) *Nature*, **363**, 603.

12 Paradise, M. and Goswami, T. (2007) *Mater. Design*, **28**, 1477.

13 Trojanowicz, M. (2006) *Trends Anal. Chem.*, **25**, 480.

14 Merkoci, A. (2006) *Microchim. Acta*, **152**, 157.

15 Smart, S.K., Cassady, A.I., Lu, G.Q. and Martin, D.J. (2006) *Carbon*, **44**, 1034.

16 He, P., Xu, Y. and Fang, Y. (2006) *Microchim. Acta*, **152**, 175.

17 Zhao, Q., Gan, Z. and Zhuang, Q. (2002) *Electroanalysis*, **14**, 1609.

18 Wang, J. (2005) *Electroanalysis*, **17**, 7.

19 Lin, Y., Yantasee, W., Lu, F., Wang, J., Musameh, M., Tu, Y. and Ren, Z. (2004) Biosensors Based on Carbon Nanotubes. in *Dekker Encyclopedia of Nanoscience and Nanotechnology*. (eds J.A. Schwarz, C. Contescu and K. Putye), Marcel Dekker, New York, p. 361.

20 Merkoci, A., Pumera, M., Llopis, X., Perez, B., del Valle, M. and Alegret, S. (2005) *Trends Anal. Chem.*, **24**, 826.

21 Gooding, J.J. (2005) *Electrochim. Acta*, **50**, 3049.

22 Gruner, G. (2006) *Anal. Bioanal. Chem.*, **384**, 322.

23 Wanekaya, A.K., Chen, W., Myung, N.V. and Mulchandani, A. (2006) *Electroanalysis*, **18**, 533.

24 Balasubramanian, K. and Burghard, M. (2006) *Anal. Bioanal. Chem.*, **385**, 452.

25 Rao, C.N.R., Satishkumar, B.C., Govindaraj, A. and Manaschi, N. (2001) *Chem. Phys. Chem.*, **2**, 78.
26 Barisci, J.N., Wallace, G.G. and Baughman, R.H. (2000) *Electrochim. Acta*, **46**, 509.
27 Barisci, J.N., Wallace, G.G. and Baughman, R.H. (2000) *J. Electroanal. Chem.*, **488**, 92.
28 Barisci, J.N., Wallace, G.G., Chattopadhyay, D., Papadimitrakopoulos, F. and Baughman, R.H. (2003) *J. Electrochem. Soc.*, **150**, E409.
29 Avouris, P. and Chen, J. (2006) *Mater. Today*, **9**, 46.
30 Saito, T., Matsushige, K. and Tanaka, K. (2002) *Physica B*, **323**, 280.
31 Hilding, J., Grulke, E.A., Zhang, Z.G. and Lockwood, F. (2003) *J. Disper. Sci. Technol.*, **24**, 1.
32 Lin, Y., Taylor, S., Li, H., Fernando, K.A.S., Qu, L., Wang, W., Gu, L., Zhou, B. and Sun, Y.-P. (2004) *J. Mater. Chem.*, **14**, 527.
33 Valcarcel, M., Simonet, B.M., Cardenas, S. and Suarez, B. (2005) *Anal. Bioanal. Chem.*, **382**, 1783.
34 Vaisman, L., Wagner, H.D. and Marom, G. (2006) *Adv. Colloid Interface Sci.*, **128–130**, 37.
35 Rink, G., Kong, Y. and Koslowski, T. (2006) *Chem. Phys.*, **327**, 98.
36 Gao, H. and Kong, Y. (2004) *Annu. Rev. Mater. Res.*, **34**, 123.
37 Song, C., Xia, Y., Zhao, M., Liu, X., Li, F. and Huang, B. (2005) *Chem. Phys. Lett.*, **415**, 183.
38 Gao, H., Kong, Y. and Cui, D. (2003) *Nano Lett.*, **3**, 471.
39 Cui, D., Ozkan, C.S., Ravindran, S., Kong, Y. and Gao, H. (2004) *Mechanics and Chemistry Biosystems*, **1**, 113.
40 Meng, S., Maragakis, P., Papaloukas, C. and Kaxiras, E. (2007) *Nano Lett.*, **7**, 45.
41 Ito, T., Sun, L. and Crooks, R.M. (2003) *Chem. Commun.*, 1482.
42 Dovbeshko, G.I., Repnytska, O.P., Obraztsova, E.D., Shtogun, Y.V. and Andreev, E.O. (2003) *Semicond. Phys. Quantum Electron. Optoelectron.*, **6**, 105.
43 Dovbeshko, G.I., Repnytska, O.P., Obraztsova, E.D. and Shtogun, Y.V. (2003) *Chem. Phys. Lett.*, **372**, 432.
44 Zheng, M., Jagota, A., Semke, E.D., Diner, B.A., Mclean, R.S., Lustig, S.R., Richardson, R.E. and Tassi, N.G. (2003) *Nature Mater.*, **2**, 338.
45 Zheng, M., Jagota, A., Strano, M.S., Santos, A.P., Barone, P., Chou, S.G., Diner, B.A., Dresselhaus, M.S., Mclean, R.S., Onoa, G.B., Samsonidze, G.G., Semke, E.D., Usrey, M. and Walls, D.J. (2003) *Science*, **302**, 1545.
46 Malik, S., Vogel, S., Rösner, H., Arnold, K., Hennrich, F., Köhler, A-K., Richert, C. and Kappes, M.M. (2007) *Compos. Sci. Technol.*, **67**, 916.
47 Baker, S.E., Cai, W., Lasseter, T.L., Weidkamp, K.P. and Hamers, R.J. (2002) *Nano Lett.*, **2**, 1413.
48 Daniel, S., Rao, T.P., Rao, K.S., Rani, S.U., Naidu, G.R.K., Lee, H.-Y. and Kawai, T. (2007) *Sens. Actuat. B*, **122**, 672.
49 Dwyer, C., Guthold, M., Falvo, M., Washburn, S., Superfine, R. and Erie, D. (2002) *Nanotechnology*, **13**, 601.
50 Ye, Y. and Ju, H. (2005) *Biosens. Bioelectron.*, **21**, 735.
51 Rubianes, M.D. and Rivas, G.A. (2003) *Electrochem. Commun.*, **5**, 689.
52 Britto, P.J., Santhanam, K.S.V. and Ajayan, P.M. (1996) *Bioelectrochem. Bioenerg.*, **41**, 121.
53 Valentini, F., Orlanducci, S., Terranova, M.L., Amine, A. and Palleschi, G. (2004) *Sens. Actuat. B*, **100**, 117.
54 Lin, X.-Q., He, J.-B. and Zha, Z.-G. (2006) *Sens. Actuat. B*, **119**, 608.
55 Wang, J. and Musameh, M. (2003) *Anal. Chem.*, **75**, 2075.
56 Pedano, M.L. and Rivas, G.A. (2004) *Electrochem. Commun.*, **6**, 10.
57 Ly, S.Y. (2006) *Bioelectrochemistry*, **68**, 227.
58 Saoudi, B., Despas, C., Chehimi, M.M., Jammul, N., Delamar, M., Bessiere, J. and Walcarius, A. (2000) *Sens. Actuat. B*, **62**, 35.
59 Qi, H., Li, X., Chen, P. and Zhang, C. (2007) *Talanta*, **72**, 1030.

60 Wang, H.S., Ju, H.X. and Chen, H.Y. (2001) *Electroanalysis*, **13**, 1105.

61 Kerman, K., Morita, Y., Takamura, Y. and Tamiya, E. (2005) *Anal. Bioanal. Chem.*, **381**, 1114.

62 Ferancová, A., Ovádeková, R., Vaníčková, M., Šatka, A., Viglaský, R., Zima, J., Barek, J. and Labuda, J. (2006) *Electroanalysis*, **18**, 163.

63 Ferancová, A., Adamovski, M., Gründler, P., Zima, J., Barek, J., Mattusch, J., Wennrich, R. and Labuda, J. (2007) *Bioelectrochemistry*, **71**, 33.

64 Ovádeková, R., Jantová, S., Letašiová, S., Štěpánek, I. and Labuda, J. (2006) *Anal. Bioanal. Chem.*, **386**, 2055.

65 Wu, K., Fei, J., Bai, W. and Hu, S. (2003) *Anal. Bioanal. Chem.*, **376**, 205.

66 Chaki, N.K. and Vijayamohanan, K. (2002) *Biosens. Bioelectron.*, **17**, 1.

67 He, P., Li, S. and Dai, L. (2005) *Synth. Met.*, **154**, 17.

68 Chechik, V., Crooks, R.M. and Stirling, C.J.M. (2000) *Adv. Mater.*, **12**, 1161.

69 Lu, Y., Yang, X., Ma, Y., Du, F., Liu, Z. and Chen, Y. (2006) *Chem. Phys. Lett.*, **419**, 390.

70 Hazani, M., Hennrich, F., Kappes, M., Naaman, R., Peled, D., Sidorov, V. and Shvarts, D. (2004) *Chem. Phys. Lett.*, **391**, 389.

71 Wang, S.G., Wang, R., Sellin, P.J. and Zhang, Q. (2004) *Biochem. Biophys. Res. Commun.*, **325**, 1433.

72 Lim, S.H., Wei, J. and Lin, J. (2004) *Chem. Phys. Lett.*, **400**, 578.

73 Guo, M., Chen, J., Liu, D., Nie, L. and Yao, S. (2004) *Bioelectrochemistry*, **62**, 29.

74 Cai, H., Cao, X., Jiang, Y., He, P. and Fang, Y. (2003) *Anal. Bioanal. Chem.*, **375**, 287.

75 Tang, H., Chen, J., Cui, K., Nie, L., Kuang, Y. and Yao, S. (2006) *J. Electroanal. Chem.*, **587**, 269.

76 Guo, M., Chen, J., Nie, L. and Yao, S. (2004) *Electrochim. Acta*, **49**, 2637.

77 Wang, G., Xu, J.-J. and Chen, H.-Y. (2002) *Electrochem. Commun.*, **4**, 506.

78 Heng, L.Y., Chou, A., Yu, J., Chen, Y. and Gooding, J.J. (2005) *Electrochem. Commun.*, **7**, 1457.

79 Chambers, G., Carroll, C., Farrell, G.F., Dalton, A.B., McNamara, M., in het Panhuis M. and Byrne, H.J. (2003) *Nano Lett.*, **3**, 843.

80 Dodziuk, H., Ejchart, A., Anczewski, W., Ueda, H., Krinichnaya, E., Dolgonos, G. and Kutner, W. (2003) *Chem. Commun.*, 986.

81 Liu, P. (2005) *Eur. Polym. J.*, **41**, 2693.

82 Wang, Z., Wang, Y. and Luo, G. (2003) *Electroanalysis*, **15**, 1129.

83 Wang, G.Y., Liu, X.J., Luo, G.A. and Wang, Z.H. (2005) *Chin. J. Chem.*, **23**, 297.

84 Yin, T., Wei, W. and Zeng, J. (2006) *Anal. Bioanal. Chem.*, **386**, 2087.

85 He, J.L., Yang, Y., Yang, X., Liu, Y.L., Liu, Z.H., Shen, G.L. and Yu, R.Q. (2006) *Sens. Actuat. B*, **114**, 94.

86 Wang, Z., Xiao, S. and Chen, Y. (2005) *Electroanalysis*, **17**, 2057.

87 Wang, Z., Xiao, S. and Chen, Y. (2006) *J. Electroanal. Chem.*, **589**, 237.

88 Ramanavičius, A., Ramanavičiene, A. and Malinauskas, A. (2006) *Electrochim. Acta*, **51**, 6025.

89 Cai, H., Xu, Y., He, P.G. and Fang, Y.Z. (2003) *Electroanalysis*, **15**, 1864.

90 Xu, Y., Jiang, Y., Cai, H., He, P.G. and Fang, Y.Z. (2004) *Anal. Chim. Acta*, **516**, 19.

91 Chen, G.F., Zhao, J., Tu, Y., He, P. and Fang, Y. (2005) *Anal. Chim. Acta*, **533**, 11.

92 Moulton, S.E., Minett, A.I., Murphy, R., Ryan, K.P., McCarthy, D., Coleman, J.N., Blau, W.J. and Wallace, G.G. (2005) *Carbon*, **43**, 1879.

93 Tkac, J. and Ruzgas, T. (2006) *Electrochem. Commun.*, **8**, 899.

94 Wang, S.F., Shen, L., Zhang, W.-D. and Tong, Y.-J. (2005) *Biomacromolecules*, **6**, 3067.

95 Xu, C., Cai, H., Xu, Q., He, P. and Fang, Y. (2001) *Fresenius J. Anal. Chem.*, **369**, 428.

96 Li, J., Liu, Q., Liu, Y., Liu, S. and Yao, S. (2005) *Anal. Biochem.*, **346**, 107.

97 Yang, Y., Wang, Z., Yang, M., Li, J., Zheng, F., Shen, G. and Yu, R. (2007) *Anal. Chim. Acta*, **584**, 268.

98 Wang, J., Liu, G., Jan, M.R. and Zhu, Q. (2003) *Electrochem. Commun.*, **5**, 1000.
99 Yang, M., Yang, Y., Yang, H., Shen, G. and Yu, R. (2006) *Biomaterials*, **27**, 246.
100 Zhu, N., Chang, Z., He, P. and Fang, Y. (2005) *Anal. Chim. Acta*, **545**, 21.
101 Jensen, A.W., Wilson, S.R. and Schuster, D.I. (1996) *Bioorg. Med. Chem.*, **4**, 767.
102 Bosi, S., Da Ros, T., Spalluto, G. and Prato, M. (2003) *Eur. J. Med. Chem.*, **38**, 913.
103 Baena, J.R., Gallego, M. and Valcarcel, M. (2002) *Trends Anal. Chem.*, **21**, 187.
104 Sherigara, B.S., Kutner, W. and D'Souza, F. (2003) *Electroanalysis*, **15**, 753.
105 Winkler, K., Balch, A.L. and Kutner, W. (2006) *J. Solid State Electrochem.*, **10**, 761.
106 Li, M.X., Li, N.Q., Gu, Z.N., Zhou, X.H., Sun, Y.L. and Wu, Y.Q. (1999) *Microchim. Acta*, **61**, 32.
107 Pang, D.W., Zhao, Y.D., Fang, P.F., Cheng, J.K., Chen, Y.Y. Qi, Y.P. and Abruna, H.D. (2004) *J. Electroanal. Chem.*, **567**, 339.
108 Bae, A.H., Hatano, T., Sugiyasu, K., Kishida, T., Takeuchi, M. and Shinkai, S. (2005) *Tetrahedron Lett.*, **46**, 3169.
109 Shiraishi, H., Itoh, T., Hayashi, H., Takagi, K., Sakane, M., Mori, T. and Wang, J. (2006) *Bioelectrochemistry*, **71**, 195.
110 Pleskov, Y.V. (2002) *Russ. J. Electrochem.*, **38**, 1275.
111 Pedrosa, V.A., Suffredini, H.B., Codognoto, L., Tanimoto, S.T., Machado, S.A.S. and Avaca, L.A. (2005) *Anal. Lett.*, **38**, 1115.
112 Cui, F.Z. and Li, D.J.A. (2000) *Surf. Coat. Technol.*, **131**, 481.
113 Ushizawa, K., Sato, Y., Mitsumori, T., Machinami, T., Ueda, T. and Ando, T. (2002) *Chem. Phys. Lett.*, **351**, 105.
114 Yang, W., Auciello, O., Butler, J.E., Cai, W., Carlisle, J.A., Gerbi, J.E., Gruen, D.M., Knickerbocker, T., Lasseter, T.L., Russell, J.N., Jr., Smith, L.M. and Hamers, R.J. (2002) *Nature Mater.*, **1**, 253.
115 Knickerbocker, T., Strother, T., Schwartz, M.P., Russell, J.N., Jr., Butler, J., Smith, L.M. and Hamers, R.J. (2003) *Langmuir*, **19**, 1938.
116 Hamers, R.J., Butler, J.E., Lasseter, T., Nichols, B.M., Russell, J.N., Jr., Tse, K.-Y. and Yang, W. (2005) *Diamond Relat. Mater.*, **14**, 661.
117 Yang, W., Butler, J.E., Russell, J.N., Jr. and Hamers, R.J. (2004) *Langmuir*, **20**, 6778.
118 Takahashi, K., Tanga, M., Takai, O. and Okamura, H. (2003) *Diamond Relat. Mater.*, **12**, 572.
119 Gu, H., Su, X. and Loh, K.P. (2004) *Chem. Phys. Lett.*, **388**, 483.
120 Apilux, A., Tabata, M. and Chailapakul, O. (2006) *Bioelectrochemistry*, **71**, 202.
121 Shin, D., Tokuda, N., Rezek, B. and Nebel, C.E. (2006) *Electrochem. Commun.*, **8**, 844.
122 Ivandini, T.A., Honda, K., Rao, T.N., Fujishima, A. and Einaga, Y. (2007) *Talanta*, **71**, 648.
123 Thostenson, E.T., Li, C. and Chou, T.-W. (2005) *Compos. Sci. Technol.*, **65**, 491.
124 Serp, P., Corrias, M. and Kalck, P. (2003) *Appl. Catal. A*, **253**, 337.
125 Takeuchi, K.J., Marschilok, A.C., Lau, G.C., Leising, R.A. and Takeuchi, E.S. (2006) *J. Power Sources*, **157**, 543.
126 Zou, G., Zhang, D., Dong, C., Li, H., Xiong, K., Fei, L. and Qian, Y. (2006) *Carbon*, **44**, 828.
127 Marken, F., Gerrard, M.L., Mellor, I.M., Mortimer, R.J., Madden, C.E., Fletcher, S., Holt, K., Foord, J.S., Dahm, R.H. and Page, F. (2001) *Electrochem. Commun.*, **3**, 177.
128 Baker, S.E., Tse, K.-Y., Lee, C.-S. and Hamers, R.J. (2006) *Diamond Relat. Mater.*, **15**, 433.
129 Vamvakaki, V., Tsagaraki, K. and Chaniotakis, N. (2006) *Anal. Chem.*, **78**, 5538.
130 Chaniotakis, N., Sotiropoulou, S. and Vamvakaki, V. (2006) Carbon nanostructures as matrices for the development of chemical sensors and biosensors. In *Book of Abstracts, ICAS-2006, International Conference of Analytical Sciences*, June 25–30, Moscow, p. 29.

131 Lee, C.-S., Baker, S.E., Marcus, M.S., Yang, W., Eriksson, M.A. and Hamers, R.J. (2004) *Nano Lett.*, **4**, 1713.

132 Baker, S.E., Tse, K.-Y., Hindin, E., Nichols, B.M., Clare, T.L. and Hamers, R.J. (2005) *Chem. Mater.*, **17**, 4971.

133 Mann, D.G.J., McKnight, T.E., Melechko, A.V., Simpson, M.L. and Sayler, G.S. (2007) *Biotechnol. Bioeng.*, **97**, 680.

134 Guth, U., Brosda, S. and Schomburg, J. (1996) *Appl. Clay Sci.*, **11**, 229.

135 Macha, S.M. and Fitch, A. (1998) *Microchim. Acta*, **128**, 1.

136 Mousty, C. (2004) *Appl. Clay Sci.*, **27**, 159.

137 Zen, J.M. and Kumar, A.S. (2004) *Anal. Chem.*, **76**, 205A.

138 Navrátilová, Z. and Kula, P. (2000) *J. Solid State Electrochem.*, **4**, 342.

139 Navrátilová, Z. and Kula, P. (2003) *Electroanalysis*, **15**, 837.

140 Huang, W., Zhang, S. and Wu, Y. (2006) *Russ. J. Electrochem.*, **42**, 153.

141 Riu, J., Maroto, A. and Rius, F.X. (2006) *Talanta*, **69**, 288.

142 Fritzsche, W. (2001) *Rev. Molec. Biotechnol.*, **82**, 37.

143 Welch, C.M. and Compton, R.G. (2006) *Anal. Bioanal. Chem.*, **384**, 601.

144 Wang, J. (2003) *Anal. Chim. Acta*, **500**, 247.

145 Ozsoz, M., Erdem, A., Kerman, K., Ozkan, D., Tugrul, B., Topcuoglu, N., Ekren, H. and Taylan, M. (2003) *Anal. Chem.*, **75**, 2181.

146 Merkoci, A., Aldavert, M., Marin, S. and Alegret, S. (2005) *Trends Anal. Chem.*, **24**, 341.

147 Sharma, P., Brown, S., Walter, G., Santra, S. and Moudgil, B. (2006) *Adv. Colloid Interface Sci.*, **123–126**, 471.

148 Costa-Fernandez, J.M., Pereiro, R. and Sanz-Medel, A. (2006) *Trends Anal. Chem.*, **25**, 207.

149 Kerman, K., Morita, Y., Takamura, Y., Ozsoz, M. and Tamiya, E. (2004) *Anal. Chim. Acta*, **510**, 169.

150 Jin, B., Ji, X. and Nakamura, T. (2004) *Electrochim. Acta*, **50**, 1049.

151 Wang, M., Sun, C., Wang, L., Ji, X., Bai, Y., Li, T. and Li, J. (2003) *J. Pharm. Biomed. Anal.*, **33**, 1117.

152 Cai, H., Wang, Y., He, P. and Fang, Y. (2002) *Anal. Chim. Acta*, **469**, 165.

153 Lu, L.-P., Wang, S.-Q. and Lin, X.-Q. (2004) *Anal. Chim. Acta*, **519**, 161.

154 Yang, J., Yang, T., Feng, Y. and Jiao, K. (2007) *Anal. Biochem.*, **365**, 24.

155 Park, S., Taton, T.A. and Mirkin, C.A. (2002) *Science*, **295**, 1503.

156 Koehne, J.E., Chen, H., Cassell, A.M., Ye, Q., Han, J., Meyyappan, M. and Li, J. (2004) *Clin. Chem.*, **50**, 1886.

157 He, L. and Toh, C.-S. (2006) *Anal. Chim. Acta*, **556**, 1.

158 Huang, X.-J. and Choi, Y.-K. (2007) *Sens. Actuat. B*, **122**, 659.

12
Metal Nanoparticles: Applications in Electroanalysis
Nathan S. Lawrence and Han-Pu Liang

12.1
Introduction

Nanotechnology has recently come to the forefront of analytical chemistry. A large number of nanomaterials – specifically nanoparticles exhibiting differing properties – have found a wide range of applications in various analytical methodologies [1]. The small size of nanoparticles (typically 1 to 100 nm) means that they exhibit unique chemical [2–4], physical [5–8], and electronic properties [9] that are distinctly different from those of bulk materials, and as such they have attracted considerable attention. This interest is highlighted by the substantial worldwide governmental funding in this area, creating enormous scientific activity. Numerous review articles and books on nanomaterials have witnessed the tremendous growth in the study and application of nanomaterials (e.g., Refs. [10–13]).

Considering the properties of these nanoparticles, it is relatively easy to see that an emerging area is their use in electroanalysis and electrochemical-based sensors [14–20]. Even a brief examination of the literature shows that there has been large growth in this area, whereby traditional macroelectrodes are being replaced with their nanoparticle analogues. The predominant advantages of using nanoparticle-modified electrodes compared to typical macroelectrodes is their large effective surface area, increased mass transport, high catalytic activity, and the ability to exert control over the local environment at the electrode surface [17].

Probably the most important application of metal nanoparticles is their use in *chemical catalysis*. Several reviews relating to catalysis, based on metal nanoparticles have highlighted their important role [21–23]. The small size of the particle endows them with a high surface area, providing more active sites to catalyze the reaction. It has long been speculated that metal nanoparticles exhibit significant improvement in detection and catalytic capabilities due to the increase in both their surface area and their ability to electronically interact with reactant molecules.

12 Metal Nanoparticles: Applications in Electroanalysis

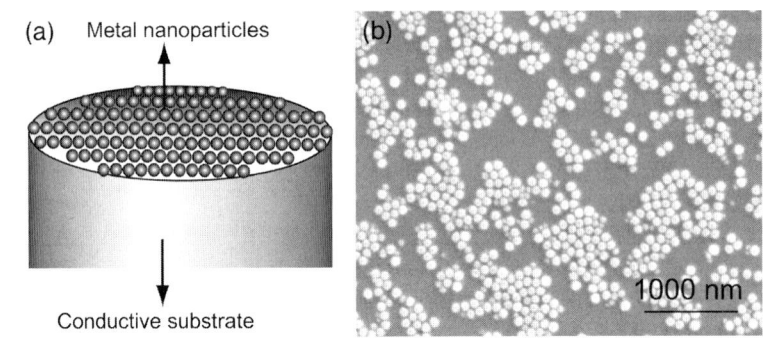

Figure 12.1 (A) Schematic representation of a nanoparticle-modified electrode. (B) An SEM image of a typical metal nanoparticle-modified surface.

A schematic representation of a nanoparticle-modified electrode is shown in Figure 12.1a, while a scanning electron microscopy (SEM) image of a metal nanoparticle-modified surface is shown in Figure 12.1b. As mentioned above, this type of electrode can offer a high surface area in comparison with a planar macroelectrode, and coupling this increase with the high catalytic activity of the particle (which lowers the overpotentials of redox species) means that such a set-up might act as a more selective and sensitive electrochemical sensor. Furthermore, such electrodes are cost-effective; typically, noble metal macroelectrodes such as gold, platinum, and palladium are expensive in comparison to manufactured nanoparticles, which is of course a significant advantage for the development of commercial sensors.

Metal nanoparticles are usually prepared via reduction of the corresponding metal precursor in the presence of a surface stabilizer (capping agent), for example phosphines, thiols, polymers, and amines [23–25]. Apart from the capping agents' ability to stabilize and disperse nanoparticles, tailoring of the capping agent can enrich the particles' chemistry further [14,15]. For example, it can introduce functional groups which may be redox active, catalytic, bioactive, or act as selective recognition sites. Thus, these tailored nanoparticles have found use in a variety of applications, ranging from bioelectronic [14–17], electronic wiring [26,27], optoelectronic [19,28] and sensoric applications [29].

In this chapter, we review the use of metal nanoparticles in electroanalysis, the aim being to produce an outline of the key areas in which each metallic nanoparticle is currently used. A summary of the current applications of each nanoparticle in electroanalysis is provided in Table 12.1. This highlights the analytes which have been detected, and gives a brief summary of how the nanoparticle has been utilized in their detection, either via a direct analytical signal or in conjunction with enzymes, as in the case of biosensors (for more detail, see Chapter 11). Within the chapter it will be shown that a major step in developing these sensors is the immobilization of the particles to the underlying inert conductive substrate; hence, these procedures are given suitable attention throughout the chapter.

Table 12.1 A comparison of the linear ranges and limits of detection achieved for a variety of analytes with a range of nanoparticles.

Nanoparticle/Reference	Analyte	Methodology	Dynamic range	LoD
Au [37]	H_2O_2	Biosensor	0.05–30.6 mM	0.02 mM
Au [38]	Glucose	Biosensor	0.001–1.0 mM	0.69 µM
Au [39]	Glucose	Biosensor	0.02–5.7 mM	8.2 µM
Au [40]	Glucose	Biosensor	Up to 60.0 mM	3.0 µM
Au [41]	H_2O_2	Biosensor	0.005–1.4 mM	0.401 µM
Au [42]	NADH	Biosensor	Up to 5 mM	5 nM
Au [44]	H_2O_2	Biosensor	2–24 µM	0.91 µM
Au [45]	H_2O_2	Biosensor	0.008–15.0 mM	2.4 µM
Au [46]	H_2O_2	Biosensor	0.01–7.0 mM	4.0 µM
Au [47]	Catechol	Biosensor	2–110 µM	0.32 µM
Au [47]	Phenol	Biosensor	2–110 µM	0.60 µM
Au [47]	p-Cresol	Biosensor	2–55 µM	0.18 µM
Au [48]	Cholesterol	Biosensor	0.075–50 µM	5 nM
Au [49]	H_2O_2	Biosensor	0.0025–0.5 mM	0.48 µM
Au [51]	H_2O_2	Biosensor	0.1–1 mM	N/A
Au [52]	Cholesterol	Biosensor	10–70 µM	N/A
Au [53]	Nitrite	Biosensor	0.3–700 µM	0.1 µM
Au [54]	Rabbit immunoglobulin G	Immuno	6.4–3200 fM	1.6 fM
Au [55]	Hepatitis B	Immuno	4–800 ng mL^{-1}	1.3 ng mL^{-1}
Au [56]	Paraxon	Immuno	1920 µg L^{-1}	12 µg L^{-1}
Au [57]	DNA	DNA	N/A	10^{-11} M
Au [58]	DNA	DNA	0.51–8.58 pM	N/A
Au [59]	DNA	DNA	N/A	1.2 mM
Au [60]	DNA	DNA	0.050–10 fM	10 fM
Au [62]	DNA	DNA	6.9–150.0 pM	2.0 pM
Au [63]	Promethazine	DNA	20–160 µM	10 µM
Au [63]	Chlorpromazine	DNA	10–120 µM	7 µM
Au [65]	Dopamine	Direct	2.5–20 µM	0.13 µM
Au [65]	Ascorbic acid	Direct	6.5–52 µM	N/A
Au [66]	Epinephrine	Direct	0.1–200 µM	60 nM
Au [67]	Glucose	Direct	Up to 8 mM	N/A
Au [69]	As(III)	Direct	0.09–4 ppm	0.09 ppb
Au [70]	As(III)	Direct	1–5 µM	5 ppb
Au [71]	As(III)	Direct	N/A	6 nM
Au [72]	As(III)	Direct	0.005–2.5 µM	0.0096 ppb
Au [73]	As(III)	Direct	Up to 15 ppb	0.25 ppb
Au [74]	Cr(III)	Direct	0.1–0.4 mM	N/A
Au [75]	Nitrite	Biosensor	0.1–9.7 µM	0.06 µM
Pt [80]	H_2O_2	Direct	0.0005–2 mM	7.5 nM
Pt [80]	Acetylcholine	Direct	N/A	2.5 fM
Pt [80]	Choline	Direct	N/A	2.3 fM
Pt [82]	H_2O_2	Direct	0.00064–3.6 mM	0.35 µM
Pt [83]	Glucose	Biosensor	0.5–5000 µM	0.5 µM
Pt [84]	Glucose	Biosensor	1–25 mM	1 mM

(*Continued*)

Table 12.1 (Continued)

Nanoparticle/ Reference	Analyte	Methodology	Dynamic range	LoD
Pt [85]	Glucose	Biosensor	0.1–13.5 mM	0.1 mM
Pt [86]	Glucose	Direct	2–14 mM	1 μM
Pt [87]	Thrombin	Nucleic Acid	N/A	1 nM
Pt [88]	Dopamine	Direct	3–60 μM	10 nM
Pt [89]	cholesterol	Biosensor	0.01–3 mM	N/A
Pt/Fe [90]	Nitrite	Direct	0.0011–1.1 mM	0.47 μM
Pt/Fe [91]	Nitric oxide	Direct	0.084–7800 μM	0.018 μM
Pt/Fe [92]	Uric acid	Direct	0.0038–1.6 mM	1.8 μM
Pt [93]	As(III)	Direct	N/A	2.1 ppb
Ag [103]	Hydroquinone	Direct	0.003–2 mM	0.172 μM
Ag [104]	H_2O_2	Biosensor	0.0033–9.4 mM	0.78 μM
Ag [105]	DNA	DNA	8–1000 nM	4 nM
Ag [106]	H_2O_2	Direct	5–40 μM	2.0 μM
Ag [107]	Hypoxanthine	Direct	1.0–100 mg mL^{-1}	1.0 mg mL^{-1}
Ag [108]	Thiocyanate	Direct	0.5–400 μM	40 nM
Ag [109]	Brilliant cresyl blue	Direct	10–210 μM	N/A
Ag [110]	H_2O_2	Biosensor	0.001–1 mM	0.4 μM
Ag [111]	H_2O_2	Biosensor	0.003–0.7 mM	1 μM
Ag [112]	Nitrite	Biosensor	0.2–6.0 mM	34.0 μM
Ag [113]	Nitric oxide	Biosensor	1–10 μM	0.3 μM
Hg/Ag [114]	Cysteine	Direct	0.4–13 μM	0.1 μM
Pd [122]	Methane	Direct	0.125–2.5% in air	0.125%
Pd [123]	Hydrazine	Direct	31–204 μM	2.6 μM
Cu/Pd [126]	Hydrazine	Direct	2–200 μM	0.27 nM
Pd [127]	Glucose	Biosensor	Up to 12 mM	0.15 mM
Cu [130]	Nitrate	Direct	N/A	1.5 μM
Cu [131]	Alanine	Direct	5–500 μM	24 nM
Cu [132]	Nitrate	Direct	1.2–124 μM	0.76 μM
Cu [133]	Amikacin	Direct	2–200 μM	1 μM
Cu [134]	H_2O_2	Direct	Up to 200 μM	0.97 μM
Cu [135]	Nitrite	Direct	0.1–1.25 mM	0.6 μM
Cu/SnO$_2$ [136]	H_2S	Direct	N/A	20 ppm
Cu [138]	Nitrite	Direct	0.05–30 mM	0.02 mM
Cu [146]	Oxygen	Direct	1–8 ppm	N/A
Cu [148]	Glucose	Direct	Up to 26.7 mM	N/A
Cu [149]	Catechols	Direct	Up to 200 μM	3 μM
Cu [149]	Dopamine	Direct	Up to 300 μM	5 μM
Cu [150]	Glucose	Direct	Up to 500 μM	250 nM
Cu [151]	Halothane	Direct	10–50 μM	4.6 μM
Cu/Au [153]	DNA	DNA	0.015–5 nM	5 pM
Ni [172]	Hydrogen sulfide	Direct	20–90 μM	5 μM
Ni [173]	Glucose	Direct	0.05–500 μM	20 μM
Ni [173]	Fructose	Direct	0.05–500 μM	25 μM
Ni [173]	Sucrose	Direct	0.10–250 μM	50 μM
Ni [173]	Lactose	Direct	0.08–250 μM	37 μM
Ni [174]	Acetylcholine	Direct	4.46–22.30 μM	N/A
Co$_3$O$_4$/Ni [175]	Carbon monoxide	Direct	10–500 ppm	N/A

Table 12.1 (Continued)

Nanoparticle/ Reference	Analyte	Methodology	Dynamic range	LoD
Co_3O_4/Ni [175]	Hydrogen	Direct	20–850 ppm	N/A
$NiFe_2O_4$ [176]	Liquefied Petroleum Gas	Direct	Up to 25 ppm in air	N/A
Fe [188]	Humidity	Direct	5–98%	N/A
Fe/In [188]	Ozone	Direct	N/A	30 ppb
Fe [191]	Glucose	Biosensor	0.006–10 mM	3.17 μM
Fe [192]	Glucose	Biosensor	Up to 20 mM	N/A
Ir [205]	Glucose	Biosensor	Up to 12 mM	N/A

LoD, Limit of detection.

12.2
Electroanalytical Applications

12.2.1
Gold Nanoparticles

Gold nanoparticle-based biosensors have received considerable attention over the past few years, due mainly to the nanoparticles' biocompatibility and ease with which they can be functionalized [30–36]. Among published reports, two variations of biosensor are predominant. Figure 12.2 illustrates, schematically, a biosensor in which a mediator species that is capable of catalyzing the oxidation or reduction of the products from the enzyme reaction, is first deposited onto the macro-electrode

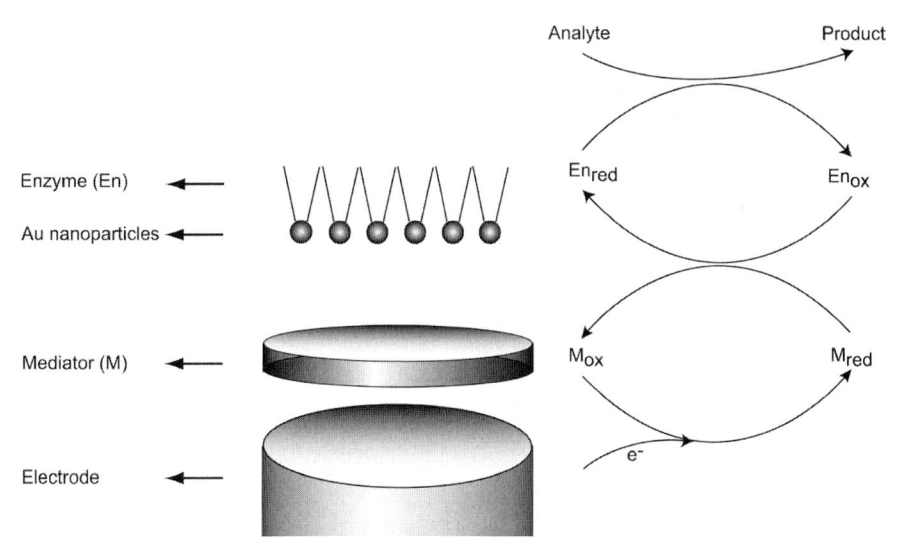

Figure 12.2 (A) Schematic representation of a nanoparticle-based mediator biosensor along with the corresponding detection mechanism.

surface. The nanoparticles are then adsorbed onto the mediator-electrode surface upon which the enzyme is finally immobilized [37–40]. The detection mechanism for the biosensor is also outlined in Figure 12.2. In these cases, it was found that the presence of the nanoparticles enhanced the sensitivity of the sensor. This enhancement was attributed to the increased concentration of enzyme upon the sensor surface due to the larger active surface area at the nanoparticle immobilized layer.

A second biosensor has also been developed in which the mediator is no longer required. In this case, the nanoparticles are directly attached to the electrode surface via certain immobilization procedures (for example, thiol-terminated sol–gels [41,42], amine–cysteamine linkages [39,43], and thiol films [44]). The enzyme is then attached to the nanoparticle layer. In this way it is the nanoparticle which can catalytically oxidize or reduce the products of the enzyme reaction. It can be foreseen that these biosensors offer significant construction simplicities compared to the mediator-based sensors. Examples of proteins and enzymes to be attached to Au nanoparticles to be used in such sensors include horseradish peroxidase [37,41,45,46], tyrosinase [47], cholesterol oxidase [48], myoglobin [49] glucose oxidase [38–40,50], microperoxidase [51], cytochrome P450scc [52], and hemoglobin [43,53].

The ability of Au to undergo facile surface modification by a host of compounds means that immunosensors and DNA sensors have also been constructed (for further details, see Chapter 11). In the case of immunosensors the surface is modified with a specific antigen or antibody; examples include anti-rabbit immunoglobulin G [54], the hepatitis B surface antibody [55], or paraoxon antibodies [56]. In order to immobilize DNA, either the Au surface or the DNA strands are typically modified with specific functionalities [57–62]. Once the DNA is immobilized it can be used in the detection of DNA and a variety of other analytes, including phenothioazine-based drugs [63], uric acid, and norepinephrine [64].

Although a large number of sensors utilizing Au nanoparticles require modification of the nanoparticle surface, there are analytes which can be detected directly at the nanoparticle layer. In these cases the interaction of Au with the analyte is utilized in a productive manner. A major task in bioanalysis is resolving and determining the concentrations of neurotransmitters in the presence of ascorbic acid, due to their similar redox potentials at bare unmodified electrodes. It has been found that Au nanoparticles can successfully and selectively determine dopamine in the presence of ascorbate. In this case, the oxidation potential of the ascorbic acid was found to shift to lower potentials due to the high catalytic activity of Au nanoparticle, whilst the electrochemical reversibility of dopamine improved significantly [65]. In addition, a nano-Au electrode has been applied to the sensing of epinephrine, when the Au electrode was found to enhance its electrochemical response, due to the increased surface area of the nano-Au electrode compared to a conventional planar electrode [66]. It was shown above that glucose oxidase, when immobilized on Au nanoparticles, can be utilized for both variations of biosensor. However, under alkaline conditions glucose can be directly oxidized and therefore detected at Au and Au/Ag alloy nanoparticle layers [67,68]. This direct oxidation technique must be used cautiously within authentic samples due to the large range of oxidizable interferents within such samples, and hence only niche applications have been investigated. It can be foreseen, however, that further applications will be possible when the direct detection techniques are used in conjunction with analytical separation techniques.

The strong interaction of gold with either arsenic or chromate has been utilized in the sensing of these compounds [69–74]. Arsenate As(III) is a common pollutant in river waters in the Third World, and therefore its detection is important. The ability of both surface-bound and electrodeposited Au nanoparticles to interact and accumulate As(III) has been used in the determination of the latter, either directly in the sample [69–72] or using an end-column detector unit [73]. Gold nanoparticles have also been used in the detection of nitrite, where the nanoparticles are modified with hemoglobin [75], and for the direct oxidation of nitric oxide [76,77].

At this point it is worth mentioning a further novel use of nanoparticles which, although slightly beyond the scope of this chapter, is of interest to the analytical chemist. In this case, gold nanoparticles are used in conjunction with chip-based capillary electrophoresis, not for detection, but to aid in separation. It was found that incorporation of the Au nanoparticles into the microchannel improved selectivity between solutes and increased the efficiency of the separation, with both resolution and plate numbers of the solutes being doubled in the presence of nanoparticles [78].

12.2.2
Platinum Nanoparticles

Platinum, like gold, is one of the most extensively studied metals in electrochemistry due to its relative chemical inertness and its ability to provide reasonable solvent windows such that oxidation, reduction and electron-transfer rates [79] can be studied. Furthermore, platinum – unlike carbon – has been shown to catalyze the redox chemistry of several compounds. Its ability to catalyze both the oxidation and reduction of H_2O_2 has meant that the Pt nanoparticles provide sensitive H_2O_2 detectors. In this case the nanoparticles were successfully deployed by the modification of carbon film electrodes [80,81], carbon fiber ultramicroelectrodes [82] and carbon nanotubes (CNTs) [83]. The fact that H_2O_2 is the product of several enzymatic reactions means that Pt nanoparticles will play a key role in potential electrochemical biosensors.

An example of such systems is the detection of glucose via the glucose oxidase enzyme. This enzyme has been shown to be easily absorbed onto Pt, Pt nanoparticle-doped carbon films [80] and CNTs [83–85], and even onto synthesized ordered Pt-nanotubular arrays [86]. All of these have been demonstrated as effective glucose sensors. Figure 12.3a shows a typical transmission electron microscopy (TEM) image of a Pt nanoparticle-embedded carbon film [80], whereby the dark spots correspond to the Pt nanoparticles and the light areas relate to the carbon film. From this it can be clearly seen that nanoparticles with diameters of 2.5 nm can be prepared. Figure 12.3b shows, graphically, a comparison of the voltammetric response of the Pt nanoparticle-embedded carbon film (responses a and c) with that of a planar platinum electrode (responses b and d) in the presence (responses a and b) and absence (responses c and d) of 1 mM H_2O_2. It can be clearly seen that the platinum nanoparticle layer produces a lower oxidation potential than that of the planar electrode, indicating a faster electron transfer rate at the nanoparticle layer.

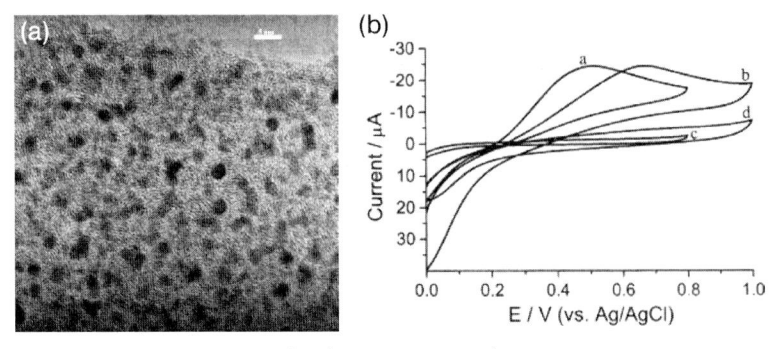

Figure 12.3 (A) A TEM image of a platinum nanoparticle-embedded graphite carbon film. (B) The cyclic voltammetric response comparing the signal obtained at the platinum nanoparticle-embedded graphite carbon film (a, c) and a planar platinum electrode (b, d) in the absence (c, d) and presence (a, b) of 1 mM H_2O_2. (From Ref. [80].)

In an analogous manner to the gold nanoparticles discussed above, the relative ease of functionalization of the Pt particles means that it is possible to prepare nucleic acid-modified Pt nanoparticles. These act as catalytic labels for the amplified electrochemical detection of DNA hybridization, and in aptamer/protein recognition [87].

The direct detection of several important species has been achieved using Pt nanostructures. These include dopamine [88], cholesterol [89], nitrite [90], nitric oxide [91], uric acid [92], and As(III) [93]. With the advantages of high surface area and activity, Pt nanostructures allow sensitive and low detection limits in comparison with a Pt bulk electrode. For example, an approximately one order of magnitude lower detection limit with a Pt nanoparticles-embedded carbon film electrode has been observed compared to a planar platinum electrode [80]. Furthermore, nanoporous platinum oxide has been investigated as a hydrogen ion/pH probe. These oxide nanostructures have been found to exhibit a near-Nernstian behavior over a wide pH range, between 2 and 12 [94]. A second platinum nanoparticle-based pH sensor to be developed relied upon the ability of the nanoparticle to be easily coated with the pH-sensitive poly(quinoxaline) species [95].

Until now, the majority of investigations utilizing the properties of Pt nanostructures have been dedicated to studying methanol oxidation, formic acid oxidation, and oxygen reduction [96–99]. These, rather than for electroanalytical purposes, are examined in an attempt to develop alternative fuel cell energy resources to those used traditionally today.

12.2.3
Silver Nanoparticles

In its pure state, silver has the highest thermal and electrical conductivity of all metals, it is highly stable in air and water, and it is more abundant than gold. These properties mean that it is of interest to the electroanalytical community. Indeed, silver has not only been used as the sensing component of electrochemical sensors, but it can often be found as the integral component of a reference electrode. The stability of the Ag/AgCl redox couple means that this couple is often used as an inexpensive alternative to the saturated calomel electrode used is a vast number of electrochemical systems. Furthermore, like gold, the ability of silver to chemically adsorb species onto its surface means that immobilization of the particles is relatively facile.

The attachment of silver nanoparticles to the sensing surface has received considerable attention. Probably the most facile method is via electrochemical techniques, in which silver ions in solution are electrochemically reduced onto the electrode substrate, to form either colloidal [100] or dispersed nanoparticles [101]. In these techniques careful control of both the deposition potentials and times are required to produce the nanoparticles layer. Chemical methods in which the nanoparticle is essentially tethered to the electrode surface have also been used. Such examples are the attachment of silver nanospheres and nanorods to indium tin oxide (ITO) [102]; in this case the attachment was successful without the need for a bridging reagent. Such electrode modifications were found to improve the redox behavior of the ferro/ferricyanide redox couple compared to the bare ITO electrode. Typically, silver nanoparticles have been covalently immobilized onto Au electrodes through either cystamine or cysteine linkers [103,104]. A novel method for the fabrication of DNA biosensors has been developed by means of self-assembling colloidal Ag to a thiol-containing sol–gel network in analogous manner to the Au particles discussed above [41,42]. The thiol groups of the sol–gel initiator served as binding sites for the covalent attachment to the gold electrode surface and silver nanoparticles [105].

Once immobilized, the silver nanoparticles can either be used in the direct determination of analytes, or as the template in a biosensor. The high catalytic activity of the Ag surface has allowed it to be used in the direct determination of hydroquinone [103], hydrogen peroxide [106], hypoxanthine [107], thiocyanate [108], and brilliant cresyl blue [109].

A host of biosensors have also been developed in which the enzyme is directly attached to the silver nanoparticles. These sensors have allowed for the indirect detection of hydrogen peroxide, using either horseradish peroxidase [104,110] or myoglobin [111], and in the sensing of nitrite or nitric oxide using hemoglobin-modified electrodes [112,113].

Before moving onto assess the capabilities of palladium nanoparticles in electroanalysis, it is of interest to note that silver nanoparticles have been used in conjunction with mercury electrodes. In this case, a mercury film electrode was deliberately doped with silver nanoparticles to enhance the detection of cysteine. It was shown that the electrode layer strongly adsorbed cysteine whilst catalyzing the electrode reaction of cysteine more efficiently than a mercury film alone [114].

12.2.4
Palladium Nanoparticles

Palladium is inert in both air and water, and is therefore a strong candidate to be used in electroanalysis. Palladium nanoparticles have been influential in understanding the electrochemical response of the well-known palladium/hydrogen system. Pd undergoes a highly specific interaction with hydrogen, as hydrogen atoms have a high mobility within the Pd lattice and therefore can diffuse rapidly through the metal. The reduction of hydrogen ions has been studied extensively at palladium macroelectrodes [115–117], with hydrogen atoms being found to absorb into the Pd lattice. This is a so-called "dissolution adsorption mechanism" in which H^+ ions first adsorb onto the Pd surface, and are subsequently reduced to form adsorbed hydrogen atoms (H_{ad}). These adsorbed hydrogen atoms finally diffuse into the bulk Pd such that they lay underneath the first few atomic layers of Pd atoms, thus forming absorbed hydrogen (H_{ab}). At macroelectrodes, only a single oxidative wave is observed which encompasses both the oxidation of H_{ab} and H_{ad} to H^+. Recently, however, it has been shown that the use of electrodeposited and homogenously synthesized Pd nanoparticles, allows the oxidation of both the H_{ad} and H_{ab} to be deconvoluted [118–120] such that two oxidative waves are observed (see Figure 12.4). In this case, nanotechnology has provided an insight into the oxidative process: the ability to form Pd nanoparticles upon Au nanoparticles provides a unique methodology to control the nanoparticles size and structure such that the Pd/Hydrogen interaction can be manipulated. It was found that careful control of the concentration of Pd salts within the synthesis solutions allowed manipulation of the surface coverage of Pd upon the Au, and hence the electrochemical signal. Figure 12.5 shows the X-ray diffraction (XRD) patterns of a range of Au/Pd samples in which the H_2PdCl_4 concentration was decreased from sample 1 to 7. It can be clearly seen that the peaks of Pd decrease, whilst those of Au increase as the H_2PdCl_4 concentration is lowered. Figure 12.6 shows typical TEM images of the various samples, and clearly demonstrates that the thickness of Pd shell – that is, Pd

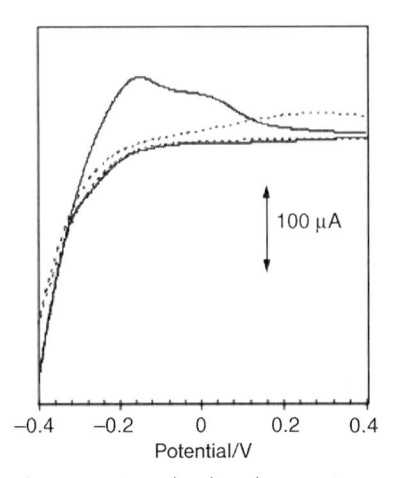

Figure 12.4 Typical cyclic voltammetric responses (scan rate = $0.1\,V\,s^{-1}$) obtained at Au/Pd dispersed layer (solid line) and a planar palladium macroelectrode (dashed line) when placed in 1 M H_2SO_4.

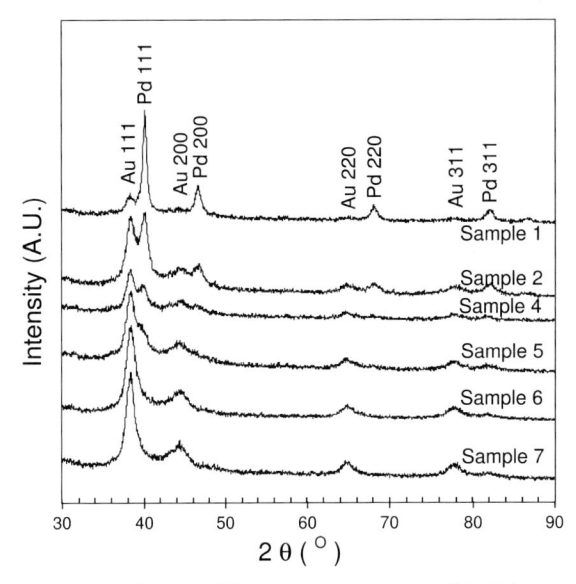

Figure 12.5 The X-ray diffraction (XRD) patterns of Au/Pd core-shell nanoparticles at different stages by deliberately decreasing the volume of H_2PdCl_4 in the feed solution. (From Ref. [120].)

coverage – gradually declines with decaying H_2PdCl_4 concentration. In order to examine the structure of the bimetallic nanoparticles, high-resolution (HR) TEM is normally used.

Figure 12.7 shows the HRTEM images of an Au/Pd sample. This high magnification allows the structure of the Pd deposit to be observed, and may in future provide a means of understanding the nucleation/growth mechanism of these particles. These images show that the nanoparticles with an irregular surface are the Au/Pd core-shell nanoparticles, with the smaller Pd particles being deposited as flat (shown within the white ellipse) rather than spherical material. The characterization techniques utilized in Figures 12.5 to 12.7 are essential in understanding the structure of the nanoparticles although, when examining their chemical properties, electrochemistry provides a niche insight. Figures 12.8a and b show, graphically, the cyclic voltammetric (scan rate = $0.1\,V\,s^{-1}$) response of both the bare Au nanoparticles and the Au/Pd particles when dispersed on a boron-doped diamond (BDD) electrode and placed in 1 M H_2SO_4. In Figure 12.8 it can be clearly seen that, at high Pd surface coverage (sample 1), two oxidative processes are observed, with lowering of the Pd coverage (sample 7) resulting in a single oxidative wave. This was rationalized as follows: lowering of the H_2PdCl_4 concentration in the starting solution produces well-dispersed flat, Pd nanoparticles on the Au. These Pd nanoparticles have a large surface area onto which the hydrogen atom can adsorb. As the surface coverage decreases, the ratio of sites on which the hydrogen atoms can either adsorb or absorb increases, and hence the analytical signal can be

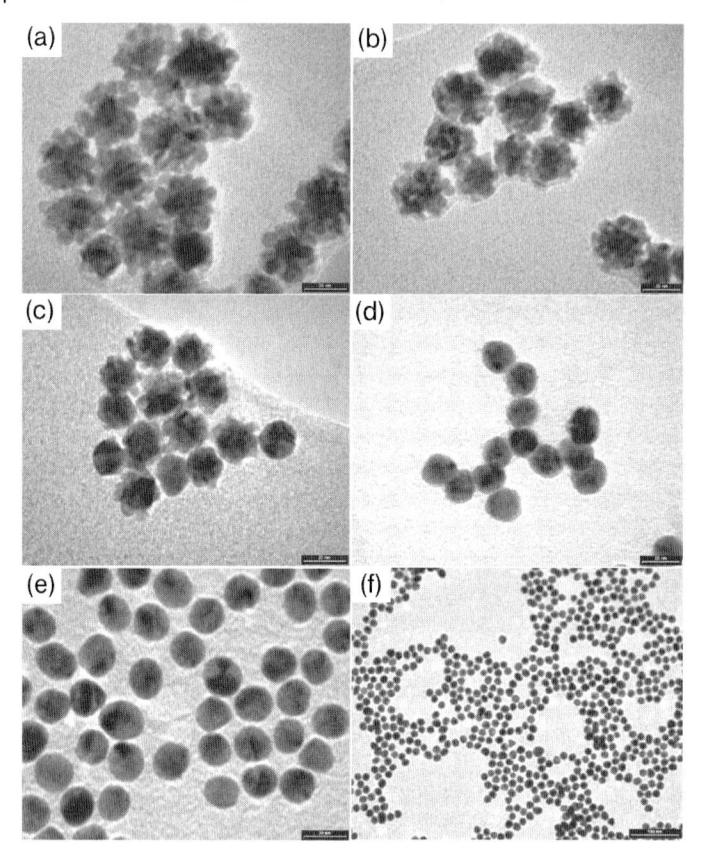

Figure 12.6 Typical high-magnification TEM image of: (a) Sample 2; (b) Sample 3; (c) Sample 5; (d) Sample 6; and (e) Sample 7. (f) A low-magnification TEM image of Sample 7. (Sample numbers relate to those given in the XRD patterns of Figure 12.5) Scale bar in (a–e) = 20 nm; scale bar in (f) = 100 nm. (From Ref. [120].)

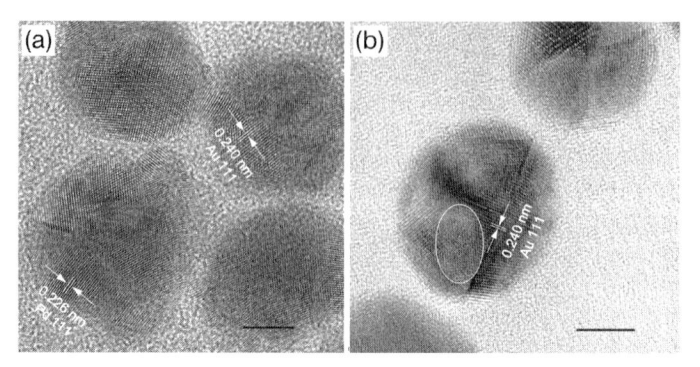

Figure 12.7 Typical high-resolution transmission electron microscopy (HRTEM) images of two Au/Pd co-shell structures. Scale bars = 5 nm. (From Ref. [120].)

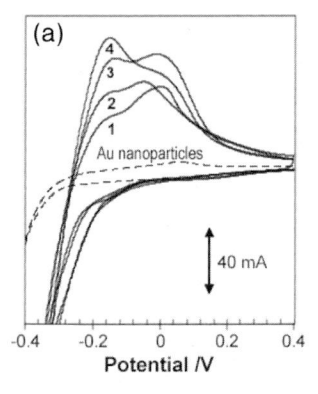

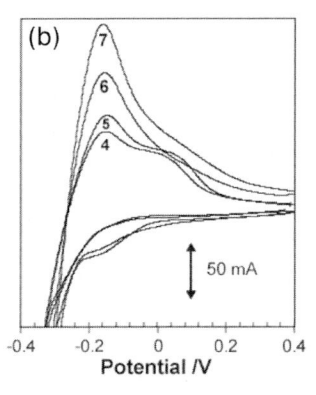

Figure 12.8 Cyclic voltammetric response of various Au/Pd core-shells structures [(a) samples 1–4; (b) samples 4–7] in which the palladium concentration is decreased in the feed solution from samples 1 to 7 (samples numbers relate to those given in the XRD patterns of Figure 12.5), when dispersed on a boron-doped diamond (BDD) electrode and placed in 1 M H_2SO_4. The Pd loading for each sample was 0.28 µg. Also shown is the response of the bare Au nanoparticles [dashed lines in (a)]. (From Ref. [120].)

manipulated. The electrodeposited nanoparticles [118,119], along with Pd mesowire arrays [121], have been subsequently used in the detection of hydrogen gas. Due to the strong interaction of Pd with hydrogen it is not surprising to see that Pd nanostructures have been reported to exhibit highly sensitive responses to other hydrogen-rich compounds such as methane [122] and hydrazine [118,122–126].

The incorporation via electrodeposition of Pd onto CNT films offers advantages in the deployment of these particles in electroanalysis. It can be envisaged that the increased electroactive surface area intrinsic in the CNT layer will further enhance analytical signals. To this end, a novel glucose sensor has been developed in which both Pd nanoparticles and glucose oxidase are co-deposited onto the CNT layer; in the presence of glucose, the Pd efficiently oxidizes and reduces the newly generated hydrogen peroxide [127].

Electrodeposited mesoporous Pd films on Pt microdiscs have been found to act as pH microsensors. The newly generated film is first electrochemically loaded with hydrogen such that it forms two phases of palladium hydride. Once formed, the potentiometric response of the electrode was examined over a wide pH range. A highly stable and reproducible Nernstian response was observed over the pH range between 2 and 12. [128]

Efforts have also been directed towards the synthesis and characterization of bimetallic nanostructures. These architectures represent a highly interesting class of materials exhibiting improved catalytic properties, which are postulated to result from both electronic and structural effects of the bimetal. For instance, Cu/Pd nanoparticles exhibit enhanced electrocatalytic activity in the detection of hydrazine [126], whilst Pd/Fe has a higher catalytic performance than pure Pd catalysts for oxygen reduction [129].

12.2.5
Copper Nanoparticles

The ease of oxidation of copper means that pure copper nanoparticles are virtually impossible to produce – which is the reason why the majority of analytical techniques deploy copper oxide nanoparticles. These can be prepared in a variety of manners, ranging from classical electrodeposition [130–132], incorporation of the particles into screen printed, carbon paste [133–135] and tin oxide [136,137] films. More elaborate processes include the development of cysteine-stabilized colloidal copper using either the Brust method [138], reverse micelles [139], microemulsions [140], or radiation techniques [141].

The most common protagonists to be detected using copper macroelectrodes are carbohydrates [142], amino acids [143], and nitrogen oxides [144,145], all of which have been shown to produce catalytic responses. The detection process at each is found to depend on the conditions of the copper species. In alkaline conditions, copper oxide nanoparticle-modified electrodes have been successfully used in the determination of amino acids, with little or no electrode passivation. This was attributed to a unique CuO/Cu_2O redox catalyst mechanism [131]. Copper nanoparticle-modified electrodes have also been used in the detection of nitrate [130,132], amikacin [133], hydrogen peroxide [134], nitrite [135,138], oxygen [146], glucose [147,148], and various catechol derivatives [149].

Copper nanoparticles have also been co-deposited with CNTs to provide enhanced detection of amino acids compared to both CNTs and the copper particles alone [150]. This construction couples the larger surface area inherent in CNT layers with the enhanced catalytic activity of the copper nanoparticles and edge plane sites in the CNT. Furthermore, rather than deliberately mixing the copper particles and CNTs, it has been shown that if, during the production of the CNT, copper is present then the "impurities" found within the CNT can catalyze the oxidation of halothane [151].

The responses of solid-state gas sensors have also been improved by the incorporation of these particles into the sensing layer [136,137]. It was found initially that a thin, uniformly distributed CuO layer deposited on a SnO_2 substrate showed enhanced sensitivity towards H_2S detection with fast response times achievable at a relatively low operating temperature (150 °C) compared to SnO_2 alone [152]. Reducing the layer thickness – and eventually the size of copper oxide particle – to the nanoscale was shown to further enhance the performance of the sensor. The SnO_2–CuO-nanosensor produced a highly sensitive response towards the gaseous species at a lower temperature compared to the thin-film sensor [136,137].

The ability of copper to form alloys has been put to significant use in the development of bimetallic nanoparticles. Examples include the development of a Cu/Au alloy nanoparticle as oligonucleotide labels for the electrochemical detection of DNA hybridization [153] and Cu/Pd nanoparticles for the enhanced electrocatalytic detection of hydrazine [126]. It was also shown that the electrodeposition of silver and copper produced bimetallic particles with two archetypes [154]. Although, to date, the Ag/Cu structures have not been used for electroanalysis, these unique bimetallic materials may offer some advantages to the two metallic particles when used separately.

12.2.6
Nickel Nanoparticles

The chemical stability of nickel electrodes and the formation of its oxide layers has been studied in depth. Several reports have been devoted to understanding the passivation phenomena and to the study of composition of the oxidation films of nickel in aqueous media (e.g., Refs. [155–157]). The formation of a nickel hydroxide film under alkaline conditions has been achieved in different ways: either via potentiostatic [158,159], potentiodynamic [160,161], or galvanostatic [162,163] anodic sweeps, or by the cathodic electrochemical deposition of Ni salts in alkaline solution [164]. Due to the nature of these oxide layers it is not surprising that nickel and nickel-based oxides are important materials in nickel-containing batteries, as well as in a wide range of technological applications, including electrochromic devices, water electrolysis, electrosynthesis, and fuel cells [155,156].

The synthesis and immobilization of nickel nanoparticles is therefore important in defining the structure of the particle and the form of the oxide layer which is present. Naturally, classical electrochemical deposition has been utilized to form Ni particles [165] as well as Ni and Cu nanowire arrays [166], although other techniques such as precipitation [167–169] and reverse microemulsions have also been developed [170,171].

Analytical uses of Ni have been developed, electrodeposited nickel oxide was found to catalyze the detection of hydrogen sulfide [172], sugars [173], and acetylcholine [174]. The incorporation of these oxide nanoparticles into Co_3O_4 porous sol–gel films, $NiFe_2O_4$ spinels and platinum catalysts has also aided in the determination of carbon monoxide [175], liquefied petroleum gas [176] and hydrogen [177], respectively.

One common non-analytical electrochemical application of Ni nanoparticles is their use in rechargeable batteries [178]. In this case, it was shown that nickel hydroxide with nano-sized, well-crystallized particles exhibited a high electrochemical capacity of up to 380–400 mA·h g^{-1}, with an excellent rate-capacity performance and long-term stability during repetitive electrochemical cycling.

Another new product to be developed is that of Ni-filled CNTs. Although, to date, no analytical uses have been reported for these materials, the tubes were shown to produce a magnetic field of 2 T at a temperature of 2 K, which is an improved ferromagnetism compared to bulk Ni [179].

12.2.7
Iron Nanoparticles

Iron oxide nanoparticles are an influential species in both industrial and environmental processes, and can occur in a range of forms including colloids, in oceans and groundwater [180], and as sludge in water and mineral processing. Naturally occurring processes involving Fe_2O_3 nanoparticles include direct electron transfer [181], surface adsorption of trace metals and organic matter [182], photoexcitation, and the photochemically induced transfer of electrons [183,184]. It is therefore important to understand the redox chemistry of these particles before they can be used in analytical tools. The direct electrochemistry of both adsorbed and solution-based Fe_2O_3 nanoparticles has been studied, with the redox chemistry being found to depend heavily on the pH of solution. This is expected due to the various iron oxides which can be formed as the pH of solution is altered [185].

Unsurprisingly, pure Fe nanoparticles have found little application within electroanalytical sensors due to the ease of oxidation of the particle surface. Nonetheless, iron oxide nanoparticles have been utilized in a diverse array of sensors, primarily as humidity sensors [186–188]. Although these are slightly beyond the scope of traditional electroanalytical sensors, they deserve a mention. In these cases the iron oxide was immobilized either in a sol–gel [186], a polypyrrole film [187], or on sepiolite powder [188]. Subsequently, the resistance of each sensor was found to vary according to the humidity. Other solid-state systems consisting of iron oxide nanoparticles in conjunction with various species have been used for the determination of ozone [189], and also in the catalytic oxidation of methanol [190].

As with virtually all of the metal nanoparticles discussed above, glucose sensing via its direct association with glucose oxidase has been achieved due to the ability of Fe_2O_3 to directly and electrocatalytically oxidize hydrogen peroxide [191,192]. In fact, it has been shown that the ability of CNTs to catalyze such oxidation is not due to the edge plane graphite defects within the CNT, but rather to the presence of Fe nanoparticles inherent from the synthesis process [193].

12.2.8
Nanoparticles of Other Metallic Species

Other metallic nanoparticles to be synthesized which have found limited applications in electroanalysis include ruthenium, rhodium, and iridium. When ruthenium nanoparticles were used in the electrochemical reduction of oxygen, the results showed that reduction take place by a multi-electronic charge transfer process with the formation of hydrogen peroxide [194]. Although this technique has not yet been used for sensing oxygen, a solid-state oxygen sensor based on a Ru–carbon nanocomposite film has been investigated. In this case, Ru nanoparticles (diameter ca. $\sim$5–20 nm) were dispersed in an amorphous carbon matrix, and the conductivities associating to the interface charge transfer between the Ru–carbon composite electrode and the Y_2O_3-stabilized ZrO_2 electrolyte were shown to be 100- to 1000-fold higher than that of Pt electrodes [195]. Although limited applications of Ru nanoparticles in electroanalysis have been reported, it has been shown that bimetallic Ru/Pt nanoparticle layers catalyze methanol oxidation (e.g., Refs. [196–199]).

Iridium nanoparticles, which are usually found in the form of iridium oxide, have been synthesized in a variety of ways, either as nanorods [200], as particles using sol–gels [201], or as colloidal suspensions containing mixed metal oxides [202]. Although several iridium structures have been prepared, their electroanalytical applications to date have primarily been in biosensors, due to their ability to catalytically oxidize and reduce hydrogen peroxide [203–206].

Although rhodium nanoparticles have found little application in electrochemical sensors, their ability to catalyze the reduction of nitrate [207] and oxidation of methanol and ethanol [208] provides some promise that they may soon find niche markets within electrochemical sensing.

12.3
Future Prospectives

The abundance of nanoparticles currently used in electroanalytical chemistry is clear from the content of this chapter. Indeed, given the large number of reports available on the analytical utility of metal macroelectrodes in electroanalysis, it can be foreseen – with relatively little forethought – that these systems will be studied at the nanoparticle layer, doubtless with significant enhancements in sensitivity due an increased electroactive surface area. As further studies of the nanotechnology and novel methods used to fabricate these nanostructures are undertaken [209–211], the application of such particles in electroanalysis will become increasingly predominant. Coupled with the growth in nanoparticle fabrication, it may be anticipated that the catalytic ability of nanoparticles, and the ease by which their surface can be chemically modified, will aid in their application to electroanalysis. The main challenge in developing these sensors will be to produce uniformly dispersed nanoparticle layers on the underlying substrate. Although this chapter has highlighted some of these procedures, emphasis must be placed on methods of preparing reproducible, stable, and low-cost sensors for practical applications. If these initial problems can be overcome, then the growth of nanoparticle-based electrochemical sensors will depend on their ability to tailor the particles' surface and structure for any analyte. In this way, the nanoparticle may be used to capture an analyte of interest before catalyzing its detection.

References

1 Penn, S.G., He, L. and Natan, M.J. (2003) *Curr. Opin. Chem. Biol.*, **7**, 609.
2 Lewis, L.N. (1993) *Chem. Rev.*, **93**, 2693.
3 Kesavan, V., Sivanand, P.S., Chandrasekaran, S., Koltypin, Y. and Gedankin, A. (1999) *Angew. Chem. Int. Ed.*, **38**, 3521.
4 Ahuja, R., Caruso, P.-L., Mobius, D., Paulus, W., Ringsdorf, H. and Wildburg, G. (1993) *Angew. Chem. Int. Ed.*, **32**, 1033.
5 Mulvaney, P. (1996) *Langmuir*, **12**, 788.
6 Alvarez, M.M., Khoury, J.T., Schaaff, T.G., Shafigullin, M.N., Vezmar, I. and Whetten, R.L. (1997) *J. Phys. Chem. B*, **101**, 3706.
7 Alivisatos, A.P. (1996) *J. Phys. Chem.*, **100**, 13226.
8 Brus, L.E. (1991) *Appl. Phys. A*, **53**, 465.
9 Khairutdinov, R.F. (1997) *Colloid J.*, **59**, 535.
10 Fahrner R. (Ed.) (2005) *Nanotechnology and Nanoelectronics, Materials, Devices,*

Measurement Techniques, Springer, New York.

11 Lockwood D.J. (Ed.) (2002) *Nanostructure Science and Technology*, Springer, New York.

12 Wang Z.L. (Ed.) (2006) *Micromanufacturing and Nanotechnology Nanowires and Nanobelts: Volume 1: Metal and Semiconductor Nanowires, Volume 2: Nanowires and Nanobelts of Functional Materials*, Springer, New York.

13 Cao, G.Z. (Ed.) (2004) *Nanostructures & Nanomaterials: Synthesis, Properties & Applications*, Imperial College Press, London.

14 Wang, J. (2005) *Analyst*, **130**, 421.

15 Wilner and Wilner B. (2002) *Pure Appl. Chem.*, **74**, 1773.

16 Luo, X., Morrin, A., Killard, A.J. and Smyth, M.R. (2006) *Electroanalysis*, **18**, 319.

17 Katz, E., Willner, I. and Wang, J. (2004) *Electroanalysis*, **16**, 19.

18 Hernandez-Santos, D., Gonzalez-Garcia, M.B. and Garcia, A.C. (2002) *Electroanalysis*, **14**, 1225.

19 Crouch, S.R. (2005) *Anal. Bioanal. Chem.*, **381**, 1323.

20 Welch, C.M. and Compton, R.G. (2006) *Bioanal. Chem.*, **384**, 601.

21 Schloglm, R. and Hamid, S.B.A. (2004) *Angew. Chem. Int. Ed.*, **43**, 1628.

22 Astruc, D., Lu, F. and Aranzaes, J.R. (2005) *Angew. Chem. Int. Ed.*, **44**, 7852.

23 Roucoux, A., Schulz, J. and Patin, H. (2002) *Chem. Rev.*, **102**, 3757.

24 Bonnermann, H. and Richards, F.J.M. (2001) *Eur. J. Inorg. Chem.*, 2455.

25 Niemeyer, C.M. (2001) *Angew. Chem. Int. Ed.*, **40**, 4129.

26 Willner, I., Helg-Shabtai, V., Blonder, R., Katz, E., Tao, G., Buckmann, A.F. and Heller, A. (1996) *J. Am. Chem. Soc.*, **118**, 10321.

27 Xiao, Y., Patolsky, F., Katz, E., Hainfeld, J.F. and Willner, I. (2003) *Science*, **299**, 1877.

28 Sheeney-Haj-Ichia, L., Wasserman, J. and Willner, I. (2002) *Angew. Chem. Int. Ed.*, **41**, 2323.

29 Lahav, M., Shipway, A.N. and Willner, I. (1999) *J. Chem. Soc. Perkin Trans.*, **2**, 1925.

30 Liu, S.Q., Leech, D. and Ju, H.X. (2003) *Anal. Lett.*, **17**, 1674.

31 Crumbliss, A.L., Perine, S.C., Stonehuerner, J., Tubergen, K.R., Zhao, J. and Henkins, R.W. (1992) *Biotechnol. Bioenerg.*, **40**, 483.

32 Xiao, Y., Ju, H.X. and Chen, H.Y. (1999) *Anal. Chim. Acta*, **391**, 73.

33 Riboh, C., Haes, A.J., Macfarland, A.D., Yonzon, C.R. and Van Duyne, R.P. (2003) *J. Phys. Chem. B*, **107**, 1772.

34 Krasteva, N., Besnard, I., Guse, B., Bauer, R.E., Mullen, K., Yasuda, A. and Vossmeyer, T. (2002) *Nano Lett.*, **2**, 551.

35 Vossmeyer, T., Guse, B., Besnard, I., Bauer, R.E., Mullen, K. and Yasuda, A. (2002) *Adv. Mater.*, **14**, 238.

36 Krasteva, N., Guse, B., Besnard, I., Yasuda, A. and Vossmeyer, T. (2003) *Sensor. Actuat. B Chem.*, **92**, 137.

37 Zhu, Q., Yuan, R., Chai, Y., Zhuo, Y., Zhang, Y., Li, X. and Wang, N. (2006) *Anal. Lett.*, **39**, 483.

38 Xue, M.-H., Xu, Q., Zhou, M. and Zhu, J.-J. (2006) *Electrochem. Commun.*, **8**, 1468.

39 Zhang, S., Wang, N., Yu, H., Niu, Y. and Sun, C. (2005) *Bioelectrochemistry*, **67**, 15.

40 Zhao, W., Xu, J.J. and Chen, H.Y. (2005) *Front. Biosci.*, **10**, 1060.

41 Xu, Q., Mao, C., Liu, N.-N., Zhu, J.-J. and Sheng, J. (2006) *Biosens. Bioelectron.*, **22**, 768.

42 Jena, B.K. and Raj, C.R. (2006) *Anal. Chem.*, **78**, 6332.

43 Downard, A.J., Tan, E.S.Q. and Yu, S.S.C. (2006) *New J. Chem.*, **30**, 1283.

44 Zhang Jiang, X., Wang, E. and Dong S. (2005) *Biosens. Bioelectron.*, **21**, 337.

45 Luo, X.-L., Xu, J.-J., Zhang, Q., Yang, G.-J. and Chen, H.-Y. (2005) *Biosens. Bioelectron.*, **21**, 190.

46 Xu, S. and Han, X. (2004) *Biosens. Bioelectron.*, **19**, 1117.

47 Liu, Z.M., Wang, H., Yang, Y., Yang, H.F., Hu, S.Q., Shen, G.L. and Yu, R.Q. (2004) *Anal. Lett.*, **37**, 1079.

48. Zhou, N., Wang, J., Chen, T., Yu, Z. and Li, G. (2006) *Anal. Chem.*, **78**, 5227.
49. Zhang, J. and Oyama, M. (2005) *J. Electroanal. Chem.*, **577**, 273.
50. Ren, X., Meng, X. and Tang, F. (2005) *Sensor. Actuat. B Chem.*, **110**, 358.
51. Patolsky, F., Gabriel, T. and Willner, I. (1999) *J. Electroanal. Chem.*, **479**, 69.
52. Shumyantseva, V.V., Carrara, S., Bavastrello, V., Riley, D.J., Bulko, T.V., Skryabin, K.G., Archakov, A.I. and Nicolini, C. (2005) *Biosens. Bioelectron.*, **21**, 217.
53. Xu, X., Liu, S., Li, B. and Ju, H. (2003) *Anal. Lett.*, **36**, 2427.
54. Liao, K.-T. and Huang, H.-J. (2005) *Anal. Chim. Acta*, **538**, 159.
55. Tang, D.P., Yuan, R., Chai, Y.Q., Zhong, X., Liu, Y., Dai, J.Y. and Zhang, L.Y. (2004) *Anal. Biochem.*, **333**, 345.
56. Hu, S.-Q., Xie, J.-W., Xu, Q.-H., Rong, K.-T., Shen, G.-L. and Yu, R.-Q. (2003) *Talanta*, **61**, 769.
57. Liu, S.-F., Li, Y.-F., Li, J.-R. and Jiang, L. (2005) *Biosens. Bioelectron.*, **21**, 789.
58. Zhang, Z.-L., Pang, D.-W., Yuan, H., Cai, R.-X. and Abruna, H.D. (2005) *Anal. Bioanal. Chem.*, **381**, 833.
59. Zheng, H., Hu, J.-B. and Li, Q.-L. (2006) *Acta Chim. Sinica*, **64**, 806.
60. Zhang, J., Song, S., Zhang, L., Wang, L., Wu, H., Pan, D. and Fan, C. (2006) *J. Am. Chem. Soc.*, **128**, 8575.
61. Lee, T.M.-H., Cai, H. and Hsing, I.-M. (2005) *Analyst*, **130**, 364.
62. Wang, J., Li, J., Baca, A.J., Hu, J., Zhou, F., Yan, W. and Pang, D.-W. (2003) *Anal. Chem.*, **75**, 3941.
63. Zhong, J., Qi, Z., Dai, H., Fan, C., Li, G. and Matsuda, N. (2003) *Anal. Sci.*, **19**, 653.
64. Lu, P. and Lin, X.Q. (2004) *Anal. Sci.*, **20**, 527.
65. Raj, C.R., Okajima, T. and Ohsaka, T. (2003) *J. Electroanal. Chem.*, **543**, 127.
66. Wang, L., Bai, J., Huang, P., Wang, H., Zhang, L. and Zhao, Y. (2006) *Electrochem. Commun.*, **8**, 1035.
67. Jena, B.K. and Raj, C.R. (2702) *Chem. Eur. J.*, **2006**, 12.
68. Tominaga, M., Shimazoe, T., Nagashima, M., Kusuda, H., Kubo, A., Kuwahara, Y. and Taniguchi, I. (2006) *J. Electroanal. Chem.*, **590**, 37.
69. Song, Y.-S., Muthuraman, G., Chen, Y.-Z., Lin, C.-C. and Zen, J.-M. (2006) *Electroanalysis*, **18**, 1763.
70. Dai, X. and Compton, R.G. (2006) *Anal. Sci.*, **22**, 567.
71. Dai, X. and Compton, R.G. (2005) *Electroanalysis*, **17**, 1325.
72. Dai, X., Nekrassova, O., Hyde, M.E. and Compton, R.G. (2004) *Anal. Chem.*, **76**, 5924.
73. Majid, E., Hrapovic, S., Liu, Y., Male, K.B. and Luong, J.H.T. (2006) *Anal. Chem.*, **78**, 762.
74. Welch, C.M., Nekrassova, O., Dai, X., Hyde, M.E. and Compton, R.G. (2004) *ChemPhysChem.*, **5**, 1405.
75. Liu, S. and Ju, H. (2003) *Analyst*, **128**, 1420.
76. Zhang, J. and Oyama, M. (2005) *Anal. Chim. Acta*, **540**, 299.
77. Yu, A., Liang, Z., Cho, J. and Caruso, F. (2003) *Nano Lett.*, **3**, 1203.
78. Pumera, M., Wang, J., Grushka, E. and Polsky, R. (2001) *Anal. Chem.*, **73**, 5625.
79. Watkins, J.J., Chen, J.Y., White, H.S., Abruna, H.D. and Maisonhaute, E. (2003) *C. Amatore, Anal. Chem.*, **75**, 3962.
80. You, T., Niwa, O., Tomita, M. and Hirono, S. (2003) *Anal. Chem.*, **75**, 2080,
81. You, T., Niwa, O., Horiuchi, T., Tomita, M., Iwasaki, Y., Ueno, Y. and Hirono, S. (2002) *Chem. Mater.*, **14**, 4796.
82. Wu, Z., Chen, L., Shen, G. and Yu, R. (2006) *Sensor. Actuat. B Chem.*, **119**, 295.
83. Hrapovic, S., Liu, Y.L., Male, K.B. and Luong, J.H.T. (2004) *Anal. Chem.*, **76**, 1083.
84. Yang, M., Yang, Y., Liu, Y., Shen, G. and Yu, R. (2006) *Biosens. Bioelectron.*, **21**, 1125.
85. Tang, H., Chen, J., Yao, S., Nie, L., Deng, G. and Kuang, Y. (2004) *Anal. Biochem.*, **331**, 89.

86 Yuan, J.H., Wang, K. and Xia, X.H. (2005) *Adv. Funct. Mater.*, **15**, 803.
87 Polsky, R., Gill, R., Kaganovsky, L. and Willner, I. (2006) *Anal. Chem.*, **78**, 2268.
88 Selvaraju, T. and Ramaraj, R. (2005) *J. Electroanal. Chem.*, **585**, 290.
89 Yang, M., Yang, Y., Yang, H., Shen, G. and Yu, R. (2006) *Biomaterials*, **27**, 246.
90 Wang, S., Yin, Y. and Lin, X. (2004) *Electrochem. Commun.*, **6**, 259.
91 Wang, S. and Lin, X. (2005) *Electrochim. Acta*, **50**, 2887.
92 Wang, S., Lu, L. and Lin, X. (2004) *Electroanalysis*, **16**, 1734.
93 Dai, X. and Compton, R.G. (2006) *Analyst*, **131**, 516.
94 Park, S., Boo, H., Kim, Y., Han, J.-H., Kim, H.C. and Chung, T.D. (2005) *Anal. Chem.*, **77**, 7695.
95 Li, X., Zhang, Z., Zhang, J., Zhang, Y. and Liu, K. (2006) *Microchim. Acta*, **154**, 297.
96 Liu, H.S., Song, C.J., Zhang, L., Zhang, J.J., Wang, H.J. and Wilkinson, D.P. (2006) *J. Power Sources*, **155**, 95.
97 Antolini, E. (2003) *Mater. Chem. Phys.*, **78**, 563.
98 Zhang, J., Sasaki, K., Sutter, E. and Adzic, R.R. (2007) *Science*, **315**, 220.
99 Ye, H.C. and Crooks, R.M. (2005) *J. Am. Chem. Soc.*, **127**, 4930.
100 Rodrigues Blanco, M.C. and Lopez-Quintela M.A. (2000) *J. Phys. Chem. B*, **104**, 9683.
101 Zoval, J.V., Stiger, R.M., Biernacki, P.R. and Penner, R.M. (1996) *J. Phys. Chem.*, **100**, 837.
102 Chang, G., Zhang, J., Oyama, M. and Hirao, K. (2005) *J. Phys. Chem. B*, **109**, 1204.
103 Fang, B., Wang, G.-F., Li, M.-G., Gao, Y.-C. and Kan, X.-W. (2005) *Chem. Anal. Warsaw*, **50**, 419.
104 Ren, C., Song, Y., Li, Z. and Zhu, G. (2005) *Anal. Bioanal. Chem.*, **381**, 1179.
105 Fu, Y., Yuan, R., Xu, L., Chai, Y., Liu, Y., Tang, D. and Zhang, Y.J. (2005) *Biochem. Biophys. Methods*, **62**, 163.
106 Welch, C.M., Banks, C.E., Simm, A.O. and Compton, R.G. (2005) *Anal. Bioanal. Chem.*, **382**, 12.
107 Zhu, X., Gan, X., Wang, J., Chen, T. and Li, G. (2005) *J. Mol. Cat. A Chem.*, **239**, 201.
108 Wang, G.-F., Li, M.-G., Gao, Y.-C. and Fang, B. (2004) *Sensors*, **4**, 147.
109 Li, M.-G., Gao, Y.-C., Kan, X.-W., Wang, G.-F. and Fang, B. (2005) *Chem. Lett.*, **34**, 386.
110 Xu, J.-Z., Zhang, Y., Li, G.-X. and Zhu, J.-J. (2004) *Mat. Sci. Eng. C*, **24**, 833.
111 Gan, X., Liu, T., Zhong, J., Liu, X. and Li, G. (2004) *Chem. Bio. Chem.*, **5**, 1686.
112 Zhao, S., Zhang, K., Sun, Y. and Sun, C. (2006) *Bioelectrochem.*, **69**, 10.
113 Gan, X., Liu, T., Zhu, X. and Li, G. (2004) *Anal. Sci.*, **20**, 1271.
114 Li, M.-G., Shang, Y.-J., Gao, Y.-C., Wang, G.-F. and Fang, B. (2005) *Anal. Biochem.*, **341**, 52.
115 Breiter, M.W. (1980) *J. Electroanal. Chem.*, **109**, 253.
116 Breiter, M.W. (1977) *J. Electroanal. Chem.*, **81**, 275.
117 Mengoli, G., Fabrizio, M., Manduchi, C. and Zannoni, G. (1993) *J. Electroanal. Chem.*, **350**, 57.
118 Gimeno, Y., Creus, A.H., González, S., Salvarezza, R.C. and Arvia, A.J. (2001) *Chem. Mater.*, **13**, 1857.
119 Batchelor-McAuley, C., Banks, C.E., Simm, A.O., Jones, T.G.J. and Compton, R.G. (2006) *Chem. Phys. Chem.*, **7**, 1081.
120 Liang, H.-P., Lawrence, N.S., Jones, T.G.J., Banks, C.E. and Ducati, C. (2007) *J. Am. Chem. Soc.*, **129**, 6068.
121 Paillier, J. and Roue, L. (2005) *J. Electrochem. Soc.*, **152**, E1.
122 Bartlett, N. and Guerin, S. (2003) *Anal. Chem.*, **75**, 126.
123 Batchelor-McAuley, C., Banks, C.E., Simm, A.O., Jones, T.G.J. and Compton, R.G. (2006) *Analyst*, **131**, 106.
124 Ji, X.B., Banks, C.E., Xi, W., Wilkins, S.J. and Compton, R.G. (2006) *J. Phys. Chem. B*, **110**, 22306.
125 Li, F.L., Zhang, B.L., Dong, S.J. and Wang, E.K. (1997) *Electrochim. Acta*, **42**, 2563.

126 Yang, C.C., Kumar, A.S., Kuo, M.C., Chien, S.H. and Zen, J.M. (2005) *Anal. Chim. Acta.*, **554**, 66.

127 Lim, S.H., Wei, J., Lin, J.Y., Li, Q.T. and KuaYou, J. (2005) *Biosens. Bioelectron.*, **20**, 2341.

128 Imokawa, T., Williams, K.-J. and Denuault, G. (2006) *Anal. Chem.*, **78**, 265.

129 Shao, M.-H., Sasaki, K. and Adzic, R.R. (2006) *J. Am. Chem. Soc.*, **128**, 3526.

130 Welch, C.M., Hyde, M.E., Banks, C.E. and Compton, R.G. (2005) *Anal. Sci.*, **21**, 1421.

131 Zen, J.-M., Hsu, C.-T., Kumar, A.S., Lyuu, H.-J. and Lin, K.-Y. (2004) *Analyst*, **129**, 841.

132 Ward-Jones, S., Banks, C.E., Simm, A.O., Jiang, L. and Compton, R.G. (2005) *Electroanalysis*, **17**, 1806.

133 Xu, J.-Z., Zhu, J.-J., Wang, H. and Chen, H.-Y. (2003) *Anal. Lett.*, **36**, 2723.

134 Zen, J.-M., Chung, H.-H. and Kumar, A.S. (2000) *Analyst*, **125**, 1633.

135 Šljukic, B., Banks, C.E., Crossley, A. and Compton, R.G. (2007) *Electroanalysis*, **19**, 79.

136 Chowdhuri, A., Gupta, V., Sreenivas, K., Kumar, R., Mozumdar, S. and Patanjali, P.K. (2004) *App. Phys. Lett.*, **84**, 1180.

137 Zhang, G. and Liu, M. (2000) *Sensor. Actuat. B Chem.*, **69**, 144.

138 Wang, H., Huang, Y., Tan, Z. and Hu, X. (2004) *Anal. Chim. Acta*, **526**, 13.

139 Kitchens, C.L., McLeod, M.C. and Roberts, C.B. (2003) *J. Phys. Chem. B*, **107**, 11331.

140 Kitchens, C.L., McLeod, M.C. and Roberts, C.B. (2005) *Langmuir*, **21**, 5166.

141 Zhao, Y., Zhu, J.-J., Hong, J.-M., Bian, N. and Chen, H.-Y. (2004) *Eur. J. Inorg. Chem.*, **20**, 4072.

142 Prabhu, S.V. and Baldwin, R.P. (1989) *Anal. Chem.*, **61**, 852.

143 Luo, Zhang, F. and Baldwin R.P. (1991) *Anal. Chem.*, **63**, 1702.

144 Davis, J., Moorcroft, M.J., Wilkins, S.J., Compton, R.G. and Cardosi, M.F. (2000) *Analyst*, **125**, 737.

145 Davis, J., Moorcroft, M.J., Wilkins, S.J., Compton, R.G. and Cardosi, M.F. (2000) *Electroanalysis*, **12**, 1363.

146 Zen, J.-M., Song, Y.-S., Chung, H.-H., Hsu, C.-T. and Kumar, A.S. (2002) *Anal. Chem.*, **74**, 6126.

147 Ren, X. and Tang, F. (2000) *Chin. J. Cat.*, **21**, 458.

148 Kumar, A.S. and Zen, J.-M. (2002) *Electroanalysis*, **14**, 671.

149 Zen, J.-M., Chung, H.-H. and Kumar, A.S. (2002) *Anal. Chem.*, **74**, 1202.

150 Male, K.B., Hrapovic, S., Liu, Y., Wang, D. and Luong, J.H.T. (2004) *Anal. Chim. Acta*, **516**, 35.

151 Dai, X., Wildgoose, G.G. and Compton, R.G. (2006) *Analyst*, **131**, 901.

152 Chowdhuri, A., Gupta, V. and Sreenivas, K. (2003) *Sensor. Actuat. B Chem.*, **93**, 572.

153 Cai, H., Zhu, N., Jiang, Y., He, P. and Fang, Y. (2003) *Biosens. Bioelectron.*, **18**, 1311.

154 Ng, K.H. and Penner, R.M. (2002) *J. Electroanal. Chem.*, **522**, 86.

155 Arvia, A.J. and Posadas, D. (1975) in *Encyclopedia of electrochemistry of the elements*, (ed. A.J. Bard) Marcel Dekker, New York, p. 211.

156 Cordeiro, G., Mattos, O.R., Barcia, O.E., Beaunier, L., Deslouis, C. and Tribollet, B. (1996) *J. Appl. Electrochem.*, **26**, 1083.

157 Epelboin, I. and Keddam, M. (1972) *Electrochim. Acta*, **17**, 177.

158 Burke, L.D. and Twomey, T.A.M. (1984) *J. Electroanal. Chem.*, **162**, 101.

159 Burke, L.D. and Whelan, D.P. (1980) *J. Electroanal. Chem.*, **109**, 85.

160 Weininger, J.L. and Breiter, M.W. (1964) *J. Electrochem. Soc.*, **111**, 707.

161 Weininger, J.L. and Breiter, M.W. (1963) *J. Electroanal. Chem.*, **110**, 484.

162 Wolf, J.F., Yeh, L.S.R. and Damjanovic, A. (1981) *Electrochim. Acta*, **26**, 409.

163 Wolf, J.F., Yeh, L.S.R. and Damjanovic, A. (1981) *Electrochim. Acta*, **26**, 811.

164 Wohlfahrt-Mehrens, M., Oesten, R., Wilde, P. and Huggins, R.A. (1996) *Solid State Ionics*, **86–88**, 841.

165 Motoyama, M., Fukunaka, Y., Sakka, T., Ogata, Y.H. and Kikuchi, S. (2005) *J. Electroanal. Chem.*, **584**, 84.
166 Singh, V.B. and Pandey, P. (2005) *J. New Mater. Electrochem. Sys.*, **8**, 299.
167 Guan, X.-Y. and Deng, J.-C. (2007) *Mater. Lett.*, **61**, 621.
168 Li, Z.-P., Yu, H.-Y., Sun, D.-B., Wang, X.-D., Fan, Z.-S. and Meng, H.-M. (2006) *Chin. J. Nonferrous Metals*, **16**, 1288.
169 Zheng, M.-B., Cao, J.-M., Chen, Y.-P., He, P., Tao, J., Liang, Y.-Y. and Li, H.-L. (2006) *Chem. J. Chin. Univ.*, **27**, 1138.
170 Zhou, H., Peng, C., Jiao, S., Zeng, W., Chen, J. and Kuang, Y. (2006) *Electrochem. Commun.*, **8**, 1142.
171 Liu, H., Zhu, L. and Du, Y. (2005) *Mater. Sci. Forum*, **475**, 3835.
172 Giovanelli, D., Lawrence, N.S., Wilkins, S.J., Jiang, L., Jones, T.G.J. and Compton, R.G. (2003) *Talanta*, **61**, 211.
173 You, T., Niwa, O., Chen, Z., Hayashi, K., Tomita, M. and Hirono, S. (2003) *Anal. Chem.*, **75**, 5191.
174 Shibli, S.M.A., Beenakumari, K.S. and Suma, N.D. (2006) *Biosens. Bioelectron.*, **22**, 633.
175 Cantalini, C., Post, M., Buso, D., Guglielmi, M. and Martucci, A. (2005) *Sensor. Actuat. B Chem.*, **108**, 184.
176 Satyanarayana, L., Reddy, K.M. and Manorama, S.V. (2003) *Mater. Chem. Phys.*, **82**, 21.
177 Matsumiya, M., Shin, W., Izu, N. and Murayama, N. (2003) *Sensor. Actuat. B Chem.*, **93**, 309.
178 Hu, W.-K., Gao, X.-P., Noreius, D., Burchardt, T. and Nakstad, N.K. (2006) *J. Power Sources*, **160**, 704.
179 Tyagi, K., Singh, M.K., Misra, A., Palnitkar, U., Misra, D.S., Titus, E., Ali, N., Cabral, G., Gracio, J., Roy, M. and Kulshreshtha, S.K. (2004) *Thin Solid Films*, **127**, 469.
180 Stumm, W. (1993) *Colloids Surf. A– Physicochem. Engineer. Aspects*, **73**, 1.
181 Mulvaney, P. (1998) in *Nanoparticles and Nanostrucured Films: Preparation, Characterisation and Applications*, (eds Fendler J.H.)VCH, Weinheim. p. 275.
182 Stumm, W., Sulzberger, B. and Sinniger, J. (1990) *Croat. Chem. Acta*, **63**, 277.
183 Wu, F. and Deng, N.S. (2000) *Chemosphere*, **41**, 1137.
184 Nikandrov, V.V., Grätzel, C.K., Moser, J.E. and Grätzel, M. (1997) *J. Photochem. Photobiol. B Biology*, **41**, 83.
185 McKenzie, K.J. and Marken, F. (2001) *Pure Appl. Chem.*, **73**, 1885.
186 Tongpool, R. and Jindasuwan, S. (2005) *Sensor. Actuat. B Chem.*, **106**, 523.
187 Tandon, P., Tripathy, M.R., Arora, A.K. and Hotchandani, S. (2006) *Sensor. Actuat. B Chem.*, **114**, 768.
188 Esteban-Cubillo, A., Tulliani, J.-M., Pecharromain, C. and Moya, J.S. (2007) *J. Eur. Ceram. Soc.*, **27**, 1983.
189 Baratto, C., Ferroni, M., Benedetti, A., Faglia, G. and Sberveglieri, G. (2003) *Proc. IEEE Sens.*, **2**, 932.
190 Rumyantseva, M., Kovalenko, V., Gaskov, A., Makshina, E., Yuschenko, V., Ivanova, I., Ponzoni, A., Faglia, G. and Comini, E. (2006) *Sensor. Actuat. B Chem.*, **118**, 208.
191 Wu, J., Zou, Y., Gao, N., Jiang, J., Shen, G. and Yu, R. (2005) *Talanta*, **68**, 12.
192 Rossi, L.M., Quach, A.D. and Rosenzweig, Z. (2004) *Anal. Bioanal. Chem.*, **380**, 606.
193 Sljukic, B., Banks, C.E. and Compton, R.G. (2006) *Nano Lett.*, **6**, 1556.
194 Duron, S., Rivera-Noriega, R., Nkeng, P., Poillerat, G. and Solorza-Feria, O. (2004) *J. Electroanal. Chem.*, **566**, 281.
195 Kimura, T. and Goto, T. (2006) *Ceramic Trans.*, **195**, 13.
196 Tsai, M.-C., Yeh, T.-K., Juang, Z.-Y. and Tsai, C.-H. (2007) *Carbon*, **45**, 383.
197 Tsai, M.-C., Yeh, T.-K. and Tsai, C.-H. (2006) *Electrochem. Commun.*, **8**, 1445.
198 Dubau, L., Hahn, F., Coutanceau, C., Leger, J.-M. and Lamy, C. (2003) *J. Electroanal. Chem.*, **554–555**, 407.
199 Zhang, X. and Chan, K.-Y. (2003) *Chem. Mater.*, **15**, 451.

200 Chen, S., Huang, Y.S., Liang, Y.M., Tsai, D.S. and Tiong, K.K. (2004) *J. Alloys Compounds*, **383**, 273.
201 Birss, I., Andreas, H., Serebrennikova, I. and Elzanowska, H. (1999) *Electrochem. Solid-State Lett.*, **2**, 326.
202 Reetz, M.T., Lopez, M., Grunert, W., Vogel, W. and Mahlendorf, F. (2003) *J. Phys. Chem. B*, **107**, 7414.
203 Luque, G.L., Ferreyra, N.F. and Rivas, G.A. (2006) *Microchim. Acta*, **152**, 277.
204 Abu Irhayem, E.M., Jhas, A.S. and Birss, V.I. (2004) *Procs. Electrochem. Soc.*, **18**, 152.
205 Elzanowska, H., Abu-Irhayem, E., Skrzynecka, B. and Birss, V.I. (2004) *Electroanalysis*, **16**, 478.
206 You, T., Niwa, O., Kurita, R., Iwasaki, Y., Hayashi, K., Suzuki, K. and Hirono, S. (2004) *Electroanalysis*, **16**, 54.
207 Tucker, M., Waite, M.J. and Hayden, B.E. (2004) *J. Appl. Electrochem.*, **34**, 781.
208 Salazar-Banda, G.R., Suffredini, H.B., Calegaro, M.L., Tanimoto, S.T. and Avaca, L.A. (2006) *J. Power Sources.*, **162**, 9.
209 Liang, H.-P., Zhang, H.-M., Hu, J.-S., Guo, Y.-G., Wan, L.-J. and Bai, C.-L. (2004) *Angew. Chem. Int. Ed.*, **43**, 1540.
210 Hu, Y.S., Guo, Y.-G., Sigle, W., Hore, S., Balaya, P. and Maier, J. (2006) *Nat. Mater.*, **5**, 713.
211 Bell, A.T. (2003) *Science*, **299**, 1688.

Index

a

AAO template-assisted fabrication
– of nanostructures 88
absorption–desorption process 353, 354
AC impedance techniques 305
Altair Nanotechnologies 388
aminophenyl-modified BDD electrode 423
aminopropyltriethoxysilane (APTES) 245
anisotropic magnetoresistance (AMR) 234
anodes
– Si-based 337
– Sn-based 335
anodic alumina membranes (AAM) 245
anodic aluminum oxide 128
anodic porous alumina 50
– miscellaneous properties of 27
– post-treatment of 81
– re-anodization of 87
– steady-state growth of 9, 12, 21, 29, 40, 48, 58, 70
– structure of 8
anodic synthesis
– anodization 119
– electropolishing 119
anodizing current density–time transients 30
asymmetric hybrid supercapacitors 394
atomic force microscopy (AFM) 1, 228, 366, 417
Auger electron spectroscopy (AES) 20, 237, 275, 367

b

ball-milling technique 371
bi-cinchoninic acid (BCA) assay 142
biosensor interface fabrication 152
biotinylated functional lipid vesicle 193
blue photoluminescence (PL) band 28
boron-doped diamond (BDD) surface 423, 445
bottom-up approach 188, 190, 349, 350
bovine serum albumin (BSA) 196
branched alumina nanotubes (bANTs) 131
Brillouin spectroscopy 85
Brunauer–Emmett–Teller (BET) surface area 391

c

capacitance–voltage measurements 85
carbon fiber ultramicroelectrodes 441
carbonic anhydrase of bovine source (CAB) 194
cathodic synthesis
– nanowires 144
chemical vapor deposition (CVD) method 3, 247, 389
chip-based capillary electrophoresis 441
chronoamperometry 201, 227, 228
CNT–DNA interaction 412
copper nanoparticles 448
Cottrell equation 214
Csokan's theory 33
current in the plane (CIP) geometry 212
current perpendicular to the plane (CPP) geometry 212
current–time transient, see potential–time transient
cyclic extrusion compression (CEC) method 350
cyclic voltammetry (CV) 155, 170

d

Delaunay's triangulations 77
differential scanning calorimetry (DSC) 361
dihexadecyl hydrogen phosphate (DHP) 416
direct detection techniques 441
dissolution adsorption mechanism 444
DNA hybridization biosensors 420, 426, 423
double-layered anodic oxides 124
dye-sensitized solar cell (DSSC)

- electron recombination 400
- TiO$_2$ role in 398

e

electrical bridge model 45
electroanalytical applications
- gold nanoparticles 425, 439
- palladium nanoparticles 443
- platinum nanoparticles 441
- silver nanoparticles 442
electrocatalytically oxidize hydrogen peroxide 450
electrochemical anodizing technique 126
electrochemical capacitors (ECs) 168
electrochemical deposition 255
electrochemical detachment method 81
electrochemical double layer capacitors (EDLC) 140
electrochemical impedance spectroscopy (EIS) 85, 251, 303, 308, 411
electrochemical methods 118
- anodic approach 118
- cathodic approach 118
electrochemical plating method 245
electrochemical quartz crystal microbalance method 411
electrochemical step edge decoration (ESED) 4, 251
electrodeposited nanowires
- physical properties of 231
electrodes
- CD film-coated 420
- electrochemical cycling 320
- fullerene-modified 422
electron-beam assisted deposition (EBAD) techniques 235
electron-beam lithography (EBL) systems 171, 189, 191
electronic-grade cold-rolled annealed polycrystalline (EG-CRA) 311
electron probe microanalysis (EPMA) 20
embedding active materials 334
encapsulating active materials, see embedding active materials
end-column detector unit 441
energy dispersive X-ray (EDX) analysis 228, 308
equal channel angular extrusion (ECAE) 350
equilibrium elasticity equation 327
ex-situ electron probe (X-ray) microanalyzer 285
extended X-ray absorption fine structure (EXAFS) 164

f

Faraday's constant 47
Faraday's law 218
fast Fourier transform (FFT) 74
ferri cyanide redox couple 443
ferrocene dicarboxylic acid 201
ferro cyanide redox couple 443
field effect transistor (FET) 260
field emission scanning electron microscopy (FESEM) images 147, 205
film anisotropic nanostructure 170
flexible TiO$_2$ photoanodes
- preparation of 401
fluoride-containing oxalic acid electrolyte 7
focused ion beam (FIB) lithography 52, 196

g

giant magnetoresistance (GMR) phenomenon 160, 212
glass–carbon electrode (GCE) 416
glow discharge optical emission spectroscopy (GDOES) 20
gold nanoparticle (GNP) 415, 425
grazing incident X-ray scattering (GIXS) 164
Griffith brittle crack theory
- application of 329
Griffith's criterion 342

h

high-energy ball milling (HEBM) 350
highly oriented pyrolytic graphite (HOPG) 4, 251
high-performance liquid chromatography (HPLC) 424
high-power Li-ion batteries 396
high-resolution ion beam lithography 190
high-velocity oxy-fuel (HVOF) 301
holographic optical lithography 139
honeycomb alumina 128
human serum albumin (HSA) 194, 207
hydride
- covalent molecular hydrides (CH$_4$) 352
- ionic hydrides (LiH) 352
- metallic hydrides 352
hydrogenation–disproportionation–desorption–recombination (HDDR) 350
hydrogen storage systems
- overview of 358
hydrothermal synthesis 123

i

impedance spectroscopy measurements 415
individual alumina nanotube (ANT) 130

inert gas condensation and *in-situ* warm compress (IGCWC)   312
in-situ spectroscopic ellipsometry   85
Institute of Materials Science and Engineering   350
International Energy Agency   349
in-vitro hydroxyapatite   123
ion beam-assisted deposition (IBAD)   235
ion-sensitive field effect transistor (ISFET)   260
iron nanoparticles   449

k

Konarka technologies   402

l

Langmuir–Blodgett film   4
$LaNi_5$-type system   366
laser-assisted direct imprinting   6
lateral microstructuring technique   138
lateral stepwise anodization (LSA) configuration   130
light-emitting diode (LED)   141
lithography patterning technique   3
low-molecular-weight molecules
– chlorpromazine hydrochloride   419
– promethazine hydrochloride   419
low-pressure chemical vapor deposition (LPCVD)   91

m

Macdonald's model   26
magnetic force microscopy (MFM), *see* scanning thermal profiling (STP)
magnetic multilayered nanowire arrays
– electrodeposition of   222
magnetic nanowires   147
– arrays   231
magnetohydrodynamic (MHD) effects   159
mechanical alloying (MA)   350
mechanically activated field-activated pressure assisted synthesis (MAFAPAS)   300
mechanochemical processing (MCP)   350
membrane-templating methods   213
– drawback of   5
mercaptopropionic acid (MPA)   258
mesoporous thin film (MTF)   5
metal hydrides (MH)   350
metal–organic chemical vapor deposition (MOCVD)   91, 205
micro electro mechanical systems (MEMS)   187
microfilm resistor fabrication   126
molecular beam epitaxy (MBE)   212

molecular-beam lithography   267
multi-walled carbon nanotube (MWNT)   410

n

nanocomposite electrode
– approach   334
nanocomposite hydride materials   371
nanocrystalline material
– behavior of   292
nanoelectrode   188
– based on chemically modified surface   249
– considerations for choosing   189
– electrochemical aspects of   248
– using top-down approach   190
nanofabrication techniques   145, 188
nano-imprint lithography (NIL)   3, 189, 199, 200
nanosecond transient absorption spectroscopy   401
nanosized titanium oxide powders
– preparation of   387
nanosphere lithography (NSL)   3, 53
nanostructured field-emission cathodes   124
nanostructured metal anodes   331
nanostructured metal hydrides
– prospects for   355
nanowire-based electronic devices   213
nanowire-based gas sensors   253
negative nanoimprint lithography (N-NIL)   201
Nernst equation   360
nickel–cadmium (NiCd) battery   350
non-aqueous asymmetric hybrid supercapacitor   394
non-high-resolution techniques   205
normal stepwise anodization (NSA) configuration   130
nuclear magnetic resonance (NMR)   420

o

one-dimensional (1D) quantum phenomena   213
on-line inductively coupled plasma (ICP)   297
open-circuit processes   118
organic light-emitting diode (OLED)   141
over-discharge reaction   358

p

Permalloy layers   160
phenothioazine-based drugs   440
phosphinothricin acetyltransferase gene   428
photoelectrocatalysis   121
photolithographic techniques   195
photonic crystals   95, 137

photon scanning tunneling microscopy (PSTM) 1
physical vapor deposition (PVD) technique 6, 211, 390
Pilling–Bedworth ratio (PBR) 45
planar electrodes
– fracture process of 320
planarized aluminum interconnection 124
plasma-enhanced chemical vapor deposition (PECVD) 202
plastic bellcore-type battery 396
platinum counter electrode 399
platinum nanowire (PtNW) 152
Poisson's ratio 327
polydimethylsiloxane (PDMS) 201
polymethyl methacrylate (PMMA) 195
pore-filling method 46
pore-widening process 85
porous alumina growth 32
– Akahori's hypothesis of 33
– phenomenological models of 41
– theoretical models of 44
porous alumina templates 212
porous anodic alumina 128
– as template 135
porous polycarbonate (PC) membrane 145
potential–time transient 28
potentiodynamic polarization curves 301
Poznan University of Technology 350
precision thin-film resistors 124
pressure injection filling technique 6
protein film voltammetric (PFV) technique 248
pulsed electrodeposition technique
– disadvantage of 223
pulse plating techniques 223
pulse voltage techniques 81
pyrolytic graphite (PG) surface 423

q
quantum conductance 255
quantum dots 119, 425
– semiconductor 2
quantum point contacts 145
quasi-Marcovian process 44

r
radar-transparent structures 131
radiofrequency magnetron sputtering system 91
Raman spectroscopy 284, 371
rapid parallel method 171
Risegang-ring process 268
ruthenium oxide nanostructures
– electrosynthesis of 147
Rutherford backscattering spectroscopy (RBS) 20

s
scanning beam lithography 191
scanning electrochemical microscope (SECM) 164, 189
scanning electron microscopy (SEM) 121, 402, 436
scanning near-field optical microscopy (SNOM) 1, 285
scanning probe microscopy (SPM) techniques 1, 52, 164, 235
scanning thermal profiling (STP) 1
scanning tunneling microscopy (STM) 1, 164. *See also* atomic force microscopy (AFM)
screen-printed carbon electrode (SPCE) 414
secondary ion mass spectrometry (SIMS) 20
selected area electron diffraction (SAED) pattern 360
self-assembled monolayer (SAM) 170, 250
self-organized AAO
– structural features of 60
self-organized anodic porous alumina
– kinetics of 28
semi-empirical models 376
silicon-based nanostructured stamping tool 202
silicon-on-insulator (SOI) technology 260
single-walled carbon nanotube (SWNT) 410
small-angle X-ray scattering (SAXS) techniques 28
sodium montmorillonite modifier 425
sol–gel additives 402
sol–gel self-assembly 390
sol–gel synthesis 123
solid electrolyte interface (SEI) passivation layer 396
sputtering deposition techniques, *see* molecular beam epitaxy (MBE)
streptavidin–biotin bonds 426
sulfonated triphenylmethane acid dye 30
superconducting quantum interference device (SQUID) 212
surface-enhanced infrared adsorption spectroscopy 413
surfactant-assisted hydrothermal process 145

t
tantala–alumina couple 126
template-based mesoporous materials 244
template method 145
thin-film capacitors 124

three-dimensional (3D) photonic crystal   2
tin-doped indium oxide layer   142
titanium-based alloys   121
titanium tetraisopropoxide (TTIP)   387, 402
track-etched polycarbonate membrane   212, 224
transmission electron microscopy (TEM)   130, 222, 331, 360, 441
Turing systems modeling   45
2D Fourier transforms   13
two-dimensional photonic crystal development   2, 95
two-step anodization method   128

u
ultimate tensile strength (UTS)   344
ultra-high vacuum deposition processes   164
ultra-violet ozone treatment   402
ultraviolet photochemical reduction method   4
underpotential deposition (UPD)   162
uniaxial anisotropy   160
uniform nanowire arrays   156

v
vapor–liquid–solid (VLS) growth mechanism   6, 145
vapor-phase hydrolysis   389
Voronoi tessellation concepts   129

w
wet chemistry routes   387

x
XPS valence band   377–380
X-ray diffraction (XRD) spectra   275, 359, 393
X-ray photoelectron spectroscopy (XPS)   20, 303, 308, 367, 413
X-ray powder diffraction pattern   225

y
Young's modulus mechanism   28

z
ZrV_2-type system   364